eries of Books in Biology

tor: Cedric I. Davern

Modes of Sp

A S

Edi

Modes of Speciation

Michael J. D. White
THE AUSTRALIAN NATIONAL UNIVERSITY

W. H. Freeman and Company
San Francisco

Library of Congress Cataloging in Publication Data

White, Michael James Denham, 1910–
Modes of speciation.

Bibliography: p.
Includes indexes.
1. Origin of species. I. Title.
QH380.W47 575 77-10955
ISBN 0-7167-0284-3

Printed in the United States of America

9 8 7 6 5 4 3 2 1

Contents

Preface

Most books on the genetics of evolutionary phenomena have dealt only incidentally with speciation, the process responsible for "cladistic" splitting of evolutionary lineages. Exceptions exist, of course, such as Ernst Mayr's *Populations, Species, and Evolution* (1970) and Verne Grant's *Plant Speciation* (1971). Nevertheless, speciation is a very important and distinct aspect of evolution, and since evidence on the genetic processes underlying it has accumulated very fast in recent years, prevailing arguments and positions need to be reexamined frequently. The viewpoint put forward here is somewhat different from those that predominate in earlier works, notably in putting less emphasis on geographic isolation (allopatry) as an absolute precondition for speciation and in attributing a much larger role to structural chromosomal rearrangements. As Dobzhansky stated in his last work, "It is becoming more and more apparent that speciation can occur in different ways" (Dobzhansky *et al.,* 1977, p. 166). Thus, the *comparative* study of speciation, in relation to the population structure and genetic architecture of living organisms, is assuming an increasing importance in evolutionary studies.

There can be no doubt that the "neo-Darwinian synthesis" of evolutionary theory during the 1940s and 1950s was a great advance in biological thinking. But it is also clear, in retrospect, that it led to some rather dogmatic oversimplification. As Lamotte (1960) put it, "Few problems have been as subject to gratuitous hypotheses and unfounded guessing as that of the mechanisms of evolution. Nowhere else have arguments based on authority played—and still play—such a large part." But that is—or should be—past history. The growth of biochemical, cytogenetic, and ethological studies of speciation, in many groups of organisms, has been so great in the past fifteen years or so that firm, testable models are beginning to take the place of earlier speculations and assertions. Thus, most of the literature referred to in this book has been published in the recent period; earlier works are referred to only when they seem to have historic importance.

The organization of the book presented some major difficulties. Many controversial topics could have been discussed equally well in any of several chapters. The evidence from plants and animals is to some extent dealt with separately, since there seem to be some overall differences between the mechanisms of speciation in these great groups of organisms. No attempt has been made to deal with problems of speciation in the prokaryotes or even in the lower eukaryotes (protozoa, fungi), since the evidence on these groups still seems inadequate.

I am greatly indebted to a number of colleagues who read chapters of the book and suggested improvements and alterations, particularly my colleague Graham C. Webb and Professors Guy L. Bush and Hampton L. Carson. Naturally, none of these are in any way responsible for what I have written. I am also grateful to a large number of colleagues who supplied information on many points and allowed me to read their unpublished manuscripts. To Mr. Howard Beckman, of W. H. Freeman and Company, I am indebted for critical editing, which improved the manuscript considerably.

M. J. D. White

April 1977 *Canberra*

1

Species and Speciation

In 1922 the British geneticist William Bateson, then a recent convert to the chromosome theory of heredity, wrote: "In dim outline evolution is evident enough. But that particular and essential bit of the theory of evolution which is concerned with the origin and nature of species remains utterly mysterious." Fifty years later we have made considerable progress, but a good deal of mystery still remains. The gradual adoption of the term "*speciation*" to designate the genetic changes whereby new species come into existence was a recognition that these processes are a distinct part of the general course of evolution, and hence a legitimate field of study. Those biologists who have been concerned with the broader aspects of evolution—whether expressed in changes in the cytochrome *c* molecule from yeast to man, or in the alterations of the skull bones of mammals—have never been much concerned with speciation. This lack of interest is clearly evident in the paleontological literature, where frequently only the names of genera are used, the species being regarded as trivial or unworthy of consideration. Similarly, population geneticists, who study the effects of natural and artificial selection and related problems, have not always seemed aware of the distinct genetic problems involved in speciation.

Modern accounts of organic evolution recognize the existence of two processes: *anagenesis,* phyletic change in the course of geological time; and *cladogenesis* or speciation, the origin of new species of organisms through splitting of pre-existing ones. Both are population phenomena. Cladogenesis compensates for the loss of species by extinction, so that no sudden or major changes in the total amount of biotic diversity occur (the question whether the number of species of organisms in the world was less in past geological periods, as has been suggested by some authors, will not be considered here).

Discussions of the nature of speciation assume that species have an objective existence. If species cannot be objectively defined, and are merely artificial constructs or subjective figments of the imagination of taxonomists, then speciation can hardly be said to be a real process.

The "orthodox" view is that species can be defined in terms of the "biological species concept." According to Mayr (1940, 1963a), "species are groups of actually or potentially interbreeding natural populations which are reproductively isolated from other such groups." Dobzhansky (1970) states that "Species are . . . systems of populations; the gene exchange between these systems is limited or prevented by a reproductive isolating mechanism or perhaps by a combination of several such mechanisms." Both these authors stress that *absolute* reproductive isolation cannot be used as the criterion of specific distinctness; occasional hybridization and even the production of fertile hybrids occurs between species that are normally genetically isolated from one another and that maintain distinct and different gene pools.

It will be obvious that the biological species concept—based on the existence of effective interbreeding between conspecific populations and absence of or ineffective interbreeding between different species—cannot logically apply to asexual organisms (in which no interbreeding occurs). This problem will be discussed in Chapter 9. For the moment it will be enough to say that the majority of asexual populations seem to have been derived rather recently from sexually reproducing forms and that consequently their genetic systems bear the imprint of their origin, so that in most cases remnants of an organization into biological species clearly persist.

Criticism of the biological species concept has come from a number of workers in the field of phenetic classification (Ehrlich, 1961; Ehrlich and Holm, 1962, 1963; Ehrlich and Raven, 1969; Sokal and Crovello, 1970). A particularly radical conclusion is that of Ehrlich and Holm (1963), based in part on studies of butterflies of the genus *Euphydryas* in California. According to these authors "the generalization that organisms exist as distinct species is largely invalid." The main utility of phenetic classification would seem to be in classifying the higher categories (genera, families, classes, etc.). Its great weakness is its application to species and infraspecific categories. On the one hand it exaggerates the significance of different phenotypes in the same interbreeding population, while on the other it cannot deal with the existence of sibling species that are reproductively isolated but exhibit minimal or no morphological differences (Mayr, 1969; Dobzhansky, 1970). As applied to the genus *Drosophila,* for example, phenetic classification would lump together *D. melanogaster* and *D. simulans,* two quite different biological species whose adaptive potential is very different (Parsons, 1975). At the same time, it would probably assign to separate genera certain Hawaiian *Drosophila* species that are very different in external morphology but so closely related genetically that they exhibit absolutely identical sequences of bands (genes) in their giant polytene chromosomes (Carson, Clayton, and Stalker, 1967; Carson *et al.*, 1970).

It cannot be emphasized too strongly that every species is at the same time a *reproductive community*, a *gene pool*, and a *genetic system*. The fact that most species are described by museum taxonomists on the basis of dead material does not affect the fundamental situation; it merely means that the taxonomist is forced to make certain reasonable assumptions about the biological behavior of his material in nature—assumptions that generally prove correct but which inevitably are erroneous in a few instances.

Ehrlich and Holm's statement, quoted above, implies that most species intergrade morphologically with other species and that the gene pools are not separate but exchange genetic material on a significant scale. Both implications can be shown to be incorrect in every instance where complexes of closely related species have been subjected to really detailed genetic study. This is certainly the case in every species group of *Drosophila* that has been extensively studied—except for those relatively few instances, such as the *paulistorum* complex, where we admittedly have species *in statu nascendi*, i.e., populations undergoing speciation at the present time, and which are races by some criteria but species by others (see Dobzhansky and Pavlovsky, 1966, 1967; Dobzhansky, Pavlovsky, and Ehrman, 1969; and Dobzhansky, 1972).

In support of their view that distinct species are largely a myth, Ehrlich and Holm (1962) state that "it is commonly admitted that species in certain groups of plants simply are not as 'good' as those in other groups," and then go on to refer to some difficulties in the taxonomy of North American butterflies. Commenting on this type of evidence, Stebbins (1963) states his impression that 70 to 80 percent of the species of higher plants "conform well to the biological species definition" and consequently show morphological discontinuity based on reproductive isolation. Most of the problematical species are due, in his opinion, to a combination of hybridization and polyploidy. Since these processes (allopolyploidy) hardly occur in animal evolution at all (see Chapter 8), we would expect to find in animals an even greater number of entirely objective, noncontroversial species once they have been thoroughly studied. Innumerable modern taxonomic monographs testify to this fact. Chromosomal studies of lepidopterans have greatly contributed to the clarification of taxonomic problems in the genera *Lysandra* and *Agrodiaetus* (De Lesse, 1952, 1953, 1954, 1957, 1959, 1960, 1969, 1970), *Erebia* (Lorković, 1958a, b; De Lesse, 1955, 1964), *Cidaria* (Suomalainen, 1963, 1965), and a number of others. It seems likely that the instances of apparent intergradation between doubtful species of Lepidoptera used by Ehrlich and Holm as counter examples to the view of species as distinct, objective entities could be more readily clarified by careful modern cytogenetic and biochemical techniques than by the application of the methods of phenetic taxonomy.

Sokal and Crovello's (1970) criticism of the biological species concept seems to be somewhat different from Ehrlich and Holm's. They recognize that there are "discontinuities in the spectrum of phenetic variation" and they specifically emphasize that reproductive barriers are neither "nonexistent or unimportant in evolution," but on the contrary are "of fundamental significance." However, they are so impressed with the difficulty of applying the criterion of interbreeding to practical taxonomic problems that they conclude that the biological species concept is "neither operational nor necessary for evolutionary theory," that it is not heuristic and in fact lacks practical value. But they do not seem to throw out the species concept altogether, and conclude that "the phenetic species as normally described and whose definition may be improved by numerical taxonomy is the appropriate concept to be associated with the taxonomic category 'species,' while the local population may be the most useful unit for evolutionary study."

It is, of course, true that local populations can and should be studied from many standpoints. But if reproductive barriers are important in evolution they should also be studied, and they will need to be studied at a higher level than the local population—at the level where they begin to appear and become important. That level can only be the level of the biological species. Sokal and Crovello do not deny that speciation mechanisms constitute a legitimate field of inquiry. However, paradoxical as it seems to this writer, they claim that abandonment of the biological species concept would lead to "few changes if any in . . . our understanding of speciational mechanisms."

The strongest argument against the theoretical position of Sokal and Crovello comes from the extremely detailed studies that have been carried out on the evolutionary cytogenetics of a large number of species of *Drosophila*. Such species as *D. pseudoobscura, D. persimilis,* and *D. miranda* conform absolutely to the biological species concept, and so do many others. There are morphological discontinuities between these species and strong reproductive barriers, and complete concordance between these two criteria of distinctness. It seems highly significant that attacks on the species concept or the "biological" definition of it have come almost entirely from workers in theoretical fields or those who have been concerned with imperfectly known material, e.g., the *Euphydryas* butterflies. They have certainly not come from any of the workers most closely concerned with evolutionary studies in the genus *Drosophila,* the only group of animals or plants that have been really adequately studied in the laboratory as well as in the wild, and in which there are no insuperable obstacles to experimental hybridization studies on a large scale. Dobzhansky (1970), a *Drosophila* geneticist who started his career as a beetle taxonomist, states that: "It is not true that species defined as reproductively isolated populations are operationally unusable because working taxonomists can only rarely obtain the pertinent information. . . ."

It may be worthwhile to consider what the practical consequences of abandonment of the biological species concept would be. If we were to give up the notion of species altogether, most discussions in such fields as ecology, ethology, population genetics, and cytogenetics (to name only a few) would simply become impossible. If we were to retain a purely phenetic concept of the species, disregarding the interbreeding criterion, we would face two dangers. On the one hand, every local race might be named a species. This actually happened in some cases before acceptance of the biological species concept became general. For example, Locard recognized 251 "species" of freshwater mussels (genus *Anodonta)* in France, based on differences in shell form; only one biological species is now considered to exist (Schnitter, 1922; Adensamer, 1937; Haas, 1940). On the other hand, there would be the danger of lumping together whole complexes of sibling species. In the case of groups such as the anopheline mosquitoes (where some sibling forms are vectors of human malaria and others are not) vast programs of disease control would be disorganized and retrogression to the quasi-scientific, purely empirical situation that existed several decades ago would occur. Workers on economically important species, whose studies of crop plants or insect pests have been intensive and many-sided, all seem to accept, implicitly or explicitly, the biological species concept (for insect pests see especially the articles in Wright and Pal, 1967).

In summary, attacks on the biological concept of the species have been based on plausible but superficial arguments and poorly studied cases, and must be judged to have failed. This is clearly the view of most geneticists. Species are genetic systems and can only be properly understood by using genetic techniques for their study. Phenetics in the narrow sense, i.e., without accompanying data from population genetics, experimental hybridization, karyotype studies, and biochemical investigations, is a very inadequate tool for understanding either species or speciation. On the other hand, if it is conceived and defined as including all these types of information (as it is by Sokal and Crovello at one point) it becomes synonymous with biology as a whole. The fact that a relative shortage of professional biologists limits really detailed evolutionary studies to a minute number of animal and plant groups should not prevent us from extrapolating the conclusions about species and speciation derived from those groups that have been investigated in depth to the great number of groups that have been studied only superficially. And if it is claimed that the biological species concept is of little practical use to the museum taxonomist, we may retort that the geneticist studying evolution can hardly do without it.

Although the species concept can be defined fairly rigorously, it does not necessarily follow that it is equally easy to define speciation. In fact, the term has probably been used in more than one sense by different authors. Even if speciation is held to include the entire process of evolutionary dichotomy—from the first appearance of a genetic isolating mechanism, however weakly developed, to the final severance of genetic continuity between two populations—it is possible to emphasize either the early or later stages in the model. Some arguments among evolutionists seem to owe their origin to the fact that one author had the initial phase of speciation in mind, while another focused on the final stage. Key (1974) states that speciation is "like the process of becoming a professor, it can be identified only when it is complete." For him the critical early stages of evolutionary divergence, in which genetic isolating mechanisms arise and initiate the entire process, constitute raciation rather than speciation. As used in this book, the term "speciation" refers always to the whole of the process. But a good deal of our attention will be concentrated on what seems to be the most critical phase, namely, the establishment of the first genetic isolating mechanism, however incomplete it may be at its origin.

Recognition of the critical importance of genetic isolating mechanisms in speciation began with the botanical work of Du Rietz (1930). But it was not until the publication of Dobzhansky's *Genetics and the Origin of Species* (1937b) that the term was actually coined and the concept gained general acceptance as one of the major aspects of evolutionary theory. Dobzhansky's original classification of isolating mechanisms is as follows (considerably abbreviated):

I. Mechanisms that prevent production of hybrids or prevent such hybrids as are produced from reaching the sexually mature stage.
  A. No meeting of parental forms.
    1. Ecological isolation.
    2. Seasonal or temporal isolation.
  B. Meeting of parental forms but no hybrids are formed or their development is incomplete.
    1. Ethological isolation.
    2. Mechanical isolation due to incompatibility of genitalia.
    3. Failure of the sperms to fertilize the eggs, or failure of the pollen tubes to grow.
    4. Hybrid inviability (hybrids die before sexual maturity).
II. Production of sterile hybrids.

The classification of isolating mechanisms was further advanced by Mecham (1961), who divided them into *premating* (those that prevent wastage of gametes) and *postmating* (those that involve wastage of gametes). Based on this distinction, Mayr's latest (1970) classification is as follows (somewhat abbreviated):

I. Mechanisms that prevent interspecific crosses (premating mechanisms).
   A. Seasonal and habitat isolation.
   B. Ethological isolation.
   C. Mechanical isolation, preventing sperm transfer during attempted copulation.

II. Mechanisms that reduce success of interspecific crosses (postmating mechanisms).
   A. Gametic mortality.
   B. Zygotic mortality.
   C. Hybrid inviability.
   D. Partial or complete hybrid sterility.

A somewhat more complicated classification of isolating mechanisms, with nine categories instead of seven, is given by Littlejohn (1969). The additional categories are "hybrid ethological isolation" and "hybrid breakdown."

Most pairs of species, even ones that are quite closely related on the basis of all morphological, genetic, and biochemical evidence, are kept apart by a combination of several of these mechanisms. In each such case the question arises: Which of various mechanisms evolved first and which developed later; that is, which was the primary cause of the initial divergence and which played a role at a later stage, reinforcing the primary isolation? This is one of the basic questions in speciation theory. Another question of great importance is whether the primary isolating factor in a particular group, such as the insects or the higher plants, is generally or always of the same type, or whether any one of the factors listed above may play the primary role in individual cases. A third question, which has been much debated, is how populations that are reproductively isolated (partially or wholly) by a primary isolating factor are distributed geographically. Must they necessarily occupy entirely distinct territories with an unoccupied zone in between (the *allopatric* model in the strict sense)? Or may they be segments of a geographic gradient or cline, or even occupy different habitats or ecological niches in the same territory (the *sympatric* model)?

In the more recent literature a number of instances have been described in which closely related taxa (races or species) occupy adjacent geographic areas with a very narrow zone of overlap, often no more than a few hundred meters wide, within which hybridization occurs. Such *parapatric* distributions are in many instances clearly "primary," i.e., not due to secondary contact between populations that were formerly separated geographically.

A large part of this book will be devoted to a discussion of these questions. We shall also examine the genetic basis of isolating mechanisms. Most of these matters have been recently discussed by Murray (1972). However, one question that neither he nor Mayr have examined in depth is the extent to which structural rearrangements of chromosomes, such as inversions and translocations, have played a special and perhaps a primary role in the origin of many or even most species. Information bearing on this question has been accumulating very fast in recent years and will no doubt result in increasingly critical evidence as the techniques of cytogenetics, and especially molecular cytogenetics, are refined and applied to particularly crucial cases. The molecular architecture of the eukaryote chromosome has been the subject of several recent conferences and symposia (see for example Cold Spring Harbor, 1973; Canberra, 1975) and is beginning to be properly understood. This gives us for the first time a real basis for understanding genetic changes in evolution. Although full resolution of all the outstanding questions in this field lies some years in the future, it is clear that most and probably all chromosomes include, in addition to the "unique" DNA coding for polypeptides, two categories of "repetitive" DNA: the moderate repetitive sequences associated with each genetic locus, and the highly repetitive satellite DNA (of which there may be several kinds) that forms the heterochromatin of classical cytogenetics.

At present there seem to be four fundamentally different types of genetic processes that occur in the evolution of higher organisms, all of which may conceivably play significant and perhaps different roles in speciation. We may distinguish between:

1. Allelic changes occurring at single genetic loci. These are "gene mutations" in the classical sense. Most of them lead to substitution of one amino acid by another in a polypeptide chain.
2. Changes in the amount of DNA associated with a genetic locus. These changes are still somewhat mysterious. They may involve a sudden doubling of the amount of DNA (see Keyl, 1965). When they occur in most or all of the genetic loci, they lead to major changes in the total "DNA value" without any change in the amount of satellite DNA.
3. Changes in the satellite DNA's (constitutive heterochromatin). The karyotypes of most species seem to include blocks of one to four different types of satellite DNA's. Evolutionary changes may involve alterations in the number, size, and position of these blocks and the origin of new satellites by nucleotide substitutions in existing satellite sequences.
4. Chromosomal rearrangements of the classical type, such as inversions and translocations.

Most discussions of genetic processes of evolution have been focused particularly on changes of types 1 and 4. It seems, however, that it is now very necessary to recognize types 2 and 3 as distinct processes, even though they may be more important in phyletic evolution than in speciation.

In general, recent discussions of speciation mechanisms in animals (e.g., Littlejohn, 1969; Maynard Smith, 1962, 1966; Murray, 1972) do not discuss the possibility of chromosomal rearrangements having a distinct role in speciation. However, Mayr, who earlier paid rather little attention to chromosomal rearrangements in his discussions of speciation, has more recently (1969a) discussed the three possible situations: (1) purely genic isolating mechanisms; (2) chromosomal rearrangements leading to cytogenetic polymorphism without affecting isolating mechanisms; and (3) chromosomal isolating mechanisms. He suggests that the third situation is especially likely to occur in sedentary animals that tend to occur in small colonies, a conclusion that seems well founded.

Although genetic isolating mechanisms clearly play an essential role in speciation, most closely related species (even so-called sibling species, which exhibit no clear-cut or only minimal morphological differences) differ in many ways that probably have nothing to do with genetic isolation as such, e.g., biometrical and biochemical, and in some cases karyotypic, differences. It would be a mistake to think that these differences have necessarily evolved since effective genetic isolation of the two species; in many instances they may have existed, between geographic races, even before speciation began.

In addition to effective genetic isolating mechanisms and the other differences referred to above, most closely related species have somewhat different ecological requirements, so that they occupy substantially different habitats and can coexist geographically without an intolerable level of competition. Mayr (1969a) has spoken of ecological exclusion as a prerequisite for successful speciation.

Although this may be true in many cases, there seem to be so many exceptions that we should be very cautious in the application of this principle. If full speciation necessarily implies ecological exclusion and the ability of the two species to coexist sympatrically, there should be no parapatric species and all parapatric populations should be regarded as less than full species, i.e., as races. It is clear, however, that a rigid application of this principle leads to many absurdities. For example, among the Australian grasshoppers of the subfamily Morabinae there are many examples of undoubtedly distinct species, with significant differences in morphology (especially of the male genitalia) and karyotype, occupying parapatric ranges (Key, 1974). This situation seems to be the usual one in such genera as *Achuraba, Achurimima, Carnarvonella,* and *Keyacris,* as well as several others. It is quite ridiculous to regard *Keyacris marcida* and *K. scurra* as conspecific simply because their distributions are parapatric (Key, 1974). Borderline cases exist as well, of course. For example, among the morabine grasshoppers there are instances of parapatric forms, especially in the genera *Vandiemenella* and *Culmacris,* that can only be regarded as full species on the basis of their cytogenetics, whereas Key (unpubl.)

regards them as races on the evidence of morphology. Other borderline cases may include *Mus m. musculus* and *M.m. domesticus* in Denmark (Selander, Hunt, and Yang, 1969; Hunt and Selander, 1973; and see p. 206) and the species formerly included in *Rana pipiens* (Littlejohn and Oldham, 1968). In the latter case seven sibling species are now recognized: *pipiens, sphenocephala, berlandieri, blairi, magnaocularis,* and two unnamed ones (Brown, 1973; Brown and Brown, 1972; Platz, 1972; Mecham *et al.,* 1973; Pace, 1974; Frost and Bagnara, 1976). The zone of overlap between *blairi* and *berlandieri* is 16 km wide in north-central Texas and 56 km wide in west Texas; that between *berlandieri* and *sphenocephala* 8 km wide in central Texas; and that between *pipiens* and *blairi* 8 km wide in Colorado. These species differ considerably in mating call. Small numbers of natural hybrids between *blairi* and *berlandieri* have been recorded in the zone of overlap, but there is no evidence of any hybridization between *blairi* and *sphenocephala* or between *blairi* and *pipiens* (Mecham *et al.,* 1973). This case thus strongly supports the view that full species can maintain parapatric ranges, i.e., they can exclude one another geographically rather than merely ecologically.

For many evolutionists, support for the biological species concept is linked with acceptance of the views on "geographic speciation" put forward by Mayr (1942, 1947, 1963a, b, 1969a, b, 1970). The general position adopted by Mayr, and which he has not substantially changed since 1942, is that virtually all speciation is allopatric, i.e., that geographic isolation of two populations is a necessary prerequisite in time to the development of genetic isolating mechanisms. However, he avoids the use of the precise term "allopatric," preferring instead to speak of "geographic speciation," a somewhat vaguer expression that seems to allow a little more latitude. In his 1942 book he discussed the possibility of what he calls "semigeographic speciation." This he defined as "the development of species gaps between populations not completely separated geographically" (p. 188). Elsewhere in the same book (p. 99) he states: "It is not known whether the slope in a zone of primary intergradation can become continuously steeper, up to the point where one species breaks into two. It seems unlikely, on the basis of our knowledge of gene dispersal, that this would happen without other isolating mechanisms becoming involved." Later (1970, p. 318), he concluded that "The occurrence of such *semigeographic speciation* seems unlikely. There is no evidence that reproductive isolation can be acquired in a zone of primary intergradation as long as the two populations are in broad contact with each other. Gene flow and the cohesion of the gene pool of the species prevent semigeographic speciation, regardless of diversifying selection."

Mayr's views on speciation have been presented with so much supporting evidence and developed with so much erudition that for many biologists they undoubtedly appear to be the ultimate truth on the matter. Nevertheless, beginning about 1962, doubts as to the universality of the allopatric model have been expressed by a number of workers, particularly Lewis (1962, 1966), Bush (1966, 1969a, b, 1974, 1975a, b), Bazykin (1965, 1969), Maynard Smith (1962, 1966), Murray (1972), Mettler and Nagle (1966), White (1968, 1969, 1970a, 1973, 1974), Craddock (1974a), and James (1970b). These authors approach the problem from a wide variety of different standpoints.

The time therefore seems ripe for a reconsideration of the whole question of the mechanisms of speciation. This seems particularly worthwhile since Mayr has never discussed in full the relative roles of chromosomal rearrangements and mutations of single genes in speciation, and there is now strong evidence, to be considered in Chapter 3, for the view that structural karyotypic changes very frequently play a special role in speciation. The theory of allopatric speciation was originally developed on the tacit assumption that chromosomal rearrangements were either unimportant in speciation or that their role was in no way different from that of ordinary gene mutations. If these assumptions can now be shown to be in error, it is unquestionably incumbent on us to take a new look at the whole subject of speciation mechanisms. It is probable that a number of modes of speciation exist, and that the simple antithesis of allopatric and sympatric modes is a false one, since most of these involve a geographic component of some kind (several of them would be called "semigeographic" by Mayr). For this reason we shall not use the term "geographic speciation" (in Mayr's sense), preferring instead to employ the more precise expression "allopatric speciation" for the situation in which new species are derived from populations that are completely isolated, geographically, from the parent species.

A study of speciation that aims to be more than superficial should in the case of each incipient species include all aspects that might have a bearing on the situation. By themselves, cytogenetic studies, ethological investigations of mating calls, or research on any single component of what is an extremely complex situation, involving many processes at different levels, cannot be expected to throw much light on the mechanism of speciation. Speciation is the result of the combined action and interaction of many processes, and any model that relies exclusively on a single process is bound to be simplistic.

A fully documented history of a single case of speciation might include the following data on *both* forms:

1. A precise map of present distributions.
2. Geological and climatological evidence suggesting past distributions.
3. Detailed morphological description (with special attention to the genitalia in the case of many groups).

4. Detailed information on geographical variation, including multivariate biometrical studies.
5. Ecological data on habitats, niches.
6. Extensive information on biochemical polymorphisms and the extent of allelic differences between the two forms.
7. Detailed descriptions of the karyotypes, based on the most modern techniques (including details of any chromosomal polymorphism in either form).
8. DNA values.
9. Information on the types of satellite DNA present, and the amount and distribution of each type of satellite.
10. Seasonal cycles, i.e., duration of mating season.
11. Results of experimental hybridization, including a cytological study of meiosis (in both sexes) of any hybrids obtained.
12. Information as to hybridization (if any) in nature.
13. Information regarding any ethological isolating mechanisms (chemical nature of any pheromones, detailed descriptions of courtship behavior, records of any mating calls used).

It is perhaps unneccesary to state that there is probably no single case of speciation for which all of this information exists.

It is a basic fact of scientific investigation that what emerges from a piece of research depends not only on the data obtained but on the interpretations of the investigator. In no field is this more true than in speciation studies. As the theory of allopatric speciation became more and more accepted (roughly in the years 1948–1968) as the *only* true one and hardened into dogma, innumerable investigations, allegedly into speciation in this or that group, were carried out by biologists who were so committed to this particular model of speciation that their studies were designed and executed in such a way as to exemplify the model rather than to test it. Obviously, such investigations greatly expanded our knowledge of the taxonomy of the groups concerned, and in many instances provided evidence of the role of particular types of isolating mechanisms, especially ethological ones. But the majority of these efforts seem to have contributed rather little to speciation theory, which they were not really designed to explore.

Nearly all of the most illuminating investigations of speciation have been those in which a number of techniques have been applied so as to explore the biochemical, ethological, cytogenetic, and ecological aspects of particular cases. Speciation studies that employ a single technique or study the role of a single type of isolating mechanism necessarily reveal only one aspect of the speciation process, and may give a very distorted view. This is especially the case if the investigator is uninterested in the other aspects. He who believes that ethological isolation is all-important in speciation will study the ethological factors in particular cases and "prove" his initial assumption; the same is equally true of the cytogeneticist who studies only the karyotypic differences between species.

Ever since Darwin studied the finches and giant tortoises of the Galápagos archipelago, biologists have paid particular attention to speciation on islands.* Archipelagos seem to be ideally suited to the allopatric model of speciation, and some zoologists who have studied insular situations have done so in the deliberate belief that insular faunas were especially important because they afforded understanding of processes that were universal but more difficult to study on continental land masses. But islands occupy only a minute fraction of the globe and contain only a very small proportion of the world's species. The vast majority of species have consequently *not* arisen under the easily understood conditions of insular isolation, but on continental landmasses. Extrapolating from a known case to an unknown one is a common method of scientific investigation. But when, as in the case of speciation in archipelagos, a highly unusual process is used to generalize to the usual, one must necessarily be doubtful of the validity of the inferences. The "species flocks" in certain ancient lakes, such as Baikal in Siberia, and Malawi and Tanganyika in Africa, have been compared by some evolutionists to the drosophilids of the Hawaiian islands and the finches of the Galápagos. We shall discuss such cases in detail in Chapter 4. Here a word of caution as to whether the processes of speciation in archipelagos and ancient lakes are really similar seems indicated.

No real understanding of speciation processes was possible until the full extent of genetic and cytogenetic polymorphism had been revealed. The true extent of polymorphism was gradually recognized during the period 1930–1960. Since then, allozyme studies and related biochemical techniques have been applied to populations of numerous sexually reproducing species. The extent of polymorphism found by these methods has almost always been greater than anyone had previously suggested. This evidence will be discussed in detail in Chapter 3. For the moment it is sufficient to point out that in studies of heterozygosity in species of *Drosophila,* rodents, man, and the horseshoe "crab" *Limulus polyphemus,* the mean heterozygosity per gene locus ranged from 0 to 20.2 percent, whereas the proportion of loci polymorphic per population varied from 0 to 86 percent (Lewontin, 1973). Some deep-sea (1,000–2,000 m) invertebrates seem to be just as genetically polymorphic as land animals (Gooch and Schopf, 1972). For the giant clam *Tridacna maxima,* 63 percent of

*The most recent major investigation has been that of the Hawaiian Drosophilidae (Carson *et al.*, 1970; Carson, 1974; Stalker, 1972; Spieth, 1974; Craddock, 1974b; and Richardson, 1974).

all loci are polymorphic and individuals are heterozygous on the average for 20 percent of their loci (Ayala *et al.*, 1973). These data refer almost entirely to the first type of cytogenetic process listed on p. 8; types 2, 3, and 4 likewise seem to be developed in many species to an extent that was not suspected 20–30 years ago.

Speciation studies are necessarily concerned largely with taxa on the borderline between races and full species. This inevitably creates problems of nomenclature because in such situations taxonomists are likely to disagree as to just which taxa have or have not attained the rank of species. Some authors use the term *semispecies* for populations on the borderline between races and species, i.e., those that are more differentiated than geographic races (subspecies) but not as completely isolated as full species.

The length of time required for speciation can only be estimated in rather special cases, which may be atypical. MacArthur (1972) has made the general statement that "the length of time it normally takes for a species to split and diverge sufficiently to be regarded as two species is a small, uncertain number of thousands of years." Most of those species for which the duration of the speciation process can be estimated are found on islands, where speciation may be unusually rapid because of the availability of numerous vacant ecological niches. One example is provided by five Hawaiian species of the moth genus *Hedylepta,* all of which feed on bananas. The banana plant is not native to the archipelago but was introduced by polynesians about 1,000 years ago (Zimmerman, 1960). The five species of *Hedylepta* are presumed to have evolved from one or two ancestral species feeding on *Pritchardia* palms. Another example of exceptionally rapid speciation in the Hawaiian archipelago is the family Drosophilidae; 650–700 species have evolved from one or two introduced species in less than 5.6 million years (the age of Kauai, the oldest island), and 20 "picture-winged" species have evolved on the island of Hawaii from nine "founder" species in less than one million years (Carson *et al.*, 1970). Data such as this do not, however, provide any basis for estimating an average duration of the speciation process itself.

Speciation is one of the main ways by which living organisms adapt in order to exploit the diversity of environments available to them. Groups like the coelacanths and the rhynchocephalians, which apparently have lost the ability to speciate, may have been able to survive for a long time undergoing periodic phyletic change in response to changing environments. But they are restricted to a single habitat at any one time, and this makes their survival extremely precarious. However, even if the possession of efficient mechanisms of speciation is usually necessary for the continued survival of a group in geologic time, it does not follow that the mechanisms must be of the same type in all groups.

Theoretically, we may postulate three main sets of variables involved in speciation. First, there are the underlying genetic mechanisms, which may consist solely of allelic changes at individual gene loci (homosequential speciation, see p. 45) or may include one or more chromosomal rearrangements. Second, genetic isolating mechanisms play a primary role in initiating the phyletic dichotomy, which may be ethological, seasonal, or of some other premating or postmating type. Finally, there is the geographic component; at least in theory, and perhaps in practice, prior geographic isolation of the diverging populations may range from complete (strict allopatry) to absent (complete sympatry).

These three sets of variables provide the basis for a large number of possible models of speciation. They are, of course, not the only variables we need to take into consideration. A most important additional one is the degree of vagility of the organisms. As far as animals are concerned, this may include two components: the ability to move actively by walking, running, swimming, or flying, and the liability of being passively transported by wind, water, or other accidental means. For plants, dispersal is invariably passive, and we may distinguish between pollen-dispersal anu the dispersal of seeds or other types of propagules. Precise definitions of vagility are discussed on p. 19.

It should be obvious that the complexities of these multiple variables are too great to be accounted for in a single model, even with today's computers. Useful computer models of particular situations have, however, been published by James (1970b), Crosby (1970), Endler (1973), and Jain and Bradshaw (1966). They throw considerable light on certain critical aspects of the process of speciation.

Mayr (1957) distinguished 12 conceivable modes of speciation, but concluded that only two of them were of any real significance: polyploidy and "geographic speciation." We shall not follow his classification of modes of speciation here because he seems to have drawn a false antithesis between instantaneous speciation through individuals and gradual speciation through populations, and because he held chromosomal rearrangements to be unimportant in speciation since they occur in single individuals. Moreover, his category of "semigeographic speciation" seems to include a number of different types of situation.

Seven models of speciation will be considered here. In addition to these, speciation by polyploidy (reduplication of chromosome sets beyond the two present in diploid organisms) is discussed in Chapter 8, and the various processes resulting in asexual or parthenogenetic populations are discussed in Chapter 9. These models certainly do not cover all possibilities, but too elaborate a classification would be self-defeating. The models are based primarily on the geographic component, and only secondarily on that of the genic or chromosomal mechanisms involved. They do not include the type of isolating mechanism primarily responsible for divergence. The models are:

1. Strict allopatry without a narrow population bottleneck.
2. Strict allopatry with a narrow population bottleneck ("founder principle").
3. Extinction of intermediate populations in a chain of races.
4. Clinal speciation.
5. Area-effect speciation (primarily genic).
6. Stasipatric speciation (primarily chromosomal).
7. Sympatric speciation.

The first three models are essentially allopatric—what Mayr calls "geographic speciation." The next three (4–6) would be called "semigeographic" by Mayr. Rather than attempt to rigorously define these models, we shall describe each of them briefly.

1. *Strict allopatry without a narrow population bottleneck.* A single ancestral species population becomes separated into two by a geographical barrier that individuals cannot cross (or cross so rarely that the effect of immigration is negligible). During the period of isolation both populations undergo extensive genetic change, so that if they later become sympatric (as a result of range extension) they will have developed sufficiently strong genetic isolating mechanisms to keep them apart. This is the "dumbbell diagram" model (Mayr, 1942, Fig. 16; Stebbins, 1966, Fig. 5-1; and included in numerous popular accounts of evolution). It is assumed that during the period of isolation both populations are large enough that genetic drift does not play any major role. If the two species remain allopatric, their genetic isolation can only be tested in laboratory experiments.

2. *Strict allopatry with a narrow population bottleneck.* In this model a daughter species arises from a very few individuals (in the extreme case a single fertilized female) that invade a new territory, such as an island, and thus serve as the founders of an immigrant population. The new population necessarily starts out with only a small fraction of the genetic variability of the original population. The "founder principle" was discussed by Mayr (1954) and has been advocated especially by Carson (1971), who has proposed that 23 species of *Drosophila* endemic to the island of Hawaii have evolved from 11 individual founders (gravid females).

3. *Extinction of intermediate populations in a chain of races.* A number of instances are known in which the terminal populations of a ring of races overlap sympatrically without hybridization. A classic case, which deserves

reinvestigation by modern methods, is that of the northern and central races of the Buckeye butterfly *Junonia lavinia,* which interbreed in northern Mexico but coexist in Cuba without interbreeding (Forbes, 1928). It is obvious that there must also exist chains of races whose terminal populations do not overlap geographically but which would be found to be genetically isolated if tested in the laboratory. It is uncertain whether incipient or weakly developed isolating mechanisms exist between the successive members of such a chain. It is fairly obvious, however, that in many cases extinction of the central populations in a chain or ring of races would leave us with two fully isolated species.

4. *Clinal speciation.* Many species exhibit a clinal type of variation along a geographical gradient. Fisher (1930) long ago suggested that a steep "step" in a cline might develop as a result of selection against individuals having a strong tendency to migrate. Murray (1972) has especially stressed the possibility of local differentiation within a cline overcoming the cohesive force of gene flow by migration to such an extent that the cline undergoes fragmentation into two or more distinct species. However, he regards this model as essentially a sympatric one and designates it as "parapatric speciation." The clinal speciation model is here regarded as essentially different from the sympatric one, and the term "parapatric speciation" will be avoided because clinal speciation seems to be only one of several mechanisms that can lead to parapatric distributions.

5. *Area-effect speciation.* In any geographically widespread species there is a tendency for local populations to undergo genetic differentiation. If vagility is restricted—as it is in higher plants, most groups of invertebrates, and such vertebrate groups as amphibians, many reptiles, small rodents, and insectivores —areas much larger than panmictic demes may be characterized by unusual alleles or frequencies of alleles. In their work on snails of the genus *Cepaea,* Cain and Currey (1963a, b) have called these *area effects.* When two or more populations characterized by unusual genetic composition expand their ranges, they may meet in a zone across which there are sharp changes in gene frequency. If populations whose gene complexes are adaptive produce less well-adapted hybrids, there may be a basis for further divergence and eventual speciation. This model has much in common with the previous one, but does not assume any regular geographic gradient or an initially clinal gradient in gene frequencies or phenotype. The differentiation of species with a continental type of distribution into a number of essentially parapatric races, e.g., the North American song sparrow *Passerella melodia* (Miller, 1956), may in many instances be the result of area effects that have not yet led to full speciation. This would occur usually because the different gene complexes are insufficiently incompatible to lead to fragmentation at the zones where the races meet, i.e., selection against interracial hybrids is not strong enough to lead to speciation.

6. *Stasipatric speciation.* In the previous model differentiation is due to numerous genic changes, but no structural change in the karyotype is assumed. In the *stasipatric* model (White, Blackith, Blackith, and Cheney 1967; White, 1968, 1974) the essential feature is a chromosomal rearrangement, originating somewhere within the area occupied by the ancestral species, which reduces fecundity when heterozygous. If such a rearrangement manages to establish itself (either by drift in a local deme or because it shows "meiotic drive") it may spread geographically throughout a part of the area occupied by the species, because of homozygote superiority, and may act as an incipient isolating mechanism between the population homozygous for it and the original population.

7. *Sympatric speciation.* A population separates into two through adaptation of two subpopulations to different ecological niches or habitats within the same geographic area.

The classification of models of speciation into the above categories seems to have the merit of including population structure and genetic mechanisms, as well as the geographic component. Grant (1971), writing primarily from the standpoint of a botanist, admits three modes of "primary" speciation (speciation that does not involve hybridization): *geographical, quantum,* and *sympatric*. The chief weakness of his scheme seems to be that his category of "quantum speciation" includes several very different models whose only common characteristic is that they are "rapid and radical" in their genetic effects (geographical speciation is conceived as "gradual and conservative," i.e., slow, by Grant).

The models for clinal speciation, area-effect speciation, and stasipatric speciation all lead to parapatric distribution patterns, i.e., the existence of narrow zones of overlap where two taxa meet and may hybridize. If there is free hybridization the zone will be one of intergradation. Such zones of coexistence or intergradation are *primary,* i.e., they are quite different in origin to *secondary* zones of coexistence that have arisen through range extensions of formerly geographically separated (allopatric) populations. Following the definition adopted here, sympatric speciation is quite different from the three models just referred to, in that it does not give rise to any narrow zones of intergradation, except insofar as thousands of these may be postulated to exist between different ecological niches or habitats within the same general environment and geographic area.

The concept of *vagility,* which is so important in discussions of speciation, does not appear to have been rigorously defined in evolutionary literature. For purposes of ecological discussion it could be regarded as the mean distance, in a straight line, between the point at which an individual is born and the point where it dies. This, however, is not very meaningful, genetically. For most evolutionary purposes it would be better to define it as the mean distance between the point at which an individual comes into being (by fertilization) and the point at which it meets a mate (or strictly speaking, gives rise to a new zygote). The distinction between mobility and vagility may be illustrated by reference to certain species of bats, where the breeding of an entire colony of many thousands of individuals may be restricted to a single cave, e.g., *Miniopterus schreibersii* (Dwyer, 1966). The *mobility* of these animals is very great, since in their foraging for food they may fly up to 64 km in a single night. But their *vagility,* in the sense defined above, may be only a few tens of meters, i.e., not much more than that of many wingless insects and molluscs. Davis and Baker (1974) have described a parapatric zone of overlap less than 50 km wide in Mexico for two species of bats of the genus *Macrotus* with different karyotypes. They express surprise that the zone should be so narrow, in view of the high "vagility" of the organisms. We do not know whether *Macrotus* spp. have the same kind of population structure as *Miniopterus*. But any degree of restriction to limited breeding areas, to which at least part of the population returns year after year (philopatry), will very drastically reduce the level of vagility (in the sense defined above), whatever the mobility of the organisms.

Although it is convenient, and indeed necessary, to discuss speciation in terms of a number of ideal models, it must be emphasized that many actual speciation events probably do not conform precisely to any one theoretical model but combine features of two or more. In particular, various intermediates between the models we call *clinal, area effect,* and *stasipatric* seem possible and probably occur.

The extent to which different types of genetic isolating mechanisms reinforce one another's action in restricting gene flow between populations in the course of speciation is undoubtedly highly variable. The stasipatric model is conceived as being one in which chromosomal rearrangements play a major role, and in which postmating isolation is primary. But it is by no means impossible that chromosomal rearrangements may play a part (but perhaps a secondary one) in some of the other models outlined above. Numerous cases are known in *Drosophila* where it is reasonable to postulate that an ancestral population was polymorphic for two chromosome sequences differing with respect to an inversion, the existing species being homozygous for the two alternative gene arrangements. Another common situation is one in which one species is polymorphic for a rearrangement for which a related species is homozygous. Clearly, not all karyotype differences between species are the result of stasipatric speciation. Many of them are relics of formerly adaptive polymorphisms that have ceased to be adaptive, at least in the territory occupied by one of the descendent species, usually because the heterozygote is no longer adaptively superior to both homozygous types.

In considering the part that these various modes of speciation may have played in different groups of higher organisms, there seem to be two extreme tendencies we should avoid. One is to assume that since all higher organisms show certain similarities at the genic level, and since the basic algebra of population genetics is the same for all organisms, therefore the mechanism of speciation must be relatively uniform, with only minor variations of little significance. The other is to assume that differences in mode of life, population structure, and perhaps genetic systems between the various groups of higher organisms are such that each group must have its own characteristic mode or modes of speciation, and that consequently fish speciation necessarily differs from bird speciation and that both differ from *Drosophila* speciation.

The viewpoint that speciation is uniform for all higher organisms, is clearly erroneous, for two basic reasons. On the one hand, speciation and evolution are not simply processes occurring at the genic level; they also involve the chromosomal level, i.e., increases and decreases in the amount of DNA associated with the individual genes, changes in the quantity and types of satellite DNA's (heterochromatin), and changes in basic cytogenetic mechanisms, such as those determining sex. This means that the genetic systems that undergo speciation are quite diverse in different groups of organisms and may also differ significantly even between closely related species. Moreover, each of the basic parameters of population structure—numbers of individuals comprised in a species, their vagility, and the area of the territory they occupy—ranges through at least two or three orders of magnitude, when all higher organisms are considered.

The other extreme, that each group has a characteristic mode of speciation, must also be rejected. There are, after all, only a limited (though large) number of combinations of the various factors influencing the mechanism of speciation. Particular combinations must hence be expected to recur, and it should not surprise us if, in the future, it is shown that certain groups, perhaps widely different in morphology and in systematic position, have very similar modes of speciation.

The problem of speciation is part of a wider nexus of problems that includes the genetic structure of natural populations and the architecture of the eukaryote chromosome. Advances in both these fields are inevitably correlated, so that new discoveries and concepts in one area are likely to lead to a deeper level of understanding in the others.

2

The Genic Structure of Species and Species Differences

Until recently, direct comparisons between different populations (or races or species) were possible only in terms of the overall morphology and physiology of the taxa. Taxonomy is essentially based on such structural data, and the vast majority of the animal and plant species that have been described are actually "morphospecies." Biometrical techniques, such as those employed in multivariate analysis, have made morphological comparisons more sophisticated, but these techniques have not overcome the basic limitations of morphological comparisons.

The introduction of gel electrophoresis, along with special biochemical staining methods, has quite changed this situation.* These techniques have led to quantitative estimates of the levels of heterozygosity and polymorphism in natural populations (Lewontin, 1974, Chapter 3) and of the extent of allelic differences between related species (Lewontin, 1974, Chapter 4). Thus far these techniques have been applied mainly to genetic variants of various enzymes (so-called allozymes) and proteins in solution in the blood or hemolymph.

Because of the great theoretical importance of the data obtained by these methods, it is necessary to point out their limitations and the very large sources of error that exist at the present time. On the one hand, it has been pointed out

*For the literature on the development of these techniques, see Hubby and Throckmorton (1965); Prakash, Lewontin, and Hubby (1969); Nair, Brncic and Kojima (1971); Ayala, Mourão, Pérez-Salas, Richmond, and Dobzhansky (1970); Ayala and Powell (1972); Ayala, Powell and Tracey (1972); Ayala and Tracey (1973, 1974); Hedgecock and Ayala (1973); Johnson and Selander (1971); Johnson, Selander, Smith, and Kim (1972); Selander, Hunt and Yang (1969) and Avise and Smith (1974a, b).

that gel electrophoresis necessarily leads to underestimates of the amount of variability, since the majority of amino acid substitutions in proteins do not lead to any change in the electrostatic charge and, hence, are generally undetectable by gel electrophoresis. Various authorities (e.g., Henning and Yanofsky, 1963; and Lewontin, 1967, 1974) have in fact estimated that standard electrophoretic techniques only detect from 25 to 77 percent of the total variability per locus. More sophisticated techniques can, of course, provide additional information. For example, Bernstein and co-workers (1973) used heat denaturation in addition to electrophoresis in their studies of the xanthine dehydrogenase allozymes of the *virilis* group of species in the genus *Drosophila;* this revealed 1.74 times as many alleles in the local populations as had been detected by electrophoresis alone. And, for all nine species of the group studied, the combined technique distinguished between 32 alleles, whereas gel electrophoresis alone had revealed only 11.

On the other hand, current methods of gel electrophoresis may in some respects overestimate the amount of variability. It has been suggested, for example, that the structural proteins may be less polymorphic than soluble enzymes and serum or hemolymph proteins (Clarke, 1974). However, the evidence on this point is conflicting. Certainly the blowfly storage protein lucilin is quite polymorphic (Thomson *et al.,* 1976), and the complexity of the genetic system coding for avian keratin (Kemp, 1975) suggests a high level of polymorphism there also.

Functionally different categories of enzymes may likewise exhibit different levels of polymorphism and heterozygosity. Thus, the enzymes involved in energy metabolism, being universally essential for life, might show less variability than other types of enzymes (Gillespie and Kojima, 1968; and Kojima, Gillespie, and Tobari, 1970). Data on polymorphism and heterozygosity for glycolytic and nonglycolytic enzymes in *Drosophila* and mammals, while not really extensive, tend to support this view (Table 2-1). Since glycolytic enzymes are generally over-represented in samples of enzymes studied by gel electrophoresis, this source of bias would tend to underestimate the total variability.

It may be concluded that the existing data almost certainly underestimate the total amount of variability, both within and between species, but the magnitude of the error is extremely difficult to estimate. It is quite likely, however, that levels of variability measured by current techniques are only about half the true values.

For outbreeding species with large panmictic populations occupying extensive geographic areas, surveys by gel electrophoresis mostly indicate high levels of variability (Table 2-2). There is a suggestion in the data, although perhaps only a slight one, that such species as *Drosophila willistoni, D. subobscura, D.*

TABLE 2-1
Polymorphism and heterozygosity for glycolytic (I) and nonglycolytic (II) enzymes in Drosophila *and mammals, determined by gel electrophoresis.* (From Lewontin, 1974.)*

Species	Polymorphism		Heterozygosity	
	I	II	I	II
D. melanogaster	36%	50%	9.4%	15.6%
D. simulans	36	100	3.0	36.4
D. willistoni	—	—	11.2	22.3
Mus musculus	24	45	8.9	10.6
Homo sapiens	21	32	4.8	7.7

*The number of loci studied ranged from seven in the case of the group II enzymes of *D. simulans* to 47 in that of the group II enzymes of *Homo sapiens*, but was generally between 10 and 20.

obscura, and *D. pseudoobscura,* which are highly polymorphic for inversions, show higher levels than species not known to show chromosomal polymorphism.

Several electrophoretic surveys have indicated much lower levels of heterozygosity and polymorphism than those given in Table 2-2. Serov (1972) found no variability at all in a sample of 138 individuals of *Rattus rattus* from Novosibirsk, which were examined for 21 loci. This is a very unexpected result for a widespread species that forms large populations. Lewontin (1973) has in fact expressed doubt as to its reliability; but Patton and Myers (1974; and see Patton, Yang, and Myers, 1975) have also found a very low level of allozyme variability in this species.* On the other hand, it is hardly surprising (although worthy of note) that Selander and co-workers (1974) found no variability in a sample of 30 individuals of the fossorial rodent *Geomys tropicalis,* since this species is restricted to a sand dune area of about 300 square kilometers on the coast of Tamaulipas, Mexico, and is roughly estimated to number only 74,000 individuals. A complete lack of allozymic variability was also found in three species of bees by Snyder (1974); the reason for this state of affairs is not known, since some other species of Hymenoptera with haploid males are known to show significant levels of allozyme polymorphism (Crozier, 1973; Johnson, Shaffer, Gillaspy, and Rockwood, 1969; Mestriner, 1969). No allozyme variability at 24 loci sampled was found in the northern Elephant seal, *Mirounga angustirostris* (Bonnell and Selander, 1974). The present population of this species is estimated at over 30,000 individuals, but it may have been reduced to as few as 20 in the 1890s as a result of human predation. The southern Elephant seal, *M. leonina,* showed five polymorphisms at 18 loci sampled.

*However, Moriwaki and co-workers (1975) report some polymorphism for transferrins in both the 42-chromosome and 38-chromosome forms of *Rattus rattus*.

TABLE 2-2
Levels of genic variation in vertebrate and invertebrate species.

	Number of populations	Number of loci	Polymorphism per population	Heterozygosity per locus	Ref.
INVERTEBRATES					
Phoronida:					
Phoronopsis viridis	1	39	48.7%	9.4%	(1)
Brachiopoda:					
Liothyrella notorcadensis				3.9	(2)
Frielesis halli		16	56	16.9	(3)
Arachnida:					
Limulus polyphemus	4	25	25	6.1	(4)
Insecta:					
Drosophila melanogaster	1		42	11.9	(5)
D. simulans	1		61	16.0	(6)
D. pseudoobscura	10	24	43	12.8	(6, 7)
D. persimilis	1	24	25	10.6	(8)
D. obscura	3 regions	30	53	10.8	(9)
D. subobscura	6	31	47	7.6[a]	(10)
D. willistoni	Varies	27	56	17.7	(11)
Mollusca:					
Tridacna maxima	1	30	63	20.2	(12)
Echinodermata:					
Ophiomusium lymani	1	15	53	16.6	(3)
VERTEBRATES					
Lepomys macrochirus (sunfish)	47	15	12.1%	2.95%	(13)
Mus. m. musculus	4	41	29	9.1	(14)
M. m. brevirostris	1	40	30	11.0	(15)
M. m. domesticus	2	41	20	5.6	(14)
Peromyscus polionotus	7 regions	32	23	5.7	(16)
P. floridanus	4	39	21	5.3	(17)
Geomys bursarius	12	23	22.6	3.8	(18)
G. personatus	5	23	18.5	4.4	(18)
G. arenarius	2	25	12.0	5.0	(18)
G. tropicalis	1	34	0	0	(18)
Thomomys bottae	5	27	22.6	7.1	(19)
T. umbrinus	1	27	11.1	3.1	(19)
Sigmodon hispidus	5	23	8.1	3.6[b]	(20)
S. arizonae	1	23	9.1	4.9[b]	(20)
Homo sapiens	1	71	28	6.7	(21)

[a] Higher values for this species were obtained by Zouros *et al.*, 1974

[b] Values adjusted for absence of esterase loci; see Selander and Johnson, 1973; and Selander, Kaufman, Baker, and Williams, 1974.

REFERENCES:

1. Ayala, Valentine, Barr, and Zumwalt, 1974.
2. Ayala, Valentine, De Laca, and Zumwalt, 1975.
3. Ayala and Valentine, 1974.
4. Selander, Yang, Lewontin, and Johnson, 1970.
5. Kojima, Gillespie, and Tobari, 1970.
6. Prakash, Lewontin, and Hubby, 1969.
7. Lewontin, 1974.
8. Prakash, 1969.
9. Lakovaara and Saura, 1971a.
10. Lakovaara and Saura, 1971b.
11. Ayala, 1972a.
12. Ayala, Hedgecock, Zumwalt, and Valentine, 1973.
13. Avise and Smith, 1974a.
14. Selander, Hunt, and Yang, 1969.
15. Selander and Yang, 1969.
16. Selander, Smith, Yang, Johnson, and Gentry, 1971.
17. Smith, Selander, and Johnson, 1973.
18. Selander, Kaufman, Baker, and Williams, 1974.
19. Patton, Selander and Smith, 1972.
20. Johnson, Selander, Smith, and Kim, 1972.
21. Harris and Hopkinson, 1972.

Low levels of genic variability were rather consistently found in populations of the kangaroo rats *Dipodomys merriami* and *D. ordii* by Johnson and Selander (1971). In the first species the proportion of loci polymorphic per population ranged from 12 to 29 percent, and in the second from 6 to 24 percent; the mean heterozygosity per locus in *merriami* was 2–7.1 percent and in *ordii* 0.4–1.7 percent.

A question of some importance for theories of speciation is whether in the case of species with continuous areas of distribution on continental land masses the levels of variability are any less in marginal or peripheral populations than they are in ones more centrally located. It is well known that in the case of certain species of *Drosophila* the amount of inversion polymorphism in peripheral populations is notably less than in central ones. This is certainly so for *Drosophila willistoni* (Da Cunha, Burla, and Dobzhansky, 1950; Da Cunha, Dobzhansky, Pavlovsky, and Spassky, 1959; Townsend, 1952, 1958; Dobzhansky, 1957). In some populations in central Brazil the average female fly may be heterozygous for over nine inversions and the average male for more than six, but among peripheral populations in the far south of Brazil, Florida, and the islands of the Antilles the level of inversion polymorphism is very much lower on the average, and on the small island of St. Kitts the average fly is heterozygous for only 0.2 inversions. However, there is no significant difference between the levels of genic polymorphism in island and mainland populations of this species (Ayala, Powell, and Dobzhansky, 1971; Ayala *et al.,* 1972). A similar lack of correlation between levels of genic and inversion polymorphism was found in *D. robusta,* another species in which peripheral populations have much less inversion polymorphism than central ones (Carson, 1965). Yet Prakash (1973) actually found slightly higher levels of genic polymorphism and heterozygosity in peripheral populations of this species. A similar situation has been demonstrated in *Drosophila subobscura* by Zouros and co-workers (1974).

There have been several interpretations of the different levels of inversion polymorphism in central and peripheral populations of such species as *Drosophila willistoni* and *D. robusta.* Dobzhansky (1951) considered the inversions to be differently adapted to the variety of ecological niches available in the environment. The more central environments were regarded as more complex, i.e., as providing a greater range of different niches for the species. Peripheral environments were ones in which a population maintained itself, perhaps precariously, in a single niche, or at any rate in a small number of different niches. By contrast, Carson (1959) argued that the stringent conditions of marginal environments require high levels of genetic recombination in order to provide new combinations of alleles. Since inversion polymorphism reduces the level of recombination, it is not favored at the periphery of a species' distribution. Carson thus contrasts "heteroselection" in the center with "homoselection" at the periphery.

Lewontin (1957, 1974) has put forward a third view, arguing that marginal environments are unstable over time because of shifting equilibria in ecotonal situations. He accepts much of Carson's interpretation but claims that the constantly fluctuating environments occupied by marginal populations make high genic polymorphism necessary for long-term survival but do not favor inversion polymorphism because this restricts recombination too much, as claimed by Carson. Lewontin's interpretation seems to be the only one that adequately explains the discrepancy between the data on genic and chromosomal polymorphism in marginal populations. However, it is necessary to point out that a decrease in inversion polymorphism in peripheral populations does not occur in all species showing this type of polymorphism, and that even where it is present it may simply have an historical explanation—certain inversions may be absent in peripheral populations because they have not reached them.

The discussion of levels of genic polymorphism in peripheral populations has been based on organisms, such as *Drosophila,* whose marginal populations are large enough to render loss of alleles due to genetic drift a negligible factor. The situation is different for species with very small peripheral populations. Several investigations have compared the amount of genic variability in populations of deer mice *(Peromyscus* spp.) on small islands and on the mainland. Selander and co-workers (1971) found much lower levels of polymorphism and heterozygosity in island populations of *P. polionotus* than in populations from the Florida mainland. The average heterozygosity per gene locus in Santa Rosa Island was 1.8 percent, compared with figures of 5.4 percent for the adjacent mainland and 8.4 percent in peninsular Florida. Similar studies were carried out by Avise and co-workers (1974) on a number of species of the same genus on islands in the Gulf of California and the adjacent mainland. In mainland populations heterozygosity was about 6 percent, and polymorphism ranged from 13 to 26 percent; populations from tiny islands showed less than one percent heterozygosity and 0–4 percent polymorphism. The reduced variability of the island species could be ascribed to genetic drift in small populations. Nevertheless, anomalies in the data suggest that this explanation, although correct, may be somewhat too simple to explain all the facts. Mainland populations of *P. eremicus* from California (from which several insular endemic species may have been derived) show very low levels of variation; so do populations of the same species from near El Paso, Texas, and populations of *P. merriami* from northern Sonora, Mexico.

Taken in conjunction with the data on *Rattus rattus,* this evidence would seem to indicate that certain species of very limited distribution may show extremely low levels of genetic variability as a result of drift, but that some other species, including ones with wide distributions, large populations, and high vagility, may have equally little variability owing to their "genetic nature," i.e., for reasons we do not understand.

In a comparison of surface and cave populations of the fish *Astyanax mexicanus* collected in a small area of eastern Mexico, Avise and Selander (1972) found a marked reduction in genic polymorphism in the latter (Table 2-3). One of the three cave populations studied includes a range of fish showing every gradation from an eyed, pigmented phenotype to an eyeless, unpigmented one; the other two cave populations are uniformly eyeless and pale. Surface populations, of course, consist entirely of eyed, pigmented fishes. In spite of these morphological differences, all the populations are considered one species. The coefficient of genetic similarity (Rogers, 1972) is .96 for the six surface populations, .83 for the three cave populations, and .82 for cave populations versus surface populations. These data, as well as the figures for levels of genic polymorphism (Table 2-3), clearly indicate the effect of drift (or of the founder effect) in determining the genetic composition of these small cave populations.

The existence of generally high levels of genetic polymorphism and heterozygosity in sexually reproducing species seems to have resolved the conflict between the so-called "classical" and "balance" theories of population structure (Dobzhansky, 1955). The classical model grew out of the Morgan school of *Drosophila* geneticists (1910–1930); its chief and most articulate protagonist was H. J. Muller (1950). According to this viewpoint, the individuals comprising natural populations are homozygous at most of their gene loci. However, mutations arise from time to time. Those that prove deleterious are eventually eliminated from the population; but until they have disappeared they constitute a "genetic load" on the population. Mutations that prove beneficial rapidly increase in frequency and soon reach fixation in the genotype of the species. Each individual in a natural population is assumed to be heterozygous at about one percent or less of its gene loci, and most of this heterozygosity is for deleterious mutations. Thus, simple heterosis is regarded as an exceptional and not

TABLE 2-3
Estimates of genic polymorphism at 17 loci in Astyanax mexicanus. *(From Avise and Selander, 1972.)*

	Mean percentage of loci		Mean no. alleles per locus per population
	Polymorphic per population	Heterozygous per individual	
Surface populations (6)	37.3 (29.4–31.3)%	11.2 (7.7–13.8)%	2.13
Cave populations (3)	13.7 (0–29.4)	3.6 (0–7.7)	1.22

a general feature of natural populations. The role of adaptive polymorphism in natural populations is therefore minimal in the classical model.

By contrast, in the balance theory, individuals in a natural population of diploid cross-fertilizing organisms are polymorphic at a large proportion of their gene loci (usually more than 10 percent, and in extreme cases tending toward 100 percent). This immense variability is viewed as beneficial, i.e., adaptive to environmental conditions. Polymorphism at each locus is believed to be maintained by balancing selection of some kind. Earlier formulations of the theory stressed heterosis almost to the exclusion of other forms of equilibrating forces, but more recently there has been increasing realization of the importance of frequency-dependent selection.

Closely related to the conflict between the classical and the balance interpretations of natural populations is the controversy regarding the adaptive significance of allele changes that become incorporated in phylogeny. One school of workers, in particular Kimura and Ohta (Kimura, 1968, 1969; Kimura and Ohta, 1971a, b, 1974), have argued that most such allele changes are adaptively neutral and that their establishment is due to random stochastic processes in populations and not to the operation of natural selection. Accordingly, most of the biochemical differences between species are the result of non-Darwinian evolution. A full treatment of this question will not be attempted here, especially because it has been discussed very fully by Lewontin (1974) and Ayala (1975b). Nevertheless, the issue whether most allele changes in phylogeny are adaptive must be faced in this book, since it is relevant to the understanding of the genetic mechanisms of speciation.

The classical model, in its original form, has been decisively disproved by the modern evidence for extensive genic heterozygosity and polymorphism—at least for the overwhelming majority of sexually reproducing species that have been investigated. In these species there is no single homozygous "wild type" individual from which "mutant" individuals deviate in various respects; all are heterozygous for a large fraction (5–20 percent) of their genes, and individual populations are polymorphic for 10–80 percent of the gene loci (these figures are minimal estimates, which take no account of amino acid substitutions that are not electrophoretically detectable). Nevertheless, present evidence suggests the definite possibility that a *few* species do approximate the classical model—there may, after all, be such a thing as a "wild type" rat *(Rattus rattus)*! And among species that reproduce by parthenogenesis (Chapter 9) we find some that have cytogenetic mechanisms that virtually enforce complete homozygosity and others that have chromosome cycles that necessarily lead to ever-increasing heterozygosity. Thus, there certainly exist species (even if not sexually reproducing ones) that conform to the most extreme versions of both the classical and balance theories—both models are biochemically possible.

The present status of Kimura and Ohta's adaptive neutrality hypothesis is less certain, but there is now a great deal of evidence against it from natural populations. If the hypothesis were true, one would expect allele frequencies in populations consisting of a large number of semi-isolated colonies between which migration is on a very minor scale to vary drastically from one colony to another. A number of workers have produced evidence that this is not so. Bullini and Coluzzi (1972) studied the phosphoglucomutase gene *(Pgm)* of the mosquitoes *Aedes aegypti* and *Ae. mariae*. In the former species, which has seven alleles, Pgm^{A1} was the most frequent in 17 out of 19 populations from tropical regions of Africa, Asia, the Pacific, and Puerto Rico. In the latter species, which has five alleles, Pgm^{B} had frequencies of 70 to 100 percent in 15 different Mediterranean populations. *Ae. mariae* is one of a group of three sibling species; in *Ae. zammitii* the Pgm^{A1} allele predominates, and in *Ae. phoeniciae* the Pgm^{C} allele.

Even more convincing evidence against the neutrality hypothesis has been provided by the work of Nevo (1973) on the frequencies of alleles at 24 polymorphic loci in 11 populations (including six chromosome races or semi-species) belonging to the pocket gopher superspecies *Thomomys talpoides*. The genic similarity between the populations, in spite of the reproductive and/or geographical isolation between them, strongly supports natural selection rather than adaptive neutrality.

Ayala (1975b) has presented six arguments against the neutrality hypothesis, based on the known facts on protein polymorphism in the *willistoni* group of *Drosophila*. These may be very briefly summarized as follows.

1. The mean frequency of homozygotes (in five species) for electrophoretically detectable alleles is 0.823. This is 23 times the value predicted by the neutrality hypothesis on the basis of the lowest acceptable estimates of population size and mutation rates.
2. The distribution of locus heterozygosities, with an observed mean of 0.177 for all the species, has a mode at 0 and a flat range from 0.02 to 0.68, whereas the neutrality hypothesis predicts a normal distribution.
3. Even if we assume a hundredfold range of mutation rates, the proportion of essentially monomorphic loci is far in excess of that predicted by the neutrality hypothesis.
4. Allelic frequencies in different populations of the same species are far more similar than would be expected by the neutrality hypothesis (the point made by Bullini and Coluzzi in the case of *Aedes*).
5. The similarities of allelic frequencies in different species of the *willistoni* group are far in excess of those expected by the neutrality hypothesis (the same point made by Nevo in the case of *Thomomys*).

6. Pairs of experimental populations of *D. equinoxialis* and *D. tropicalis* showed similar changes in the frequency of allele 0.94 at the malate dehydrogenase-2 locus, but the changes were in opposite directions in the two species, corresponding to the frequencies in wild populations. This experiment is very conclusive, but of course only for this particular allele.

Quite a different line of evidence has been brought to bear on this question by Maynard Smith (1970), who points out that if the neutrality hypothesis were true, there should be (at equilibrium between mutation and selection) a higher proportion of polymorphic loci in large populations than in small ones. Applying this principle to the human hemoglobin variants, he concludes that if the neutrality hypothesis is true there must have been a strongly marked "bottleneck" in population numbers at some time during the past million years (which seems inherently improbable, although perhaps not impossible).

Lewontin (1974) has criticized the kind of argument relied upon by Bullini and Coluzzi on the ground that a migration rate between populations of the order of one individual per thousand per generation would be sufficient to prevent differentiation, as a result of genetic drift, between populations of moderate size. But this objection does not apply to Nevo's results, where the populations considered are strongly isolated reproductively as well as geographically, and are for the most part separate biological species. In fact, Lewontin himself cites numerous cases of sibling species of *Drosophila* in which allele frequencies at many loci are far too similar for the neutrality hypothesis to be credible.

A major argument in favor of the neutrality hypothesis seems to arise from the fact that if all the variation discovered in natural populations in the last 10 years consisted of balanced polymorphisms, the total "genetic load" (segregation load) would be intolerable and the degree of inbreeding depression far in excess of what is actually observed. Various ways out of this dilemma have been pointed out (King, 1967; Milkman, 1967; Sved, Reed, and Bodmer, 1967; Lewontin, 1974). However, all calculations on this question depend rather directly on estimates of the total number of loci in the genome. It is quite probable that these totals have been consistently overestimated. Some authors, in order to arrive at a figure, have simply taken the total DNA value of a species and divided this figure by the amount of DNA needed to code for a protein of average size—a calculation that ignores the fact that an unknown but large fraction (perhaps as much as 90 percent in many species) is "noninformational."

Lewontin (1974) has based some calculations on a figure of 15,162 polytene chromosome bands in *Drosophila melanogaster,* assuming that each band contains one structural locus. But this figure is an error and should be 5,161. Even this number seems to be a maximum estimate (White, 1973); the true figure may be as low as 3,441. In the newt, *Triturus cristatus,* the number of chromomeres in the lampbrush chromosomes is at least 3,300 in the haploid genome, according to the estimate of Mancino (1963), and this estimate might be increased somewhat by improved techniques. If the total number of structural genes in such organisms as insects and urodeles is only 4,000 to 5,000—rather than the 10,000 or 20,000 that many recent workers have postulated—the dilemma posed by high levels of balanced polymorphism, although it still has to be faced, is much less serious.

There is a significant difference in the levels of genic heterozygosity between vertebrate and invertebrate species, the latter having levels two or three times those usual in vertebrates (see Table 2-2). Assuming the existence of 5,000 loci in both cases, we might suggest that the average vertebrate is heterozygous at about 250 loci, the average invertebrate at 600–700. Dobzhansky (1972), who accepts an estimate of 100,000 loci in *Drosophila,* is forced to the conclusion that the average population contains 25,000 polymorphic loci and that the average fly is heterozygous for some 8,000 genes. Many authorities will consider the estimates accepted in this book as too low, but Dobzhansky's are certainly too high, by at least an order of magnitude.

A full discussion of the present evidence on levels of polymorphism and heterozygosity and its significance would be outside the scope of this book; the whole subject has been very thoroughly dealt with by Lewontin (1974). It has been necessary, however, to present a general and simplified account of the data as a preliminary to the main topic of this chapter, namely the extent of the genic differences between related species and, especially, between populations in various stages of speciation.

The view of such evolutionists as Mayr (1954, 1963a, 1970) and Dobzhansky (1970) has been that speciation is normally accompanied by a major reorganization of the genotype. Mayr emphasizes that this is especially the case in species that arise from geographically isolated populations descended from a small number of founder individuals: the incipient species passes through a period in which it loses a great deal of its variability, and afterward undergoes a so-called "genetic revolution," from which a new "integrated genotype" is created. Dobzhansky's (1970) view, while less specifically stated, is that even closely related species ". . . must differ in thousands of genes."

Mayr starts his argument with somewhat extreme views on the role of epistasis in population genetics. An opponent of "beanbag genetics," he claims that "continuous populations of species are held together by such close ties of genetic cohesion that one can scarcely conceive of this essentially single gene pool being divided into two." And, "The real problem of speciation is not how differences are produced but rather what enables populations to escape from the cohesion of the gene complex and establish their own independent identity. No one will comprehend how formidable this problem is who does not understand the power of the cohesive forces in a coadapted gene pool."

It is certainly impossible to deny the existence of many types of epistatic interactions between different gene loci and the importance of epistasis as a factor in the genetics of evolving populations. But it is clear that the newer knowledge of levels of polymorphism in natural populations has thrown considerable doubt on the concept of an extremely powerful "cohesive force" preventing speciation from occurring. An elementary model may help clarify what sort of "force" this might be. Let us imagine a population that is polymorphic for two alleles at a *single* locus. If three genotypes A_1A_1, A_1A_2, and A_2A_2 are approximately equal in fitness, we can imagine that the population might split into subpopulations homozygous for A_1 and A_2, respectively. In this case there is no cohesive force, or only a minimal one. On the other hand, if A_1A_1 and A_2A_2 both have a much lower fitness than A_1A_2, i.e. if there is strong heterosis, then it is highly unlikely that homozygous populations will be produced by the original polymorphic one. Similarly, we may imagine that if there are epistatic interactions between two loci A and B such that $A_1A_1B_1B_1$ and $A_2A_2B_2B_2$ are adaptive combinations but genotypes containing A_2B_1 and A_1B_2 are not, we may expect that the population may split into homozygous subpopulations. However, this would be less likely if there were adaptive heterozygous combinations, the extreme case being the double heterozygote $A_1A_2B_1B_2$ having by far the highest fitness.

In view of the known facts regarding levels of genetic polymorphism in natural populations, Mayr's estimate of the strength of cohesive forces seems to imply the existence of completely intolerable genetic loads. There can be no doubt that some kind of cohesive force exists, but it must be seriously questioned whether it is strong enough to act as a barrier to speciation. This was also the view of Haldane (1964): "Mayr devotes a good deal of space [in *Animal Species and Evolution*] to such notions as 'genetic cohesion,' 'the coadapted harmony of the gene pool,' and so on. These apparently become explicable 'once the genetics of integrated gene complexes has replaced the old beanbag genetics.' So far as I can see, Mayr attempts to do this in his Chapter 10. . . . This chapter contains a large number of enthusiastic statements about the advantages of large populations which, in my opinion, are unproved and not very probable." Felsenstein (1975) has characterized the overemphasis on genetic interactions—the use of such expressions as "ubiquitous," "infinitely strong," "infinitely complex," and "not susceptible of analysis"—as "positively medieval."

Since, even if they exist, the role of cohesive forces in preventing speciation cannot be measured, even approximately, by present-day techniques, it seems legitimate for the time being to regard the origin of genetic isolating mechanisms rather than subdivision of the gene pool as the prime cause of speciation, and this is the position that will be taken throughout this book. It is true that in Chapter 7 we shall discuss some instances of "host races" in which there has been a subdivision of the gene pool before the origin of genetic isolating mechanisms, and this may even be the usual course of events in allopatric speciation (Chapter 4). But host races and geographic subspecies are not to be regarded as species, although they may be incipient ones.

In order to compare the genetic constitution of two populations of the same or different species, for which the allelic frequencies at a number of loci are known, some kind of index is required. A number of statistics have been proposed, of which perhaps the most satisfactory is the distance coefficient developed by Rogers (1972):

$$D = \frac{1}{L} \sum_{i=1}^{L} \left[\frac{1}{2} \sum_{j=1}^{A_i} (P_{ijX} - P_{ijY})^2 \right]^{1/2},$$

where L is the number of loci; A_i is the number of alleles at the ith locus; and P_{ijX} and P_{ijY} are the frequencies of the jth allele at the ith locus in populations X and Y, respectively. The range of D is from zero, when the two populations have identical gene frequencies, to 1.0, when they have reached fixation for alternative alleles at all loci. In some cases it may be more convenient to use the similarity coefficient, $S = 1 - D$.

An alternative statistic measuring genetic differentiation between taxa is the one proposed by Nei (1972):

$$D = -\ln I,$$

where $I = J_{XY} / \sqrt{J_X J_Y}$; J_X, J_Y, and J_{XY} are the arithmetic means, over all loci, of Σx_i^2, Σy_i^2, and $\Sigma x_i y_i$ (where x_i and y_i are the frequencies of the ith alleles in populations X and Y, respectively). Because of the different methods of calculation, values of Rogers's and Nei's coefficients are not directly comparable; for the same set of data, the Nei coefficient is generally higher.

More is known about the population genetics of *Drosophila pseudoobscura* and its sibling *D. persimilis* than any other species. Therefore, we shall begin the discussion of genetic differentiation between populations and species with this example.

The distribution of *D. pseudoobscura*—from British Columbia southward to Texas, Mexico, and Guatemala, with an isolated population around Bogotà, Colombia, which may be of relatively recent origin—is well known. *D. persimilis* has a much smaller range, being essentially confined to mountainous areas from British Columbia to central California (Figure 1). The phylogenetic relationships of the third-chromosome inversions in these species, and the distribution of the third-chromosome sequences in the natural populations, are likewise classical knowledge in the field of evolutionary genetics (Figure 2; see also

FIGURE 1

Distribution range of *Drosophila pseudoobscura* and *D. persimilis* and location of populations sampled for allozymes to determine levels of heterozygosity (see Table 2-4). [Based on Lewontin, 1974]

p. 176). A new era in the understanding of the genetic architecture of *D. pseudoobscura* was initiated by Hubby and Lewontin (1966; and see Lewontin and Hubby, 1966) with their gel-electrophoretic study of the natural variation of allozymes and larval hemolymph proteins. This early work was followed by a more extensive electrophoretic study by Prakash, Lewontin, and Hubby (1969) in which 24 genetic loci were analyzed in 12 populations (Table 2-4).

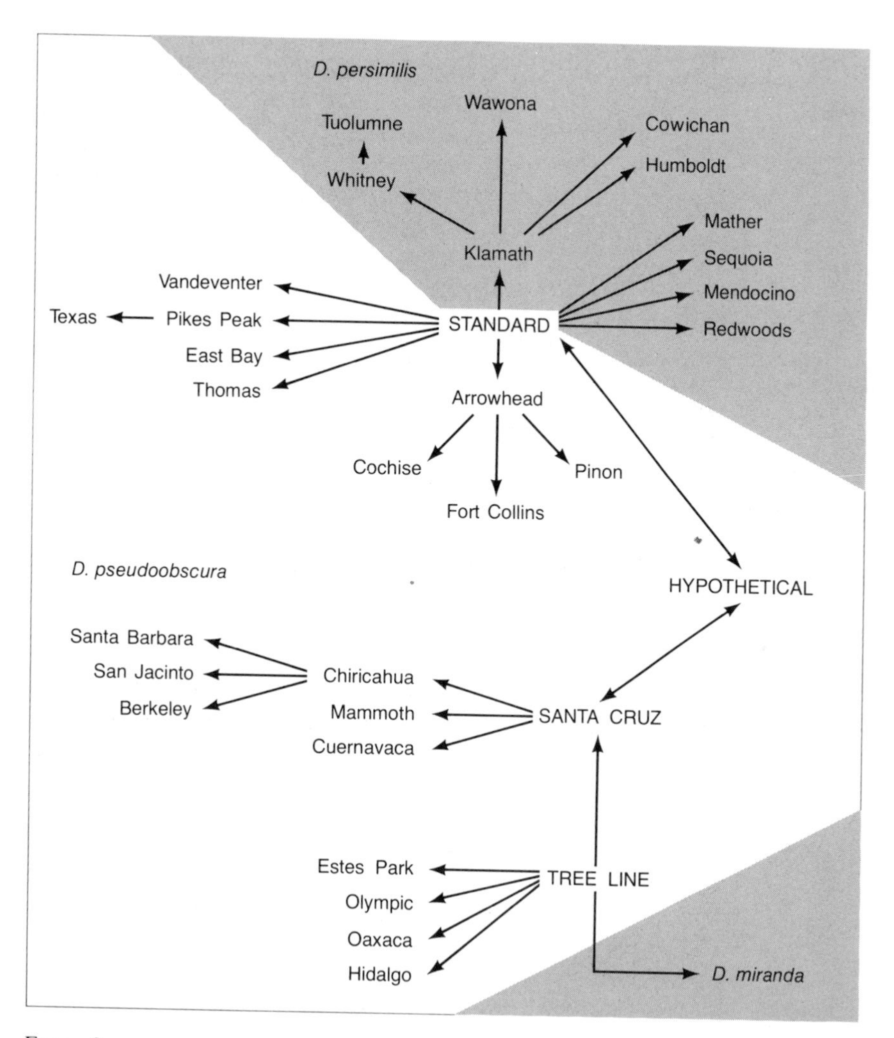

FIGURE 2

Phylogenetic relationships between arrangements on the third chromosome of *Drosophila pseudoobscura* and *D. persimilis*. [After Lewontin, 1974; based on Dobzhansky and Epling, 1944]

The results showed that in populations an average of 40 percent of the gene loci are polymorphic, and in individuals an average of 12 percent of the loci are heterozygous. Of the 24 loci examined, 11 are invariant in all populations. All populations in the main area of distribution of the species have very similar

TABLE 2-4
Genic polymorphism in populations of Drosophila pseudoobscura. *(From Lewontin, 1974).*

Population	Loci polymorphic	Loci heterozygous	Heterozygotes for third-chromosome inversions
Strawberry Canyon, Calif.	50%	16.1%	71%
Wild Rose, Calif.	38	12.9	67
Charleston, Nevada	46	12.6	47
Sheep Range, Nevada	42	12.5	52
Cerbat, Arizona	46	11.2	58
Mesa Verde, Colorado	46	11.7	4
State Recreation, Colorado	42	13.1	69
Hardin Ranch, Colorado	33	12.3	69
Nelson Ranch, Colorado	42	14.0	79
Austin, Texas	46	12.6	38
Guatemala	21*	8.1	49
Bogotà, Colombia	25	5.1	47
Mean:	40	11.9 ± 0.8	
Mean, excluding Guatemala and Bogotà:	43	12.9 ± 0.6	

*A spuriously low value, due to small sample size.

levels of genic polymorphism and heterozygosity. This contrasts with the striking differences in the degree of polymorphism for third-chromosome inversion sequences. The Mesa Verde population, from an area where almost the only third-chromosome sequence is Arrowhead, actually has a higher than average proportion of its loci polymorphic. It is only in the geographically isolated populations in the mountains of Guatemala and Colombia that we find a markedly lower level of genic polymorphism. The level of structural chromosomal heterozygosity is not particularly low in these populations, but the number of different third-chromosome sequences (not shown in Table 2-4) is much reduced—there are only three sequences in Guatemala and two in Colombia, as compared with eight at Strawberry Canyon and six at Austin.

Some of the gene loci investigated lie on the third chromosome, and particular alleles may be associated with certain inversion sequences of that chromosome. Thus, at the protein-10 locus the 1.04 allele is present in all Standard third chromosomes and in most Arrowhead ones (except at Mesa Verde, where the 1.02 allele predominates in Arrowhead chromosomes). The 1.04 allele is present in almost all Pikes Peak chromosomes, but is absent from Santa Cruz chromosomes, which have allele 1.06. At Mather, California, half the Chiricahua chromosomes carry allele 1.04, the other half carry 1.06; but at Strawberry Creek all of them carry 1.06. Summarizing an enormous quantity of data, Lewontin (1974) stresses four features of the genetic architecture of *D.*

pseudoobscura populations as follows: "(1) Monomorphic loci are identically monomorphic over all populations; (2) most polymorphic loci show no pronounced geographical differentiation; (3) loci associated with inversions show patterns of geographic differentiation characteristic of the inversions; and (4) a completely isolated population [i.e. Bogotà] is more homozygous than populations in the continuous range of the species."

At 24 gene loci in *D. persimilis* and *D. pseudoobscura* studied by Prakash (1969), the allele frequencies were remarkably similar in the two species. The same eleven loci are monomorphic in both, and at 10 out of the 13 polymorphic loci the predominant allele was the same in the two species. There was not a single instance of fixation or near-fixation for alternative alleles in the two siblings. Commenting on these results, Lewontin points out that if there are, as there must surely be, species-specific genes in *pseudoobscura* and *persimilis,* they were not found in a random sample of 24 loci, and must hence be less than 10 percent of the total. However, there is a serious source of error in this argument. What is referred to as an "allele" is really a band on an electrophoretogram. Using the technique of heat denaturation in addition to gel electrophoresis, Bernstein, Throckmorton, and Hubby (1973) found the proportion of "alleles" possessed in common by species pairs of the *virilis* group of *Drosophila* to be considerably less than that found by electrophoresis alone (Table 2-5). Thus far this kind of comparison has only been carried out for the xanthine dehydrogenase locus in this one species group of *Drosophila,* and it is possible that this locus is not typical. Nevertheless, as far as the discussion in this chapter is concerned, all estimates of genetic similarity between populations and species are regarded as overestimates and, conversely, estimates of genetic differences are underestimates.

TABLE 2-5
Proportion of the alleles of the xanthine dehydrogenase locus possessed in common by Drosophila virilis *and other members of the* virilis *group. (From Bernstein* et al., *1973.)*

D. virilis and:	Electrophoresis	Electrophoresis and heat denaturation
D. a. americana	0.60	0.31
D. a. texana	0.67	0.14
D. novamexicana	0.67	0
D. lummei	0.20	0.13
D. littoralis	0.17	0.08
D. ezoana	0	0
D. montana	0.13	0.07
D. flavomontana	0	0
D. borealis	0	0
D. lacicola	0.14	0.08

The early work of Hubby and Throckmorton (1965) on the members of the *virilis* group led to an overall assessment of the proteins that were identical from species to species. It was concluded that speciation involved a change in 20 percent of the gene pool. Altogether, 29 to 42 different proteins in 10 different members of the *virilis* group were studied. The proportion of these proteins unique to the species ranged from 2.6 percent in *virilis* to 28.2 percent in *littoralis,* with a mean of 14.3 percent. On the assumption that the existing species of the *virilis* group have arisen from four immediately ancestral species, Lewontin (1974) estimates that a minimum of 8.5 percent of the proteins in the present-day species have arisen since speciation. He points out, however, that it is impossible to determine how much of this divergence occurred during the actual speciation process and how much has taken place in subsequent phyletic evolution.

In later work, Hubby and Throckmorton (1968) studied nine "triplets" of species, each composed of two morphologically indistinguishable (or almost so) sibling species, and a third, related but clearly distinguishable, species. The percentage of identity between siblings ranged from 85.7 in the *D. victoria–D. lebanonensis* pair down to 22.5 in *D. willistoni–D. paulistorum*. However these results were based on only one strain, usually homozygous, of each species, which makes them of limited value for evolutionary discussions.

The genetic differentiation between the sibling species *D. melanogaster* and *D. simulans* was studied by Kojima, Gillespie, and Tobari (1970) on the basis of 17 enzyme loci. At no locus were alternative alleles fixed in the two species, but at one locus *melanogaster* has undergone fixation for an allele that has a frequency of less than one percent in *simulans*.

One general conclusion that emerges very strongly from all this work is the fact that alleles fixed in a species are almost invariably present, even if in low frequency, in closely related species. As Lewontin (1974) puts it, "The overwhelming preponderance of genetic differences between closely related species is latent in the polymorphisms existing within species." Whether this rule applies to the genes directly responsible for genetic isolating mechanisms we do not know—probably it does not.

The geographically isolated Bogotà population of *D. pseudoobscura* was probably derived from flies that reached Colombia from Guatemala some thousands of years ago (Dobzhansky, 1975). The idea that it reached Colombia only about 25 years ago, accepted by Lewontin (1974), is certainly wrong. The Bogotà population has acquired a partial genetic isolation from the rest of the species, i.e., sterility of the F_1 males, but its genic evolution has consisted mainly in becoming homozygous for alleles that are the most common ones in other populations of *pseudoobscura*.

Particularly high coefficients of similarity were obtained by Yang, Wheeler, and Bock (1972) for four species of the *bipectinata* group of *Drosophila* from southeast Asia and the Pacific. Two of these are each subdivided into pairs of subspecies; the coefficient of similarity between one pair was .92 and .82 between the other. For the four species the mean coefficient of similarity was .76 (range .66 to .91). None of them would be regarded as sibling species, but genically they are more similar than sibling species in many other groups. The similarity coefficients of .91 for the pair *D. bipectinata–D. malerkotliana pallens* and .89 for *D. bipectinata–D. m. malerkotliana* are among the highest that have been recorded for full species. They constitute strong evidence for the view that a major "genetic revolution" does not necessarily accompany speciation. An average of 9.3 fixed interspecific chromosomal inversions separates each pair of species in this complex, an unusually high figure for the genus *Drosophila*. There is some reason therefore to suppose that these inversion differences may have played a special role in speciation in the *bipectinata* group, and that the mechanism of speciation may in fact have been somewhat different to that which has prevailed in some other groups of the genus *Drosophila* (such as the *willistoni* group, where the genetic similarities between the species are very much less).

Some data on the extent of genetic divergence in the Hawaiian species of *Drosophila* has also been published by Rockwood and co-workers (1971). Of particular interest are the data on "homosequential" species pairs, i.e., ones that have no differences in the banding pattern of their polytene chromosomes. The divergence turns out to be highly variable, ranging from 26.3 to 80.0 percent, with a mean of 54.7 percent. Obviously, identity of gene sequence does not necessarily imply a close genic similarity, although in the "picture-winged" group of *Drosophila* the average genetic divergence between non-homosequential species is slightly higher than the average between the homosequential species pairs.

In the *willistoni* group the "semispecies" of the *D. paulistorum* complex (Dobzhansky and Pavlovsky, 1967; Dobzhansky, Pavlovsky, and Ehrman, 1969; Ehrman, 1965; Ehrman and Williamson, 1969; Kernaghan and Ehrman, 1970) have attracted particular attention because they seem to be precisely on the threshold of species status. (Most subspecies are probably nowhere near the point of speciation divergence, while even the most closely related sibling species may have undergone a great deal of independent genetic evolution *since* speciation.) There are six of these semispecies: the Centroamerican; Transitional, Amazonian, Andean–Brazilian, Orinocan, and Interior populations. A seventh, originally named the Guianan, is now regarded as a full species, *D. pavlovskiana*. There is, in general, rather strong ethological isolation between the semispecies. Laboratory crosses between different semispecies sometimes give no progeny at all. If progeny are obtained, they may consist of sterile males and fertile females. Occasionally (and especially in crosses involving the Transitional population) fertile F_1 individuals of both sexes can be bred. Sterile male hybrids seem to be the result of infection by a microorganism (a mycoplasma) transmitted from the mother via the egg cytoplasm. It is thus possible that one factor in the incipient speciation of these semispecies is an adaptive interaction, or at least tolerance, between the microorganism and the genes of the fly that is liable to break down in the hybrids.

Each semispecies has some characteristic chromosomal inversions of its own, but there are also some inversion polymorphisms that are shared between two or even three semispecies. There is a great deal of sympatry between the semispecies; the Andean–Brazilian, Amazonian, and Interior populations coexist over a wide area without losing their identity. Sexual isolation is virtually complete between some semispecies, but is generally incomplete between the Transitional population and the others, so that the Transitional population in Colombia and Venezuela renders possible a limited gene flow between semispecies. The genetic differences between the six semispecies have been studied by Richmond (1972) and Ayala, Tracey, Hedgecock, and Richmond (1974). The results are summarized in Table 2-6.

Several recent authors, in discussing the amount of genetic change that takes place in the course of speciation, have distinguished between early and late stages of the process. This is an advance from earlier discussions, in which speciation was considered as a single event. However, the significance of any such division of speciation into stages depends on the precise model of speciation being considered. Thus, when Ayala and co-workers (1974) speak of a "first" and "second" stage of speciation (on the basis of their work on the *willistoni* group of *Drosophila*) they are referring to the orthodox allopatric model. They conclude that much genetic differentiation takes place during the first stage (corresponding to the formation of geographic subspecies) and that little genetic differentiation occurs during the second stage ("development of complete reproductive isolation"). Yet Lewontin (1974), also thinking in terms of the allopatric model, concludes from a review of earlier evidence on the

TABLE 2-6
Genetic similarity (above the diagonal) and genetic distance (below the diagonal) between pairs of semispecies of Drosophila paulistorum. *Distance is calculated using Nei's statistic (see p. 33). (From Ayala* et al., *1974.)*

	Centroamerican	Transitional	Andean–Brazilian	Amazonian	Orinocan	Interior
Centroamerican	—	0.957	0.932	0.859	0.811	0.880
Transitional	0.043	—	0.935	0.838	0.820	0.862
Andean–Brazilian	0.071	0.067	—	0.844	0.848	0.881
Amazonian	0.152	0.176	0.170	—	0.827	0.875
Orinocan	0.209	0.198	0.165	0.189	—	0.929
Interior	0.127	0.149	0.127	0.134	0.074	—

willistoni group, the subspecies of *Mus musculus*, and the Bogotà population of *Drosophila pseudoobscura* that the first stage of speciation ("the acquisition of primary reproductive isolation in geographical solitude") does not involve a major genetic revolution and "may result from chance changes in a few loci." During the second stage ("when the isolated populations have again come into contact") there is some additional genetic divergence, but it is during the "third stage of species evolution" that major genetic change occurs independently in the two species.

It is fairly clear that Lewontin's "third stage" is not really part of the speciation process itself, but rather phyletic evolution following on speciation. Lewontin has, however, greatly clarified the discussion by pointing out that major genetic divergence occurs during this stage. In other words, we cannot measure the genetic divergence that has occurred between species (even sibling species) and conclude that this amount of genetic change has occurred during speciation—most of it, in the majority of cases, will have arisen after speciation has been completed.

But what of the apparent divergence of views between Ayala and Lewontin as to what happens in stages 1 and 2? Clearly, both consider that geographic subspecies result from the first stage. However, these can be either "old" or "recent" pairs of subspecies. The subspecies of *D. equinoxialis* and *D. willistoni* considered by Ayala show no sexual isolation, but in the former the hybrid males are always sterile and in the latter they are sterile when the male parent is *D. w. willistoni* and the female *D. w. quechua*. In both these examples the evolutionary separation probably took place hundreds or even thousands of years ago. Ayala rightly says that much genetic differentiation occurred during subspeciation ("stage 1" of speciation), since the genetic distance (D) is 0.23. But the further step from subspecies to semispecies does not seem to have involved any further increase in genetic distance (Table 2-7). The progression

TABLE 2-7
Mean genetic similarity ($\bar{I}$) *and genetic distance* ($\bar{D}$) *between taxa at various levels of evolutionary divergence in the* willistoni *group of* Drosophila. *The values are calculated using Nei's method. Standard errors are not given but are small in all cases. (From Ayala* et al., *1974.)*

Level of divergence	$\bar{I}$	$\bar{D}$
Local populations	0.970	0.031
Subspecies	0.795	0.230
Semispecies	0.798	0.226
Sibling species	0.563	0.581
Non-sibling species	0.352	1.056

from semispecies to sibling species, however, seems to have necessitated at least a doubling of the genetic distance between the evolving populations (from .23 to .58 in the case of the *willistoni* group).

Genetic similarity indexes for a number of animal groups are given in Table 2-8. The highest values for what are undoubtedly different species were obtained by Turner (1974) for five relict species of the pupfish genus *Cyprinodon* inhabiting Death Valley, California and river systems that were formerly con-

TABLE 2-8
Mean genetic similarities between taxa at different levels of evolutionary divergence (mostly from Ayala et al., *1974, simplified). Indexes are calculated from the formula of Rogers (1972) except for those with asterisks, which have been calculated from Nei's formula (1972).*

	Number of species	Number of loci	Taxonomic level			Ref.
			Subspecies	Semispecies	Species	
Mammals:						
Dipodomys	11	18	—	—	.61 (.31–.89)	(1)
Sigmodon	2	23	—	—	.76 (.76–.77)	(2)
Mus musculus	1	41	.84 (.82–.88)*	—	—	(3)
Peromyscus	16	21	—	—	.66 (.34–.99)	(4)
Thomomys	1	27	.84 (.83–.86)	—	—	(5)
Reptiles and Amphibia:						
Anolis	4	23	—	—	21 (.16–.29)	(6)
Sceloporus[a]	1	20	—	.79 (.73–.84)	—	(7)
Taricha	3	18	.87 (.77–.90)*	—	.63 (.50–.77)*	(8)
Fish:						
Lepomys	10	14	—	—	.53 (.37–.79)	(9)
Cyprinodon	5	31–38	—	—	.90 (.84–.97)	(10)
Drosophila species groups:						
virilis	9	37	.65	—	44 (.28–.79)[b]	(11)
obscura	11	22	—	—	.29 (.15–.74)[b]	(12)
bipectinata	4		.91 (.90–.92)	—	.79 (.68–.96)*	(13)
mesophragmatica	6	24	—	—	.50 (.30–.77)	(14)
willistoni	7	36	.80 (.78–.81)*	.80 (.77–.82)*	.45 (.27–.79)*	(15)

[a]These populations of *Sceloporus* were considered subspecies by Ayala *et al.*, but are designated "independently evolving biological species" by Hall and Selander. In the above table they have been regarded as semispecies.

[b]Calculated as the proportion of loci at which the most common allele is identical in the two taxa.

REFERENCES:
1. Johnson and Selander, 1971.
2. Johnson *et al.*, 1972.
3. Calculated by Ayala *et al.*, 1974, from Selander *et al.*, 1969.
4. Avise *et al.*, 1974.
5. Patton *et al.*, 1972.
6. Webster *et al.*, 1972.
7. Hall and Selander, 1973.
8. Hedgecock and Ayala, 1974.
9. Avise and Smith, 1974a, b.
10. Turner, 1974.
11. Hubby and Throckmorton, 1965.
12. Lakovaara *et al.*, 1972.
13. Calculated by Ayala *et al.*, 1974, from Yang *et al.*, 1972.
14. Nair *et al.*, 1971.
15. Ayala *et al.*, 1974.

nected with Death Valley. These species, which are morphologically quite diverse, have been cited as an example of very rapid post-Pleistocene evolution (Miller, 1948, 1950; Hubbs and Miller, 1948). In spite of striking differences in size, shape, dentition, fins, scalation, ecology, and behavior, the genetic similarity indexes range from .84 to .97, values only slightly less than those obtained for populations of the lizard *Anolis carolinensis* and the cotton rat *Sigmodon hispidus* (.957–.986 in the first case, .970–998 in the second).

The results for *Cyprinodon* contrast strongly with those for the sunfishes of the genus *Lepomis* in the eastern and central United States (Avise and Smith, 1974a, b). Despite the propensity of these species to hybridize, they show major allelic differences at about half the gene loci tested ($\bar{S}$ = .53, ranging from .37 to .79). It seems likely that the evolutionary separation of the *Lepomis* species took place at a much earlier period than that of the Death Valley pupfishes.

Summarizing the results that have been obtained thus far on genic differences at different levels of evolutionary separation, we can say that there is no evidence for the accumulation of major differences during that particular critical stage of speciation when genetic isolating mechanisms are acquired. In this sense, Mayr's theory of a "genetic revolution" occurring during speciation has not been confirmed by modern biochemical work. Allelic differences accumulate in the course of phyletic evolution, regardless of whether this takes place before or after the acquisition of genetic isolating mechanisms. Dobzhansky (1972) is certainly correct in stating that ". . . reproductive isolation may sometimes follow and at other times precede the adaptive divergence of gene pools of populations." Thus, subspecies that are isolated geographically for long periods of time may build up considerable genetic differences even if they have not acquired genetic isolating mechanisms. Conversely, some semispecies or incipient species may acquire significant genetic isolating mechanisms but rather minimal levels of allelic divergence, e.g., the Bogotà population of *D. pseudoobscura*. The degree of morphological divergence, which is what classical taxonomists are concerned with, is not closely correlated with genetic divergence, e.g., in the Death Valley pupfishes. Thus, man and the chimpanzee are separated by a remarkably small genetic distance, comparable to that found between sibling species of *Drosophila* and rodents (King and Wilson, 1975). Most of the genetic divergence between so-called closely related species (even sibling species) is due to phyletic evolution that has occurred subsequent to evolutionary separation, and has little or nothing to do with speciation as such. Minimal genic divergence may occur in spite of considerable morphological differentiation, especially when "explosive" or very rapid speciation has occurred. It is unfortunate that as yet we have no really adequate comparisons between the degree of allelic and karyotypic divergence in any one group of organisms. It might be particularly useful to have comparisons of allelic divergence between species pairs with large-scale differences in quantities of heterochromatin and those without.

Studies of allelic variation by gel electrophoresis have thus far been largely confined to animals. In the future we may expect highly significant data for a number of plant groups. The understanding of polyploidy, in particular, will be greatly increased by gel electrophoresis.

In theory, there is a much more satisfactory technique than gel electrophoresis for studying species differences at the molecular level, namely the direct sequencing of the amino acids in the various proteins. Immensely laborious, this technique has thus far been applied only to a very limited range of proteins, such as cytochrome *c*, the hemoglobins, and the fibrinopeptides. The results, on widely different species, are more relevant to problems of long-term phyletic evolution than they are to the question of genetic changes during speciation. Nevertheless, it is clear that the present state of our knowledge is a transitional one and that soon the evidence from protein-sequencing will begin to have an impact on the speciation problem and remove some of the imprecision inherent in estimates based on electrophoretic techniques.

3

Chromosomal Differences Between Species

Not all species differ in their karyotypes, but most of them certainly do. Only in the genus *Drosophila* do we have certain evidence of the existence of a small number of instances where from two to five closely related species have karyotypes that are indistinguishable in the polytene salivary gland chromosomes, the banding pattern being precisely the same. Two such cases—a pair of sibling species and a trio, barely distinguishable, if at all, in external phenotype—were reported by Wasserman (1962, 1963). They were both in the *repleta* species group, in which the number of chromosomal rearrangements that have established themselves is lower than in other *Drosophila* species groups. Carson, Clayton, and Stalker (1967) described several groups of Hawaiian species with identical banding patterns in their polytene chromosomes and introduced the term *homosequential complexes* for such groups. There are now 13 known homosequential complexes in the genus *Drosophila,* 10 of them in the Hawaiian fauna (Table 3-1). Many of the homosequential Hawaiian species are strikingly different in external phenotype, so that their close relationship would hardly have been suspected if they had not been studied cytogenetically. Altogether, the 13 complexes include 38 species (out of about 500 species of *Drosophila* whose polytene chromosomes have now been analyzed). It is not certain, however, that all homosequential species have precisely the same distribution of satellite DNAs in their heterochromatic segments, or even the same amount and distribution of heterochromatin, which cannot easily be analyzed in the polytene chromosomes. In fact, Ward and Heed (1970) showed that two Mexican species of *Drosophila* considered homosequential actually differed in their heterochromatic segments.

TABLE 3-1
Homosequential species complexes in the genus Drosophila.*

Complex	Geographical distribution
1. *Mulleri, aldrichi,* and *wheeleri*	Western U.S.A., Mexico
2. *Meridiana* and *meridionalis*	Texas, Mexico, Brazil
3. *Bostrycha, disjuncta, grimshawi orphnopeza,* and *villosipedis*	Hawaiian archipelago
4. *Glabriapex, pilimana,* and *vesciseta*	Hawaiian archipelago
5. *Limitata, sejuncta* and *ochracea*	Hawaiian archipelago
6. *Balioptera* and *murphyi*	Hawaiian archipelago
7. *Hawaiiensis, musaphilia, recticilia,* and *silvarentis*	Hawaiian archipelago
8. *Heteroneura, silvestris* and *planitibia*	Hawaiian archipelago
9. *Neopicta* and *obscuripes*	Hawaiian archipelago
10. *Hanaulae* and *oahuensis*	Hawaiian archipelago
11. *Adiastola, cilifera* and *peniculipes*	Hawaiian archipelago
12. *Ocellata, paucipuncta, punalua,* and *uniseriata*	Hawaiian archipelago
13. *Guaramunu* and *griseolineata*	South America

*Complexes 1 and 2 according to Wasserman, 1962, 1963; complexes 3–12 according to Carson *et al.*, 1970; complex 13 according to Kastritsis, 1969.

The existence of these homosequential complexes proves very clearly that chromosomal rearrangement is not a *sine qua non* for speciation. Nevertheless, homosequential complexes are very rare in *Drosophila* and have not been found at all among the numerous species with polytene chromosomes in other Dipteran genera that have been extensively studied *(Sciara, Rhynchosciara, Simulium, Prosimulium, Twinnia, Gymnopais, Anopheles, Chironomus,* and *Polypedilum).* In groups that lack polytene chromosomes or whose polytenes are not suitable for detailed banding studies (and this includes the overwhelming majority of animal groups and all higher plants) there are innumerable examples of closely related species that differ in their visible metaphase karyotypes and hence cannot possibly be homosequential. But in such groups the metaphase karyotypes of two or more species sometimes, though rarely, appear identical.

Eventually most of these instances of apparently identical karyotypes will prove to be spurious as the techniques for karyotypic analysis are improved. Already the newer chromosome banding techniques have demonstrated some structural differences between karyotypes that formerly were indistinguishable. These techniques are of several different types (Hsu, 1973). Staining with quinacrine produces bands that fluoresce in ultraviolet light (Q-banding). Staining with the Giemsa stain after treatment with acid and alkaline solutions reveals the extent and location of blocks of constitutive heterochromatin, including those around the centromeres (C-banding). Staining with Giemsa stain after treatment of the preparation with trypsin and a variety of other agents yields a more elaborate cross-banded pattern (G-banding). The banding patterns revealed by all these techniques clearly indicate aspects of the underlying architecture of the chromosomes. They help us to interpret chromosomal rearrangements, such as inversions and translocations, and to identify and understand differences between the karyotypes of related species. Six levels of karyotypic analysis may currently be found in the literature:

Alpha karyology. Only chromosome numbers and approximate sizes have been determined. At present this is the only practicable level of analysis in yeast and most protozoa, fungi, and algae.

Beta karyology. Chromosome numbers and lengths of chromosome arms are known, i.e., centromere positions have been accurately located. Sex chromosomes, if present, have been identified. Most karyotypic data are of this type.

Gamma karyology. Giemsa and fluorescent banding techniques have been carried out and maps showing locations of the main C-, G-, and Q-bands are available.

Delta Karyology. Locations of satellite DNAs, nucleolar organizers, and 5-s rRNA loci have been determined.

Epsilon karyology. On the basis of lampbrush chromosome analysis, the main distinctive loops and other landmarks have been mapped.

Zeta karyology. On the basis of polytene chromosome analysis, thousands of bands and other landmarks (puffs, nucleolar organizers, etc.) have been mapped.

These six types of karyology represent increasing degrees of detailed knowledge. But they are also, to some extent, alternatives based on different materials and techniques. Thus, delta karyology provides detailed information on aspects of chromosome architecture not revealed by epsilon or zeta karyology. We shall repeatedly refer to these different types of karyology. Those readers who are not cytogeneticists must bear in mind that each type brings very different amounts and types of information to problems of evolution and speciation. In particular, we must not expect beta karyology, which is all that we have for most groups of organisms, to provide anything like the critical evidence that is available from delta, epsilon, or zeta karyology.

It is very difficult at the present stage to hazard a guess as to what proportion of related species of eukaryote organisms may have homosequential karyotypes. For the genus *Drosophila* 10 percent would seem to be an upper limit, bearing in mind that there appear to be far more homosequential complexes in the Hawaiian fauna than in the continental faunas, which have been less extensively studied. Ten percent would also seem to be an absolute upper limit for other groups of animals, with one percent a more probable figure. It is more difficult to arrive at an estimate for plants, due to the prevalence of polyploidy in many genera and the fact that there have so far been few fluorescent- or Giemsa-banding studies on plant chromosomes. All we can really say at the present stage is that there are very few pairs of plant species that could possibly be homosequential.

However, even if we accept an estimate of one percent homosequential species (in the case of animals), this does not necessarily imply that 99 percent of all speciation events involve karyotypic changes. In most groups many species have become extinct, so that some surviving species that differ in karyotype are not closely related and may have no really close relatives. So-called sibling species, however similar morphologically, are not always related as parent and offspring species or as direct descendants of a third species (extinct or not). They may in fact be related only through several extinct species. Thus, the fraction of speciation events during which chromosomal rearrangements have occurred is bound to be somewhat less than 99 percent. However, there can be little doubt that in many groups that have undergone a great deal of recent speciation (so-called "explosive" speciation) all the species that have ever existed still survive, i.e., there has been no extinction of any species. In such groups the number of heterosequential species pairs would truly reflect the incidence of chromosomal rearrangements in speciation. Many genera of Australian morabine grasshoppers, the European species of mole crickets *(Gryllotalpa),* and some species groups in Drosophilidae, Culicidae, Chironomidae, and Simuliidae are probably ones in which all the species still survive and all are karyotypically unique, there being no homosequential species. However, even if one or more chromosomal rearrangements have occurred during a speciation process, there can in most cases be no certainty that they played a primary role in initiating genetic isolation and divergence. It can certainly be argued that in some cases conditions incidental to speciation, e.g., a small localized initial population, facilitated the establishment of a rearrangement, rather than that the occurrence of the rearrangement facilitated speciation.

Some types of chromosomal rearrangements are very much easier to detect than others, and thus relatively superficial studies are apt to lead to erroneous or biased conclusions. In organisms with polytene chromosomes *all* structural rearrangements of the euchromatic part of the karyotype, except ones at the submicroscopic level, should theoretically be analyzable. In the polytene chromosomes, however, rearrangements in the heterochromatic segments are difficult to analyze—they may in fact be more easily studied in ordinary mitotic chromosomes.

Two general principles concerning chromosomal rearrangements should be briefly mentioned here. First, in many groups there is a parallelism between the types of "fixed" and "floating" rearrangements. For example, in the genus *Drosophila,* considered as a whole, so-called paracentric inversions (ones that do not include the centromere) are common polymorphisms, whereas other types of rearrangements (e.g., pericentric inversions, translocations, and changes of chromosome number) are almost unknown as population polymorphisms, i.e., they do not occur in the "floating" state. We find exactly the same picture for the "fixed" cytotaxonomic differences between related species of *Drosophila:* paracentric inversion differences are common, while other types of structural rearrangements are very rare. Similarly, in certain acridid grasshoppers belonging to the tribe Trimerotropi (in both North and South America) we find that pericentric inversions (ones that straddle the centromere) are common, both as floating polymorphisms and as fixed cytotaxonomic differences.

We shall discuss the significance of this type of parallelism later. It is certainly not a universal phenomenon. For example, in the Australian morabine grasshoppers at least 61 evolutionary changes of chromosome number are known to have occurred (White, 1974), yet not a single such chromosomal fusion or dissociation is known to exist in a polymorphic state in a natural population (apart from narrow hybrid zones between parapatric races or species, and in such cases we are clearly not dealing with balanced polymorphism).

The second general principle governing the occurrance of chromosomal rearrangements is the one that has been called *karyotypic orthoselection* (White, 1973, 1975). This is the tendency for the same type of rearrangement to occur over and over again in different chromosomes of the same species. Thus, in certain evolutionary lineages one fusion after another will occur, producing a progressive decrease in chromosome number, while in other lineages repeated increases of chromosome number or repeated conversions of rod-shaped (acrocentric) chromosomes into V- or J-shaped (metacentric) ones by pericentric inversion are the rule. This very widespread tendency extends down to many minute rearrangements. For example, in the genus *Vandiemenella,* of the Australian morabine grasshoppers, the short limbs of all the acrocentric chromosomes are very minute in the 19-chromosome race of *V. viatica,* but very much longer in the closely related species "P45c" (White, Blackith, Blackith, and Cheney, 1967).

Some possible causal explanations of the principle of karyotypic orthoselection have been discussed by White (1975). It is almost certainly not due to particular types of rearrangements simply occurring more frequently in some groups than in others (the *differential origin* explanation). Three other factors, acting singly or in combination, seem more likely to have played a role in determining karyotypic orthoselection: (1) the fact that similar rearrangements may have similar effects on the phenotype; for example, in the mollusc *Nucella lapillus,* studied by Staiger (1954), five different chromosomal dissociations apparently help the species to adapt to a particular type of habitat (see p. 214); (2) the need for the sizes, shapes, and numbers of the chromosomes to be in harmony with all the dimensions of the cell, and especially with those of the mitotic spindle, in all the tissues of the body; and (3) possible regularities in the architecture of the interphase nucleus.

It is fairly certain that in many species of animals various types of chromosomal rearrangements are (for reasons we do not completely understand) directly adaptive to certain types of habitats and ecological niches. This fundamental fact was first clearly realized and stated by Mayr (1945). The case of *Nucella lapillus* is only one example. A rather special type of geographical and ecological polymorphism of the X chromosome has been described in the collembolan *Neanura monticola* by Cassagnau (1974). Salivary gland nuclei of this species show a typical polytene structure in the autosomes, but the X consists of a looser bundle of chromonemata (Figure 3). Populations in the Pyrenees show five main types of X that seem to result from an interaction of four variables: extent of separation of the chromonemata; parallel orientation of chromonemata or radial orientation from a mass of α-heterochromatin; development of cross-banding; and increase in the amount of α-heterochromatin. Type 1 occurs in damp forests at low elevation (900–1000 m); types 2 and 3 are found in cold forests (1600–1800 m); while types 4 and 5, with extra heterochromatic segments, are characteristic of sub-alpine environments above 2000 m. There is no suggestion that speciation is occurring, since heterozygotes occur in all but the most extreme environments. But clearly a polymorphism such as this, one so closely adaptive to an altitudinal gradient of different biotypes, is potentially divisive and could eventually lead to speciation. *Neanura monticola* lacks a Y-chromosome, so there is no question here of X-Y coadaptations such as exist in Simuliidae (p. 58); any coadaptations that exist must be between the various types of Xs and genetic variation in the autosomes.

A very different type of chromosomal variation has been shown to exist in two species of burrowing rodents, *Thomomys bottae* and *T. umbrinus,* in which the number of acrocentric chromosomes is highly variable (0–38 in the first species, and 2–56 in the second), in spite of virtual constancy in chromosome number (76–78 in both species). Berry and Baker (1971) ascribe the differences between acrocentric and metacentric or submetacentric chromosomes in this case to pericentric inversions, but the possibility that it may be due to variation in the amount of heterochromatin does not seem to have been definitely excluded. They claim that populations living in deep soils in mesic habitats show low numbers of acrocentric chromosomes, while ones inhabiting arid areas with shallow stony soils have many.

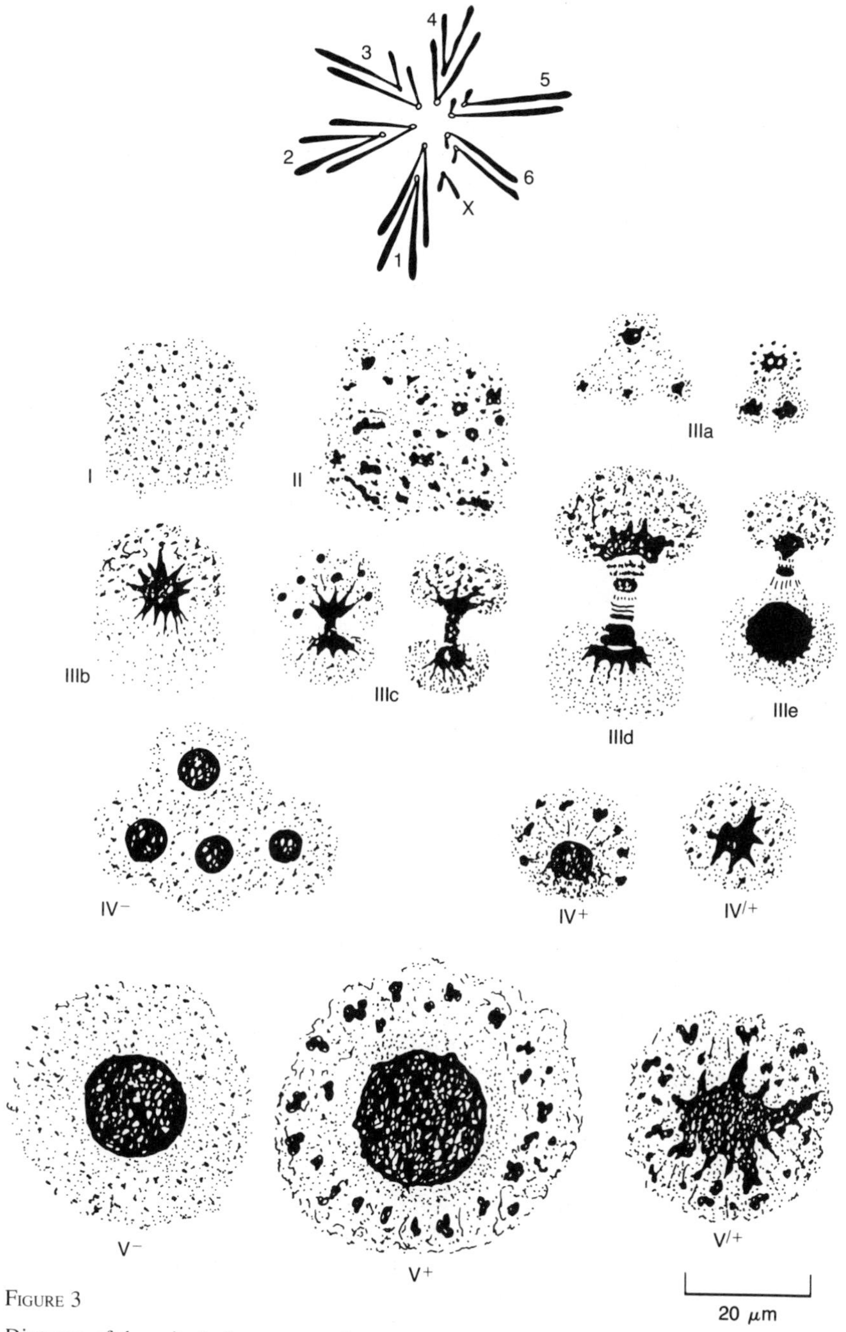

FIGURE 3

Diagram of the mitotic karyotype of a male *Neanura monticola* (top) and drawings of 13 different types of polytene X chromosomes from salivary gland nuclei (IIIa–e are subtypes of "type III," and similarly for the subtypes of IV and V.) [After Cassagnau, 1974]

Individuals heterozygous for paracentric and pericentric inversions are at least potentially capable of forming "reversed loops" in their chromosomes at meiosis, as a result of accurate synapsis between all homologous regions. If they do so, and if crossing-over occurs within the loop, chromatids carrying duplications and deficiencies will be produced. In the case of paracentric inversions these abnormal chromatids will carry either two centromeres or no centromere at all. *Drosophila* species generally escape paying the penalty for paracentric inversion heterozygosity because in the male there is no crossing-over and in the oocyte the dicentric and acentric chromosomes are shunted into the polar body nuclei and do not get into the functional egg nucleus. No such protective mechanism operates in the case of pericentric inversions, so these chromosomal rearrangements are virtually unable to establish themselves in *Drosophila* populations and do not show up as differences between species.

In species of *Chironomus* there are also many floating and fixed paracentric inversions, but crossing-over does occur in the male. A combination of factors protects the heterozygotes against producing sperms carrying duplications and deficiencies. The number of reversed loops formed at meiosis is considerably less than the potential maximum (Martin, 1967), and if dicentric and acentric chromatids are produced they lead to the formation of large diploid spermatids that do not develop into functional sperms (Beerman, 1956). Heterozygosity for pericentric inversions does not occur in wild populations of *Chironomus,* any more than in *Drosophila*.

In certain species of grasshoppers (but only a small minority) pericentric inversions exist in both the floating (polymorphic) and fixed conditions. Formation of reversed loops at meiosis does not seem to occur, or is rare, in these species.

Mutual translocations, when heterozygous, lead to chains or rings of chromosomes at meiosis. A single interchange between two chromosomes leads to a ring or chain of four chromosomes, two of which are the originals, and the other two the translocated ones, e.g., AB–BC–CD–DA. Several sequential interchanges may lead to chains or rings of 6, 8, 10 . . . chromosomes, e.g., AB–BC–CD–DE–EF–FA. The fecundity of such heterozygotes will not be reduced if at first anaphase the alternate members of the chain or ring regularly segregate to opposite poles. But if adjacent chromosomes pass to the same pole, aneuploid, i.e., inviable or lethal, gametes will be produced. Consequently, translocation polymorphism is rare in natural populations. Examples are known in a few special cases, such as the plant *Paeonia californica* (Walters, 1942; Grant, 1964b) and some cockroaches of the genera *Periplaneta* and *Blaberus* (John and Lewis, 1957, 1958, 1959; Lewis and John, 1957; Rajasekarasetty and Ramanamurthy, 1964). A few species of plants belonging to the genera *Oenothera, Gayophytum, Rhoeo, Isotoma,* and *Hypericum* are "fixed heterozygotes" for translocations (Cleland, 1962, 1964; Sax, 1931; Renner, 1925; James, 1965). Some species of scorpions are also heterozygous for translocations and show rings or chains of chromosomes at meiosis, but their genetic systems are not well understood (Piza de Toledo, 1943, 1947; Sharma, Parshad, and Joneja, 1959).

In general, however, mutual translocations seem to diminish the fecundity of the heterozygotes too much to establish themselves in natural populations or to provide a basis for speciation. However, this is not always so. For example, interchanges between minute regions at the tips of chromosome limbs may occur; aneuploid gametes will then be formed but they will carry duplications and deficiencies of only a few gene loci that may not be essential for life. Translocations of whole chromosome arms may also take place; quite a number of these have occurred in the phylogeny of the midge genus *Chironomus* (Keyl, 1962; Wülker, Sublette, and Martin, 1968) and may have played a role in speciation.

A great many cytotaxonomic differences between closely related species are of the so-called Robertsonian type,* where a metacentric, V-shaped chromosome in one species is represented by two acrocentric, rod-shaped chromosomes in the other. This may be due either to *centric fusion* between two acrocentrics (leading to a decrease in chromosome number) in one species or *dissociation* (producing an increase in chromosome number) in the other. Both are probably special types of translocations: the former involves loss of minute regions of the ends of the chromosomes that undergo fusion, while the latter are essentially translocations between metacentric chromosomes and minute chromosome fragments additional to the normal karyotype (White, 1957, 1973). However, some cytogeneticists believe that simple end-to-end fusion between acrocentrics can occur without loss of any material—which seems unlikely. Others (e.g., Todd, 1970; Webster, Hall, and Williams, 1972) believe that functional telocentric, rod-shaped chromosomes can arise in evolution as a result of a simple break in the centromere of a metacentric. The latter interpretation implies not only that half-centromeres can be functional—which is probably true—but also that not all functional chromosome ends are specialized DNA sequences known as telomeres (which are probably DNA palindromes; see Cavalier Smith, 1974).

The controversy over *dissociation* versus simple *centric fission* has not been settled. Conceivably both processes have occurred, in different groups. In the present work the term "dissociation" will be used for all evolutionary changes

*After W. R. B. Robertson, Kansas cytogeneticist, who in 1916 described some instances of centric fusions in grasshoppers. He never studied any cases of "dissociation" or "centric fission," i.e., evolutionary increases in chromosome number.

in which a metacentric chromosome has been replaced by two acrocentrics (or telocentrics), regardless of the precise mechanism involved.

The occurrence of Robertsonian rearrangements in chromosome evolution makes it necessary when describing karyotypes to cite the number of major chromosome arms (not counting the very short "second arms" of acrocentrics, which are probably heterochromatic in all cases). This is Matthey's "nombre fondamental" (N.F.), which is cited following the chromosome number.

The overwhelming majority of cytogeneticists confronted with species pairs in which one species has two acrocentric chromosomes replaced by a metacentric in the other species have assumed that the former was the primitive condition—the karyotype with the metacentric chromosome having acquired an evolutionary fusion of the two acrocentrics by their proximal ends, i.e., a centric fusion. Certainly cytogenetic mechanisms for fusion are easier to envisage than mechanisms leading to dissociation of a metacentric into two acrocentrics (basically, the loss of a centromere is easier to conceive than the acquisition of an additional centromere, although the matter is really more complex than this, and depends in part on the still controversial nature of "telomeres"). In the long run, however, as many increases as decreases in the number of chromosomes must have occurred in evolution. At least this must be so unless the "evolutionary potential" (the probability of leaving descendent species) is significantly less in organisms with lower numbers of chromosomes (those that have acquired many centric fusions in their past phylogeny) than in those with higher numbers. In about 200 species of Australian grasshoppers of the subfamily Morabinae, in which the direction of karyotype evolution is exceptionally clear, 39 chromosomal fusions and 22 dissociations have occurred (White, 1974). Histograms of chromosome numbers for some groups, like Lepidoptera (see Figure 5, p. 73), also suggest more fusions than dissociations (although some lepidopteran species have undergone spectacular increases in chromosome number, presumably by multiple dissociations).

The fecundity of individuals heterozygous for any particular centric fusion or dissociation depends on three factors. A metacentric chromosome AB must undergo meiotic pairing with two acrocentrics, A and B. Following this, chiasmata must form, i.e., genetic crossing-over must occur, between acrocentric A and the A limb of the AB chromosome, and between acrocentric B and the B arm of the AB. If synapsis fails to occur or chiasmata are not formed, meiosis will result in univalents, which will then segregate randomly, frequently leading to aneuploid gametes. But even if both synapsis and chiasma-formation do occur, the trivalent must orientate on the spindle of the first meiotic division in such a manner that acrocentrics A and B pass to one pole at anaphase, while the metacentric AB passes to the opposite pole. If the chain of three orientates in a linear manner, so that the acrocentrics A and B pass to opposite poles, the result will inevitably be aneuploid sperms or ova. Whether trivalents of this kind regularly orientate in a zigzag (disjunctional) or linear (nondisjunctional) manner is highly unpredictable and does not seem to be related in any simple manner to the lengths of the chromosome arms. For example, in the shrew *Sorex araneus,* whose populations are polymorphic for six fusions or dissociations, the trivalents seem to be regularly disjunctional (Sharman, 1956; Ford, Hamerton, and Sharman, 1957; Meylan, 1964, 1965; Ott, 1968; Kozlovsky, 1970; Ford and Hamerton, 1970), and the same is probably true of a number of species and races of the African mouse genus *Leggada,* sometimes regarded as a subgenus of *Mus* (Matthey, 1964, 1966). But in laboratory hybrids between *Mus musculus* (house mouse) (2n = 40) and *Mus poschiavinus* (Swiss "tobacco mouse") (2n = 26; homozygous for seven centric fusions), linear orientation of the trivalents is very common in both male and female meiosis (Gropp, Tettenborn, and von Lehmann, 1970), so that the fecundity of these hybrids is sharply reduced. Ford and Evans (1973) estimate the frequency of nondisjunction of different trivalents as 5.7–33.5 percent.

It follows from the above that fusions (or dissociations) are able to maintain themselves in a floating polymorphic condition in populations of *Sorex araneus* and certain species of *Leggada,* but that this was not possible in the speciation of *Mus poschiavinus* (which must have arisen from *Mus musculus).* In the latter case it seems almost certain that the centric fusions had a divisive effect, and that they (or at any rate one of them) played a primary causative role in the origin of *Mus poschiavinus.*

A special kind of chromosomal difference between related species, one that is probably much more widespread, was discovered by Keyl (1965, 1966) in two sibling species of midges, *Chironomus thummi* and *C. piger* (Keyl regarded them as subspecies, but the genetical and biological evidence shows that they are specifically distinct). When their polytene chromosomes are compared in the nuclei of laboratory hybrids, it is found that although some bands of both species have the same quantity of DNA, in many *thummi* bands the amount of DNA is two, four, eight, or 16 times the amount in the corresponding *piger* band (Figure 4). There are no exceptions to the above statement; the amount is never greater in *piger* than in *thummi,* and the relationship is always a power of two, there being no cases where the amount in *thummi* is five or six times that in *piger.* In some bands the increase is variable, e.g., some individuals of *thummi* show a fourfold increase, others an eightfold one.

Some independent evidence suggests that *piger* is the ancestral species from which *thummi* has been derived (and the relationship between the bands is cer-

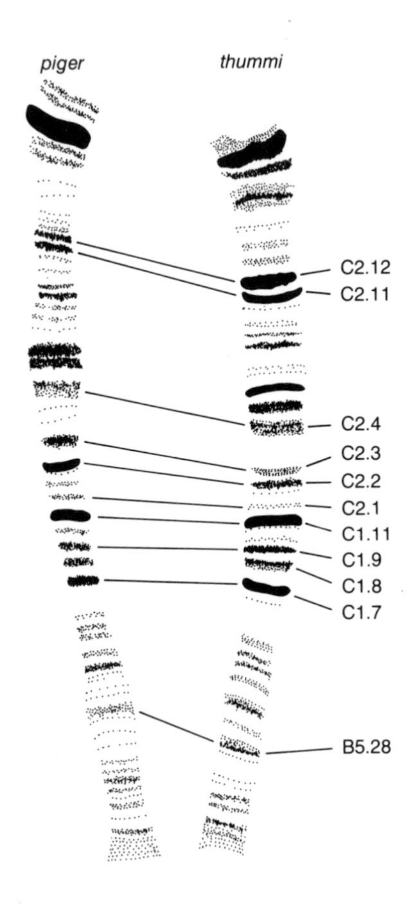

FIGURE 4

Comparison of the polytene banding patterns of chromosome arm IIR of *Chironomus piger* and *C. thummi* in the hybrid between them. Bands C2.11 and C2.4 show a twofold increase in DNA content in *thummi;* band C2.12 shows a fourfold increase; bands C2.2, C2.1, and C1.9 show the same amount. The increase of bands C2.3, C1.11, C1.8, C1.7, and B5.28 in *thummi* is variable. [After Keyl, 1965]

tainly easier to understand if this is the case). The total DNA per polytene (or spermatocyte) nucleus is 27 percent greater in *thummi*. The DNA increase in *thummi* is clearly not due to differential polyteny or lateral reduplication—

longitudinal reduplication of gene loci has occurred in the course of the past evolution of *thummi*.

A further and most striking example of the "Keyl phenomenon" has been found in black flies (Simuliidae) by Rothfels (unpublished). The genus *Gymnopais* includes two diploid species, *holopticus* and *dichopticus,* in Alaska. There are also two triploid parthenogenetic forms, which have presumably arisen by hybridization. The first, which is widespread in the Arctic, has two sets of *holopictus* chromosomes and one of *dichopticus* chromosomes, while the second, which occurs in the Rocky Mountains, has two sets derived from *dichopticus* and one from *holopticus*. A study of the polytenes of these forms has shown that *dichopticus* has at least 28 bands whose DNA content has been duplicated, quadruplicated, etc., compared with the corresponding bands in *holopticus*. Like the above case in *Chironomus,* there are no bands in which the relationship is the other way round, i.e., no bands of *holopticus* have a higher DNA content than those of *dichopticus*.

It would be digressive here to discuss the molecular mechanism of DNA reduplication in these cases. Natural selection has probably favored repeated duplication of gene loci controlling a large number of metabolic pathways in such species as *Chironomus thummi* and *Gymnopais dichopticus*. But, even if the structural genes are located in the bands—and this is not absolutely certain, since they may lie in the interbands—they clearly form only a small part of the DNA of the band. It is therefore possible, although perhaps improbable, that Keyl-type duplications do not involve structural genes. It would seem highly desirable to carry out studies on the general adaptive systems, e.g., rate of development, width of ecological niches occupied, etc., of species such as *Chironomus thummi* and *Gymnopais dichopticus* and compare these with those of the presumed ancestral species having less DNA. Another important line of investigation would seem to be a study, in depth, of the role and significance of the natural polymorphisms in *C. thummi* for certain band duplications.

In the remainder of this chapter an attempt will be made to describe the nature and extent of the chromosomal differences between species of a number of groups on which extensive general studies have been carried out. Additional information on geographic distribution and genetic isolating mechanisms, suggesting general models of speciation, will be considered in Chapter 6.

The existence of giant polytene chromosomes in the somatic nuclei of some families of Diptera makes it possible to carry out zeta karyology, providing a more detailed cytotaxonomic picture of these families than of other groups. We shall therefore first examine the dipteran families Simuliidae (blackflies), Chironomidae (midges), Culicidae (mosquitoes), and Drosophilidae, and then discuss the evidence for other groups—including vertebrates and higher plants—in which polytene chromosomes do not exist.

Simuliidae

The evolutionary cytogenetics of simuliid species has been studied especially by Rothfels and his students in Canada (for a general review see Rothfels, in press) and by a number of workers in the USSR. The chromosome number is low in all the species (the primitive number is n = 3, secondarily reduced to n = 2 in a number of different lineages). Most species have chiasmata in the male meiosis, but a fairly large number have developed achiasmatic spermatogenesis, thus further reducing the level of genetic recombination, which is in any case very low because of the low chromosome number.

Most of the classical morphospecies of Simuliidae turn out, when studied by modern cytogenetics, to be complexes of several sibling species. An extreme example is *"Simulium damnosum,"* a complex of great medical importance in Africa, and which has been shown by Dunbar (1972) to include about 16 species, only some of which bite man. In the holarctic region several species names, such as *Simulium tuberosum* and *Prosimulium onychodactylum,* include as many as 11 siblings (Rothfels, unpublished; Newman, unpublished). *Eusimulium aureum* includes at least seven (Dunbar, 1959) and other "species" have been shown to consist of small numbers of sibling forms. Many of the morphospecies are wide-spread in the holarctic region, and have not been studied cytogenetically throughout their whole range. In all probability there are many sibling species still awaiting discovery. These difficulties prevent, for the time being, a full interpretation of the speciation processes, especially since it is not practicable to breed simuliids in the laboratory.

The family Simuliidae is characterized by male heterogamety; however, two species in Tahiti, have developed female heterogamety. The X and Y sex-determining segments are single bands, or at any rate very short; usually they are borne on chromosome 2. The proliferation of sibling species in Simuliidae seems to be due to a tendency to frequent evolutionary change in the sex chromosome mechanism. Apparently, the species from time to time pass through a stage in which they are polymorphic for two kinds of X segments, X and X′, which may be located in different regions of the same or different chromosomes, and two kinds of Y′ segments, Y and Y′. If X and Y form an adaptive combination and X′ and Y′ do likewise, but XY′ and X′Y are for some reason poorly adapted, speciation is likely to occur. This seems to have happened in *Simulium tuberosum*. Landau (1962) distinguished five sibling forms of this superspecies in southern Ontario (a total of 11 are now known from the entire range) on the basis of seven types of chromosome arm 2S, which differ in respect of a number of inversions. Four of these, A, AB, FG, and CKL, are X chromosomes, while the standard and CDE are Y chromosomes; the seventh, which is structurally FGH, apparently may carry either X or Y sex loci. These second chromosomes occur as follows:

Sibling	Males	Females
1	$X_A Y_{CDE}$	$X_A X_A$
2	Mainly $X_{AB} Y_{ST}$	$X_{AB} X_{AB}$
3	Mainly $X_{FG} Y_{ST}$	$X_{FG} X_{FG}$
4	$X_{FGH} Y_{FGH}$	$X_{FGH} X_{FGH}$
5	$X_{CKL} Y_A$?	$X_{CKL} X_{CKL}$

Apart from the inversion differences in the 2S arm, no fixed inversion differences between the siblings are known. However, Landau identified 83 different inversions in a floating polymorphic condition, and some inversion polymorphisms were present in two or three of the siblings, a situation that is not unknown, but distinctly rare, in *Drosophila*.

Sibling species in Simuliidae tend to be extensively sympatric, but there is no really critical evidence that they have arisen by a process of sympatric speciation, although this is most likely the case (see Chapter 7), especially since the closest siblings (on cytogenetic criteria) *always* seem to be sympatric. In *Prosimulium onychodactylum* as many as seven sibling species have been found in the same creek in Oregon (to some extent these emerge at different periods of the year, so that interbreeding is avoided).

Simulium vittatum has the distinction of possessing the most elaborate structural chromosomal polymorphism known—at least 127 simple inversions and seven complex ones (Pasternak, 1964). It is closely related to *S. argus,* from which it differs by several fixed inversions and especially by structural modifications of chromosome limb 2L (the sex-determining arm).

The situation in the *Prosimulium magnum* complex is more complicated (Ottonen and Nambiar 1969). In this group the third chromosome pair is sex determining. Two sibling species, *P. magnum* and "form 2," have the X_0Y_0 condition, i.e., they have no cytologically recognizable sex chromosomes; they differ in respect of a fixed inversion in chromosome 2. Form 3 has an X_cY_c mechanism, where the two sex chromosomes differ from the primitive X_0 (or Y_0) condition by different inversions; Y_c in this species may be replaced by more complex inversion derivatives Y_1, Y_3 or Y_5. The closely related *Prosimulium multidentatum* has a sex-determining mechanism located on chromosome arm 2L, i.e., it is not homologous to the one present in *P. magnum*. Only one type of X is known, but there are four structurally different kinds of Y chromosomes (Y_1, Y_2, Y_3, and Y_4), each of which is characteristic of a different geographic area.

In general, chromosomes X_0 and Y_0 are coadapted and form an efficient sex determining mechanism in *P. magnum* and form 2. Similarly, X_c is coadapted with Y_c in form 3. Some hybridization between *P. magnum* and form 1 occurs in New York State. Hybridization between *P. magnum* and form 3 also occurs in some areas and gives rise to X_0Y_c populations. Individuals with X_cY_0 have never been found, and this combination is assumed to be lethal. A few hybrids between *P. multidentatum* and *P. magnum* and between the former and form 3 have been found in New York state.

Three mechanisms of speciation seem to have been at work in this group. Form 2 apparently differentiated from *P. magnum* when an inversion in chromosome 2 became fixed in the homozygous condition. The origin of Y_c was the first step in the separate evolution of form 3 and was followed by the transformation of X_0 into X_c (populations with X_0Y_c, instead of being secondarily hybrid, as assumed above, may be relicts of a former intermediate condition). *P. multidentatum,* on the other hand, clearly arose as a segregate from a population that, like the midge *Chironomus tentans* in northern Germany and southern Sweden (Beerman, 1955), simultaneously possessed two different sex-determining mechanisms, located on different chromosome pairs. In all simuliid species except the two Tahitian ones mentioned on p. 58, sex determination depends on "dominant Y" mechanisms, so that if there are Y factors located on different chromosome pairs, any one of these will lead to maleness. Whether the sex locus in *P. multidentatum* was transferred from chromosome 3 to chromosome 2 is not known (the alternative being that it arose *de noyo*).

A special type of evolution of the sex-determining mechanism has taken place in the Norwegian simuliid *Cnephia lapponica,* which includes two sibling forms. The more primitive one has n = 2, the X and Y being on chromosome pair 1. The "derived" form is a permanent heterozygote, in the male, for a translocation between chromosomes 1 and 2. It has two kinds of sex-determining loci, each on a different chromosome, so that the constitution may be represented as $X_1Y_1X_2Y_2$(♂): $X_1X_1X_2X_2$(♀), there being no autosomes. At meiosis in the male, which is achiasmatic, the four chromosomes form a cross-shaped configuration, and segregation at anaphase is such that X_1 and X_2 pass to one pole and Y_1 and Y_2 to the other. The North American species of *Cnephia* have three pairs of chromosomes. In *C. mutata* and an undescribed species from British Columbia chromosome 1 is the Y (i.e., it has male-limited inversions) while in another undescribed species from Arizona chromosome 2 is the Y (Madahar, 1969).

Speciation in Simuliidae thus appears to be mainly due to the development of new sex-determining mechanisms. No evidence exists that these are directly adaptive to particular habitats, climatic zones, or other aspects of the environment, although of course they may be. The most probable interpretation is that new coadapted sex-determining mechanisms arise by mutation and then function as isolating mechanisms. If they then acquire some ecological diversification, they may be able to maintain themselves sympatrically with the population carrying the original sex mechanism. The main gap in our knowledge would seem to be the matter of the existence and nature of ethological isolating mechanisms in multiple-sibling super-species inhabiting the same environment, such as the *Prosimulium onychodactylum* complex inhabiting the same streams in Canada. We may be reasonably confident, however, that they exist.

Chironomidae

We have already referred to Beermann's work on sex chromosome mechanisms in different European populations of the midge *Chironomus tentans*. In North American populations of this species two further sex-determining mechanisms are known (Thompson, 1971; Thomson and Bowen, 1972). In Wisconsin and Ontario the females are heterogametic for a dominant female-determining factor at the tip of chromosome 1. In Iowa, however, males are heterogametic for a primary male-determining factor linked with an inversion. In laboratory crosses it is possible to obtain midges carrying both male- and female-determining factors; such individuals are fertile males.

The closely related Australian species *Chironomus australis*, *C. duplex*, and *C. oppositus* have male heterogamety for sex-determining factors on chromosome arms C, B, and A, respectively (Martin, in press). Obviously, there is now a good deal of evidence that replacement of one sex-determining mechanism by another has rather frequently preceded or accompanied speciation in Chironomidae, as it undoubtedly has in Simuliidae.

Culicidae

Speciation has been studied in considerable detail in certain groups of mosquitoes (Culicidae), but there is less reason to believe that mutational changes in the sex-determining loci have played a role in mosquito speciation. Four cases will be considered here, although in only two of these is the chromosomal evidence really adequate.

The Anopheles maculipennis complex. The *Anopheles maculipennis* complex of sibling species is historically important in that work on it established the idea that genuine biological species could exist without the development of morphological differences, or with only minimal or cryptic differences. The

Palearctic species *A. maculipennis (typicus* in the earlier literature), *A. labranchiae, A. l. atroparvus, A. messeae, A. melanoon melanoon, A. melanoon subalpinus,* and *A. sacharovi* comprised the first clear example of such a complex to receive widespread recognition, following the pioneering work of Falleroni (1926), who showed that they could be distinguished on the basis of differences of the egg surface and floats, and of Missiroli (1939) and Bates (1940), who demonstrated marked ecological differences between some of them.

Chromosomal differences between the members of this complex were studied by Frizzi (1947, 1949, 1950, 1951, 1953). As one might expect, the two races of *labranchiae* show no karyotypic differences, although in hybrids between them some long chromosome segments are unsynapsed in the polytene nuclei, perhaps indicating some differences in the architecture of the DNA, even if there is no difference in banding pattern. Similar unsynapsed regions are seen in hybrids between *maculipennis* and *atroparvus,* in which there is also a fixed inversion difference in chromosome 3. The X chromosomes fail to synapse in almost all hybrids obtained between the Palearctic members of the *maculipennis* group (Kitzmiller, Frizzi, and Baker, 1967), indicating either that extensive minor structural rearrangements occurred in this chromosome during speciation or that other types of changes, perhaps in repetitive DNA sequences, inhibit synapsis. *Labranchiae, atroparvus, maculipennis,* and *subalpinus* are said to have the same banding pattern in their polytene X chromosomes, but the mapping has not been detailed enough to exclude the existence of minor rearrangements.

Frizzi (1951) reported two forms of *messeae* in the vicinity of Pavia, Italy, one with the standard sequence in the X chromosome (as in *atroparvus*), the other with an X that has undergone extensive rearrangement. Some heterozygous females were found in nature, but far fewer than the number expected if the two forms were mating at random. It is probable that the second form was actually *A. melanoon subalpinus.* It is not known whether there is any difference between the Y chromosomes accompanying the two kinds of X chromosomes.

Some crosses between Palearctic members of the *maculipennis* complex cannot occur in nature because of ethological barriers, but may be obtained in the laboratory by the technique of "induced copulation." In the case of some combinations the larvae die, but in others F_1 hybrids may be obtained; both sexes may be sterile, or the females may be fertile and the males sterile (in the cross between *labranchiae* and *atroparvus,* regarded taxonomically as races rather than distinct species).

Six North American species of the *maculipennis* group have been studied: *A. freeborni, A. quadrimaculatus, A. punctipennis, A. earlei, A. aztecus,* and *A. orientalis.* In general, the results of crossing these species are similar to those obtained with the Palearctic forms; but the North American species do seem to have diverged morphologically, genetically, and karyotypically to a greater extent than the European forms. Hybrids between the various North American species—when they can be obtained and reared to a late larval stage—show almost complete failure of pairing of the polytene chromosomes. This has been ascribed by Kitzmiller and co-workers to "genic changes which are not reflected in the banding pattern." It is unlikely, however, that such changes are simply ordinary allelic changes that lead to variant allozymes or structural proteins; changes of this kind would not be expected to lead to polytene asynapsis. More probably, profound changes have occurred in the moderately repetitive DNA, i.e., not in the satellite DNA, which is underreplicated in the polytene chromosomes and probably does not affect synapsis.

Crosses between North American and Palearctic mosquitoes of the *maculipennis* group generally produce inviable zygotes that fail to hatch from the egg or die as larvae. But Frizzi and De Carli (1954) obtained both male and female adults from the cross *A. atroparvus* × *A. freeborni,* and the females were even fertile in backcrosses.

A number of other North American, Neotropical, and Mediterranean species are either members of the *maculipennis* complex or related to it, but the cytogenetics of these species has not been explored in depth. In general, it appears that speciation in the *maculipennis* complex has been accompanied by moderately extensive chromosomal rearrangement, but the role of changes in the sex chromosomes in initiating speciation is less clear than in Simuliidae.

The Anopheles gambiae Complex. The African *Anopheles gambiae* complex includes four freshwater sibling species, A, B, C, and D (the last being adapted to mineralized waters), and two saltwater species, *A. merus* in East Africa and *A. melas* in West Africa (Davidson *et al.,* 1967; Coluzzi, 1964, 1966, 1970; Coluzzi and Sabatini, 1967, 1968b, 1969; Davidson and Hunt, 1973). Morphological differences between these species are minimal; the two saltwater species can be distinguished from the freshwater forms, but the latter are virtually indistinguishable morphologically. Studies of the polytene chromosomes have shown that the autosomes of species A and B are very similar, but that the X chromosomes have very different sequences, due to multiple rearrangements. Species A and B share four autosomal inversion polymorphisms. Species C seems to be intermediate between A and B. *A. melas* has a very long heterochromatic segment in the X chromosome that is not present in *merus* or in the freshwater species. Species A and B are sympatric over a large part of Africa, but B is better able to survive in drier savannah areas. These species bite man and are efficient malaria vectors, whereas species C, found mainly in southeast Africa, bites animals other than man and is of no importance in malaria transmission. In the laboratory, viable hybrids can be obtained between all the species; male hybrids are sterile and female ones fertile. In some crosses there is a marked excess of males in the progeny. The two saltwater species are not particularly closely related; *melas* is phylogenetically close to species C, *merus* to species A.

The Aedes mariae Complex. The polytene chromosomes of other genera of mosquitoes are unsuited for the type of detailed analyses possible in *Anopheles*. Detailed studies on the *mariae* complex of *Aedes* have been carried out by Coluzzi and Sabatini (1968a), Coluzzi and Bullini (1971) and Coluzzi, Gironi, and Muir (1970). These are Mediterranean forms that breed in saltwater rock pools and have a littoral distribution. There are three allopatric sibling species: *mariae* (western Mediterranean), *zammitii* (Adriatic, Aegean and Ionian Seas), and *phoeniciae* (eastern Mediterranean). Male hybrids between *phoeniciae* and the other two have abnormal spermatogenesis and are sterile. The ovaries of hybrid females are partially atrophied, although some such females will lay a few fertile eggs. The sterility barriers between *mariae* and *zammitii* are less strong than those between *phoeniciae* and the other two.

The Genus Culex. Mosquitoes of the genus *Culex* likewise do not have polytene chromosomes favorable for detailed analysis. Here, however, an unusual mode of speciation dependent on cytoplasmic factors is said to occur in some species (Laven, 1959, 1967). In crosses between different European strains of the cosmopolitan *Culex pipiens,* Laven found that if one strain was the female parent, offspring could frequently be obtained, whereas if it was the male parent, no progeny resulted. Thus, the cross Hamburg ♀ × Oggelshausen ♂ produced offspring of both sexes, but Og ♀ × Ha ♂ yielded only inviable embryos. By repeatedly backcrossing hybrid Ha/Og females to Og/Og males it was possible to obtain mosquitoes with Og chromosomes and Ha cytoplasm. But males of this type still failed to produce viable offspring when crossed with Og females, thereby proving that a cytoplasmic factor is involved.

Laven has shown that in some instances two strains of *Culex pipiens* show incompatibility in both directions, presumably because each has a cytoplasmic factor that operates against the other. This raised the question whether active speciation due to mutations of cytoplasmic factors is actually happening. Unfortunately, although the general distribution of the cytoplasmic races of *C. pipiens* in Europe is known, no data are available on what happens where they meet. Caspari and Watson (1959) showed mathematically that with random mating in a mixed population of Ha and Og genotypes the Og cytoplasm will rapidly be eliminated. One might argue from this that zones of hybridization should advance rapidly across the landscape and that one form should soon disappear. Obviously, this is *not* what happens. Mayr (1963a) criticized Laven's claims that in *Culex* new species were arising by a form of sympatric speciation due to mutation of cytoplasmic incompatibility factors. The whole matter now has to be looked at again in the light of Yen and Barr's (1974) discovery that the "cytoplasmic factors" are in fact strains of a rickettsia-like microorganism, *Wolbachia pipientis,* which exists as a symbiont in all individuals of the *C. pipiens* complex. Whether the same or a related species of *Wolbachia* is responsible for cytoplasmic incompatibility in some other mosquitoes, such as *Aedes aegypti* and *A. albopictus* (Bonnet, 1950; Toumanoff, 1950) and *A. scutellaris* (Smith-White and Woodhill, 1954, 1955), is not known, but appears probable.

The question whether Laven's "crossing types" (a total of 17 were recognized) are leading to true speciation is still unclear. The probability is that *Wolbachia* strains or mutants that cause minimal damage in *Culex* populations adapted to them have severely adverse effects when they are introduced (by fertilization) to cells that are not adapted to withstand them. Males without the symbiont (produced by tetracycline treatment in the experiments by Yen and Barr) were compatible with females of all strains tested. A number of "subspecies" of *C. pipiens* have been distinguished, partly on the basis of morphological characters, partly on the biological character of autogeny, i.e., the ability to oviposit without having had a blood meal. Currently, three subspecies are recognized: *C. p. pipiens, C. p. fatigans,* and *C. p. australicus*. But the fact that these taxa show no relationship to the "crossing types" of Laven suggests that *Wolbachia* has played little or no role in *Culex* speciation. Thus hybridization between *C. p. pipiens* and *C. p. fatigans* occurs extensively in North America and no incompatibility was found in a cross between *C. p. pipiens* from Egypt and *C. p. fatigans* from Eritrea. A general analogy between the host–symbiont relationship in *C. pipiens* and the cases of *Drosophila paulistorum* (see p. 68) and tsetse flies (Pell and Southern, 1975) certainly exists. But it does not follow that the relation to race- and species-formation is the same in all these cases.

Drosophilidae

A vast literature now exists on the chromosomal evolution of *Drosophila* species. We shall consider the speciation mechanisms of the Hawaiian species in the next chapter, as a particularly well-studied example of allopatric speciation in an oceanic archipelago. Mechanisms of ethological isolation in *Drosophila* will be dealt with in Chapter 10. Here certain general problems of the relationship of chromosomal rearrangements to *Drosophila* speciation will be discussed.

Apart from the "homosequential" complexes referred to earlier, which include only a very small fraction of the known species, closely related *Drosophila* species usually differ in respect of one or more fixed paracentric inversions and, frequently, in the size and distribution of the heterochromatic segments in the karyotype. The earlier literature merely describes, often in a rather sketchy manner, the general distribution of the heterochromatin. It is now known, however, that this material includes from three to six different kinds of satellite DNA's that are not distributed in a random manner but in organized blocks, frequently adjacent to one another in the karyotype. The overall problem of the evolution of satellite DNA's will be considered at the end of this chapter. As far as *Drosophila* is concerned, it can be stated quite definitely that the classical accounts of chromosomal evolution in this genus, e.g., those of Patterson and Stone (1952; and Stone, 1955, 1962), which merely list a few of the more striking and obvious additions of heterochromatin (Stone cites 38 cases in *Drosophila* species studied up to that time), enormously underestimated both the amount and complexity of the evolutionary changes in the heterochromatin. Although the role of such changes in speciation is almost entirely unknown, it is hard to believe that they have been of no significance, especially when it is found that there are rather radical differences between the satellite DNA's of such closely related species as *D. melanogaster* and *D. simulans*.

In addition to the inversions and changes in the heterochromatic segments, there have of course also been a number of "Robertsonian" whole-arm changes in the genus *Drosophila*. Stone (1962) identified 58 fusions in 215 species studied, which suggests that about 500 had occurred in the total world fauna of about 2,000 species. He did not believe that any increases in chromosome number (whether by dissociation or "centric fission") had occurred, but it is difficult to be certain that they have not. Nevertheless, with one exception that may be a rather special case, no *Drosophila* populations polymorphic for fusions are known to exist. The exception is a hybrid population of *D. a. americana* and *D. a. texana* in Missouri, studied by Carson and Blight (1952). In *D. a. americana* the X is fused to autosome 4, so that the sex-chromosome mechanism in the male is effectively neo-XY_1Y_2, Y_1 being the original Y and Y_2 the unfused member of the fourth chromosome pair; in *D. a. texana* the X is unfused.

The predominant category of chromosomal rearrangements in *Drosophila* is of course paracentric inversion. Stone (1962) estimated that 592 floating (polymorphic) inversions were known in 42 species of *Drosophila* that had been carefully studied cytogenetically. This is an average of 14 per species, although *D. willistoni* and *D. subobscura* are both known to be polymorphic for about 50 different inversions. Obviously, some species and species groups are far more polymorphic for inversions than others. In fact, no inversion polymorphism has been recorded in *D. repleta* or *D. mulleri,* although they have been extensively sampled. The *virilis* group has an average of 5.67 polymorphic inversions per species (Stone, 1962), while the *repleta* group has 1.13 per species (Wasserman, 1963). No doubt both these figures are underestimates—future work would be more likely to reveal new inversions than new species—but the figures do make the point that there are at least fivefold differences in the level of inversion polymorphism between the different species groups of the genus *Drosophila*. These differences are correlated with the number of paracentric inversions that have been fixed in phylogeny, e.g., far more inversion differences exist between species of the *virilis* group than between those of the *repleta* group. It is thus clear that floating inversions may become fixed in the course of speciation, but there is rather little evidence in *Drosophila* to suggest that inversions frequently play a primary role in speciation. A few closely related pairs of species share the same inversion polymorphism, i.e., both have the same two alternative sequences; for example, *D. pictura* and *D. pictilis* both have the polymorphism $2n^3/+$ (Wasserman, 1963). Nevertheless, such situations are not common. Some inversion polymorphisms are undoubtedly ancient. Carson (1974) has presented evidence that one X chromosome polymorphism in the Hawaiian *D. neopicta* is older than 11 species in which the "derived" sequence has undergone fixation, i.e., that it is at least 1.5 to 2.0 million years old.*

The distribution of both floating and fixed inversions throughout the karyotype is strongly asymmetrical. For example, in the *repleta* group 103 out of the 144 inversions that have established themselves (and 39 out of the 52 floating ones) are in the second chromosome (Wasserman, 1963). In a few instances a large number of inversions have undergone fixation in the X chromosome; in the *virilis* group the species *ezoana, littoralis, montana,* and *lacicola* all have highly distinctive X chromosomes that have undergone such a degree of repatterning, as a result of multiple inversions, that no precise analysis has ever been carried out. Clearly, there is a *prima facie* case for believing that radical reorganization of the X chromosome has played a primary role in the speciation of one section of this group (those derived from the hypothetical ancestor "Primitive III" of Stone, 1962). If so, such reorganization may be analogous to the changes in sex chromosomes in Simuliidae and Chironomidae that seem to have been extremely important in speciation in these families. The analogy should perhaps not be pushed too far, however, since sex

---

*It was formerly argued, on biogeographic grounds, that some extant inversion polymorphisms in *D. pseudoobscura* date from the mid-Tertiary (about 30 million years ago), but this now seems very doubtful.

determination in these families seems to depend on localized "dominant male-determining" and "dominant female-determining" factors, while in *Drosophila* it depends on a genic balance mechanism, with most or all regions of the X chromosome female-determining. Wasserman (1963) put forward the hypothesis that 12 species in the *mulleri* complex of the *repleta* group had arisen by segregation from a polytypic species having seven geographic subspecies differing with respect to fixed inversions. Alternatively, they could have arisen from a single evolving population that was polymorphic, in different geographic areas and at various times, for the several inversions.

The total number of paracentric inversion differences that have established themselves in the phylogeny of the approximately 2,000 species of the genus *Drosophila* was estimated by Stone, Guest, and Wilson (1960) to be 22,000 to 56,000, with perhaps 18,000 to 28,000 floating inversions in the living species. While the latter estimate may (in the light of subsequent work) be somewhat excessive, the estimate for the number of fixed inversions is probably about right, or at any rate is the right order of magnitude.

Karyotypes and speciation have been studied in great detail in the *Drosophila paulistorum* complex of the *willistoni* group (Dobzhansky and Spassky, 1959; Dobzhansky and Pavlovsky, 1962, 1967, 1971; Dobzhansky, Ehrman, Pavlovsky, and Spassky, 1964; Dobzhansky, Pavlovsky, and Ehrman, 1969; and Spassky *et al.*, 1971). This complex, which inhabits the vast area between Guatemala and southern Brazil, includes seven taxa, most of which are probably best regarded either as geographic races or semispecies. The latter designation seems appropriate where two or three forms have reached a level of reproductive isolation that allows them to coexist sympatrically in areas of overlap, i.e., without merging. One form originally included in the *paulistorum* complex that seems to be more strongly isolated, ethologically, than the others (Kessler, 1962) has been regarded as a distinct species, *D. pavlovskiana,* by Kastritsis and Dobzhansky (1967). The distal end of chromosome 3 of *Pavolovskiana* has undergone a rather radical repatterning during or since its divergence from the other members of the complex, and hybrids with *pavlovskiana* show little chromosomal synapsis in the salivary gland chromosomes, a characteristic of interspecific rather than intraspecific hybrids in this group.

The remaining taxa of the *paulistorum* complex (Centro-American, Amazonian, Andean–South Brazilian, Orinocan, Interior, and Transitional) are genetically isolated from one another by a combination of premating, ethological factors and the usually complete sterility of the hybrid males. All these forms are highly polymorphic for inversions; some polymorphisms are common to several of the taxa while others are confined to a single one. The sterility of the hybrid males is apparently due to a mycoplasma, which is inherited maternally and is probably carried by all strains of *D. paulistorum*. The interaction between the mycoplasma and the genotype, which seems to be harmonious in the individual semispecies, is liable to be upset in hybrids, thus causing sterility of the males (Williamson and Ehrman, 1968; Ehrman and Williamson, 1965, 1969; Kernaghan and Ehrman, 1970). Changes in the relationship between the mycoplasma and the genotype have been observed in laboratory stocks kept for a year or more. A strain from Llanos, Colombia, which produced fertile male hybrids when crossed with Orinocan forms in 1958, no longer did so from 1963 onward (Dobzhansky and Pavlovsky, 1967, 1971). Some wild strains intermediate between Orinocan and Interior, and which originally produced fertile male hybrids with both, lost the ability to produce fertile male hybrids with the Interior semispecies after about a year in the laboratory (Dobzhansky and Pavlovsky, 1975).

The relationship between the mycoplasma and the *paulistorum* genotype is clearly labile, and is probably the main cause of the evolutionary divergence of the semispecies in this complex. It is still unclear, however, whether the genetic changes responsible for divergence in *paulistorum* are occurring in the genotype of the fly, in that of the mycoplasma, or in both. It seems unlikely that inversion polymorphisms have played a primary role in initiating divergence in the complex, although the changes at the distal end of chromosome 3 in *D. pavlovskiana* may have done so. Because ethological isolation is stronger between sympatric strains of different semispecies of *paulistorum* than between allopatric strains of the same pairs of semispecies (Ehrman, 1965), selection for ethological isolation may have occurred in areas of geographic overlap.

Speciation in flies of the genus *Musca* has been discussed by Saccà (1953, 1957, 1958, 1967); Saccà and Rivosecchi (1958); Milani (1967); Rubini (1964); and Rubini and Franco (1965). The superspecies *Musca domestica* (housefly) includes a number of taxa of uncertain status that have been distinguished by taxonomists and medical entomologists. Saccà (1967) states that *domestica, vicina,* and *nebulo* together form a cline in domestic habitats in hot and temperate climates outside the Ethiopian region, while *calleva* and *curviforceps* represent a more advanced stage of speciation in central and southern Africa. The whole problem is complicated, however, by an extraordinary diversity in the sex chromosome system. Normally, the females are XX and the males XY, the metacentric Y chromosome being considerably smaller than the metacentric X. However, some strains appear to have a segment of chromosome 2 translocated to the Y and show holandric inheritance for certain loci. Some individuals have a single sex chromosome. Those that are YO are certainly males; there has been some controversy as to whether XO individuals exist, but if they do they are female, and in certain strains may be inviable. Other individuals may carry more than one X or more than one Y. Boyes (1967) found the following combinations in a sample of *M. domestica curviforceps* from South Africa:

Males:	4 XX, 13 XY, 4 XO, 5 YO, 1 YY, 4 doubtful
Females:	6 XX, 5 XY, 4 XO

He concluded that sex determination in this population was no longer controlled by the X and Y, but instead by autosomal factors. Other interpretations are possible, however, and identification of a particular heteropycnotic chromosome as an X or a Y, solely on the basis of size and arm-ratio would seem risky in this case, since different types of Xs and Ys may exist. Paterson and James (1973) have reported that most Australian populations of *M. domestica* lack a Y; some of them have male-determining loci on several autosomes, while in others only chromosome 3 is male-determining. In the United States, many female houseflies collected in nature produce progenies with aberrant sex ratios (Absher, 1975). The relationship (if any) of changes in the sex chromosome mechanism and the origin of the various races or semispecies that comprise *Musca domestica* has not been seriously investigated. In order to do so it would seem necessary to know the precise nature and distribution of the various satellite DNAs that make up the heterochromatic Xs and Ys of the various taxa.

The "ordinary" grasshoppers of the families Acrididae, Pyrgomorphidae, Pamphagidae, and some others that comprise the superfamily Acridoidea are frequently cited as a group with karyotypic conservatism, i.e., minimal cytotaxonomic differences between related species. Indeed, chromosome numbers and the general pattern of the karyotype tend to be rather constant in this group—far more so than in the grasshoppers of the tropical and warm-temperate family Eumastacidae, some subfamilies of which (especially the Australian Morabinae) exhibit great karyotypic variability (see p. 173). Thus, the great majority of species of Acrididae have 2n = 23 acrocentric chromosomes in the males (which are XO), while those of Pyrgomorphidae and Pamphagidae show 2n = 19 acrocentrics in the males. Deviant karyotypes result mainly from centric fusions (producing metacentrics); these may be between autosomes or between an autosome and an X chromosome (leading to a neo-XY sex chromosome mechanism). In a small minority of species (mainly but not exclusively members of the North American tribe Trimerotropi: Acrididae) pericentric inversions have converted originally acrocentric chromosomes into metacentrics.

This karyotypic uniformity is superficial, however, being superimposed on an underlying diversity, as John and Hewitt (1966, 1968) have rightly emphasized. Very considerable differences exist between species in the amount and distribution of the heterochromatin, and a more than twofold range of DNA values exists in Acrididae (Kiknadze and Vysotskaya, 1970). The relative lengths of the chromosomes are far from constant and the lengths of the short arms of the acrocentrics vary greatly from species to species, sometimes even within species. All these facts suggest that chromosomal rearrangements may well have played an important role in speciation in this group, even though they are less conspicuous than in some other groups. Abnormalities of meiosis in grasshopper species hybrids have been recorded by Klingstedt (1939) and Helwig (1955), and although they are somewhat varied, there is no suggestion (with one doubtful exception) that translocations occurred in (or since) the evolutionary divergence of the parental species. John and Lewis (1965) found some unequal bivalents and multivalents in the meiosis of hybrids between the African grasshoppers *Eyprepocnemis plorans meridionalis* and *E. p. ornatipes* (which may be distinct biological species rather than subspecies). They regarded these abnormalities as evidence of translocation heterozygosity, but pointed out that other associations of chromosome ends exist that cannot be accounted for by translocation heterozygosity alone. Although some of their interpretations have been criticized in detail (Nankivell, 1967; White, 1973), their main conclusion is well established: the karyotypes of the two parental forms differ with respect to several and perhaps numerous structural rearrangements. According to John and Lewis these include translocations, but White considers the rearrangements mainly changes in the length of terminal heterochromatic segments (centromeric and perhaps also telomeric heterochromatin). Hybrids of sibling species of the *Cimex pilosellus* (bed bug) complex also show unequal bivalents, due to differences in the length of heterochromatic segments of several chromosome pairs (Ueshima, 1963). The general proposition that the occurrence of multivalents in the meiosis of species hybrids necessarily indicates that the parental forms differ in respect of translocations is undoubtedly false. Hybrids between subspecies or semispecies of the European newt *Triturus cristatus,* which show chains of three or four chromosomes at male meiosis (Callan and Spurway, 1951), are almost certainly not translocation heterozygotes. In many species of animals—the evidence for plants is less clear—the tips of most or all chromosomes (telomeric heterochromatin) are homologous, and can synapse and even form chiasmata in the somewhat disturbed meiosis of hybrids.

In groups of organisms that have "holocentric" chromosomes, i.e., ones without single localized centromeres, one might expect that the role of chromosomal rearrangements in speciation and evolution would be somewhat different. Yet the evidence, even though somewhat contradictory, indicates that the principles of karyotype evolution are less divergent in such groups than might be expected. Some cytogeneticists in the past claimed that evolutionary chromosomal fragmentation is theoretically far easier in these organisms because every chromosome fragment is able to attach to the spindle. However, this viewpoint ignores the telomere problem. Simple breakage of a holocentric chromosome would be likely to yield two fragments with ends that would then undergo "sister strand union," giving isochromosomes. However, this might not happen if the chromosomes contained interstitial palindromic nucleotide sequences capable of conversion into telomeres (as probably happens during the fragmentation of chromosomes in the nematode *Parascaris,* which occurs during the development of each individual. Clearly, however, the chance of translocations establishing themselves in phylogeny is somewhat greater in species with holocentric chromosomes, since all the products of reciprocal translocations are transmissible at meiosis instead of half of them being acentric and dicentric chromosomes. It is possible for fusions of holocentric elements to result in lower chromosome numbers, but probably only if each element loses a telomere.

The main groups in which holocentric chromosomes occur are the insect orders Hemiptera (Heteroptera plus Homoptera), Trichoptera (Caddis flies), and Lepidoptera, and the higher plant families Cyperaceae and Juncaceae. The magnitude of the karyotypic differences between related species in Heteroptera and Homoptera seems to be similar to that found in insect orders with monocentric chromosomes. Some genera and families show great stability of chromosome number ($n = 7$ in the great majority of the species of Pentatomidae), but the total amount of DNA in the karyotype varies greatly from species to species. This variation was interpreted by Hughes-Schrader and Schrader (1956) in terms of the now discredited hypothesis of evolutionary polynemy (changes in the number of DNA strands in the chromatid). We must now regard it as due to multiple duplications and deletions of short DNA sequences in a mononeme (single-stranded) chromosome, as in so many other groups (see p. 100).

Although there exist a few species of Heteroptera and Homoptera in which the chromosome number has undergone reduction to $n = 3$ or $n = 2$, there are no species with spectacularly high chromosome numbers, such as occur in some Lepidoptera (see p. 74). It would seem, therefore, that some restrictions must exist that prevent increases in chromosome number beyond about $n = 25$.

Little information exists concerning the distribution of satellite DNAs in karyotypes with holocentric chromosomes, but Lagowski and co-workers (1973) have shown that in the bug *Oncopeltus* the heterochromatin is dispersed throughout the karyotype in the form of many short segments, rather than being concentrated in major blocks. This is probably, but not certainly, a general feature of holocentric chromosomes.

The order Lepidoptera is in a sense one of the best known groups of animals cytogenetically; the karyotypes, or at any rate the chromosome numbers, of approximately 740 species (mostly butterflies) have been recorded. However, the small size and generally isodiametric form of the chromosomes have precluded any really detailed analysis, and most of the data are simply the results of alpha karyology. Nevertheless, some interesting general principles have emerged. There is now convincing evidence that the chromosomes of both Lepidoptera and Trichoptera (universally admitted to be very closely related insect orders) are holocentric, so that we cannot speak of chromosome arms, and in the absence of any knowledge of the distribution of heterochromatin only the overall dimensions of the chromosomes can be measured.

If we examine a histogram of chromosome numbers in Lepidoptera (Figure 5), we find that there is a strongly marked concentration of the species around

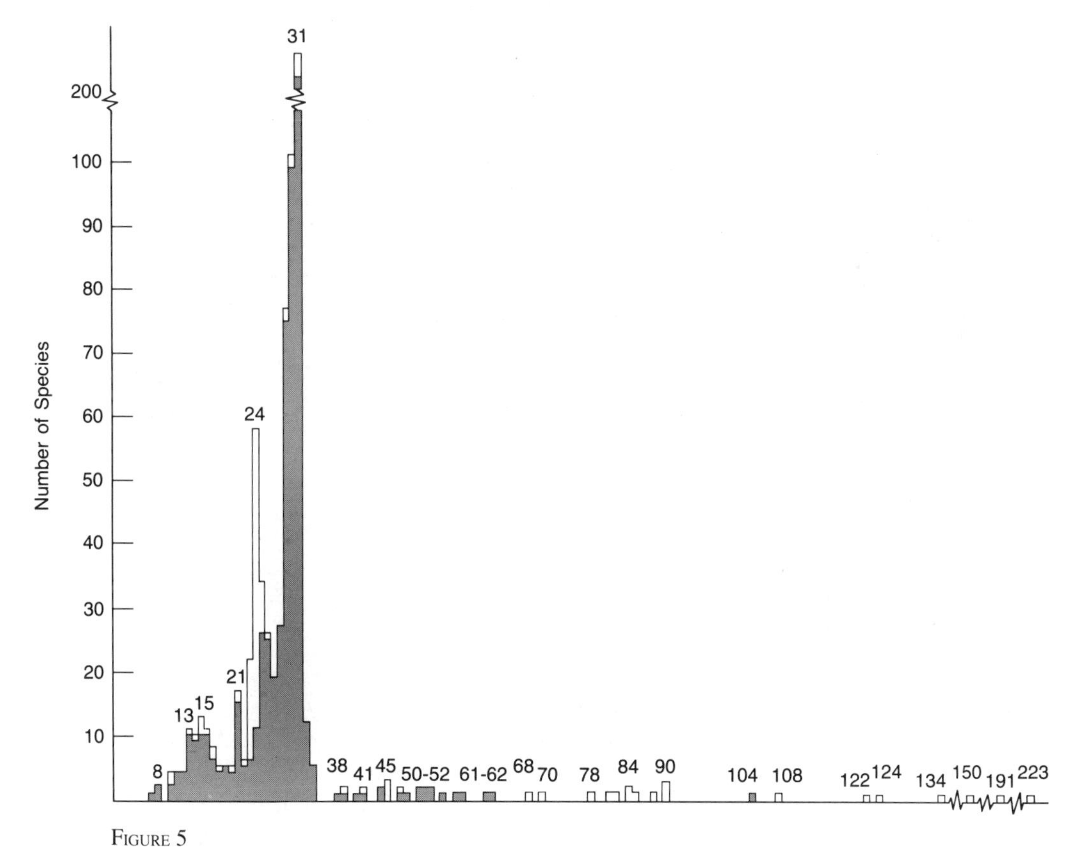

FIGURE 5

Histogram of the known haploid chromosomes in 738 species of butterflies. Haploid numbers are indicated above the bars. Data for the superfamily Lycaenoidea (families Lycaenidae and Riodinidae) are shown in outline. [After White, 1973]

n = 29, 30, and 31, with over a quarter of the species having 31 (the distribution of chromosome numbers in butterflies and moths is similar). Nevertheless, there are species with much smaller numbers (down to n = 7) and much larger ones (up to n = 223). The shape of the histogram is strongly skewed, with far more species having numbers below 31. We may take this as an indication that, in general, mechanisms for reducing the chromosome number have been far more efficient, or more successful, than ones leading to increases in chromosome number.

The only family of Lepidoptera that has a well-marked "type number" of its own is Lycaenidae (blue and copper butterflies) in which the numbers n = 23 and n = 24 are common. Spectacular increases and decreases of chromosome number have occurred in certain lineages of this family. Thus, in the Mediterranean genus *Agrodiaetus* there are species (such as *posthumus* with n = 10 and *araratensis* with n = 13) in which chromosomal fusions must have occurred, and others (*phyllis,* n = 79 – 82) in which humerous fragmentations of some kind must have established themselves. In the closely related genus *Lysandra, L. syriaca* has the primitive number n = 24, but species with 45, 82, 84, 88, 90, 124, 131–134, 147–151, 190–191, and 217–223 also occur; there is no reason to believe, as some early workers did, that polyploidy has played any part in generating these very high chromosome numbers. The karyotypes of the genus *Lysandra* include about three chromosomes that are much larger than the others and clearly represented only once even in the species with the highest numbers, a situation incompatible with polyploidy (these larger chromosomes may be sex chromosomes, immune from the fragmentation process that has affected the other chromosomes). Lycaenidae of the New World and Japan show the same tendency to develop aberrant chromosome numbers.

Although Lycaenidae show particularly striking variations in chromosome number, many other sibling or very closely related lepidopteran species have widely different karyotypes. The species of *Erebia* have been the subject of detailed studies by Lorković (1958a, b) and De Lesse (1955, 1960, 1964), while 40 species of the moth genus *Cidaria* (n = 13–32) have been studied by Suomalainen (1963, 1965).

Some lepidopteran genera appear to show rather constant karyotypes, however. Species of *Colias* all show n = 31 and all species of *Papilio* except three show n = 30 (Maeki and Remington 1960a, b; Maeki and Ae, 1966). Although variations in chromosome number have been recorded in many species of Lepidoptera, the interpretation of these cases is unclear, and it must be admitted that not a single instance of karyotype evolution in Lepidoptera is thoroughly understood. Speciation in such genera as *Lysandra, Agrodiaetus, Erebia,* and *Cidaria* seems to have been usually or always accompanied by (frequently extensive) changes in karyotype. But in such genera as *Colias* and *Papilio* it appears to have occurred without chromosomal rearrangements, or if it did, with ones that are not evident at the level of alpha karyology.

In plants, holocentric chromosomes have been reported in some algae and in the rushes and sedges (families Juncaceae and Cyperaceae). Karyotype evolution has been studied in the genus *Luzula* by Nordenskiöld (1951) but a number of points remain obscure. In some species, such as *L. campestris,* a type of chromosome fragmentation has apparently occurred; some populations have 2n = 12, while others show 2n = 24 chromosomes half the size of those in the biotype with the lower number. A few plants have intermediate chromosome numbers and mixtures of large and small chromosomes, e.g., 10 large and 4 small; they may be of hybrid ancestry. The term *agmatoploidy* (Malheiros-Gardé and Gardé, 1951) has been introduced to designate increases in chromosome number due to fragmentation of holocentric chromosomes, and is used by Grant (1971). However, at the present stage it merely serves to conceal our ignorance of the actual mechanism of this process. True polyploidy has apparently also occurred in some Juncaceae and Cyperaceae; thus in *Luzula* there is a polyploid series based on n = 6. In the genus *Carex* there is a great range of chromosome numbers, from n = 6 to n = 56, with every number from 12 to 43 being represented (Davies, 1956). This has been referred to as an "aneuploid" series, which would imply that it was due to stepwise reduplication of whole chromosomes. The situation is probably much more complicated than this, involving structural rearrangements of various kinds. It seems that these groups of plants will have to be reinvestigated by modern cytogenetic techniques in order to understand what has actually happened in their karyotype evolution. Grant rightly states that "our understanding of agmatoploidy is still in its early exploratory stages" and that "the relation between agmatoploidy and plant speciation is unclear."

Although much information now exists on the cytogenetics of a wide range of the invertebrate phyla and classes of arthropods other than insects, most of the published data do not seem to illustrate any major principles not already encountered in our discussion of karyotype evolution in the insect groups that have been more thoroughly studied. Certain special regularities seem to occur in the karyotype evolution of the spiders (White, 1973, 1975). The great majority of the approximately 250 species whose karyotypes have been determined have a peculiar sex chromosome mechanism that may be symbolized X_1X_2O (♂) : $X_1X_1X_2X_2$ (♀). A rather special meiotic mechanism seems to ensure that X_1 and X_2 regularly pass together to the same pole at the first meiotic division in the male, so that they are inherited together. Since the X_1X_2O system exists in members of three different suborders of spiders, it has presumably been inherited from their Palaeozoic ancestors. However, on at least six occasions in the phylogeny of the group lineages with only a single X, i.e., XO:XX, have arisen (probably by fusions of X_1 and X_2), and in five or more lineages $X_1X_2X_3O : X_1X_1X_2X_2X_3X_3$ mechanisms have evolved (White, 1973).

In contrast to the insect groups previously discussed, the spiders manifest a great conservatism in the sex chromosome mechanism; in most lineages it has presumably persisted in a virtually unchanged condition for perhaps 300 million years. There must have been a strong barrier to the establishment of fusions between X_1 and X_2 (breached on only a few occasions) and an absolute barrier to evolutionary fusions between X chromosomes and autosomes (which are not known to have occurred at all in this group). The chromosomes of spiders are mostly acrocentric, and rather strict proximal localization of chiasmata is a common feature of the male meiosis, probably acting as a barrier to the evolutionary success of fusions between chromosomes (trivalents with proximal chiasmata are particularly liable to undergo nondisjunction). Nevertheless, there is clear evidence that in a number of species of spiders fusions of acrocentric chromosomes to give metacentrics have led to multiple reductions of chromosome numbers, e.g., from an ancestral $2n♂ = 41$ to $2n♂ = 21$ in *Heteropoda sexpunctata,* and down to $2n♂ = 7$ in *Ariadna lateralis*.

Modern analytical studies of karyotype evolution in vertebrates have been carried out mostly on amphibians, reptiles, and mammals. Because of the relatively large number of chromosomes and their generally small size, investigations of the karyotypes of fishes and birds have been for the most part less detailed.

The comparative cytogenetics of amphibians has been discussed by Morescalchi (1973). The karyotypes of closely related species in this group frequently appear indistinguishable when studied at the alpha level, but reveal differences when more sophisticated techniques are applied. The DNA values of Urodeles are particularly high and the giant "lampbrush" chromosomes in the oocyte nuclei of many species have been studied in detail. Thus, in the genera *Triturus, Salamandra, Euproctus,* and *Pleurodeles* the chromosome number is always $n = 12$ and the relative lengths of the chromosomes are similar. Nevertheless, the DNA values in the family Salamandridae, to which these genera belong, range from 43.6 to 60.6 pg (Olmo and Morescalchi, 1975). "Maps" of the main landmarks in the lampbrush chromosomes of a number of species have been published (Callan and Lloyd, 1960; Mancino and Barsacchi, 1965, 1966, 1969; Barsacchi, Bussotti, and Mancino, 1970; Mancino, Barsacchi, and Nardi, 1969; Nardi, Ragghianti, and Mancino, 1972; and LaCroix, 1968). These show, diagrammatically, the location and general appearance of the more conspicuous pairs of loops and other structures arising from the chromosomes and, in some instances, the positions of the centromeres (Figure 6). They are hence not really comparable to the maps of the polytene chromosomes in Diptera, which show a wealth of detail not indicated in the lampbrush

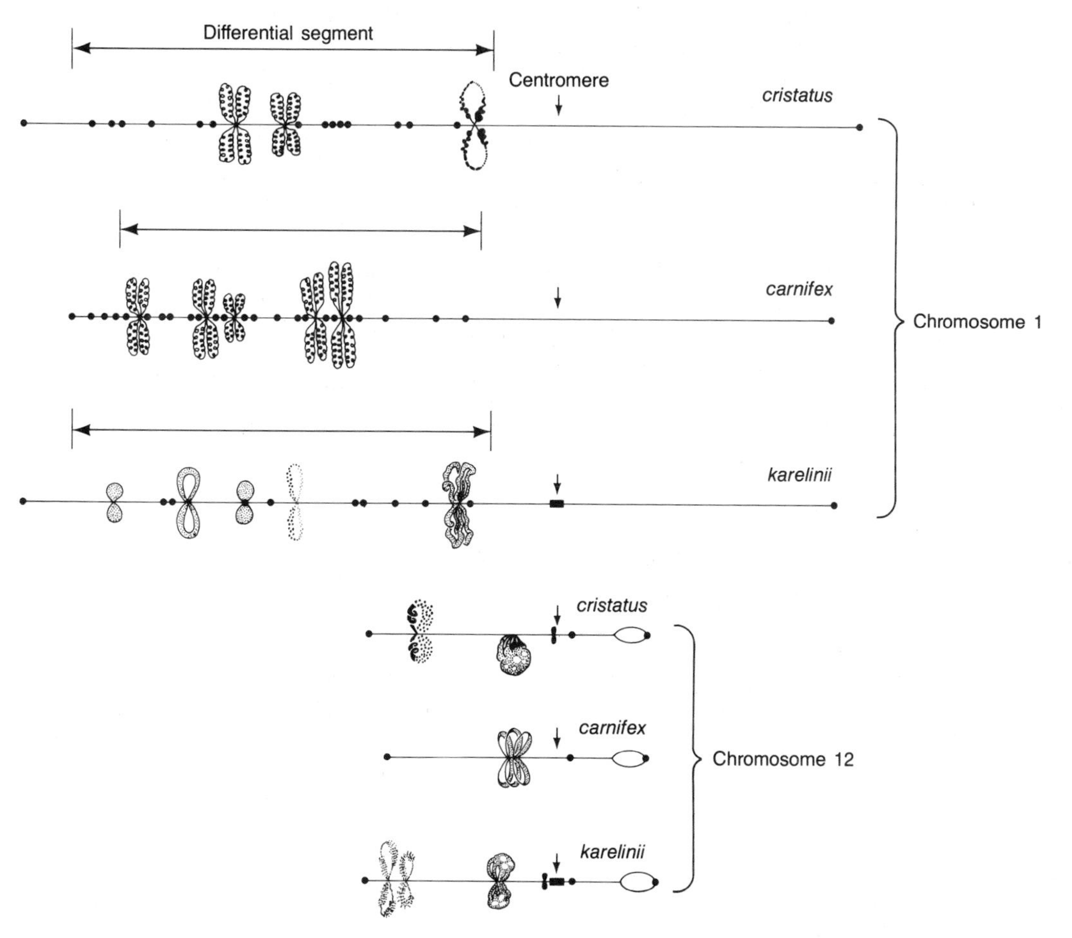

FIGURE 6

Diagrams of the main loops and other features of chromosomes number 1 and 12 in lampbrush nuclei of *Triturus cristatus cristatus, T. c. carnifex,* and *T. c. karelinii*. These "races" are probably better regarded as distinct species. Chromosome 1 is the X chromosome. [After Callan and Lloyd, 1960]

maps. More detailed lampbrush maps undoubtedly will be published in the near future, perhaps at first for short easily identified chromosome segments. But until such time it is a little difficult to state categorically what a comparison of the lampbrush "maps" of related species reveals. If a pair of conspicuously long loops are present in one race or species but are not shown in another taxon, does this mean that it is totally lacking, or that it is represented by a pair of shorter and hence inconspicuous loops? There has been a general tendency to ascribe differences in loop morphology between closely related taxa to "allelic" changes, but it seems likely that such differences, especially if conspicuous, usually or always imply that there have been changes in the amount of DNA present. Thus evolutionary interpretations of lampbrush chromosomes really await more precise models of their molecular architecture, with detailed information on the extent and distribution of unique and repetitive nucleotide sequences.

Certain things are already clear, however. The semispecies of the *Triturus cristatus* complex clearly differ in the morphology of the centromere region (Callan and Lloyd, 1960), but this is uniform for all the chromosomes in the karyotype of a particular taxon. For example, in *T. c. karelinii* from Turkey all the centromeres are flanked by thick bars of DNA, while in *T. c. carniflex* they are visible as small chromomeres without flanking bars and without loops; the centromeres of *T. c. cristatus* are similar but smaller (Figure 6), and those of *T. c. danubialis* smaller still.

Differences between the loop patterns of some Urodele taxa have led to the reclassification of some as sibling species. *Triturus italicus* was formerly considered by some authors to be merely a race of *T. vulgaris*. It and *T. vulgaris meridonalis* overlap in the Abruzzi mountains, but no natural hybrids have been encountered (Mancino, 1968) and the few obtained by laboratory crossing were sterile. The mitotic karyotypes are distinguishable, especially since there is a difference in the arm ratio of chromosome 12. The lampbrush chromosomes show substantial differences; those of *italicus* show distinct centromeric chromomeres, while those of *vulgaris* do not, and there are numerous differences in the principal landmarks.

In the *T. cristatus* complex the females are XY and the sex bivalent has a long segment in which the loop pattern is different in the two homologs; however, the loop patterns of these segments are highly polymorphic. In *T. marmoratus* and *Pleurodeles poireti* the specialized differential segments of the X and Y are unequal in length (Mancino and Nardi, 1971; LaCroix, 1970).

The frogs and toads (suborder Anura) do not possess lampbrush chromosomes of a size suitable for detailed analysis. They have much lower DNA values than Urodela. The karyotypes of Anura have been generally regarded as very conservative—those of many groups of related species and even whole genera are frequently indistinguishable (Maxson, Wilson and Sarich, 1974; Wilson, Sarich, and Maxson, 1974). However, even if alpha karyology does not reveal karyotypic differences, it by no means follows that the species are really homosequential. In his extensive studies on the toads of the large genus *Bufo,* Bogart (1972) found that the species often differed in the location of chromosomal "constrictions" (in reality, probably heterochromatic segments).

He found (1970a) much more considerable differences between the karyotypes of five Chilean species of the genus *Eupsophus* and argued from this that *Eupsophus* was an "unnatural assemblage of species." Bogart's conclusion may be the result of a bias derived from his work with genera like *Bufo, Rana,* and *Hyla,* which show much less karyotypic diversity at the "alpha" level. Certainly, the karyotypic diversity of *Eupsophus* would not be regarded as unusual in a genus of lizards or rodents. Another South American genus of leptodactylid frogs, *Eleutherodactylus,* shows even more karyotypic diversity, with chromosome numbers ranging from 2n = 18 to 36 (Bogart, 1970b). *Leptodactylus,* with 2n = 22 in the majority of the species but 2n = 24 and 26 in some aberrant ones, shows fairly extensive karyotypic differences in the 19 species studied by Bogart (1974). Even in the tree frogs of the genus *Hyla,* most of which show a considerable degree of karyotypic uniformity (2n = 24), *H. brunnea* from Jamaica has 2n = 34, clearly the result of five "dissociations" or "centric fissions" of metacentrics (Cole, 1974). A few species of Anura have become polyploid (see p. 280).

A considerable amount of detailed and reliable cytogenetic information on reptiles is now available (Gorman, 1973; Hall, 1973; King and King, 1975). The karyotypes of a large number of species are known, especially in certain groups of lizards, so that the contribution of chromosomal rearrangements to speciation processes should be relatively easy to assess. Unfortunately, lizards are rather unsuitable material for experimental hybridization, and knowledge of their genetic isolating mechanisms, biochemical polymorphisms, and basic population genetics is very inadequate as yet. The data summarized by Wilson, Bush, Case, and King (1975) indicate that karyotypic changes per evolutionary lineage per 10^8 years have been far fewer in lizards than in rodents or primates but rather more than in frogs (see Table 3-2, p. 82).

Gorman (1973) has summarized karyotypic data on about 300 species of lizards. The chromosomes of this group tend strongly to fall into two size classes: microchromosomes (*a* or *m* chromosomes, according to whether they are acrocentric or metacentric) and macrochromosomes (*A* or *M* chromosomes). The former normally occupy the central part of the spindle at mitosis, while the latter are attached to the outside of the spindle, with their long arms lying freely in the cytoplasm. In the Gecko family there is no sharp distinction between the two categories, but in the other families the dichotomy is well marked and the two classes of chromosomes seem to obey, to a considerable extent, different rules and principles. Unfortunately, the distribution of satellite DNAs over the karyotype is not known for any species of lizard.

The karyotypes of the iguanid lizards are now especially well known, and other evolutionary studies on members of this family have thrown some light on their speciation mechanisms. The largest genus is *Anolis,* with more than 200 species in the Caribbean and Central America, divided taxonomically into "alpha" and "beta" sections (Gorman and Atkins, 1968, 1969).

In the alpha section of *Anolis* the basic and presumably primitive karyotype is 2n = 36 (12*M* = 24*a*). In several species, however, the number of microchromosomes has been reduced to 22 or 20. One aberrant species, *A. monticola,* has 2n = 48 (24 acrocentric *A* chromosomes, 24 *a* chromosomes); presumably six chromosomal dissociations or centric fissions have occurred. Morphologically distinguishable sex chromosomes have not been found in the karyotypically primitive species. The general tendency in the alpha section has been toward a reduction in chromosome number. A number of species have an X_1X_2Y sex chromosome mechanism that may have evolved independently more than once. *A. eversmanni* has an XY sex chromosome pair that may have been derived by a fusion of X_1 and X_2.

In the beta section of *Anolis* the primitive karyotype seems to be 2n = 30 (14*M* + 16m, with no visible sex chromosomes). Many species have increased their chromosome number to 2n = 40 (20*A* + 4*M* + 16*m*, or a similar karyotype). Some beta anoles may have a distinguishable XY pair in the male, but *A. biporcatus* has an X_1X_2Y system.

Relatively precise phylogenetic trees of the species, based on genetic distance data as well as a fairly accurate scale of absolute evolutionary time and reasonably adequate karyotypic data, have been constructed for only a few groups of organisms. The eight α *Anolis* species related to *A. roquet,* which together inhabit the Lesser Antilles islands (Yang *et al.,* 1974; Nei, 1975) are one such group. Most of these species have the primitive karyotype (2n = 12*M* + 24*a*), but one branch (including *A. aeneus, A. extremus,* and *A. roquet*), which broke away from the main stem of the phylogenetic tree about 1.5 million years ago, has reduced the number of microchromosomes to 22, presumably by a fusion. In this case nine species appear to have evolved with only one major chromosomal rearrangement, i.e., one speciation event may have been due to a structural chromosomal change while the other seven took place without major rearrangements.

Sceloporus is another species-rich iguanid genus that shows a very considerable variety of karyotypes (Lowe, Cole, and Patton, 1967; Cole, 1971, 1972; Hall, 1973). This genus is discussed at length in Chapter 6 because the *S. grammicus* complex seems to provide strong evidence for the "stasipatric" model of speciation. Therefore, other lizard genera will be discussed here. In general, lizards seem to possess two mechanisms of speciation, only one of which involves karyotypic change. However, it must be emphasized that to date the only karyotypic analysis of this group has been of the beta type; if gamma karyology were carried out, the situation might be found to be more complex. In particular, in the numerous species with six pairs of large autosomes the chromosome arms that have united to form metacentrics are not necessarily the same.

The chameleons are the ecological equivalents of the anoles in Africa and Madagascar. The cytotaxonomy of Chamaeleontidae has been studied by Matthey (1957, 1961; Matthey and van Brink, 1956, 1960) who distinguishes two main types of karyotypes: the "continental" (African) and the "insular" (Madagascan and Comoran). The continental karyotypes, which mostly show $2n = 12M + 24a$, exhibit a sharp separation between macro- and microchromosomes. The insular karyotypes have lower chromosome numbers and show less of a distinction (or none at all) between the two classes of chromosomes, the result of fusions between originally separate microchromosomes. The correlation with geographic distribution is imperfect, since some species from Madagascar show a continental karyotype while, conversely, a few mainland species have insular karyotypes.

The karyotypes of about 1,500 species of mammals have now been studied. Great diversity exists in most orders, so that there seems to have been a far greater tendency for mammalian speciation to be accompanied by major chromosomal rearrangements than speciation of a group such as the frogs, with little karyotypic diversity (Wilson *et al.*, 1974). Obviously, only a very small part of the available evidence on the role of karyotypic changes in mammalian speciation can be considered here. Wilson and co-workers (1975; and Bush *et al.*, in press) have estimated the number of "karyotypic changes," i.e., changes in arm number or chromosome arm, per lineage per million years in various groups of placental mammals. Their data are summarized in Table 3-2. These authors have very plausibly suggested that the generally high rates of evolutionary changes in the karyotypes of placental mammals are the result of social structuring in populations of these animals, which arises ultimately from the dependence of the young on the mother during lactation. This dependence, together with the prevalence of polygamy and dominance hierarchies among the males, leads to high levels of inbreeding, thereby facilitating fixation of chromosomal rearrangements. Wilson and co-workers regard the stasipatric model of White (1968) as having been the main mode of speciation in placental mammals, in contrast to frogs and toads, in which various allopatric types of speciation have been prevalent.

The whales and carnivores show significantly less karyotypic diversity than other mammals. The high vagility of the whales is undoubtedly a partial explanation for this group (see p. 19), but an explanation for the carnivores is less obvious. It seems probable that karyotypic changes of types not included in the survey of Wilson and co-workers may have been important in carnivore speciation.

The karyotypes of most of the known species and races of macropodid mar-

TABLE 3-2
Rates of karyotype evolution in vertebrate genera. (From Bush et al., *in press.)*

	Number of genera examined	Average age of genera (million years)	"Karyotypic changes" per lineage per million years
Horses	1	3.5	1.395
Primates	13	3.8	.746
Lagomorphs	3	5.0	.633
Rodents	50	6.0	.431
Artiodactyls	15	4.2	.561
Insectivores	7	8.1	.187
Marsupials	15	5.6	.176
Carnivores	10	12.9	.078
Bats	15	9.0	.059
Whales	2	6.5	.025
	Average for mammals:	6.5	.295
Snakes	14	12.1	.048
Lizards	16	20.1	.058
Turtles and crocodiles	14	45.2	.022
Frogs	15	26.4	.023
Salamanders	11	23.4	.014
Teleost fish	12	5.7	.029

supials (kangaroos and wallabies) have been studied most carefully by Hayman and Martin (1974), who also determined the DNA values of the various taxa of this group. However, these workers did not use the modern Giemsa-banding techniques, which would undoubtedly have added something to the interpretation. In the genus *Petrogale,* two species *(P. xanthus* and *P. rothschildi)* have virtually indistinguishable karyotypes, but three forms *(hacketti, pearsoni* and *inornata)* traditionally included in *P. penicillata* have quite different ones. In three species of *Thylogale* the autosomes are very similar, but the X chromosomes differ and the DNA values are significantly different. The genus *Macropus* (10 species) and the monotypic genera *Wallabia* and *Megaleia* are very closely related, and both *Wallabia bicolor* (swamp wallaby) and *Megaleia rufa* (red kangaroo) can be hybridized with various species of *Macropus*. But in this entire group only two species, *Macropus fuliginosus* and *M. giganteus,* have karyotypes that are indistinguishable at the level of beta karyology. The sex chromosomes of most of the species are highly distinctive. The remaining genera and species of macropodids all have unique karyotypes, except for two species of *Dendrolagus* (tree kangaroos), which appear to be chromosomally indistinguishable. As in so many other groups, it is clear that the macropodids have undergone more structural changes in the sex chromosomes than in any one pair of autosomes, although not as many as in the entire set of autosomes.

As shown in Table 3-2, rodents are one of the most karyotypically variable orders of mammals. This is no doubt due to a combination of low vagility and a type of population structure that encourages inbreeding. Since we have very extensive karyotypic data for several large rodent genera, we shall discuss a few of these here.

The numerous species of *Gerbillus* and *Dipodillus* of Israel and North Africa have been studied in detail (Wahrman and Zahavi, 1955, 1958; Wahrman and Gourewitz, 1973; Zahavi and Wahrman, 1957; Wassif, Lutfy, and Wassif, 1969; and Cockrum, unpublished). Some of the data are summarized in Table 3-3. A particularly interesting species complex is the *G. pyramidum* group. There is good reason to believe that the ancestral stock of *Gerbillus* had $2n = 76 - 78$ acrocentrics, and that different centric fusions established themselves in the various lineages leading to the modern species and races. The "Tel Aviv" form of the *pyramidum* group, with 12 to 13 fusions (it is polymorphic for one fusion), occupies a very short and narrow strip of territory along the coast of Israel. Further south, in the Sinai Peninsula and the Negev, a more widespread race with only five or six fusions occurs (it is also polymorphic for one fusion, probably the same one as in the Tel Aviv race). There is an extensive zone of hybridization on the coast of Israel, at least 40 km wide, in which most of the individuals are heterozygous for chromosomal fusions and show trivalents at meiosis. The North African race that is *pyramidum sensu stricto,* whose distribution seems to include Algeria, Egypt, Ethiopia, and French Somaliland, has $2n = 38$. Hybrids between the forms from the Sinai Peninsula and North Africa generally show three bivalents (including the XY), 10 trivalents, and three chain associations of 5–7 chromosomes each. The existence of these chain associations shows that the ancestral *G. pyramidum* had far more acrocentric chromosomes than any of the existing taxa and that different fusions have occurred in Israel and North Africa. On the other hand, all the fusions in the Sinai Peninsula race, together with seven additional ones, seem to be present in the Tel Aviv race. The existence of the chains of 5–7 chromosomes in the hybrids referred to above proves that we are dealing with fusions rather than dissociations (because an acrocentric can fuse with any other acrocentric, but a metacentric can only dissociate into the same two acrocentrics).

The two races in Israel may be regarded as semispecies; the North African form (which should be regarded as specifically distinct) is obviously genetically isolated from them to a considerable extent, since F_1 hybrids must be largely sterile, due to irregular segregation of long chains of chromosomes. It would be interesting to know what happens in the zone of contact that must exist near the Suez Canal between the populations with 64–66 and 38 chromosomes. Clearly, chromosomal fusions have had a major part in speciation in gerbils.

Chromosome evolution in *Acomys* (spiny mice) seems to have followed a

TABLE 3-3
Karyotypes of three genera of gerbils.

Species	Origin	2n♂	Sex chromosomes	No. of metacentric (♂)	No. of chromosome arms (♂)
Gerbillus:					
G. gerbillus	Morocco, Algeria, Tunisia, Egypt, Negev	43	XY_1Y_2	35	78
G. tarabuli	Algeria, Tunisia, Morocco	40	XY	38	78
G. pyramidum	Egypt	38	XY	38	76
G. sp. (near *pyramidum)*	(1) Sinai Peninsula, Negev	64–66	XY	10–12	76
	(2) near Tel Aviv, Israel	50–52	XY	24–26	76
G. andersoni	Tunisia, Egypt, Israel	40	XY	40	80
G. aquilus	Iran, Pakistan	38	XY	38	76
G. cheesmani	Iran	38	XY	36	74
G. perpallidus	Egypt	40	XY	36	76
G. occiduus	Morocco	40	XY	40	80
G. gleadowi	Pakistan	51	XY_1Y_2	22	73
G. hesperinus	Morocco	58	XY	22	80
G. hoogstraali	Morocco	72	XY	8	80
Dipodillus:					
D. dasyurus	Egypt, Israel	60	XY	9	70
D. campestris	Morocco, Algeria, Tunisia, Egypt	56	XY	11	67
D. simoni	Tunisia, Egypt	60	XY	8–10	68–69
D. henleyi	Morocco, Tunisia, Egypt, Negev	52	XY	13	65
D. nanus	Morocco, Algeria, Tunisia, Egypt, Negev, Iran, Pakistan	52	XY	9	61
Meriones:					
M. lybicus	Algeria, Iran	44	XY	30	74
M. shawi	Israel	44	XY	34	74
M. crassus	Israel	60	XY	14	74
M. sacramenti	Israel	46	XY	28	?74
M. tristani	Israel	72	XY	2	?74

SOURCE: Data from Wahrman and Zahavi, 1955; Zahavi and Wahrman, 1957; Wassif, Lutfy, and Wassif, 1969; Nadler and Lane, 1967; and E. L. Cockrum, personal communication. Cockrum's data provide many additions and corrections to earlier data, both with regard to cytology and nomenclature of this taxonomically confused group.

similar course. Ten species of this genus have chromosome numbers ranging from 36 to 66; the number of major chromosome arms (the *nombre fondamental* or NF of Matthey) ranges from 66 to 76 (Matthey, 1965, 1968b). Two chromosomal races of *A. cahirinus,* differing in karyotype by a single fusion or dissociation, have a zone of hybridization about 30 km wide on the coast of the Gulf of Aqaba (Wahrman and Goitein, 1972). In the laboratory, hybrids between a Palestinian form and one from Cyprus showed chains of up to 28 chromosomes at meiosis, indicating (as in the case of the *Gerbillus pyramidum* complex) that different centric fusions have occurred in two evolutionary lineages (Wahrman, 1972).

By way of contrast to these rodent genera, in which chromosomal evolution seems to have been predominantly of the "Robertsonian" type, with centric fusions of acrocentrics especially frequent, we may consider the situation in the genus *Peromyscus* (North American deer mice). All the species in this genus appear to have 2n = 48 (Hsu and Arrighi, 1966, 1968). However, the number of major chromosome arms ranges from 56 (in *P. crinitus* and *P. boylei)* to 96 (in *P. eremicus* and *P. collatus).* This situation originally was interpreted as indicating that reciprocal translocations and pericentric inversions had occurred. Later it was shown (Duffey, 1972; Pathak, Hsu, and Arrighi, 1973) that the short arms of all the metacentric chromosomes are composed of constitutive heterochromatin. This rules out pericentric inversions. The length of the short arms may have increased (by duplication) and decreased (by deletion of repetitive satellite DNA regions) in particular lineages. But there may also have been translocations by which longer and shorter heterochromatic arms were interchanged between different members of the karyotype. A detailed study of the karyotypes of *P. maniculatus* has been carried out by Bradshaw and Hsu (1972). This is a very widespread, complex species, whose range extends from the Arctic Circle in Canada to southern Mexico. Taxonomists have described as many as 65 subspecies, which fall into two main groups, according to whether they occupy forests or grassland habitats. In the eastern United States the grassland and forest *maniculatus* mice apparently do not interbreed (Blair, 1953), but in the western region limited interbreeding presumably occurs in some areas. Almost all the 15 subspecies investigated by Bradshaw and Hsu showed some karyotypic polymorphism, manifested as variation in the number of "bi-armed" chromosomes, i.e., submetacentric ones with a heterochromatic short arm. But the mean (or modal) number of such chromosomes also differes between the subspecies. Most individuals of the prairie subspecies have 34–36 bi-armed chromosomes, but *P. m. blandus* showed only 16. Two presumably very similar subspecies, "*rufinus* I" and "*rufinus* II," were found in the region of Arizona–New Mexico–Colorado, with modal numbers of 16 and 36 bi-armed chromosomes, respectively. However, it is probable that "*rufinus* I" (from the mountain ranges of southern Arizona) actually belongs to the Mexican species *P. melanotis,* rather than to *maniculatus* (Bowers *et al.,* 1973).

Peromyscus maniculatus has been called a "nightmare to the taxonomist." Blair (1950) considered that *maniculatus* was ancestral to four geographically restricted peripheral species: *polionotus* (Florida, Georgia, South Carolina, Alabama), *sitkensis* (islands off the coast of Alaska and British Columbia), *sejugis* (islands off the Gulf of California), and *melanotis* (high elevations in Mexico and southern Arizona). Bowers and co-workers have accepted this interpretation and developed it in relation to the "centrifugal speciation" model of Brown (1957), according to which the central part of the distribution of a species is regarded as the main source of evolutionary changes leading to "potential new species." The "center" in this case is assumed to be a population or populations of *P. maniculatus* with a high number of acrocentric chromosomes, i.e., a low number of bi-armed chromosomes. Brown's model of speciation is discussed at length in Chapter 4; in its application to the *Peromyscus maniculatus* case it rests on the assumption that a large number of acrocentric chromosomes is the primitive condition. The "peripheral" populations (whether subspecies of *maniculatus* or the four species mentioned above) mostly do have large numbers (up to 18 or 19 in populations of *maniculatus* from the State of Washington and southern Ontario, 17 in California populations of the same species, a uniform 30 in all populations of *melanotis,* 24 in *polionotus*). However, *P. sitkensis* is exceptional in showing only six acrocentrics (Hsu and Arrighi, 1968).

These views have been criticized by Lawlor (1974) chiefly on the ground that there is insufficient evidence that a karyotype with a large number of acrocentric chromosomes is primitive for the genus *Peromyscus* as a whole or for the *maniculatus* group. While he is probably correct in criticizing the general tendency among vertebrate cytogeneticists to assume too easily that acrocentric chromosomes are necessarily primitive and that the metacentrics are derived (because "centric fusion" is assumed to be an easier mechanism than "centric fission" or dissociation of metacentrics), there are some sound reasons for believing that in the particular case of *Peromyscus* Bowers and co-workers are correct in their assumption. Centric fusions are a type of rearrangement in which some centric heterochromatin is lost. Thus the most plausible explanation for the fact that no centric fusions seem to have established themselves in the phylogeny of the genus *Peromyscus* (a highly unusual and perhaps unique feature of this genus in comparison with other species-rich rodent genera) is that there has been fairly strong selection against losses of centric heterochromatin.

The *Peromyscus* populations from a chain of 13 islands off the coast of British Columbia have been studied by Thomas (1973). The southern islands (Georgia Straits group) are inhabited by a form of *P. maniculatus* with 26–28 bi-armed chromosomes while the northern Scott Islands have populations assigned to *P. sitkensis* with 40 bi-armed chromosomes. In the intermediate islands (the Gordon and Goletas groups) some islands have one type and some the other—no island seems to support both types of mouse.

The karyotypes of two "subspecies" of *P. maniculatus, P. m. ozarkiarum* and *P. m. pallescens* (north Texas and Oklahoma), have been studied by Caire and Zimmerman (1975). There are some karyotypic differences between these forms and an apparent absence of interbreeding where they overlap. A hybrid male obtained in a laboratory cross showed two univalent chromosomes at meiosis in 53 percent of the cells, suggesting that its fertility was reduced by about 25 percent. On the basis of morphometric studies, Caire and Zimmerman conclude that this is a case of "circular overlap" of two noninterbreeding forms that show no habitat separation; according to their interpretation, karyotypic differences established themselves without prior allopatry.

Many other traditional species of rodents, defined by the techniques of classical taxonomy, have been shown to be complexes of morphologically very similar species by modern karyotype studies. Thus, *Rattus rattus* (black rat) is now known to consist of two or three allopatric sibling species (referred to as "karyotypic morphs" by Patton and Myers, 1974). The form with 2n = 42 occurs in eastern and southeastern Asia and most of the Indian subcontinent. Another form, with 2n = 38, occurs in southern India, western Asia, Europe, Africa, North and South America, and Australasia. Giemsa-banding studies suggest that the two forms differ with respect to two centric fusions (for which the 38-chromosome form is homozygous) and one or two pericentric inversions. A form with 2n = 40, which seems to form a link between the 42- and 38-chromosome forms, occurs in Ceylon (Yosida, 1973).

The karyotypes and heterochromatin of 22 species of the kangaroo-rats belonging to the genus *Dipodomys,* which inhabit the western U.S. and northern Mexico, have been investigated by Stock (1971), Mazrimas and Hatch (1972), and Hatch and co-workers (1976). The ancestral chromosome number for the genus is believed to be 2n = 72, most of the chromosomes being acrocentric (the number of major chromosome arms is 74–98 in different races of the most primitive species, *D. spectabilis). D. ordii* is an aberrant 72-chromosome species with 144 major chromosome arms. In a number of species the chromosome number has been reduced to 70, 64, 62, 60, 54, and 52, no doubt by fusions. Rather naturally, the species with lower chromosome numbers have high numbers of major arms, e.g., the species with 2n = 60 have 120 major chromosome arms and the widespread *D. merriami,* with 2n = 52, has 104.

The amount of heterochromatin in the karyotypes of *Dipodomys* species shows a wide range of variation, but all species possess the same satellite DNAs (see p. 103), with buoyant densities of 1.707 and 1.713. In the primitive *D. spectabilis,* with little heterochromatin, the former constitutes 11 percent and the latter 3 percent of the total DNA. The aberrant *D. ordii,* with the same chromosome number has 26 percent of each kind of satellite DNA, i.e., more than half of the total DNA is satellite.

Giemsa-banding techniques are beginning to be used in the study of rodent karyotypes, i.e., this field is moving from the level of beta karyology to that of gamma karyology. The evidence thus far seems to suggest that inversions and translocations in the euchromatic chromosome limbs have been very infrequent in this group, and that the great diversity of the karyotypes has been due mainly to "Robertsonian" changes and changes in the amount and distribution of the heterochromatic blocks (Mascarello, Stock, and Pathak, 1975; Mascarello, Warner, and Baker, 1975).

Next to the horses, the primates seem to show the second greatest karyotypic diversity among the mammalian orders (Table 3-2). Lemuridae, Tupaiidae, and Cebidae especially show much variation in karyotypes. In Cercopithecidae the genera *Macaca* and *Papio* show apparently invariant karyotypes (2n = 42 in all species), but in *Cercopithecus* the known chromosome numbers (2n) range from 42 to 72, and structural rearrangements have certainly played a part in speciation. A comparison between the karyotypes of *Macaca mulatta* (rhesus monkey; 2n = 42, 84 chromosome arms) and *Cercopithecus aethiops* (green monkey; 2n = 60, 120 chromosome arms) was carried out by Stock and Hsu (1973). The former can be derived from the latter by assuming loss of heterochromatic arms, centric fusions, and some tandem fusions; there is no evidence of earlier inversions, either paracentric or pericentric. Within the genus *Macaca* two species have been shown to have identical banding patterns by De Vries and co-workers (1975). The human species seems to have been derived from a stock whose speciation pattern was characterized by karyotypic conservatism, i.e., one that was more like *Macaca* than *Cercopithecus* in its speciation mechanisms.

The four living species of Hominoidea *(Pan troglodytes, Gorilla gorilla, Pongo pygmaeus,* and *Homo sapiens)* have very similar karyotypes, and this similarity is confirmed by modern banding studies (Dutrillaux *et al.,* 1973; de Grouchy *et al.,* 1972; Turleau *et al.,* 1972; Egozcue, 1975). Man has a centric fusion that has produced chromosome 2, which is absent in the other species, all of which have 2n = 48. At least *Pan, Gorilla,* and *Homo* all have chromosomes 13, 14, 15, 21, and 22 bearing "satellites," i.e., small segments separated from the rest of the short arm by a nuclear organizer. Certain of the chromosomes in the karyotypes of these three species have extremely similar banding patterns, e.g., chromosomes 11, 13, 16, 20, 21, and 22. Some pericentric inversions have occurred in chromosomes 4, 5, 7, 9, 17, 18, and 19 (and perhaps also in 1 and 12) of one or more species of the Hominoidea, and there are also differences in the Y chromosome. However, it seems that in the hominid lineage both chromosomal repatterning and speciation have been rather minimal.

Among some of the "lower" primates, however, speciation has normally been accompanied by karyotypic changes. The chromosomal evolution of the lemurs of Madagascar has been studied by Chu and Bender (1961, 1962), Chu and Swomley (1961), Rumpler and Albignac (1969*a, b,* 1970), and Albignac, Rumpler, and Petter (1971). It is instructive to compare the situation in this group of large forest-dwelling mammals with that found in the various genera of rodents previously considered.

According to the hypothesis of Chu and Bender, the primitive lemur karyotype would have consisted of 2n = 66, a karyotype which persists in *Microcebus murinus* and *Cheirogaleus major*. The other lemurs with lower chromosome numbers would have acquired these as a result of various centric fusions.

The most interesting and complex species is *Lemur macaco,* which, according to the most recent views, includes the various races of what was formerly called *Lemur fulvus*. Seven subspecies of *macaco* may be called *L. m. macao, L. m. flavifrons, L. m. sanfordi, L. m. albifrons, L. m. fulvus, L. m. collaris,* and *L. m. rufus*. Collectively, these inhabit almost the whole of Madagascar, except the arid southwest, where they are replaced by the species *L. catta;* the boundaries between the subspecies are probably parapatric. *Sanfordi, albifrons,* and *rufus* show the karyotype that was probably ancestral for the species (2n = 60, 2 pairs of submetacentrics, X-chromosome acrocentric). *Collaris,* which inhabits a restricted area of forest between Fort Dauphin and Manakara, exhibits no less than four different karyotypes: 2n = 60, as before; 2n = 52, X acrocentric; 2n = 52, X metacentric; 2n = 48, X acrocentric. *Fulvus* shows two kinds of individuals: 2n = 60 and 2n = 48. Finally *macaco,* which inhabits a very small area around Ambilobe, has 2n = 44.

Since no structural heterozygotes have been found among the relatively small number of individuals that have been karyotyped, it seems likely that each chromosomal subspecies occupies its own territory and that these overlap little or not at all (they are probably parapatric in most cases). A number of hybrids between the subspecies of *L. macaco* have been obtained in zoos and laboratory colonies. When examined cytologically they had the expected karyotypes in somatic cells. Unfortunately, nothing has been published concerning the meiosis of these hybrids, although in a few cases they were fertile, so that F_2 or backcross hybrids could be obtained.

The general picture that emerges from the cytogenetic work on *L. macaco* is that of a species undergoing fragmentation into a number of parapatric chromosomal races that are probably incipient species. It seems very likely that the mechanism of speciation follows the stasipatric model (see Chapter 6); but the data are insufficient for a firm decision to be reached on this point. The great forests of eastern and northern Madagascar are (or were until the last few hundred years) essentially continuous and there is little reason to believe that the chromosomal races have been separated by any significant geographic barriers. Martin (1972) takes the orthodox view that for speciation to occur there must be "physical barriers of some kind which can effectively isolate subpopulations for periods of time sufficient for speciation to occur. Without such barriers, one would expect no more than cline-formation in Madagascar, with each species forming a continuously interbreeding chain around the island." He does not consider the chromosomal evidence, which is clearly unfavorable to his conclusions, particularly in the case of the four populations of *L. m. collaris* in the Fort Dauphin–Manakara area, which clearly represent an instance of incipient speciation in a continuously forested area.

In contrast to the rodents and lemurs, a number of mammalian groups seem to be karyotypically conservative and show either no cytotaxonomic differences between their species or only minimal ones. One such group is Camelidae (Taylor *et al.,* 1968). The two living species of *Camelus* and four neotropical species in the genera *Lama* and *Vicugna* all show 2n = 74 and apparently have indistinguishable karyotypes. However, no banding studies have been carried out on this group, and with such a high chromosome number it is not easy to be certain that no structural chromosomal rearrangements have occurred in the phylogeny of the family. The conclusion that its six living species were actually "homosequential" in the same sense as the *Drosophila* complexes previously referred to would certainly be unwarranted. Camelids have a high degree of vagility, of course, but probably no higher than the species of *Equus* (horses, donkeys, zebras), which show great karyotypic diversity (see p. 92).

Two other groups of mammals characterized by relative karyotypic uniformity are Pinnipedia (seals, sea lions, and walruses) and Cetacea (whales). The data of Arnason (1972, 1974a, b, c) are unusually complete and include C-, G-, and Q-banding studies on a number of the species. In Pinnipedia 2n ranges from 32 to 36; except for the walrus, the karyotypes are extremely similar. In the whales 2n = 44, except for the sperm and pigmy sperm whales, which have 2n = 42 and quite different karyotypes from those of the other species. The amount of C-heterochromatin varies from 10 to 30 percent of the total chromosome length in the whales; some species are polymorphic for length of C-heterochromatin segments.

It is obvious that neither of these groups of marine mammals has complete karyotypic uniformity. Chromosomal rearrangements have occurred, but their number may have been relatively small; in Pinnipedia these may have been mainly fusions or dissociations, and in Cetacea interstitial deletions and duplications of heterochromatin. Arnason (1972, 1974c) contrasts the relative karyotypic uniformity in Cetacea and Pinnipedia with the much greater karyotypic variability in the rodents and insectivores. He ascribes the different patterns of karyotypic evolution to differences in the mode of life, as follows:

Pinnipedia and Cetacea	Rodentia and Insectivora
Low prolificity	High prolificity
High vagility	Low vagility
Environment without delimited niches	Environment with delimited niches

According to Arnason, these differences have led to mainly allopatric speciation in Pinnipedia and Cetacea, while in Rodentia and Insectivora speciation has been mainly stasipatric.

The cats (family Felidae) have frequently been cited as a group showing karyotypic uniformity. The great majority of the species show 2n = 38, but a few South American ones show 2n = 36. In spite of the considerable degree of karyotypic conservatism that undoubtedly exists in this group, Wurster-Hill (1973) was able to assign 30 species to five groups that are karyotypically different but within which the karyotypes could not be distinguished by conventional cytology. Using Giemsa-banding techniques, several of these groups could be further subdivided (Wurster-Hill and Gray, 1973). Several species, such as *Felis caracal, F. temmincki,* and *F. yagouaroundi,* now appear to be karyotypically unique. However, *F. lybica, F. chaus,* and *F. catus* are karyotypically indistinguishable even with the banding technique, as are the two species pairs *F. viverrina–F. bengalensis* and *F. geoffroyi–F. pajeros*. These cases *could* be homosequential groups in the sense of the *Drosophila* geneticists, but there is no certainty that they are. The resolution provided by the Giemsa-banding technique is coarser by at least two orders of magnitude than that available in the polytene chromosomes.

The horse family (Equidae) obviously has a far higher level of vagility than most groups of rodents, yet it shows very considerable karyotypic variation. At the level of alpha karyology, all the living species show different chromosome numbers; the total range is n = 16–33 (Table 3-4). At the level of gamma karyology, Biemont and Laurent (1974) have demonstrated rather considerable differences between the banding patterns of horse and donkey chromosomes, suggesting that there have been numerous chromosomal rearrangements in addition to the one responsible for the difference in chromosome number. But their work is not sufficiently precise or detailed to demonstrate the approximate number of these rearrangements or to determine whether they have been mainly inversions, translocations, or changes in the length of heterochromatic segments.

TABLE 3-4
Chromosome numbers (2n) in Equidae.

Equus przewalskii	66
E. caballus	64
E. asinus (several races)	62
E. onager	56
E. grevyi	46
E. burchelli (two races)	44
E. zebra hartmannae	32

The most closely related species of equids are undoubtedly *Equus przewalskii* (Przewalski's horse) and the domestic horse *E. caballus*. Hybrids between these two show, as we should expect, a trivalent at meiosis, but this orientates disjunctionally and the hybrids are fully fertile (Short *et al.*, 1974). On the other hand, mules (donkey × horse F_1 hybrid) and hinnies (horse × donkey F_1) are invariably sterile (Chandley *et al.*, 1974).

It has been pointed out by Bush (1975a; and see Wilson *et al.*, 1975) that although equids are large, mobile mammals, their populations are fragmented into small family groups, consisting of a stallion, several mares, and their offspring. Consequently, levels of inbreeding are probably high, facilitating the fixation of chromosomal rearrangements and leading to a mode of chromosomal speciation that would not otherwise be expected on the basis of the high level of vagility of this group.

Especially interesting from the standpoint of chromosomal mechanisms of speciation are those cases of really spectacular increases or decreases of chromosome number (although it is to be regretted that so few of these have been investigated in depth.) We shall first consider a few instances of large-scale reductions.

The mammal species with the lowest chromosome number is the Indian muntjac deer, specifically the Assam subspecies *Muntiacus muntjak vaginalis*. This has 2n♂ = 7 (XY_1Y_2), 2n♀ = 6 (XX) (Wurster and Benirschke, 1970). The karyotype is shown in Figure 7. Specimens from Malaysia assigned to *M. m. muntjak* show 2n♀ = 8 (Wurster and Atkin, 1972); the karyotype of the male has not been studied, but presumably 2n♂ = 9. The Chinese muntjac, *M. reevesi*, has the much more normal mammalian (and artiodactyl) karyotype 2n♂♀ = 46 acrocentrics. Presumably, at least 19 chromosomal fusions and 8–17 pericentric inversions (depending on whether the later fusions were centric or tandem) occurred during the immediate ancestry of *M. m. muntjac*. A further fusion (which was probably a tandem fusion of an acrocentric chromosome "3" onto the end of the short arm of chromosome "1") must have taken place in the evolution of the subspecies *vaginalis*. The "condensation" of the karyotype in the latter implies that each of its three chromosomes includes the material of seven or eight of the original *reevesi* chromosomes. If these in-

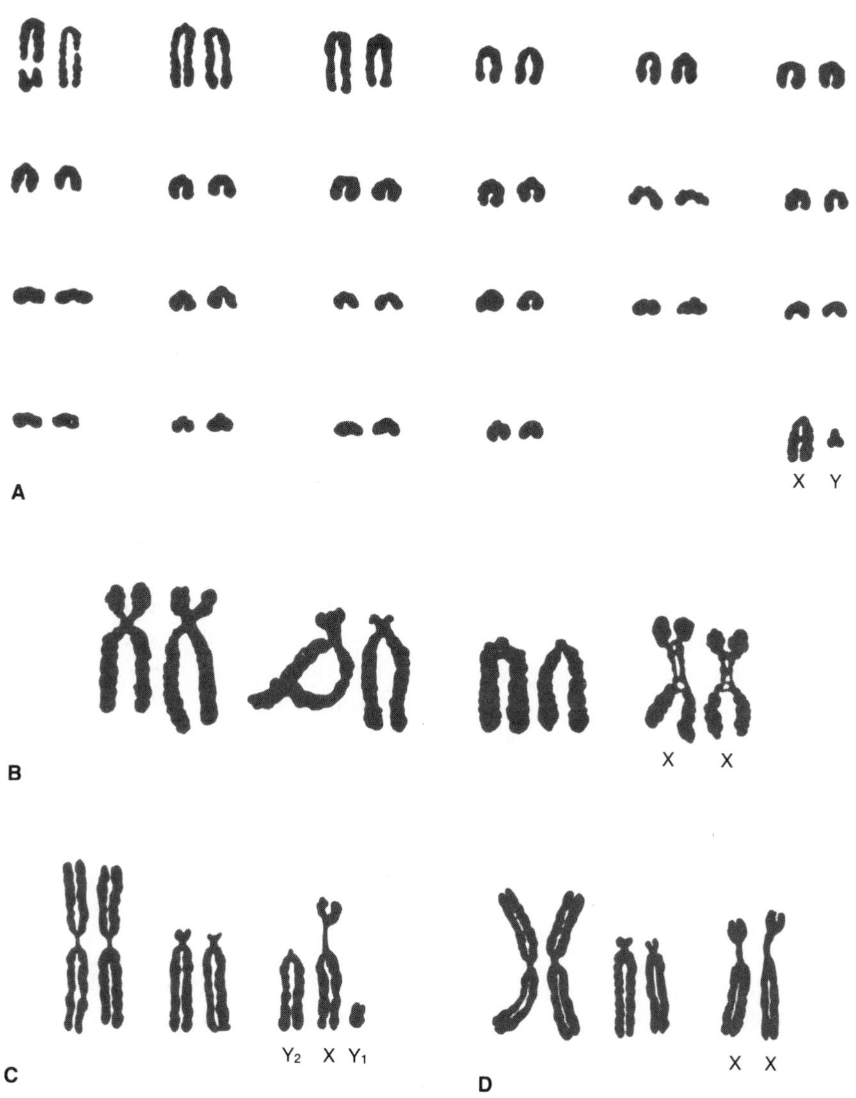

FIGURE 7

Karyotypes of Muntjac deer. **A,** *Muntiacus reevesi* (2n = 46); **B,** *M. muntiacus* female (2n = 8); **C** and **D,** *M. muntiacus vaginalis* male (2n = 7) and female (2n = 6). [After Wurster and Benirschke, 1967; and Wurster and Atkin, 1972]

terpretations are correct, the extra fusion in *vaginalis* would be expected to act as a fairly potent genetic isolating mechanism, i.e., hybrids between *M. m. muntjac* and *M. m. vaginalis* should have their fecundity severely reduced.

There are cases on record of successful hybridization (in zoos) between *M. muntjak* and *M. reevesi* (although not necessarily between the same subspecies as those that have been karyotyped) (Zuckerman, 1953). In view of the great difference in karyotype, earlier claims that these hybrids are sometimes fertile should be regarded as unreliable. The karyotypes of a number of species and subspecies of Muntjacs have not been studied—conceivably, some are cytogenetic intermediate links between *M. reevesi* and *M. m. vaginalis* that might help in the interpretation of this remarkable case of the condensation of a karyotype.

A second case of extreme karyotype condensation is the South American grasshopper *Dichroplus silveiraguidoi,* with 2n♂♀ = 8 (Saez, 1956, 1957). Some species of this genus show the normal karyotype (for an acridid grasshopper) of 2n♂ = 23, 2n♀ = 24 acrocentrics, with an XO condition in the male. *D. bergi* has acquired an X autosome fusion, so it has 2n = 22 in both sexes (male neo-XY), and some populations of *D. pratensis* have 2n♂♀ = 18; both species presumably represent intermediate stages on the way to the *D. silveiraguidoi* karyotype. If we assume that tandem fusions have not been involved, the latter would have acquired one X autosome fusion, seven fusions between autosomes, and six pericentric inversions in the evolution of its karyotype from the ancestral 2n♂ =23. If one postulates that tandem fusions have been able to establish themselves in this case, the minimum number of rearrangements would have been one X–autosome tandem fusion, three centric fusions between autosomes, and five tandem fusions between autosomes (nine rearrangements instead of 14). But whereas each centric fusion may well have been able to maintain itself in a floating (polymorphic) condition, the tandem fusions could not because of the low fecundity of the heterozygotes. Thus, each of the postulated six tandem fusions would imply a speciation event. Unlike the muntjac, where there is some evidence that one tandem fusion has actually occurred, there is no real reason for believing that tandem fusions have taken place in the ancestry of *D. silveiraguidoi*.

Except for the prevalence of polyploidy in plants, which is dealt with in Chapter 8, the cytotaxonomic picture seems to be essentially similar in higher plants and animals—the same spectrum is found, from genera with minimal chromosomal differences between related species to ones with large-scale differences. However, because of the widespread occurrence of polyploidy and the small size of the chromosomes in many groups of plants, most of the really significant studies on the role of chromosomal rearrangements in plant speciation have been carried out on such genera as *Crepis, Haplopappus,* and *Brachycome* (Compositae), in which some species have very low numbers, or on ones such as *Trillium, Lilium,* and *Tradescantia,* in which the chromosomes are very large.

Ever since the pioneering work of Navashin (1932) and Babcock and his collaborators (Babcock and Stebbins, 1938; Babcock and Jenkins, 1943; Babcock, 1942, 1947), it has been clear that three modes of evolutionary differentiation have occurred in the genus *Crepis*. Certain lineages have developed polyploid species, some forms have adopted asexual mechanisms of reproduction, and in a number of sexually reproducing species there have been phylogenetic reductions in chromosome number from n = 6 to lower numbers and ultimately to n = 3. The latter process seems to have been accompanied by a change in the life cycle, from perennial herbs to annuals.

A particularly significant case is that of *Crepis neglecta* and *C. fulginosa,* studied by Tobgy (1943). The former (n = 4) is widespread in south-central Europe, while the latter (n = 3) is restricted to Greece. The ranges of the two overlap in northern Greece, where some natural hybrids occur. The *neglecta* chromosomes are stated to be "four times as great in chromatin volume as the *fuliginosa* chromosomes" (Figure 8). This kind of situation is found in many plant and animal genera, and is discussed below (p. 99). It is interesting that in this case the species with the derived karyotype has the lower chromosome volume (i.e., DNA value). There is thus a *prima facie* case in this instance for a phylogenetic decrease in DNA value (by multiple deletions of DNA segments). But it is not certain that the immediate ancestor of *C. fuliginosa,* even if it had an outward phenotype that would have caused a taxonomist to include it in *C. neglecta,* had as much DNA as modern *C. neglecta*. Part of the difference in DNA value between the two species is undoubtedly due to differences in length of the heterochromatic blocks (segments composed of highly repetitive satellite DNA). This difference is immediately evident in the interphase nuclei, where *C. neglecta* shows numerous large chromocenters, i.e., heterochromatin, and about a dozen small ones (the total volume of chromocentral material is clearly greater in *C. neglecta*). Most of this heterochromatin is proximal, i.e., located on either side of the centromere, in chromosomes A, B and D; chromosome C seems to be entirely heterochromatic.

Chromosome pairing in the hybrid indicates that *C. fuliginosa* is homozygous for a translocation between the A and D chromosomes and for paracentric inversions in most or all of the long chromosome arms. The reduction in chromosome number was the result of a translocation between the distal segment of the long arm of the B chromosome and one limb of the C chromosome, with loss of one of the products of the translocation (the one consisting mainly of heterochromatin from the C chromosome). Tobgy's study, carried out long before modern techniques for studying the genetic architecture became available, remains one of the most careful analytical studies of karyotypic changes in plant speciation.

A karyotype with n = 3 seems to have evolved independently three times in

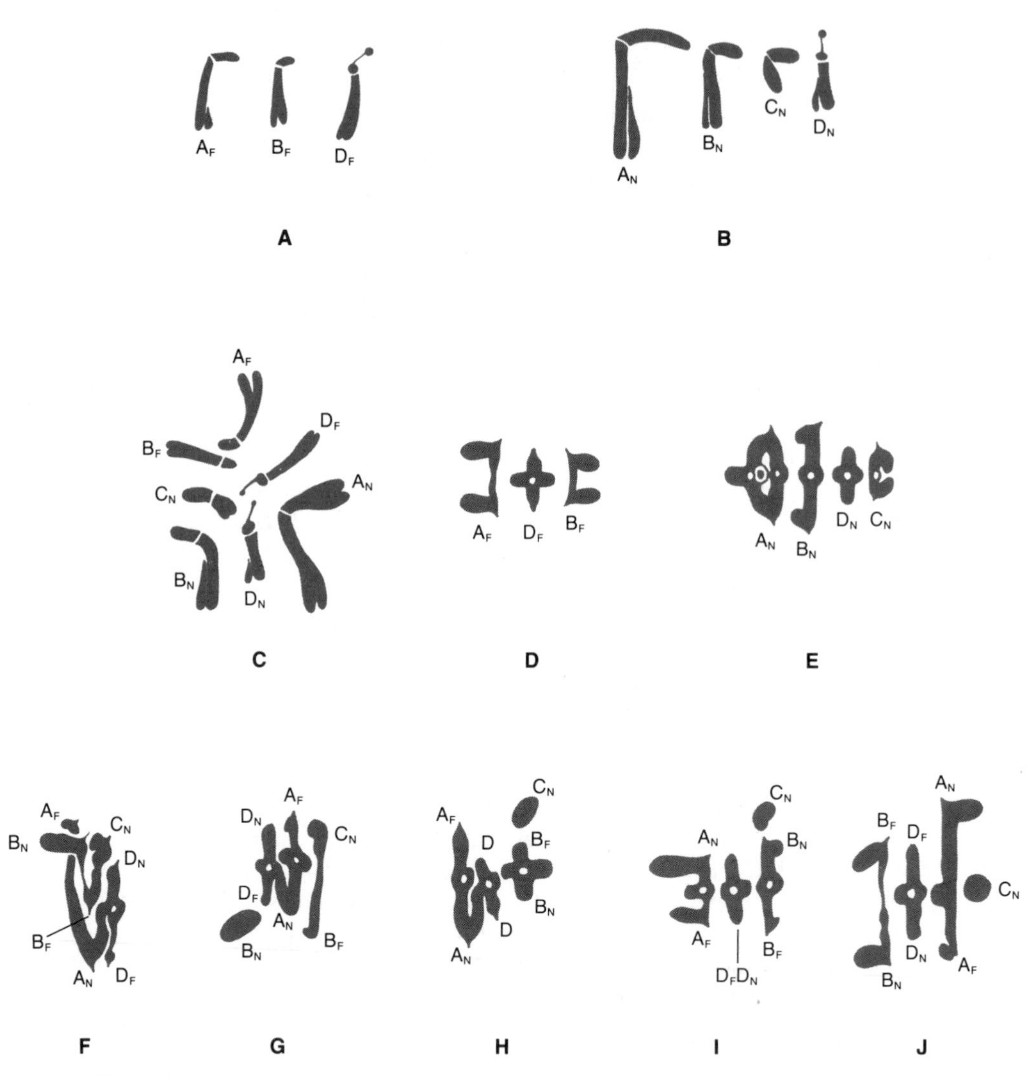

FIGURE 8

Chromosomes of *Crepis fuliginosa, C. neglecta,* and their F_1 hybrid. **A,** Haploid karyotype of *fuliginosa;* **B,** that of *neglecta;* **C,** that of the F_1 hybrid (note that the size of the parental chromosomes is conserved). **D,** Side view first metaphase of *fuliginosa;* **E,** that of *neglecta;* **F-J,** that of the F_1 hybrids. [After Tobgy, 1943]

the genus *Crepis:* in *C. fuliginosa, C. capillaris,* and *C. zacintha.* Stebbins (1950) believes that the primitive karyotype in the genus was $n = 6$, and that the species with $n = 7$ are a specialized offshoot in which there has been an increase in chromosome number. Karyotypes with n = 5 and n = 4 have probably evolved repeatedly. Sherman (1946) showed that the karyotype of *C. kotschyana* (n = 4) was derived from an n = 5 karyotype by the same kind of mechanism as Tobgy demonstrated for *C. fuliginosa*. All species of *Crepis* seem to be karyotypically unique and it seems impossible to avoid the conclusion that chromosomal rearrangements have been responsible for their speciation.

The genus *Trillium* (Liliaceae) includes diploid species with n = 5, tetraploids, and hexaploids. Japanese workers have made very detailed studies of the diploid *T. kamtschaticum*. The size and distribution of heterochromatic segments revealed by cold treatment of the chromosomes is highly variable (Kurabayashi, 1958, 1963; Haga, 1969, 1974). A total of 71, 19, 17, 16, and 20 different types of chromosomes A, B, C, D, and E, respectively are known in natural populations. Reciprocal translocations seem unable to establish themselves in this species; four individuals heterozygous for such changes were found in the course of very extensive work and were highly sterile. The different distributions of heterochromatin in the karyotype of this species result from structural rearrangements, such as inversions, insertions, duplications, and deletions. Most populations are highly polymorphic for different chromosome types; thus, in a sample of 50 plants from one large population every plant had a different karyotype. In these large populations, in which reproduction is apparently sexual and panmictic (Kayano and Watanabe, 1970), the frequencies of homo- and heterokaryotes approach Hardy–Weinberg frequencies. Nevertheless, populations that are chromosomally monomorphic, but with different homozygous combinations of chromosomes, exist in three localities.

The same type of population structure seems to exist in the American *T. sessile*. Most populations are polymorphic for their heterochromatic segments, but at one locality (Thurston County, Washington) all plants had the same homozygous combination. Darlington and Shaw (1959) have described the distribution of the heterochromatic segments in about 10 species of diploid North American *Trillium* (some of uncertain species status). But their material was obtained from dealers and botanic gardens and provides little basis for understanding the variation pattern in nature. These authors seem to regard the reproduction of the North American species as largely clonal (i.e., vegetative), which is apparently not the case with *T. kamtschaticum*. The role, if any, of variation of heterochromatic segments in the speciation of the diploid *Trillium* and the closely related genus *Paris* must be regarded as obscure, but Darlington and Shaw's data do indicate considerable differences between the species they studied with respect to the amount and distribution of heterochromatin.

In general, higher plants have metacentric or submetacentric chromosomes—acrocentric chromosomes are rare. Thus "Robertsonian" changes seem to have taken a less prominent part in the karyotypic evolution of plants than in most groups of animals. There are, however, some peculiar or archaic species of plants with acrocentric chromosomes (in some cases these have been claimed to be actually telocentric, i.e., with strictly terminal centromeres). The gymnosperm *Podocarpus nivalis* (n = 18) has 17 acrocentric chromosomes and only one metacentric (Hair and Buizenberg, 1958), while the extraordinary *Welwitschia mirabilis,* from Namibia, has a karyotype of 21 acrocentric or telocentric chromosomes (Khoshoo and Ahuja, 1963). Some cycads also have karyotypes consisting partly or mainly of acrocentric chromosomes (Khoshoo, 1969). It is not usual for plants to show the kind of chromosome dimorphism seen in most lizards (microchromosomes and macrochromosomes), but this is definitely the situation in the tribes Yucceae and Agaveae of Liliaceae (Sharma, 1969). Clearly, there are regularities of karyotype evolution in some higher plants that exemplify the general principle of karyotypic orthoselection (see p. 49).

A few species of plants have developed a type of genetic system based on fixed heterozygosity for translocations. Consequently, they show rings of chromosomes at meiosis, rather than bivalents. The most famous examples of this type of genetic mechanism are, of course, certain species of *Oenothera.* Other ring-forming plants are the Central American *Rhoeo discolor* (the only species of its genus), the West Australian *Isotoma petraea* (James, 1965, 1970a) and *Hypericum punctatum* (Hoar, 1931).

It is only in *Oenothera* that the relationship of these systems to speciation processes has been seriously explored (Cleland, 1950, 1962, 1964). All species of the North American subgenus *Euoenothera* have 14 metacentric chromosomes and all are capable of self-fertilization. There are three groups of taxa. The first, cytogenetically the most primitive, consists of populations included in *Oe. hookeri* (western U.S.A., northwestern Mexico), *Oe. grandiflora* (Gulf coast of Alabama), and *Oe. argillicola* (Appalachian Mountains). These forms are commonly outcrossed; they usually have seven bivalents at meiosis, although rings of four or six chromosomes are present in some individuals, as a result of translocation heterozygosity.

The second group in *Euoenethera,* the *irrigua* alliance described by Cleland, occurs in New Mexico, Colorado, Utah, Oklahoma, and northern Mexico. These plants are cross-pollinating and structurally heterozygous—forming rings of four, six, or eight chromosomes at meiosis. However, their translocation heterozygosity is not fixed, so that they are chromosomally polymorphic; because they lack the balanced lethal mechanisms of the third group, they are not true breeding and always give rise to some structurally homozygous progeny that form seven bivalents.

The third group consists of plants that are also structurally heterozygous—forming rings of 14 chromosomes at meiosis—but mainly self-fertilizing. Plants in this group possess a balanced lethal system of either the gametophytic or zygotic type, which prevents the occurrence of structural homozygotes. Variation in these normally highly inbred populations is produced by crossing, which gives rise to new hybrid derivatives and occasional crossing-over in proximal or differential chromosomal segments, that leads to entirely new combinations of chromosome ends. These ring-forming *Oenothera* "species," which include *strigosa, biennis,* and *parviflora,* are hence a vast complex of taxa in a stage of active evolution, constantly generating new forms from their reservoir of hidden variability and not really comparable to the species of normal cross-fertilizing diploid organisms.

In contrast to the euoenotheras, other ring-forming complex heterozygotes in plants seem to be mostly evolutionary dead ends, incapable of further speciation. *Isotoma petraea* (2n = 14) includes a range of populations—from ones always exhibiting seven bivalents, through ones with a ring of six chromosomes, two rings of six and four each, a ring of 10, two rings of six, a ring of 12, two rings of eight and six each, to ones with a single ring of 14. Essentially self-fertilizing, the species still has a residual capacity for crossing. The complex heterozygosity of *I. petraea* is very inefficient in individuals with large rings, since about 80 percent of the meiotic divisions are nondisjunctional. Certain species of scorpions are the only animals that seem to have ring-forming translocation heterozygosity of the *Euoenothera* type, but their genetic systems are not really understood.

Other forms of fixed heterozygosity may occur in a few plant species. The Australian *Calectasia cyanea,* the only species of the family Calectasiaceae, with a relict type of distribution, may be a permanent heterozygote for a rearrangement (or rearrangements) that restricts or suppresses the formation of chiasmata in the longest chromosome arm of the karyotype (Anway, 1969). In those plants that show chiasmata in this chromosome arm, pollen grains with recombinant chromosomes seem to be sterile. *Calectasia* is a self-pollinator that appears to have lost the ability to speciate, since eastern and western populations, separated by approximately 2,000 km since the Miocene, are only weakly differentiated. Meiosis in the megasporogenesis of *Calectasia* has not been studied, and the actual nature of the chiasma-suppressing mechanism is somewhat obscure.

It is now known that in a number of plant genera large-scale differences exist between species with respect to the total amount of DNA present in the karyotype—differences that are certainly not due to polyploidy. One such case in *Crepis* has already been referred to (p. 95). The data have been discussed in a general review by Rees and Jones (1972). In 22 diploid species of Ranunculaceae there is a fortyfold range, and within the single genus *Anemone* a twofold one (Rothfels *et al.,* 1966). Marks and Schweizer (1974) have studied C-banding in the karyotypes of several species of *Anemone* and the related genus *Hepatica*. They found that, in general, species with high DNA values had more C-banded material (presumably satellite DNA); yet this was not enough to explain the whole of the variation in total DNA.

The karyotypes and DNA values of members of the genus *Vicia* have been studied by Chooi (1971a); here there is a sixfold range in DNA values, with *V. faba,* which has a very aberrant karyotype for the genus, having by far the largest amount. In some cases the extra DNA appears to have been added to all or almost all the chromosome arms; thus *V. meyeri,* which has 2.8 pg of DNA, has all but one of its 14 chromosome arms longer than the corresponding arm in the related *V. hirsuta,* which has 1.8 pg. But several species with very different DNA values have a small metacentric chromosome that is the same size in all of them. *V. faba,* with n = 6 and one pair of very large metacentrics, has presumably acquired a centric fusion that is lacking in related species with n = 7. Its five acrocentric chromosomes do not appear to have gained extra material in their short arms, but they have undoubtedly done so in their long arms. Chooi's later results (1971b) do not clarify the situation, and it is still uncertain to what extent the increase in the total amount of DNA in a species like *V. faba* has been achieved by the acquisition of extra segments of highly repetitive satellite DNA, and to what extent it is due to increase in the amount of DNA in each gene locus, or to an actual increase in the number of gene loci. Many early discussions of this problem concentrated on the evidence for and against the polyneme (many stranded) model of chromosome structure. However, this model has now been decisively disproved and the alternative unineme interpretation is generally accepted (except in the special case of the polytene chromosomes). Therefore, evolutionary changes in the total amount of DNA (in such genera as *Anemone* and *Vicia*) must be seen as due to numerous duplications of short sequences (either unique or repetitive) or to losses of such segments. Nevertheless, there is as yet no critical evidence for deciding whether such evolutionary changes have been preponderantly in one direction or equally in both. Rees and Hazarika (1969) claim that the threefold range in DNA values in *Lathyrus,* a genus closely related to *Vicia,* has been due to DNA losses in the more specialized species (predominantly annual inbreeders, as opposed to the perennial outbreeding species, which have the higher DNA values). But, whether the changes have been in one direction or the other, their relation to the actual process of speciation is quite obscure.

The situation in a number of animal genera seems to be broadly similar. A particularly striking example is the flatworm genus *Mesostoma,* where *M. ehrenbergi* has about 10 times as much DNA as *M. lingua* (Göltenboth, 1973). In the North American salamander genus *Plethodon,* Mizuno and Macgregor (1974; and see Macgregor *et al.,* 1973) have shown that although all species have 14 metacentric or submetacentric chromosomes, the 2C DNA values range from 36 to 138 pg. In general, the Western species have much higher DNA values than the Eastern ones. The difference is due mainly to the "moderately repetitive" DNA fraction, i.e., neither the unique sequences that correspond to the structural genes nor the "highly repetitive" satellite DNA. Although *P. cinereus* (Eastern) and *P. vehiculum* (Western) are so similar in outward appearance that they can hardly be distinguished by an inexperienced observer, the former has about 45 percent of its total genome represented by fast-reassociating (repetitive) DNA, while in the latter about 80 percent of the DNA is repetitive. Salamanders belonging to the same species group had between 60 and 90 percent of the repetitive sequences in common; Eastern "large" and Eastern "small" species had 40–60 percent in common, while Eastern and Western species had less than 10 percent identical. It seems as if the genomes of the Western species increased in size as a result of repeated saltatory replication of sequences that were present in their common ancestor, after it had diverged from the Eastern group.

The total number of chromomeres in the lampbrush chromosomes of *P. cinereus, P. vehiculum,* and *P. dunni* has been estimated by Vlad and Macgregor (1975) on the basis of direct counts of selected segments. *Vehiculum* (73.6 pg DNA) and *dunni* (77.6 pg DNA) were estimated to have respectively 5,067 and 5,792 chromomeres in the entire haploid karyotype, compared with an estimated 3,458 in *cinereus,* which has only 40 pg of DNA. Clearly, increases in DNA value have been due to the duplication of preexisting chromomeres, possibly by the "Keyl mechanism," although Mizuno and Macgregor (1974) believe there is a difference between "tandem duplication" of the Keyl type and "saltatory replication" as proposed by Britten and Kohne (1967). But because the amount of unique-sequence DNA is "not conspicuously greater" in the species with high DNA values than in ones with lower values, Vlad and Macgregor reject the idea of a direct relationship between chromomeres and structural genes. It seems doubtful, however, that they are entirely justified in this conclusion; if most of the structural genes in such species as *vehiculum* and *dunni* have *not* undergone duplication, those that have done so will most likely be represented 4, 8, 16 . . . . times, and this will affect estimates of the amount of unique-sequence DNA.

Undoubtedly *some* differences in DNA values between related species are due to differences in the amount of highly repetitive satellite DNA present; species like *Drosophila nasutoides* (Cordeiro *et al.,* 1975) have long heterochromatic segments that are lacking in related species. But it emphatically does not seem to be the explanation for differences in DNA values in such genera such as *Vicia, Anemone,* and *Plethodon.* Studies on these genera by such workers as Rothfels, Chooi, Macgregor, and their collaborators seem to have been restricted to "full" species. The significance of changes of this type in relation to speciation will only become clear when populations in the process of speciation (i.e., isolated populations, geographic subspecies, and semispecies) are studied. It is conceivable that they are mainly or entirely post-speciation events. But this is unlikely if we are correct that *Chironomus thummi* and *C. piger* have only very recently undergone speciation (and Keyl himself regards them as subspecies rather than species). Perhaps there is some kind of cascade phenomenon by which certain newly arisen species acquire many scores or hundreds of duplications of gene loci at a stage when the total species population is still fairly small. The animal species with the highest DNA values are the three living species of lungfishes (Dipnoi) and the members of the four families of neotenous Urodeles: Cryptobranchidae, Amphiumidae, Proteidae, and Sirenidae (Pedersen, 1971; Olmo and Morescalchi, 1975).* These are extremely specialized and rather archaic types that, although still surviving, seem practically to have lost the ability to speciate.

If we are correct in supposing that such species as *Plethodon vehiculum* and *P. dunni* have undergone scores or hundreds of duplications of individual gene loci (chromomeres), it is certain that these did not all occur simultaneously but must have been spaced over a considerable period of time. Most duplications probably had little or no direct role in the actual speciation process and thus would have been much less important in initiating evolutionary divergence than the major rearrangements, such as inversions, translocations, etc. Indeed, it is rather difficult to envisage duplication of individual genes giving rise directly to genetic isolating mechanisms. On the other hand, it is clear that the role of duplications in phyletic evolution (as opposed to speciation) has been very considerable in certain groups of plants and animals.

The problem remains whether all these changes in DNA values really represent evolutionary increases (duplications) or changes in the opposite direction (deletions). Most of the evidence certainly seems to favor increases as the predominant type of evolutionary change, with forms that have undergone many such increases having a decreased potentiality for further evolution, and especially for speciation.

The DNA values in a large number of vertebrate species have been tabulated by Callan (1972). Manfredi-Romanini (1973), and, for cyclostomes and fishes, by Ohno (1974). Determinations for a number of reptile species are given by Olmo (1976). The 2C amounts range from 0.8 pg in the puffer fish *Tetraodon fluviatilis* (Hinegardner, 1968) to 248 pg in the South American lungfish *Lepidosiren paradoxa* (Ohno and Atkin, 1966)—a 300-fold difference between organisms that must be regarded as having approximately the same level of

*More recent evidence suggests that a plethodontid salamander from Costa Rica, *Bolitoglossa subpalmata,* which is not a neotenous species, has more DNA than any of the above (Macgregor, unpublished).

morphological and biochemical complexity. It seems clear that the qualitative and quantitative potentials of such very different genetic systems for future changes must be almost totally unlike. There are, however, vertebrate groups in which the DNA amounts are relatively constant. An example is the bats, in which the reported range is 3.0–3.9 pg (Capanna and Manfredi-Romanini, 1971).

The existence of chromosome segments made up of highly repetitive DNA sequences, the so-called satellite DNAs, has already been referred to several times, and it has been suggested that in some instances changes in the amount, chemical constitution, and distribution of these satellites over the karyotype may have played a role in speciation. The satellite DNAs seem to correspond to the heterochromatin of classical cytogenetics, or, strictly speaking, to the constitutive heterochromatin, since nonheterochromatic chromosomes may exist that at certain stages of the life cycle or in certain tissues exhibit heterochromatic behavior (so-called facultative heterochromatin). Satellite DNAs possibly do not contain any structural genes coding for proteins and are not transcribed in the usual way. On the other hand, their universal presence in the karyotypes of all eukaryote organisms that have been investigated and their relative constancy (despite some variability) within the species show that they must serve some important general function.

The satellite DNAs have clearly undergone an evolution of their own, the general outlines of which are beginning to be partly understood in the case of a few species groups of *Drosophila* and some rodents. Most species seem to have several different satellites, but these are frequently rather similar in structure, as if one had arisen from another or both had evolved from a common ancestral sequence by a small number of base substitutions. Each kind of satellite consists of a sequence of a small number of nucleotides (usually less than 10) repeated in tandem many thousands of times. Satellites that contain mainly adenine and thymine nucleotides are "AT rich" or "light," since they have a buoyant density less than the rest of the DNA when studied by density-gradient centrifugation. Conversely, ones consisting largely of cytosine and guanine bases are "GC rich" or "heavy." Some highly repetitive DNAs ("cryptic satellites") have a buoyant density so close to that of the main-band DNA that they cannot be separated from it by simple density-gradient centrifugation and special techniques are needed to demonstrate them.

The satellite DNAs of several species of *Drosophila* are beginning to be fairly well known. Those of *D. virilis* have been studied by Gall and Atherton (1974). There are three of them:

I: poly d (ACAAACT), density 1.692
II: poly d (ATAAACT), density 1.688
III: poly d (ACAAATT), density 1.671

Obviously, II and III each differ from I by a single base-pair change, and from one another by two such changes. The closely related forms *D. novamexicana, D. americana americana,* and *D. americana texana* all have satellite I but lack II and III; they do, however, have small amounts of two other satellites with buoyant densities of 1.721 and 1.676, whose sequences have not yet been determined. The simplest evolutionary interpretation is to suppose that the common ancestor of all these species possessed satellite I, and that in the lineage leading to *virilis* this gave rise to II and III, while in the lineage leading to the other species two other satellites with densities of 1.721 and 1.676 were acquired. Since these latter have not yet been sequenced, it is not known whether they also have been derived from satellite I by simple base-pair changes. The members of the "*montana* branch" of the *virilis* group *(montana, borealis, lacicola, flavomontana, littoralis,* and *imeretensis)* appear to lack satellite I.

The satellite DNAs of *D. melanogaster* have now been thoroughly characterized. There are five of these, with buoyant densities of 1.672, 1.686, 1.688, 1.697, and 1.705 (Peacock *et al.*, 1974; Brutlag and Peacock, 1975; Peacock *et al.*, 1977). The 1.672 satellite apparently contains two different DNA "subspecies," one with the repeated sequence ATAAT, the other with the repeat ATATAAT. The sequence of the 1.686 satellite is apparently AATAACATAG, i.e., it is a 10-mer with the general formula $A_6T_2C_1G_1$, but it includes minor components having the formulas $A_7T_1C_1G_1$ and $A_5T_3C_1G_1$ (about five percent of each). Both these satellites resemble one another in possessing an A-rich and T-rich strand, but there are obviously considerable differences between them. The 1.705 sequence, like the 1.672 one, includes two sequence isomers (satellite "subspecies"), AAGAG and AAGAAGAGAG. The 1.688 and 1.697 satellites have not yet been sequenced.

The five satellites of *D. melanogaster* together constitute about 20 percent of the total DNA in the genome (each of them forming about four percent). All of them are represented in the material of the chromocenter in the polytene nuclei and three of them (1.672, 1.686, and 1.705) are also present in the 21 C-D region of chromosome arm IIL (Goldring, Brutlag, and Peacock, 1975). Satellite 1.672 is present mainly in the centromeric heterochromatin of chromosome II and in the Y chromosome; smaller blocks occur in the proximal heterochromatin of chromosomes III, IV, and the X. It is interesting that it is represented mainly by the ATAAT "subspecies" in the Y and by the ATATAAT one in chromosome II. Accordingly, the heterochromatic segments of the *D. melanogaster* karyotype seem to be built up of alternating blocks of different satellites in a complex but regular manner. Thus the nucleolar organizers on both the X and Y chromosomes appear to be located between blocks of the 1.688 and 1.686 satellites. The major satellites of *D. simulans* appear to be different from those of the sibling species *D. melanogaster,* although the main *melanogaster* satellite sequences are present in the *simulans* heterochromatin as minor components. Several other sibling species of this complex are now known, from Africa and the island of Mauritius, but their satellite DNAs have not yet been studied.

There seems to be no evident similarity between any of the *D. melanogaster* satellite DNAs and those of *D. virilis* (which belongs to a different subgenus). The three satellites of *virilis* are located in the proximal heterochromatin of all the chromosomes and make up about 50 percent of the total DNA in the *virilis* genome. But their detailed arrangement in the heterochromatin is not known. Calculations show that the *virilis* genome contains about 1.2×10^7 copies of the seven-nucleotide sequence of satellite I and approximately 0.4×10^7 repeats of satellites II and III. *Drosophila nasutoides,* which has one pair of huge metacentric heterochromatic chromosomes, has four different satellite DNAs, which together form about 60 percent of its genome (Cordeiro *et al.,* 1975).

Satellite DNAs have been studied in a number of species of marsupials and rodents. The wallaby *Macropus rufogriseus* has one main satellite (buoyant density 1.708), which is represented in all the autosomes but not in the X or Y (Dunsmuir, 1976). A minor satellite also occurs on all the chromosomes, including the X and Y. Sequences identical or related to the 1.708 satellite occur also as minor components in the DNAs of seven other species of macropod marsupials.

Rodent species seem to show the same kinds of differences in satellite DNAs at both the intraspecific and interspecific levels. The satellites of a number of species of mice (genus *Mus*) have been studied by Sutton and McCallum (1972). *M. castaneus,* from Thailand, shows a satellite (density 1.690) indistinguishable from that of *M. musculus,* but *M. caroli, M. cervicolor,* and *M. famulus cookii* lack this satellite. *M. caroli* has three satellites, while *M. cervicolor* and *M. f. cookii* have one that is different from all of these. Cross-reassociation studies have shown that the five light (i.e., AT-rich) satellites in these species of asiatic *Mus* are a family of related sequences; the one plausible explanation is that they are phylogenetically descended from a common ancestral satellite sequence.

The molecular mechanisms for the origin and replication of both satellite DNAs and the moderately repetitive DNA are the subject of much discussion and speculation at the present time. Earlier it was believed that the repetitiveness of the satellite DNAs was imperfect, due to the fact that a number of mutational changes (base substitutions) had occurred since their origin. If this were so, there would be significant sequence heterogeneity in most satellites, depending on their relative age, e.g., one might expect more heterogeneity in satellite I of *Drosophila virilis* than in the presumably derived satellites II and III. But Brutlag and Peacock have found that the sequence heterogeneity that does occur is probably due to the presence of more than one molecular "subspecies" of a satellite (as in the case of satellite 1.672 of *D. melanogaster*). If, as now seems probable, the sequences of the satellite DNAs are highly conserved, there must be a specific mechanism for ensuring this, perhaps one that depends on interchromosomal exchanges.

Since the major satellite DNAs of related species are frequently, and perhaps always, different—although a major component in one species may be present as a minor one in a related species—the origin of new satellites and the loss or transformation of old ones are clearly relevant to the whole question of speciation mechanisms. To what extent these changes play a causative role in speciation is entirely unknown at the present time. Until the real function of satellite DNAs is understood, their relation to speciation, whether causative or consequential, is bound to remain mysterious. Our understanding of the architecture of eukarytotic chromosomes has progressed very greatly in the last few years. But it is still very incomplete. In particular, we do not really know how many significantly different architectures exist in eukaryotic organisms. At the moment it appears as if the frog *Xenopus* differs quite radically from *Drosophila* in the length of the 'unique' sequences (presumably structural genes), which are much shorter in the former (Davidson *et al.*, 1975). Apparently, most Metazoa have the *Xenopus* type of architecture, but it is still possible that most insects have the *Drosophila* type. The implications of these different patterns for evolution and speciation are still totally obscure.

4

Allopatric Speciation

The genetic and chromosomal differences between related species considered in previous chapters clearly indicate that speciation is a genetic and cytogenetic process. But speciation does not take place in a vacuum. Since it occurs in a complex and ever-changing world, we must also examine its geographical and ecological aspects.

The essential feature of the various allopatric models of speciation, briefly defined in Chapter 1, is that they all suppose that genetic isolating mechanisms are acquired under conditions of complete geographic isolation. There can be absolutely no doubt about the reality of allopatric speciation; if populations are geographically isolated for a long enough period of time they *will* evolve into different species. This is the one type of speciation whose actual existence is uncontroversial. Controversy has developed, however, as to whether all (or almost all) speciation conforms to the allopatric model (or models, since there are a number of variants on the basic theme). And even in cases of speciation that are undoubtedly allopatric, there may be considerable room for argument as to the exact course that it has followed and the relative role of different types of genetic processes involved.

The view that geographic isolation is an essential prerequisite for speciation has been supported most strongly by vertebrate zoologists. Commenting on Mayr's (1963a) statement that "it is quite evident that one isolating mechanism or several must be acquired in geographical isolation before contact is established," Blair (1964) considered this "so obvious a requirement as to need no further discussion." Such was the current view a decade ago. It is expressed in countless elementary textbooks and popular accounts of evolution.

In the present chapter a number of instances of undoubted allopatric speciation will be discussed. But a number of more doubtful cases will also be considered. These are mostly instances of speciation that earlier authorities (either the original investigators or subsequent reviewers) claimed had followed the allopatric model, but where subsequent objective and dispassionate reconsideration raises serious doubts, even if the facts are insufficient to firmly support one scheme rather than another.

Most published accounts of allopatric speciation assume that two populations that have achieved species status as a result of geographic isolation may extend their ranges to the point that they become sympatric, but without being able to hybridize effectively (since they have become genetically isolated). Such sympatry would clearly be a postspeciation phenomenon, according to the model, since speciation is expected to be completed under conditions of geographic isolation. A different situation would arise if two geographically isolated populations, expanding their ranges, came into parapatric contact, with a very narrow zone of overlap. In such a case, geographic isolation may have only initiated speciation, which would then become complicated by genetic interaction between the two incipient species. There can be no doubt that parapatric contacts between species or "semispecies" are quite common; they have been described in insects, frogs and toads, lizards, rodents, and even bats. Believers in the universality, or near universality, of allopatric speciation necessarily consider them as secondary contacts between populations that were formerly separated geographically. One trouble with this interpretation is that in many cases there is no plausible geographic or paleogeographic reason for believing that the two populations were ever geographically isolated. Moreover, there are innumerable instances of parapatry between geographic races that have by no means reached the level of separate species and there is certainly no reason to regard the zones of intergradation as due to secondary contacts. It would seem that an unbiased approach to this problem necessitates (on the principle of Occam's razor) that one should assume parapatric zones of coexistence to be primary, i.e., between populations that have never been out of contact with one another, unless there is independent evidence for a former disjunct distribution—belief in a theoretical model should not be regarded as "evidence."

As briefly indicated in Chapter 1, there seem to be several rather different modes of speciation that may be regarded as allopatric. One of these conforms strictly to the "dumbbell" model of Mayr (1942), which involves three stages: (1) range extension of an original population into new, unoccupied territory; (2) development of a geographic barrier (ocean, mountain range, desert, etc.) between the two areas; and (3) genetic modification of the separated populations, such that if they again come into contact they will be kept apart by genetical isolating mechanisms. In cases of insular speciation the geographic barrier may exist from the beginning, the range extension (stage 1) being the colonization of an island previously uninhabited by the species. In the case of organisms possessing a high degree of vagility, it is easy to imagine similar colonization of a new and geographically isolated area occurring on a continental landmass.

The main feature of the classical dumbbell model is that a sharply decreased population size (a population "bottleneck") is not assumed at any stage. The variability of the population is always maintained at a high level, with some 30–40 percent of the gene loci polymorphic. On the other hand, colonization of new areas by "founder" individuals (in the extreme case, a single gravid female) may involve a drastic reduction in the variability of the population—this is the "founder principle" of Mayr (1942). Obviously, founder individuals cannot carry with them all the alleles present in the original species population. However, if there are say 10–100 of them, they may carry a rather large proportion of the alleles present in the particular deme from which they were derived. Even a single founder female—if she is heterozygous at 10–15 percent of her gene loci and has been fertilized by a male that is equally heterozygous (largely for different alleles)—will carry with her much more genetic variability than could have been imagined when the founder principle was first put forward (when levels of heterozygosity were assumed to be much lower). Theoretically, a newly founded colony, whether derived from a small number of immigrants or a single gravid female, will at the start be more homozygous and less polymorphic than the original population. If the colony survives, however, mutation is expected to restore the level of polymorphism rather soon to approximately the original value, although the alleles may not be exactly the same as the original ones, and their frequencies would certainly be different. Moreover, in the initial stages of its existence, unless the new colony exhibits "inbreeding depression," the population size is likely to increase rapidly, as a result of the temporary absence of competitors and controlling agents. During this phase of expansion the chances of survival of new mutations are much greater than in a population of stable size.

Some plausible examples of the founder effect are provided by populations of races or subspecies of mammals on small islands. Berry (1969, 1973, 1975) has carried out a biometrical analysis of skeletal characters in the vole *Apodemus sylvaticus*. Whereas populations in Great Britain are relatively uniform, 15 taxonomic races have been described from Iceland, the Shetland Islands, and the Hebrides group. With one conceivable exception *(A. s. hirtensis* on St. Kilda Island) these island populations cannot be preglacial survivors; they must have originated from early postglacial founders, possibly brought by the Vikings, and must owe their distinctive phenotypes at least in part to founder effects. The large islands of the English channel have populations that resemble those on the mainland; these populations clearly represent relicts from the time when the islands were connected to the mainland, approximately 15,000 years ago. Certain small islands in the English Channel, however, have distinctive forms that are believed to be derived by postglacial recolonization (founder events), as in the islands off the coast of Scotland. It is unfortunate that the extremely suggestive work of Berry has not been followed up with biochemical studies of the genetic polymorphism in these interesting vole populations.

Mayr (1963a) and Carson (1959, 1965, 1971) have postulated that a new, isolated colony derived from very few individuals (even one) may initiate speciation. Carson (1968) has developed this idea in studies of the *Drosophila* fauna of the Hawaiian archipelago. He has suggested that a sudden increase ("flush") in population size followed by a "crash" may be a particularly significant factor in speciation by the founder principle. If the flush occurs in a species with a high level of genetic polymorphism and is the result of an improvement in environmental conditions, it may lead to a sudden relaxation of natural selection and "the release of enormous genetic variability as the population reaches its crest." Widespread dispersal of the abnormally large number of individuals is followed by a population crash, which, because it will be geographically nonuniform, will lead to numerous isolated demes, some of which are new colonizations for the species. These may possess gene pools quite different from the original one. In Carson's words, "The new isolates are forced to inbreed in small populations, and natural selection may first favor reciprocal adjustments of the sexes in reproduction. This, together with the well-known profound effects of random drift on the gene frequencies in the new deme, may result in reproductive isolation from the parent group should another flush occur. Thus a new species is born under circumstances where adaptation is not the guiding force."

It is difficult to estimate just how important the alleged effects of such population flushes really are in speciation. Carson emphasizes that ordinary seasonal cycles in population number cannot be regarded as flushes and do not have the genetic consequences he postulates for flushes. Most species of animals never seem to exhibit flushes and nothing really similar occurs in plants. In those animal species that notoriously exhibit occasional flushes, e.g., various species of lemmings and voles and several species of locusts, no relationship to speciation has been proved to exist, nor does any seem apparent. In the case of the locusts the situation is especially clear. The major locust pests in Africa, Asia, and Australia either belong to monotypic genera or have related species entirely outside the geographic area in which they periodically flush. *Locustana pardalina* (brown locust) is largely confined to the Karroo, but spreads out when swarming into large areas of Natal, the Transvaal, and Southwest Africa (Lea, 1969); there is not the slightest evidence that it is speciating in any of these peripheral areas. The same may be said of *Nomadacris septemfasciata* (red locust), which occurs throughout tropical Africa and Madagascar, with major "outbreak areas" in Tanzania and Zambia. *Chortoicetes terminifera* (plague locust) occurs throughout Australia except for a few coastal and tropical areas; it too swarms periodically over large areas. Key (1954) recognizes an "eastern" race, which is widespread, and a "southwestern" race, which occupies a relatively small area. The latter is weakly differentiated, morphologically, and there is no suggestion that it is an incipient species.

The situation regarding *Schistocerca gregaria* (desert locust) is somewhat different. This is the only Old World species of a genus that has a number of species in both North and South America, and thus may represent a relatively recent range extension of the genus eastward across the south Atlantic. It and the form from Southwest Africa referred to below are both regarded by Dirsh (1974) as subspecies of the widespread New World species *S. americana*. *S. gregaria* occurs from Bangladesh to Turkey and Arabia and throughout most of Africa north of the equator. A morphologically distinct race occurs in Southwest Africa and the western half of the Republic of South Africa, separated from the main area of distribution by a wide equatorial belt. This southern-hemisphere race is fully interfertile with the northern one (Nolte, 1968).

The migratory locust, unlike the others we have mentioned, consists of a number of geographic subspecies that are sufficiently distinct to have received taxonomic names: *Locusta migratoria migratoria, L. m. migratorioides, L. m. capito, L. m. cinerascens,* and *L. m. manilensis*. It occurs over a vast area that includes the whole of Africa, Madagascar, most of Europe, Asia, New Guinea, Australia, and New Zealand. The existence of distinct geographic races is not surprising in view of the great range of the species.

The history of the flushes of *Locusta migratoria* since 1871 has been documented by Betts (1961). There was a major flush in 1890–1904 and another in 1928–1942, the intervening periods being ones of "recession." The evidence suggests that flushes develop quite suddenly in this species, and that crashes are equally sudden. The recessions last on the average at least 25 years, i.e., about 100 generations.

A similar historical survey of the upsurges and recessions of populations of *Schistocerca gregaria* was carried out by Waloff (1966). The periodicity of populational fluctuations is much shorter; there seem to have been four flushes in the period 1908–1964 with recessions (each of about five years) between them.

It would seem, then, that the spectacular population expansions and outbreaks of five species of locusts provide absolutely no support for Carson's views on the role of flushes in speciation. Instead of being surrounded by peripheral or incipient species, these geographically widespread insects show only rather weakly developed geographic differentiation; the few peripheral races that have been recognized do not seem to be genetically isolated. Since the details of population structure and the periodicity of outbreaks are somewhat different in all these species, they would seem to argue strongly against the view that flushes and crashes in any way favor speciation.

The evidence from mammalian flushes seems to point in the same direction. The well-known lemming *(Lemmus lemmus)* of Lapland and the mountains of Scandinavia has related species in arctic America *(L. trimucronatus)* and in northern Russia and Siberia *(L. obensis, L. chrysogaster, L. amurensis,* and *L. paulus),* but there is no particular reason to suppose that speciation has been due to the flushes and crashes characteristic of this species (geographic isolation during the Pleistocene glaciations would seem a more probable cause).

Although Carson has, to a considerable extent, combined his concepts of founder events and population flushes in a single model of speciation, they seem to be logically independent. It seems clear that, under insular conditions, founder events can indeed lead to speciation, but the evidence that population flushes and crashes have any causative role in speciation appears to be very much weaker.

Most writers on allopatric speciation appear to believe that a speciating population must be separated from the ancestral one by a fairly major geographic barrier that persists for a considerable period of time. Mackerras (1970) espouses the orthodox view in writing about the speciation of insects in Australia: "Essentially we look for the occurrence of barriers—of various kinds, depending on the kinds of animals we are working with—that could have divided populations into discrete sections for long enough periods to permit the evolution of reproductive isolation. Existing barriers may be producing incipient speciation, for which evidence can often be found in the existence of subspecies; but we must find barriers that existed in the past and lasted, in general, for many thousands of years in order to account for the definitive species that we see around us today."

An essentially similar standpoint is taken by Key (1970): "Barriers must be adequate to prevent gene exchange almost completely, i.e., they must constitute unacceptable territory that can rarely be crossed. It is generally considered that a difference in the preferred habitat of two portions of a population does not in itself constitute effective isolation." This view is also expressed by some workers whose views on other evolutionary questions are less orthodox. Thus Croizat, Nelson, and Rosen (1974) state: "Allopatric species (vicariants) arise after barriers separate parts of a formerly continuous population, and thereby prevent gene exchange between them," and, "The earliest stages (races and subspecies) of differentiation (vicariance), separated by complete or incipient barriers to gene exchange, are entirely allopatric."

It should be noted that all of the opinions just quoted specifically exclude the possibility that widely separated geographic races of a species might evolve into separate species in the event they are connected by intermediate populations. A *gap* in the distribution, caused by a *barrier,* is held to be essential, regardless of the degree of vagility of the species.

Although some cases of speciation interpreted earlier in terms of the dumbbell model may indeed be allopatric, the particular geographical sequence of events postulated is so speculative that it does not inspire confidence. In the case of continental species, numerous former discontinuous ranges interrupted by glaciation or mountain ranges have been hypothesized. Adherents of what we may call the extreme allopatric view have encountered the greatest difficulty in postulating major geographic barriers on the continent of Australia, where there have never been major mountain ranges and where Pleistocene glaciation was minimal. Some of them have, however, made use of the Nullarbor Plain in constructing ingenious, even if completely imaginary, models of past distributions and speciation patterns. These geographic speculations are reminiscent of the hypothetical land-bridges imagined by early biogeographers to explain apparently anomalous distributions. Thus, Lee (1967) describes the dilemma faced by A. R. Main, Lee, and Littlejohn (1958) in their attempt to interpret speciation in the frog genera *Crinia, Neobatrachus,* and *Heleioporus* as "the incongruous situation of a multiplicity of species . . . in the topographically featureless southwestern pocket. . . ." Now, once major topographic barriers are regarded as a *sine qua non* for speciation, one is forced, particularly in the case of a continent such as Australia, to postulate the most surprising sequences of events. Main, Lee, and Littlejohn state their investigation was carried out "in order to see whether special factors other than geographical barriers operated so as to initiate speciation processes in this group. Needless to say none was revealed."

The hypothesis by A. R. Main, Lee, and Littlejohn is that southeastern Australia was inhabited by two climatically adapted stocks or lineages of each of the three genera: a "Bassian" one (cool mesic climate) and an "Eyrean" (hot arid climate). During every Pleistocene pluvial period these six stocks migrated across the Nullarbor Plain to colonize the southwest of the continent; and during every interpluvial period in the Nullarbor barrier isolated the southeastern and southwestern populations so that typical allopatric speciation occurred. But in the case of *Crinia* and *Neobatrachus* "all traces of earlier Bassian invaders were erased, so that at the present time only the most recent Bassian invader occurs." There is, however, quite simply no evidence of such "earlier Bassian invaders"—they are postulated merely to preserve the symmetry of the model. Of the "Eyrean" species of *Crinia,* the stock that gave rise to the southeastern *C. parainsignifera* is alleged to have produced the southwestern *C. pseudoinsignifera, C. insignifera,* and *C. subinsignifera* during the first, second, and third pluvial cycles followed by an arid period.

The frog genus *Heleioporus* includes six species (Lee, 1967). *H. australiacus* occurs in New South Wales and Victoria, in high rainfall areas in the Great Dividing Range and to the east of it. *H. albopunctatus* is widespread in southwestern Australia; at a number of localities north and east of Perth it is sympatric with the less widespread species *H. eyrei, H. psammophilus,* and *H. inornatus*. All five species are present at one locality and two or three occur together at a number of others. All the species make use of individual mating burrows in which copulation, oviposition, and embryonic development occur. Later, the burrows are flooded. In four of the species larval development (3–5 months) takes place in ponds, but in *H. inornatus* the larvae inhabit the collapsed burrows or shallow pools adjacent to them, while in *H. barycragus* the tadpoles live in flowing creeks. *H. psammophilus* and *H. eyrei* seem to be especially closely related, and some presumed natural hybrids were detected by Lee on the basis of the male call. *H. barycragus* from the Darling Scarp near Perth is the Western Australian species most closely related to the eastern *H. australiacus*. The sympatric populations appear to maintain their distinctness primarily because of differences in male mating calls, but there are also substantial postmating isolating mechanisms (abnormalities of hybrid development).

Lee has postulated for *Helioporus* the same elaborate scheme of successive westward migrations during three different pluvial periods as for *Crinia*. There are supposed to have been the same "Eyrean" and "Bassian" stocks in eastern Australia; *H. australiacus* is the direct eastern descendant of the former, while no eastern descendants of the latter have survived. Lee supposes that during the first pluvial period the Eyrean form crossed the continent to give rise to *H. albopunctatus*. During the second pluvial period it (or its descendants) again traversed the continent to give rise to *H. eyrei,* and at the same time the "Bassian" population did the same to produce the western *H. inornatus*. During the third pluvial period a double invasion of the west again occurred, as a result of which the "Eyrean" stock gave rise to *H. psammophilus* while the "Bassian" one produced *H. barycragus*.

It should be obvious that there is not the slightest direct evidence for this quintuple westward crossing of the central part of the Australian continent. It seems much simpler to suppose that *Heleioporus* is essentially a Western Australian genus—one that evolved in that area—but that at some stage it gave rise to an offshoot (having no doubt a close common ancestry with *H. barycragus*) that crossed Australia in an eastward migration to give rise to *H. australiacus*. The situation with regard to *Crinia* is somewhat different, since that genus has equal numbers of species in the southeast and southwest of the continent, so that there is no way of telling whether it evolved originally in one area rather than another.

B. Y. Main has discussed the origin of five species of mygalomorph spiders that belong to the closely related genera *Aganippe* and *Idiosoma* and inhabit Western Australia. In her first account (1957) these were interpreted in a straightforward manner as having arisen by the fragmentation of a formerly widespread western population. Later (1962), and clearly under the influence of the arguments and speculative interpretations that had developed in studies of the frog genera previously mentioned, she rejected her previous interpretation—because "it is difficult to visualize the complete geographic isolation, necessary for speciation, of these populations in the region"—and built up a hypothetical scheme based on four successive westward migrations across the Nullarbor region (the first, in the Pliocene, giving rise to the *Idiosoma* stock; the second to *A. raphiduca;* the third to *A. cupulifex;* and the fourth to *A. occidentalis).* The source of all these migrations was alleged to be a population in the Flinders–Mount Lofty Ranges of South Australia, of which *A. subtristis* is the modern descendant.

As in the case of the frogs, there is not the slightest direct evidence for this elaborate hypothesis, which seems to have been produced solely to satisfy the *a priori* postulate that speciation is only possible where populations are completely separated by major topographic barriers. The logical conclusion of this line of argument would be that none of the many thousands of animal species endemic to southwestern Australia could have evolved *in situ,* but must have been derived from successive invasions from eastern Australia. The hypothesis becomes even more ridiculous when applied to the flora of Western Australia, which is exceptionally rich in endemic species of many genera.

The above criticism of the views of A. R. Main, B. Y. Main, Lee, and Littlejohn is not meant to imply that the speciation of such genera as *Crinia, Heleioporus,* or *Aganippe* in the "topographically featureless" area of Western Australia was necessarily sympatric or followed the "stasipatric" model. The area in which they must be presumed to have evolved does not, it is true, include major geographic barriers, but it is ecologically quite varied and thus provides many diverse habitats. It is quite possible that incipient species of these genera could have been separated geographically, to a considerable extent, by ecological restriction to particular habitats. Speciation could have been according to the clinal or area-effect models (Chapter 5), or it might even have been of the classical allopatric type, but without the major barrier of the Nullarbor Plain. No cytogenetic or biochemical evidence on the above genera is available to throw additional light on the past course of events.

The real answer to the "incongruous situation" of many related species of frogs inhabiting the "topographically featureless" area of southwestern Australia comes from studies of many insect groups that have undergone extensive speciation in this area. In the vast majority of these studies there is not the slightest reason to suppose that major topographic barriers were responsible for speciation, or that any of these groups had representatives in eastern Australia that periodically migrated across the Nullarbor barrier during climatically favorable periods. The situation is perhaps even clearer in the equally "featureless" desert area of central Australia, in which active speciation of many groups of insects and small vertebrates has occurred. A particularly striking example is the grasshopper genus *Warramunga,* whose 23 species (Key, 1976) inhabit clumps of spiny grasses belonging to the genus *Triodia* throughout the Australian desert.

The clearest cases of allopatric speciation have occurred on archipelagos remote from continental land masses, such as the Galápagos and Hawaiian Islands. Because of its historical association with Darwin and the part that it played in the development of his ideas on speciation, the fauna of the Galápagos has always attracted the attention of evolutionists. In recent years, however, certain elements of the Hawaiian and Tahitian faunas have been investigated in far greater depth than any of the Galápagos groups. The following account of allopatric speciation on islands will focus on the finches and giant tortoises of the Galápagos, the drosophilids of the Hawaiian Islands, and the land snails of Moorea in the Society Islands (French Polynesia).

Patterns of speciation in archipelagos depend on many factors—on the distance of the island group from the mainland; the likelihood of accidental transport of organisms from rafting, hurricanes, and other causes; and the sizes and ages of individual islands and their distance from one another. Clearly, we must also take into account whether the climate of the islands is tropical, temperate, or cold. There is hence no single pattern of speciation on islands, and the allopatric model is likely to be manifested in a variety of modified forms.

Of the three groups of oceanic archipelagos considered here, the Galápagos probably date from Pliocene times, while the Hawaiian archipelago is significantly younger (p. 121). The age of the Society Islands does not seem to be known at all accurately. All these archipelagos are of volcanic origin. It is clear that the main Hawaiian islands (Kauai, Oahu, and Hawaii) have never been connected with one another and that each has acquired its biota by trans-ocean colonization from the mainland, from other islands of the group, or from other Pacific islands. However, the islands of Maui, Molokai, and Lanai, which are very close together, may well have been connected by land bridges in the Pleistocene.

Galápagos Islands: Geospizidae and *Geochelone*

There are two rival geological theories for the Galápagos. One is that the major islands have never been connected, while the other holds that a continuous land mass existed in the past and has undergone subsidence. Bathymetric considerations (Thornton, 1971, Fig. 3) suggest that Fernandina, Isabela, Santa Cruz, Baltra, and perhaps San Salvador and Rábida Islands were connected during the Pleistocene glaciation but that the other islands have always been isolated by water (including Pinzón, which lies more or less in the middle of a horseshoe formed by those named above and has an "anomalous" fauna). The earlier view that the Galápagos were at one time connected to the South American continent has not been supported by any recent evidence. Obviously, views about modes of speciation in the Galápagos biota depend considerably on whether one assumes that all the major islands were originally separated or continuous, or takes some intermediate position.

Some writers have asserted that islands have special ecological characteristics that produce unusual adaptations. Thornton (1971) writes: "Considering the array of circumstances which may act to promote divergence on islands, there is no wonder that it is on islands that the most unusual adaptations are often found. Ecological shifts and changes in behavior, as well as unusual structural alterations, are particularly prevalent on oceanic islands." He then gives a number of examples from the Hawaiian and Galápagos archipelagos.

Subjective judgments of this kind are obviously very unreliable. "Unusual adaptations" are prevalent on the great continental land masses as well. Thus, some cecidomyid midges have acquired three compound eyes (one on top of the head) and other species of the same family have developed pedogenesis (parasitization of a mother larva by her own offspring); some continental flies have developed eyes on long stalks; and in Australia one species of cockroach has transparent windows through the thorax. But a mere enumeration of bizarre examples will prove nothing. A great diversity of ecological niches, whether in a continental or an island situation, is likely to lead to specialized adaptations. An initially sterile, newly arisen volcanic island provides opportunities for a variety of plant species to establish themselves. As the numbers of individuals and species of plants increases, more and more ecological niches suitable for occupation by insects and other terrestrial animals are created. But it would be quite unscientific to claim that an area of Hawaii or the Galápagos (before human interference) contained more ecological niches than an equivalent area of the Amazon Basin, the Congo, or Malaysia. In fact, the reverse may very well have been the case, although we may never be able to prove it.

Related to this concept of the unusual character of island species and their adaptations is the idea that "an unusual degree of speciation is . . . a feature of archipelago evolution" (Thornton 1971, p. 234). This seems to be partly true and partly false. Oceanic islands and archipelagos have acquired their biota by chance immigration from larger continental land masses or from other islands. Their faunas and floras are thus highly "unbalanced," i.e., many genera and families found on the mainland are lacking, either because they were (and still are) incapable of crossing wide expanses of ocean, or simply because they have never done so. Thus, when immigrant species genetically capable of speciation reach an archipelago, they are likely to encounter a very large range of vacant ecological niches and opportunities, free of competitors belonging to related genera and families. The result may be "explosive" speciation (the prime example being the Hawaiian Drosophilidae, discussed below). However, to ascribe such a phenomenon entirely to geographic isolation (if this is taken to mean the saltwater gaps between the islands of the archipelago) is an oversimplification. Although isolation is undoubtedly a factor in explosive speciation on archipelagos, to regard it as the *main* cause seems to go somewhat beyond the evidence presently available.

Many popular writers (e.g., Moorehead, 1969) have asserted that Darwin's ideas on evolution were significantly influenced by his observation that species of the geospizid finches differed from island to island (Figure 9). The actual situation, however, is more complex than this. Of the 13 currently recognized species of Geospizidae, only one is confined to a single island *(Camarynchus pauper* on Santa María). The small island of Española has three species, while Isabela has 11 and San Salvador, Santa Cruz, and Fernandina each have 10 species (Lack, 1969). The very small islands of Culpepper and Wenman, remote from the rest of the archipelago, have about three species each. On the other hand, a number of endemic "subspecies" (distinguishable mainly by beak size and shape and by coloration) are confined to single islands. Thus, *Certhidea olivacea,* the only species of its genus, is present on all 16 islands and has eight subspecies, five of which are confined to single islands. On the other hand, some widespread species, such as *Geospiza fortis* (on 12 of the islands) and *C. fuliginosa* (on 13 of them) do not seem to be differentiated into subspecies. All the Geospizidae except *Certhidea* are extremely similar allozymically, and have indistinguishable karyotypes.

A critical evaluation of this classical case of speciation is hardly possible in the absence of a proper understanding of the degree of evolutionary and genetic divergence of the subspecies. It is clear that they may occupy varied ecological niches and have quite diverse diets on different islands (the beak shapes are adaptive to these diets, as was especially stressed by Bowman, 1961). And the analysis of Hamilton and Rubinoff (1963, 1964, 1967) has shown that there is a

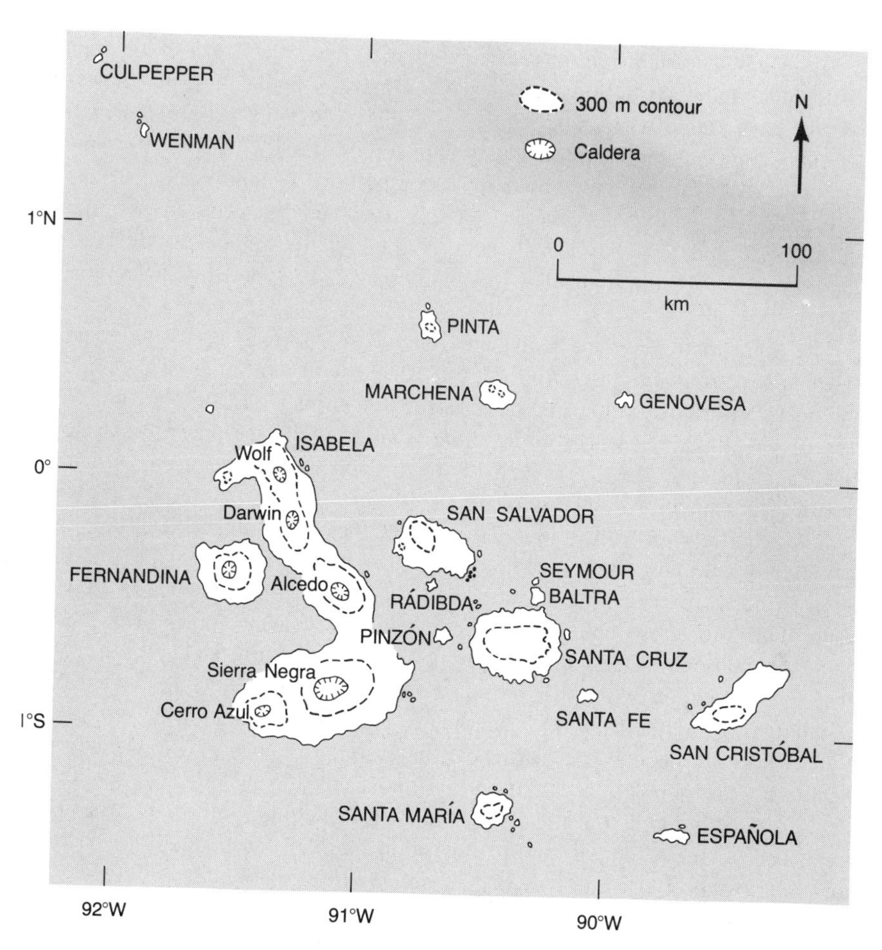

FIGURE 9

Map of the Galápagos Archipelago, showing the 300-meter contour and the *calderas* (volcanic craters).

strong correlation between geographic isolation (i.e., from the other islands) and subspecific endemism. Nevertheless, the islands are generally quite close to one another and a considerable amount of interisland wandering of the finches must occur. There seems to be no information on the genetic isolating mechanisms that keep the species from merging. A number of peculiar individuals that could be hybrids, possibly even intergeneric ones, have been collected.

The general conclusion that can be drawn from the modern taxonomic, ecological, and distributional studies of "Darwin's finches" is that new subspecies have arisen when a population was able to establish itself in a more or less unoccupied niche on a previously unoccupied island, i.e., when a diet was available that was not preempted by a competitor. The next stage in speciation—the acquisition of genetic isolating mechanisms between semispecies or incipient species—has hardly been studied in the Galápagos finches, but there seems to be no doubt that it was allopatric, in spite of some interisland genetic leakage.

The giant tortoise of the Galápagos Islands is regarded as a single species *(Geochelone elephantopus)* with 15 "subspecies" that differ in size, length of the neck, and shape, color, and thickness of the carapace (Hendrickson, 1966). Ten of the subspecies occur on separate islands, while the other five occur on the largest island, Pinta, on each of the five principal volcanoes, which are situated in a chain at intervals of about 30 km.

It is extremely difficult to imagine just how these subspecies could be related genetically, and how those on Pinta could have arisen. One theory is that Pinta was formerly five separate islands, and that separate subspecies evolved on each. This still leaves unanswered the question as to what forces keep the five Pinta populations distinct at the present time. It is claimed (Thornton, 1971) that the phenotype with a high-domed carapace and a short neck (e.g., the population on southern Pinta) is adapted to areas of grass and lush, low vegetation, while the phenotype with a low, flared-out carapace and long neck and legs (e.g., the population of the northernmost volcano on Pinta) is adapted to arid areas of tall *Opuntia* and other shrubs. No adequate information has been published on the extent (if any) of migration between the populations on Pinta and the possibility of interbreeding in nature. Presumably, the differences between them are genetic (as has been assumed by all investigators); but one should not lose sight of the possibility that they may be due to differences in growth rate, depending on the amount and type of the food supply, as in the case of some freshwater fishes (Hile, 1937).

If both the allopatric and sympatric types of speciation are possible (see Chapter 7), it follows that, although only sympatric speciation could occur on very small, isolated islands, both types should occur (at least in certain types of organisms) on archipelagos. Nevertheless, since archipelagos have provided so many of the classical examples of the allopatric process, there has been a tendency to look only for evidence of this type of speciation on archipelagos. Moreover, there have been few studies of archipelago speciation in groups where one would most expect to find evidence of the sympatric process.

Hawaiian Archipelago: Drosophilidae

The most extensive, detailed, and significant study of speciation in an archipelago is undoubtedly the investigation of all aspects of the evolutionary biology of Hawaiian Drosophilidae that has been carried out over the past 12 years. Part of its importance is derived from the fact that the Hawaiian archipelago is comparatively young, having been produced by relatively recent volcanic action. The ages of the six main islands have been rather firmly established from potassium-argon measurements (Macdonald and Abbott, 1970). The oldest island, Kauai, which is also the most northwesterly, is about 5.6 million years old; Oahu came into being about 3.4 million years ago; the islands to the southeast are considerably younger (Molokai has a maximum age of 1.8 million years, Maui 1.3 million years, and Hawaii, the largest island, less than one million years).

The number of drosophilid species in the Hawaiian archipelago is estimated at 750–800, of which somewhat over 500 have been described (Hardy, 1974). Except for 22 species that are recent immigrants, introduced as a result of human activities, the remainder are all endemic. About 73 percent of the native species are members of the genus *Drosophila;* these account for well over a quarter of the 1,260 known *Drosophila* species in the entire world. Most of the remaining native species belong to the genus *Scaptomyza;* the number of Hawaiian *Scaptomyza* species is twice the number of known *Scaptomyza* species in the rest of the world. The distinction between *Drosophila* and *Scaptomyza,* which is sharp elsewhere in the world, tends to break down in the Hawaiian fauna and it has even been suggested that *Scaptomyza* arose from a drosophiloid ancestor in the Hawaiian archipelago. (If this is true, *Scaptomyza* species in the rest of the world would all be derived from one or more emigrations of flies from Hawaii to other Pacific Islands or the mainlands of Asia and America.)

It is uncertain how many ancestral species of drosophilids reached the young islands at the northwestern end of the Hawaiian chain 5–6 million years ago (by immigration from some other land mass). Perhaps there was one ancestral *Scaptomyza* and one ancestral *Drosophila;* but the possibility that there was a single species ancestral to both genera cannot be excluded at present, and there are sound reasons for believing that there cannot have been more than two ancestral species (Throckmorton, 1966). Consequently, the adaptive radiation of the Hawaiian drosophilids must have involved "explosive" speciation on a grand scale in a remarkably short space of time. In particular, the numerous species endemic to the island of Hawaii must all be less than 700,000 years old, and many of them must be considerably younger.

The great majority of the Hawaiian drosophilids (perhaps 98 percent) are endemic to single islands (Carson 1971). Because there is a great lack of native fleshy fruits in Hawaii (which provide food and shelter for the larvae of many continental drosophilids), the Hawaiian species have acquired very varied and sometimes bizarre feeding habits, and may be said to occupy extremely specialized and narrow ecological niches in the tropical rainforest. Some species breed on leaves, others feed as larvae on decaying bark or even on spider eggs. There are numerous examples of very closely related species whose distributions are essentially or entirely sympatric. In a number of instances these species are also homosequential (see p. 45). However, in such cases, especially in the group of "picture-winged" species with conspicuous patterns on the wings, the males of closely related species are easily distinguishable. These differences are almost certainly the evolutionary result of an important component in the biology of these flies known as lek behavior (Spieth, 1968, 1974). Mature males select and then defend a small territory, on which they advertise their presence by species-specific forms of behavior (attitudes, waving the wings), which serve to attract sexually receptive females.

Two such species are *D. heedi* and *D. silvarentis* (Kaneshiro, *et al.,* 1973), which coexist on a xeric volcanic area of the island of Hawaii that has only two main trees, *Myoporum sandwicense* and *Sophora chrysophylla*. Adults of both species feed on slime fluxes exuded from *Myoporum*. But the larvae inhabit separate ecological niches; those of *D. silvarentis* live in fluxes on the tree trunks well above ground level, while those of *D. heedi* inhabit caked volcanic soil wetted by flux dripping on it from above. The species are essentially homosequential, although there is a heterochromatin difference and a puff difference in the polytene chromosomes (Raikow, 1973). The niche separation between these species is clearly a necessary condition for their successful coexistence, and depends on a very specific choice of oviposition site by the females. It seems clear that a shift in oviposition site by one of them was an essential step in their evolutionary divergence. Whether this adaptive shift occurred under conditions of sympatry or allopatry is not known. There is certainly no evidence for earlier allopatry, and on the principle of parsimony it would seem advisable to regard this as a case of sympatric speciation for the time being.

It is clear, however, that a great deal of the explosive speciation of the Hawaiian drosophilids has been due to founder individuals accidentally transported by air currents (or other means) from one island to another. On the evidence of polytene chromosome maps, Carson and Stalker (1969) and Clayton, Carson, and Sato (1972) have suggested that the entire picture-winged group of species, including four subgroups (the *grimshawi* subgroup with 56 species, the *planitibia* subgroup with 16, the *adiastola* subgroup with 14, and the *punalua* subgroup with 8) arose from a fifth subgroup of flies, without wing markings, which consists at present of only two species, *D. primaeva* and *D. attigua,* endemic to wet forests on Kauai, the oldest island of the archipelago. These two are sibling species that differ in the male genitalia and with respect to 13 fixed inversions. They carry a recognizable section of bands in the X chromosome, which also occurs in 82 other picture-winged species, and is believed by Carson (1974) to be more primitive than the "Xt" sequence. In fact, the band sequences of *D. primaeva* and *D. attigua* seem to show some resemblances to those of certain continental species, notably *D. robusta* (Stalker, 1972). It is reasonable to suppose that all four subgroups of picture-winged species arose on Kauai, possibly in the Pliocene. Of the species presently found on Kauai, the *grimshawi* subgroup is represented by six and each of the other subgroups by one. At least seven migrants from the original *Drosophila* fauna of Kauai reached the more recent islands to the east and speciated extensively, giving rise to the large number of picture-winged species existing today.

The origin of the picture-winged *Drosophila* species of the most recent island, Hawaii, has been discussed by Carson (1970a, b, 1973a, 1974). Most of these seem to have been derived from the older island of Maui, which is separated from Hawaii by the Alenuihaha Channel, which is 50 km wide and 1,950 m deep. Two particularly interesting picture-winged species on Hawaii are *D. silvestris* and *D. heteroneura* (Craddock, 1974c). These are both sympatric and homosequential. They inhabit the same habitats and have the same satellite DNAs (Craddock, unpublished). *D. planitibia* on Maui and *D. differens* on Molokai are members of the same homosequential species complex. *D. planitibia* (or a population directly ancestral to it) must have given rise to *D. silvestris* and *D. heteroneura* on Hawaii, but it is uncertain whether there was a single crossing from the Maui complex to Hawaii (which gave rise to a population that subsequently split to produce the two Hawaiian species) or two separate immigrations (each of which gave rise to one of the Hawaiian species). If there were two immigrations, it is not certain whether they came from the same *planitibia* population (presumably on Maui) or from different *planitibia* populations (on Maui and Molokai).

D. planitibia and *D. differens* are chromosomally monomorphic species, whereas the two Hawaiian species are polymorphic; *D. silvestris* (the more abundant of the two) is polymorphic for seven inversions, one of which also exists polymorphically in *D. heteroneura*. The fact that the two species share this polymorphism speaks in favor of the "single founder event" hypothesis. A study of allozyme polymorphism at 12 different loci showed that the commonest allele was usually the same in all four species, and that the two Hawaiian ones were especially alike. However, in the case of the *ADH-1* gene, different local populations of *silvestris* showed strongly marked differences in allelic frequencies; at one locality the two species showed different major alleles of this gene. Surveying the differences between populations at various localities, differences in inversion frequencies were far greater than differences in allelic frequencies. Populations on two *kipukas* (islands of vegetation isolated by lava flows) 1 km apart showed differences in inversion frequencies; one inversion present on "*kipuka* 9" was absent on "*kipuka* 14." Little if any migration seems to have taken place between *kipukas* separated by lava flows during the eruption of 1855.

A few hybrids between *silvestris* and *heteroneura* were found in nature at one locality. They certainly do not hybridize frequently, probably because of a powerful ethological barrier, i.e., differences in premating display and behavior. Also, the males of the two species differ significantly in morphological characters; these must have been subject to selection because they are important in sexual recognition and pair-formation. Males of *heteroneura* can inseminate females of *silvestris* in the laboratory (although the reciprocal cross usually fails, due to a premating barrier) and F_1 and subsequent hybrids of both sexes are fully fertile. Clearly, under conditions of sympatric coexistence premating isolation evolved earlier than postmating mechanisms.

A conspicuous exception to the general condition of single-island endemism in Hawaiian drosophilids is *D. grimshawi,* found on Kauai, Oahu, Molokai, Lanai, and Maui. The species is not differentiated into subspecies, but each local population has its own characteristic frequency of the chromosomal inversion 4a. Among populations in the eastern part of the island of Maui the frequency ranges from zero to 100 percent (Carson and Sato, 1969). We have already noted marked variation in the frequency of some inversions in *D. silvestris;* it also occurs in *D. bostrycha* (Molokai) and *D. disjuncta* (Maui) and is probably a very general phenomenon among those Hawaiian drosophilids that have inversion polymorphism. Carson and Sato ascribe it to isolation between local colonies, but it must also be maintained by natural selection. That some inversions certainly have considerable adaptive value in Hawaiian *Drosophila* species is shown by the strong tendency for closely related species to have nonidentical inversions extending over almost, but not quite, the same chromosome segments (Carson, 1969). The fact that inversion frequencies tend to vary between local populations, in some cases far more than the frequencies of allozyme alleles, suggests that they serve as a "fine-focusing" genetic mechanism of populations, as opposed to the "coarse focusing" of variations in the major polymorphic gene loci.

A minimum of 22 inter-island colonizations have been postulated by Carson and co-workers (1970) to account for the phylogeny of the Hawaiian picture-winged flies (Figure 10). All except four of these involved migration in a south-

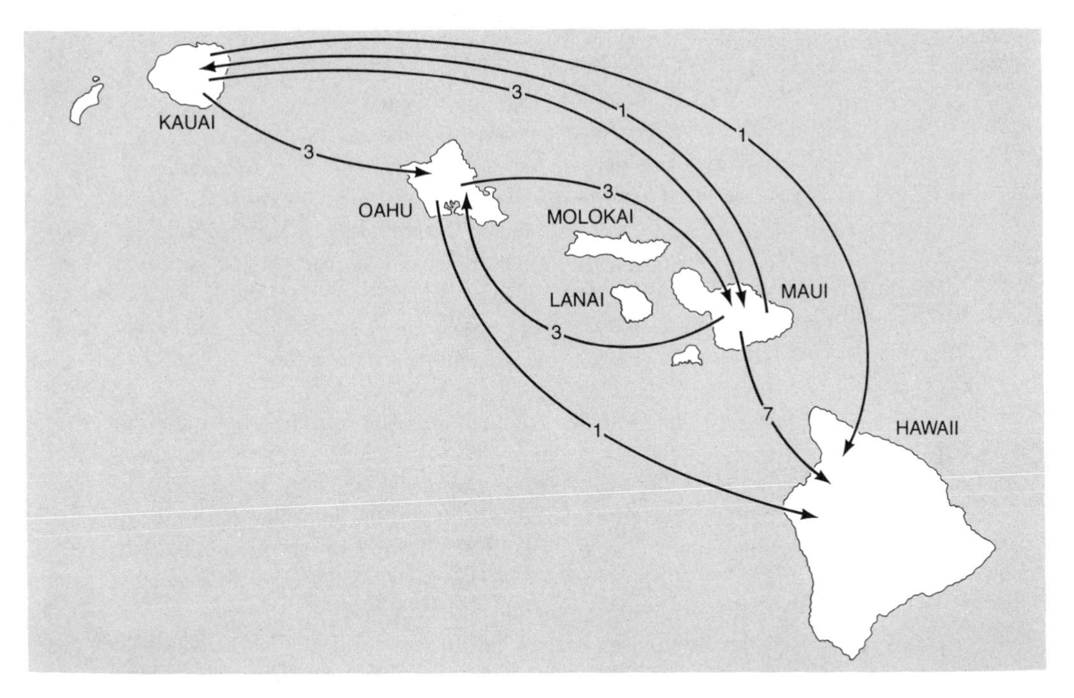

FIGURE 10

Minimum number of inter-island founder events among Hawaiian *Drosophila* species. [From Dobzhansky *et al.*, 1977]

easterly direction, i.e., recipient islands were geologically more recent than the donor islands. This is certainly not due to the trade winds, which blow from the northeast throughout the year. It must reflect the fact that new volcanic islands are biologically sterile at first, and have numerous unoccupied niches when young. Migrations from younger to older islands may have been limited in part by the small number of species (and hence of individuals) on very young islands; it is also likely that many migrations occurred but the migrant populations soon died out because no suitable empty ecological niche existed. Of the 22 founder species, only three can be proved to have been polymorphic for inversions; most founder species were hence cytologically monomorphic (Carson, 1970b). A particularly instructive example is *D. murphyi* (endemic to Hawaii), which has two sequences for the third chromosome, 3 and 3o; the same two are also found in *D. orphnopeza* (endemic to Hawaii). *D. murphyi* carries the Xg sequence in the sex chromosome; this sequence is not present in *D. orphnopeza* but does occur in a closely related Maui species, *D. balioptera*. It appears that the common ancestor of the two Maui species must have been polymorphic for X,Xg and for 3,3o and that it gave rise, by an inter-island founder event, to *D. murphyi,* which gave rise on Hawaii to two other species, one monomorphic for sequence 3, the other for sequence 3o.

Another interesting member of the picture-winged group of species is *D. neopicta,* which occurs on Molokai and on the east and west volcanoes of Maui. It is chromosomally polymorphic for the "primitive" sequences X, 2, and 4 and the "derived" ones Xt, 2m, and $4f^3$ (Table 4-1). According to Carson (1974), *neopicta* or some other population that was carrying these chromosomal polymorphisms gave rise to species homozygous for Xt, 2, and 4; Xt, 2m, and 4; Xt, 2m, and $4f^3$; and to *D. neoperkinsi,* on Molokai, which has the 2/2m and $4/4f^3$ polymorphisms but is homozygous for the "derived" Xt sequences (Figure 11).

There were apparently nine founder events in the colonization of the island of Hawaii by *Drosophila* species. In seven of these the founder came from Maui, in one from Oahu, and in one directly from the oldest and most distant island, Kauai (see Figure 10). Carson makes the very important point that in the great majority of cases invasion of a new island is followed by speciation, which must be quite rapid. Thus, on the island of Hawaii no "subspecies colonizers" are known, all the invaders being endemic "full" species.

There are, however, two cases among the picture-winged flies in which a species has colonized an island but not evolved beyond the subspecies level. *D. grimshawi* is common on the islands of the Maui complex, and is represented by flies of a slightly different phenotype on Oahu and Kauai. Flies from different islands can produce fertile hybrids, suggesting that they are indeed at the subspecific level. *D. crucigera* appears to have originated on Kauai; there it differentiated into two groups of populations characterized by somewhat differ-

TABLE 4-1
Summary of the distribution of ancestral (X, 2, 4) and derived (Xt, 2m, $4f^3$) gene arrangements in species of the Drosophila planitibia *subgroup. (From Carson, 1974a.)*

picticornis,[a] *setosifrons,*[e] *substenoptera,*[b] *obscuripes*[d]	X	2	4
neopicta[c, d]	X/Xt	2/2m	$4/4f^3$
neoperkinsi[c]	Xt	2/2m	$4/4f^3$
ingens,[d] *melanocephala*[d]	Xt	2m	$4f^3$
nigribasis[b]	Xt	2m	4
cyrtoloma,[d] *hanaulae,*[d] *hemipeza,*[b] *heteroneura,*[e] *oahuensis,*[b] *planitibia,*[d] *silvestris*[e]	Xt	2	4

[a]Kauai; [b]Oahu; [c]Molokai; [d]Maui; [e]Hawaii.

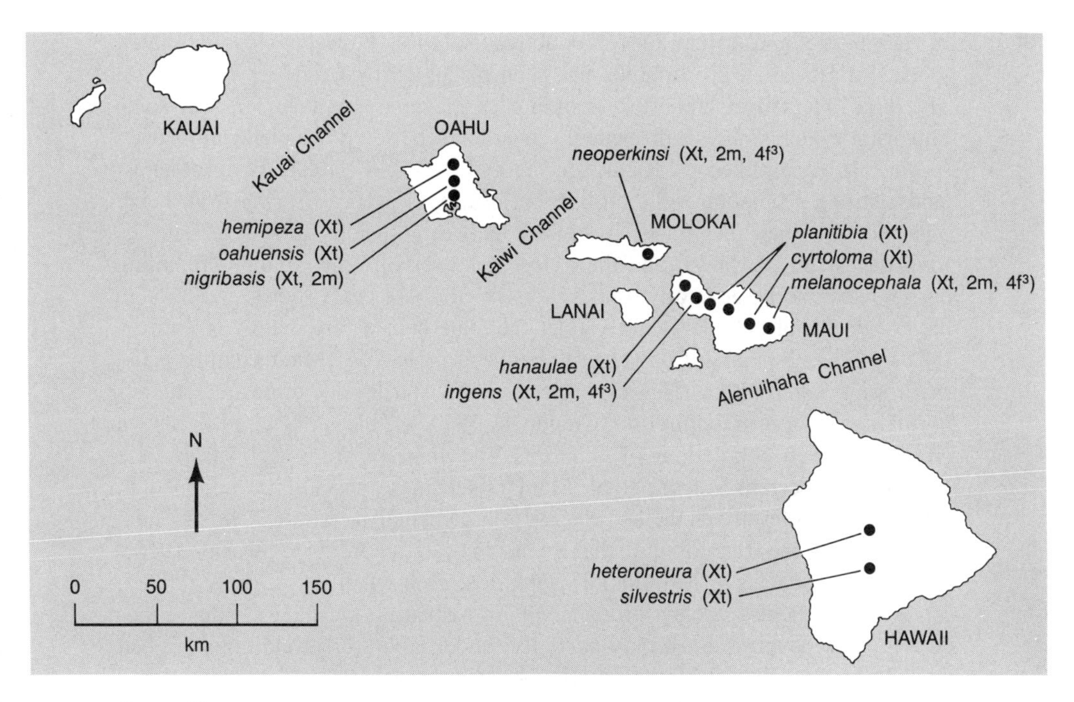

FIGURE 11

Distribution of the members of the *planitibia* subgroup of Hawaiian *Drosophila* (other than *neopicta*) that possess one or more derived chromosomes (Xt, 2m, $4f^3$). [From Carson, 1974]

ent cytogenetic constitutions (one group lacks the Xab gene arrangement of the other). The species seems to have colonized the island of Oahu twice; the Waianae range population is lacking Xab, while the Koolau range one has this sequence in high frequency (88–100 percent).

The common association of speciation with colonization in Hawaiian drosophilids is explained by Carson (1973b) on the basis of the founder principle and his own model of population flushes, which we discussed earlier. That the founder principle has played a role in drosophilid colonizations is reasonably certain on general principles. But we do not have (and are never likely to have) any evidence as to whether most or all of the colonizations involved a single gravid female or several individuals. The fact that the homosequential species *planitibia* (Maui), *differens* (Molokai), *silvestris* and *heteroneura* (Hawaii) have the same allozyme alleles and even show similarities in the frequency of those alleles seems to argue against overemphasizing the founder principle in these colonizations.

The nature and extent of reproductive isolation between closely related species of Hawaiian drosophilids has been discussed by Craddock (1974b, c). The data obtained in laboratory crossing experiments are complex, but some important conclusions seem evident. In crosses between "ancient allopatric" species, as exemplified by Kauai species and their close relatives on other islands, strong premating isolation barriers usually prevent offspring from being obtained. On the other hand, in crosses between "recent allopatric" species from the island of Hawaii and their close relatives on other islands premating barriers seem to be generally weak or absent. In these cases, however, although the F_1 hybrids of both sexes are viable, the male hybrids are mainly or entirely sterile. Thus, F_1 males from crosses between *planitibia* (Maui complex) and *silvestris* or *heteroneura* (Hawaii) are completely sterile, due to postmeiotic abnormalities of sperm maturation (Craddock, 1974c). This is likewise the case in crosses between populations of *planitibia* and *differens* from Molokai and the eastern and western volcanoes on Maui (since these species are confined to relatively high elevations, the two volcanoes are virtually separate "islands," and may have been separate for more than a million years according to Stearns, 1966).

Sympatric species of Hawaiian drosophilids, even when very closely related, are generally kept apart by strong premating isolating mechanisms that either do not break down or do so only partially (under laboratory conditions). When they do break down at all, postmating isolating mechanisms frequently come into play (inviability of the hybrids at the pupal stage, sterility of F_1 hybrid males, etc.). However, there are some exceptions to this rule; we have already noted the case of *silvestris* and *heteroneura,* whose evolutionary separation may have been extremely recent.

The obvious conclusions to be drawn from the above summary of what is admittedly a very complex situation seem to be as follows:

1. Allopatric species on different islands rapidly acquire postmating isolating mechanisms (or, at any rate, sterility of F_1 males). This must be an incidental consequence of other evolutionary changes, since it is not adaptive.
2. Such species later acquire premating (ethological) isolating mechanisms. These must also be the result of general evolutionary divergence.
3. Where closely related species occur sympatrically, they rapidly acquire both premating and postmating isolating mechanisms; premating isolation may be directly selected for as adaptive and evolve earlier than postmating isolation.

It is clear that several different modes of speciation have operated in the explosive differentiation of the Hawaiian drosophilids (Craddock, 1974b). Founder events have undoubtedly played an important role. We have stressed especially those that involved colonization of major islands previously unoccupied, because these are more easily identified. But within the "Maui complex" (Maui, Lanai, and Molokai, probably connected in the Pleistocene), and perhaps on Oahu and Hawaii as well, there were undoubtedly numerous colonizations of one volcanic peak from another. Where these led to speciation they must be regarded as conforming to the general allopatric model, the territory between the volcanoes being unsuitable for occupation because of its low elevation. Craddock (1974c) speaks of this type of speciation as "geographic speciation via adaptive differentiation of allopatric populations," and contrasts it with speciation resulting from a founder event. The distinction seems to be a rather dubious one, but if it could be shown that a population existing at low elevations had been split into two by the separation of Molokai from Maui, due to a rise in sea level, then we would have an example of the classical dumbbell model of allopatric speciation. But in addition to speciation by founder events (the allopatric model with a narrow population bottleneck, considered briefly in Chapter 1), it is clear that much *Drosophila* speciation on the Hawaiian archipelago has involved adaptation to very specialized ecological niches, and was achieved without geographic separation of the speciating populations (i.e., sympatrically) or with very minimal geographic segregation. It is certainly possible (but unproven) that the explosive speciation of the Hawaiian drosophilids has been due in part to the availability of large numbers of specialized ecological niches in rainforest habitats that in continental rainforests are occupied by flies of other families.

A detailed argument for sympatric speciation has been made by Richardson (1974) for the species *D. mimica* and *D. kambysellisi*. The two occur together on certain *kipukas* on the island of Hawaii. They are closely related members of the so-called "modified mouthparts" group of species, but differ cytologically with respect to a fixed inversion and in the fact that the *mimica* karyotype has more extensive heterochromatic blocks. Ecologically, *mimica* breeds primarily in rotting fruits of *Sapindus saponaria,* while *kambysellisi* breeds in rotting *Pisonia* leaves. Other, undescribed, species related to these two occur on the older islands of Oahu and Kauai. One interpretation of the evolution of these flies is that they arose from separate invasions of Hawaii by an ancestor of *mimica* and one of *kambysellisi*. The alternative interpretation (and the one favored by Richardson) is that only one invasion of Hawaii took place, by a *mimica* ancestor, followed by sympatric speciation based on habitat separation, which gave rise to *kambysellisi; kambysellisi* is considered the "derived" species on the basis of its very restricted distribution (possibly only two *kipukas*) and more specialized habitat. Although it is clear that Richardson has made a strong case for sympatric speciation in this instance, one would have to know more about the members of this group of species on the other islands before reaching a firm conclusion. The two species on Hawaii seem to have diverged in karyotype much more than the four members of the *planitibia* complex considered earlier, which are homosequential and have the same satellite DNAs.

In summary, it now seems possible to conclude from the very extensive studies of Hawaiian Drosophilidae that the founder model (rather than the classical dumbbell model of allopatric speciation) has been highly significant in the evolution of this group, but that most Hawaiian *Drosophila* species have not resulted from founder events. Clearly, we need to consider next the possibility of sympatric speciation as a response to niche diversity, even though it is obviously extremely difficult to prove that sympatric speciation has occurred in any particular case. If the explosive speciation of the Hawaiian drosophilids was due primarily to the diversity of ecological niches available to them, then conversely, the failure of the genus to speciate in an equally explosive manner in continental tropical areas can only be due to the great majority of the ecological niches in those areas being "closed," i.e., occupied by competitors.

The genetic components of the explosive speciation of the Hawaiian drosophilids clearly include gene mutation and changes in the amount, distribution, and nucleotide sequences of the satellite DNAs. As in the case of the continental species of *Drosophila,* there is little evidence that chromosomal inversions have played a significant direct role in speciation. And, in complete contrast to the grasshoppers, lizards, rodents, and some other groups discussed in Chapters 3 and 6, there is no evidence at all that chromosomal rearrangements of other types have been primary causes of speciation.

There are a number of other groups of invertebrates on the Hawaiian islands in which speciation seems to have been especially rapid. We have referred elsewhere (p. 14) to the moth genus *Hedylepta,* in which several endemic species have apparently arisen during the past 1,000 years (the time during which their host plant, the introduced banana, has been present in Hawaii). Unfortunately this most interesting case of speciation has not been studied by the sophisticated techniques now available. The speciation of Drepanididae (a family of birds known as honey creepers) in the Hawaiian islands parallels that of Geospizidae in the Galápagos (Amadon, 1950), but in this case many of the original species are now extinct, making a detailed analysis difficult.

Society Islands: *Partula*

It is interesting to compare the Hawaiian drosophilids with the land snails of the genus *Partula* on the island of Moorea in the Society Islands (Crampton, 1916, 1932; Clarke, 1968; Clarke and Murray, 1969, 1971; Murray and Clarke,

1966, 1968). Studies on *Partula* speciation are not nearly so advanced as those on Hawaiian *Drosophila*. Nevertheless, the very obvious differences in the evolutionary regularities of the two groups are presumably due, in the main, to a very different vagility, mode of life, and population structure, although the possibility that they are in part due to inherent differences in the genetic system cannot be excluded. *Partula* snails are hermaphrodites that normally cross-fertilize, although self-fertilization occurs occasionally. Vagility has not been studied critically but must be very low; populations only 20 m apart may show significant differences in genetic composition.

There are about a hundred species of *Partula* known from the mountainous volcanic islands of the Marianas, Marquesas, Society, and Austral archipelagos. Most of the species are apparently confined to single islands. Presumably founder events occurred, as in the case of the Hawaiian drosophilids, but since only the species of the island of Moorea have been studied in any depth, the number, direction, and chronology of these colonizations is not known. The classical work of Crampton was concerned largely with the genetic polymorphism of *Partula,* as revealed by variations in shell color and banding pattern and the dextral or sinistral coiling of the animals. Crampton considered these phenotypic characters "neutral," i.e., uninfluenced by natural selection, a conclusion which is difficult to accept today, even though no definite evidence for an adaptive significance of these characters exists as yet.

The more recent work of Clarke and Murray is much more relevant to mechanisms of speciation in this group. There appear to be about nine species of the genus *Partula* on Moorea (an extremely mountainous volcanic island of about 115 square kilometers); several of these include more than one geographic subspecies. Since these snails are relatively conspicuous animals it is unlikely that any more species remain to be discovered. All the species show a considerable amount of inter- and intra-populational variation. There have been no cytogenetic studies, so it is quite uncertain to what extent chromosomal rearrangements have played a role in speciation and in differentiation of populations. The species on Moorea fall into two complexes: the *P. suturalis* group *(P. suturalis,* with several subspecies, *P. t. tohiveana, P. t. olympia, P. mooreana,* and *P. aurantia)* and the *P. taeniata* group *(P. taeniata,* with several subspecies, and *P. exigua,* and probably also *P. mirabilis, P. solitaria,* and *P. diaphana).*

P. suturalis shows considerable geographic variation. It occurs throughout the main mountainous area of the island and is represented in the small isolated Rotui range by the subspecies *dendroica*. Usually dextrally coiled, it is replaced in the southeastern part of its range by the sinistrally coiled race *strigosa;* there is a zone 400–1,600 meters wide in which the populations are polymorphic for direction of coiling (sinistrality is dominant to dextrality in crosses, but the effect of the gene is delayed by a generation, i.e., the phenotype of an individual is determined by the genotype of its mother). *P. tohiveana,* and its race *olympia,* is a sinistral form restricted to high elevations on Mount Tohivea, Mount Mouaputa, and the ridge between them. At some localities *tohiveana* or *olympia* hybridize with *suturalis,* either to a limited extent or sufficiently to produce full intergradation; in most areas where they coexist they remain quite distinct. *P. aurantia,* which occurs on the slopes of Mount Tearai and Mount Teahroa in the northeast of the island, differs from *suturalis* in the structure of the penis, the length of the shell, its pattern and color, and the size of the parietal tooth. But extensive hybridization between the two forms seems to be occurring in some areas, while at other localities they are at least partially isolated. Murray and Clarke (1968) suggested that this was a case of "incomplete speciation," i.e., that the two incipient species had diverged while allopatric, to the point of acquiring some degree of reproductive isolation, and then after having become sympatric had partially merged again.

The evolution of the *taeniata* species group seems to have been similar in most respects to that of the *suturalis* one. The species *taeniata* and *exigua* appear to hybridize in nature in some areas, but in others they remain distinct. These forms are morphologically more distinct than *suturalis* and *aurantia.*

Clarke and Murray (1969) studied the distribution of color and pattern morphs in *P. taeniata nucleola* and found "area effects" similar to those observed by Cain and Currey (1963a) in *Cepaea* in Britain. They were inclined to invoke these as a stage in the acquisition of partial genetic isolation between incipient species, according to the "semi-sympatric" (parapatric) model of Jain and Bradshaw (1966).

Snails of the genus *Partula* do not seem to occupy narrow specialized ecological niches, in the way that many of the Hawaiian drosophilids do. Thus endemic speciation on individual islands (as opposed to speciation by founder events, which must occur as well) is more likely to be micro-allopatric than sympatric in the strict sense. The differences in genetic composition of the populations, sometimes over distances of only 10–20 meters, renders this plausible. But the actual genetic processes involved (whether genic or chromosomal) are totally unknown, and the nature of the genetic isolating mechanisms that prevent hybridization, in the areas where it does not occur, are entirely unknown. There can be no doubt that a conservative taxonomic treatment of these snails would recognize only two species on Moorea, corresponding to what we have called the *suturalis* and *taeniata* species groups. But this would leave us in the position of having to recognize a number of sympatric or partially sympatric subspecies, and it seems more in accordance with the genetic and evolutionary realities of the situation to adopt the concept of groups composed of a number of species, even though a considerable amount of hybridization is occurring in some areas.

The pattern of geographic variation shown by *Partula* on Moorea may be compared with that shown by the land snails of the genus *Cerion* in Cuba and the Bahamas (Mayr and Rosen, 1956; Mayr, 1963a). Whereas *Partula* is distributed in two dimensions (or three, if elevation above sea level is included), *Cerion* has an essentially linear type of distribution, since it inhabits a narrow strip of coastal vegetation just above the high-tide line, extending less than 200 meters inland. Along 50 kilometers of the Banes Peninsula in eastern Cuba seven populations that differ profoundly in shell shape replace one another sequentially (Figure 12). Three areas in which species would otherwise come into contact are areas unsuitable for habitation by *Cerion,* i.e., the sequence is discontinuous; but in the other areas of contact there is a narrow hybrid zone.

In their study of the *Cerion* populations of the Bimini Islands in the Bahamas, Mayr and Rosen were particularly concerned to explain the high variability for certain characters, such as degree of sculpturation of the shell and its ground color, in a few of the local colonies sampled. This variability contrasted with the relative uniformity of other colonies. Mayr and Rosen ascribed it to a hybrid origin, while leaving open the question whether the hybridization was between distinct species or resulted merely from crossing between populations of intraspecific status. Nevertheless, hybridity is not a really plausible explanation for one of Mayr and Rosen's examples, where a highly variable colony is located between two very uniform colonies, which are very similar to each other. Their work was carried out before the extent and universality of genetic polymorphism in natural populations of sexually reproducing organisms was fully appreciated, and today it does not seem necessary to invoke a special explanation for the unusually variable *Cerion* colonies, which in any case show relative uniformity for shell size. The taxonomy of *Cerion* is in a hopelessly confused state; how many species should be recognized is quite unclear. In general, only one *Cerion* population seems to exist at each locality, but there are a few areas where two presumed species coexist sympatrically without hybridization.

Gould, Woodruff, and Martin (1974; and Woodruff, 1975) have attempted, by allozyme studies and morphometric investigations, to bring some kind of order out of this chaos. After detailed study of the *Cerion* population at Pongo Carpet, a locality on Great Abaco in the Bahamas, they concluded that although by traditional taxonomic criteria it would be regarded as a distinct species, it is merely a well-marked local variant of *C. bendalli*. They believe that the number of species in *Cerion* has been greatly exaggerated by taxonomic "splitters," and that the peculiarity of this genus of molluscs is the tendency of its local populations to develop special morphological traits without acquiring re-

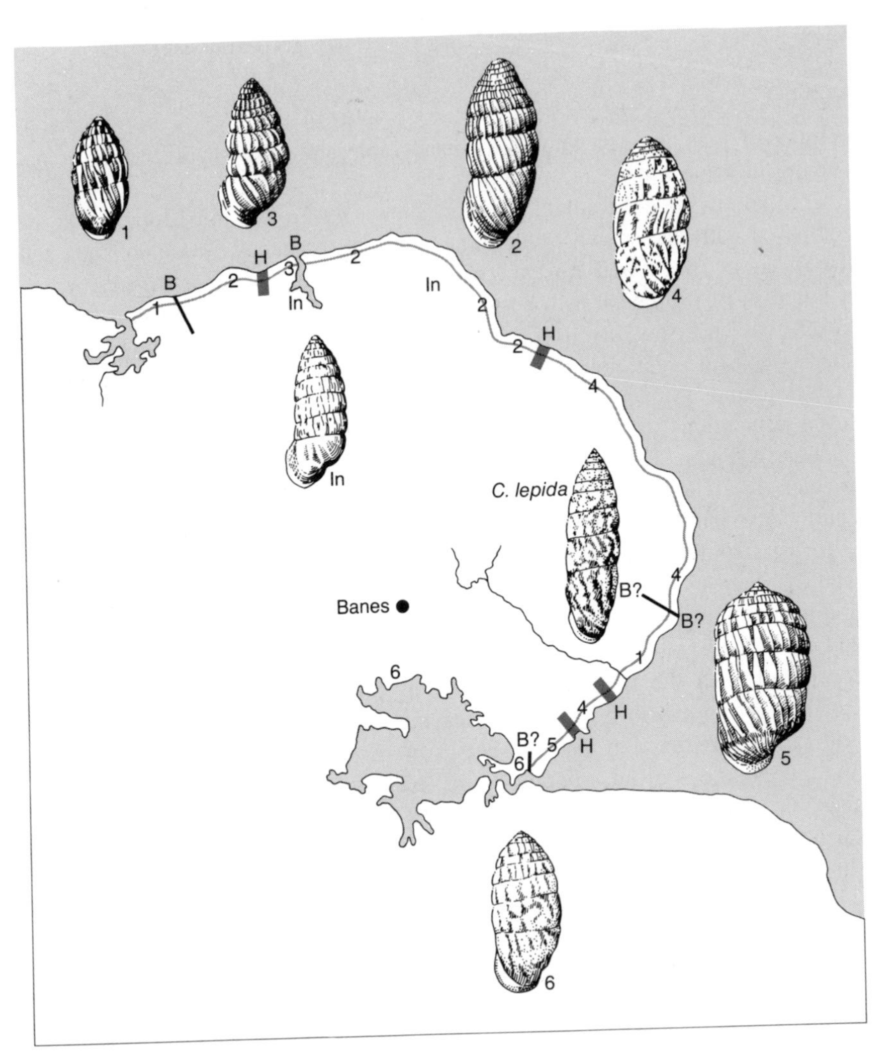

FIGURE 12

Distribution pattern of land snails of the genus *Cerion* on the coastline of the Banes Peninsula, eastern Cuba. Numbers indicate distinctive "races." With one exception, populations hybridize *(H)* where they come into contact, regardless of the degree of phenotypic difference. In some places, contact is prevented by an extrinsic barrier *(B)*. An isolated inland population is indicated by *In*. The very different species *C. lepida* does not interbreed with the other forms. [After Mayr, 1970]

productive isolation. Unfortunately, as yet there have been no studies on the reproductive isolation that must surely exist between the genuine species of *Cerion,* neither have there been any cytogenetic studies.

The probability of the classical dumbbell model of allopatric speciation occurring would seem to be maximal in groups of organisms that are very widely distributed as species but in which the vagility of individuals is not too great. Some wide-ranging species show chains or rings of geographic races. The terminal populations of such a ring may overlap; in certain cases they fail to hybridize or do so only to a very limited extent. The failure of such overlapping populations to hybridize has been regarded by Mayr (1942, 1963a) as "the perfect demonstration of speciation," "a perfect demonstration of speciation by distance," and "although only one of many proofs for geographic speciation . . . a particularly convincing one."

Such circular overlaps have been described in certain species of wide-ranging birds, such as the gulls of the *Larus argentatus* complex, but an insect example is that of the butterfly *Junonia lavinia* in Cuba, where the two end populations overlap without hybridization (Forbes, 1928). Although rare, such cases present a formal problem in that we have a taxon that is a single species by most criteria but whose terminal races behave as separate biological species. Obviously, extinction of a population (or populations) in the middle of the chain would lead to a type of allopatric speciation.

However, things are seldom quite as simple as this, as Mayr (1963a) admits in the case of the *Larus* gulls. It seems that no instance of nonhybridizing overlap has been subjected to a really searching investigation of the isolating mechanisms involved. We do not know how the terminal populations of *Junonia* are prevented from interbreeding, nor is it understood how numerous small and perhaps ineffective barriers to interbreeding between populations in a chain add up to a complete barrier between the terminal populations.

A very special dilemma for believers in the universality of allopatric speciation has always existed with regard to the "species flocks" of certain groups of animals in the ancient lakes of the world, particularly Lake Baikal in Siberia, Lakes Tanganyika and Malawi (Nyasa) in Africa, and a few others (Titicaca in the Andes, Ochrid in Albania). It is characteristic of such lakes (all of which are very deep) that, like oceanic islands, their faunas are "unbalanced," i.e., certain groups are entirely absent or underrepresented while others have undergone speciation on a mass scale. The species flocks of gammarid crustacea in Lake Baikal (240 species) and cichlid fishes in Lake Malawi (200 species) are comparable in the number of species to the drepanid birds and drosophilids of Hawaii. As shown for Lake Baikal in Table 4-2, and for the ancient African lakes in Table 4-3, the great majority of the species are endemic.

The existence of a very large number of closely related species in a small area necessarily raises the question whether sympatric speciation of some kind may not have occurred. It is important to point out that our knowledge of speciation in ancient lakes is very deficient since in no case is there adequate evidence from cytogenetics, experimental hybridization, or biochemical studies.

TABLE 4-2
*Fauna of Lake Baikal. (From Kozhov, 1963.)**

	Total number species	Number endemic in Baikal
Protozoa	317	90
Porifera	10	3
Coelenterata *(Hydra)*	2	1
Turbellaria	90	90
Trematoda	17	6
Cestoda	12	0
Nematoda (free)	10	5-6
Nematoda (parasitic)	81	3-4
Acanthocephala	3	2
Rotifera	48	5
Bryozoa	5	1
Polychaeta	1	1
Oligochaeta	62	45
Hirudinea	17	10
Copepoda–Calanoidea	5	1
Copepoda–Cyclopoidea	25	16
Copepoda–Parasitica	13	2-3
Copepoda–Harpacticoidea	43	38
Ostracoda	33	31
Cladocera	10	0
Malacostraca (Bathynellidae)	2	2
Isopoda *(Asellus)*	5	5
Amphipoda (Gammaridae)	240	239
Acarina	6	3
Tardigrada	1	?
Trichoptera	36	13
Plecoptera	2	?
Chironomidae	60	11
Anoplura	1	0
Gastropoda	72	53
Pelecypoda	12	3
Pisces (Cottoidei)	25	23
Other fishes	25	0
Mammalia	1	1
Total:	1219	708

*The figures in this table are somewhat different from those given by Brooks (1950), which were based on the earlier data of Vereshchagin.

The whole problem of speciation in ancient lakes was reviewed by Brooks (1950), who concluded that all cases could be accounted for by the classical allopatric model. Brooks' review, which seems to be based on a careful study of the literature but without benefit of original investigation, has been widely quoted. But as far as the fauna of Lake Baikal is concerned, there has been much further work, summarized by Kozhov (1963) whose conclusions are very different.

TABLE 4-3
Fish of the great lakes of Africa. (From Fryer and Isles, 1972.)

	Cichlid species		Non-cichlid species	
	Total	Endemic	Total	Endemic
Malawi	200+	All except 4	42+2?	26+2
Tanganyika	126	126	67	47
Victoria	170	All except 6	38+1?	17

The situation has been most thoroughly studied in the case of the Lake Baikal gammarids (Figure 13). The great majority of the species, which have been assigned to 35 genera, are bottom dwellers, but they occupy all zones of Baikal, from the littoral to the abyssal depths; there are some burrowing forms in both the littoral and abyssal zones. According to Kozhov they have probably evolved from four or five ancestral species that populated the original lake.

There are several theoretical possibilities regarding the speciation of the Lake Baikal endemics. In the first place this might have taken place outside the lake, in river systems draining into it, or in smaller lakes now or formerly connected with it. Alternatively, at various times in the past the lake may have been fragmented, smaller lakes on its edges being separated from the main body of water. Both these models are essentially allopatric ones; they are not sharply distinguishable, and various combinations of them or intermediate situations are conceivable. The third possibility, in view of the depth gradient (0 to 1,741 meters), is that the speciation process was intralacustrine, i.e., occurred in Lake Baikal itself. While intralacustrine speciation is not necessarily sympatric, (it may occur over a very wide area), nevertheless if it occurs in a small enough area and within a sufficiently limited time span, there is a strong possibility that it is at least largely sympatric. Mayr (1942, 1947) originally argued in favor of multiple colonization of one type or another as the explanation for the evolution of the gammarid and other species flocks of Lake Baikal. Brooks, however, while still clinging faithfully to the dogma of the universality of allopatric speciation, understood that this was unrealistic and not in accordance with the evidence. He emphasized that many of the gammarid species are restricted to geographically isolated and environmentally peculiar regions of Baikal and concluded: "The only reasonable explanation of their distribution pattern is that populations of these benthic organisms have been subdivided within the lake through the influence of geographical features. During their isolation these subpopulations have undergone genetic changes often of sufficient magnitude to effect reproductive isolation when the geographical isolation breaks down" (Brooks, 1950, p. 170).

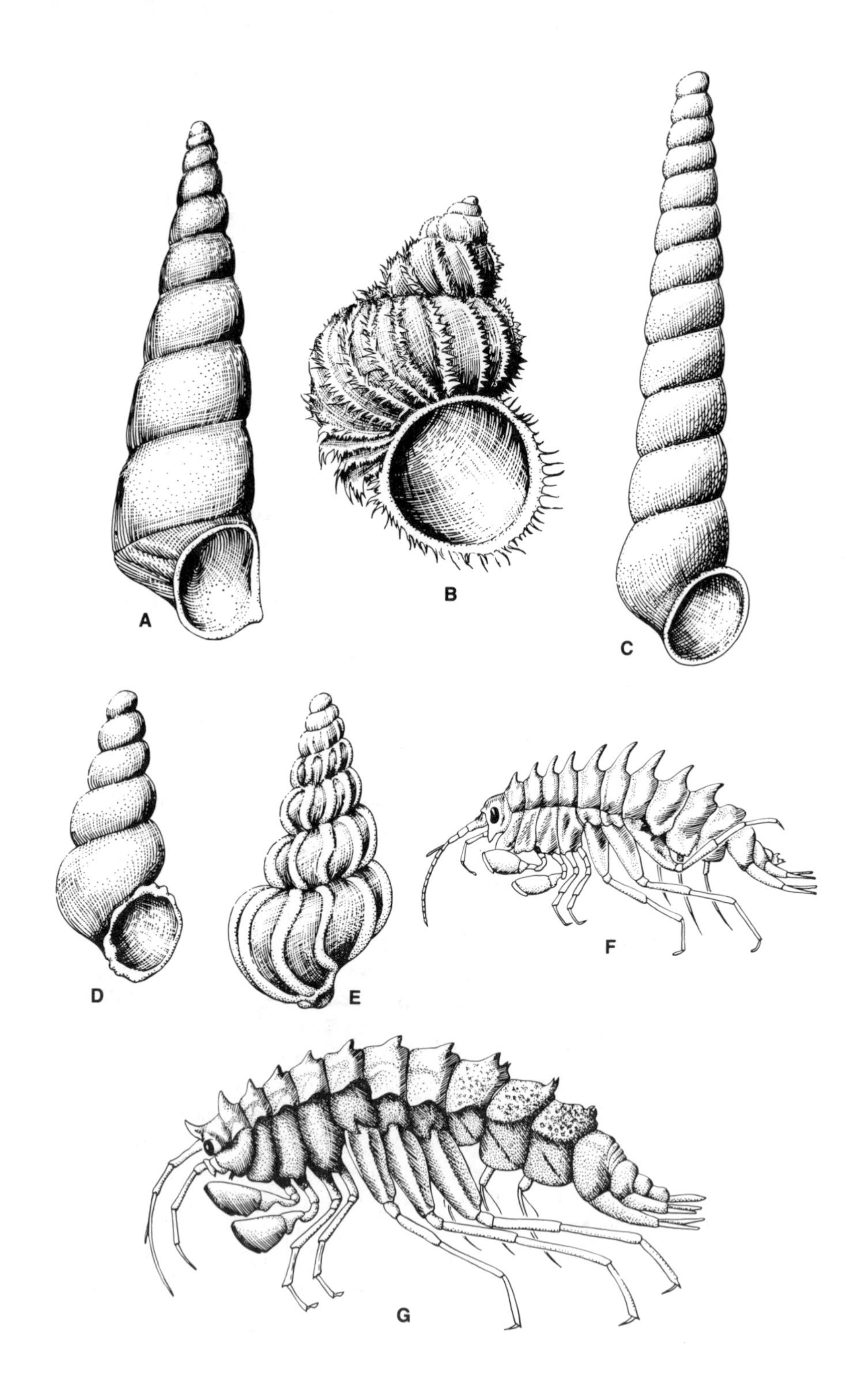
A
B
C
D
E
F
G

Kozhov (1963), who had a long direct association with Lake Baikal and its fauna, believes that both allopatric and sympatric speciation have occurred. He stresses the large number of deep-water and abyssal species and the frequent close relationship of littoral and deep-water species pairs and concludes:

> The development of forms in Baikal on the basis of isolation did and does take place. . . . Allopatric speciation must also have been favored by the fact that at one time Baikal was the centre of the giant Baikal system of tectonic troughs. . . . But a deeper analysis of the composition and distribution of the Baikalian fauna in the lake itself and outside of it leads one to believe that the main part of its evolution from relatively few parental forms was played by the gradual deepening of the lake, the resulting appearance of new biotypes, their colonization and, as a consequence of it, the gradual splitting of species into littoral and deep-water forms without any marked participation of spatial geographic isolation of the diverging populations, i.e., by the sympatric way. [p. 294]

Kozhov argues further that "spatial geographic isolation could play only a secondary part in the evolution of Baikal" because the species of such lakes as those of the Gyda watershed and Lake Taimyr (formerly connected with Baikal) differ very little from the Baikal species and in many instances are identical. He notes also that species found in the "vestiges of old lakes of the Baikal system," and isolated even longer, differ from Baikal species "only within the limits of very closely related species and varieties." By contrast with this situation, one finds a great wealth of endemic Baikalian species existing at different depths. Thus, of the gammarids, 17.1 percent are found at 0–5 meters, 49.8 percent at 5–70–150 meters, 14.8 percent at 70–150–300 meters, and 18.7 percent at 300–500 meters.

Kozhov's views must command respect, both on account of his obviously deep knowledge of the Baikal fauna and his judicious approach. Nevertheless, it is clear that he does not use the term "sympatric" in precisely the sense used in this book. Clearly, what he has in mind is the gradual formation of species, at different depths below the surface, from an originally continuous population. The actual processes may have involved the segregation of species out of a population forming a cline down the gradient or it may have involved area effects of the type discussed in Chapter 5. In neither case would the process have been allopatric, but it should probably also not be regarded as sympatric, at least in the restricted sense.

FIGURE 13

Endemic species of molluscs and gammarid crustaceans from Lake Baikal. **A,** *Baicalia carinata:* **B,** *B. ciliata;* **C,** *B. godlewskii;* **D,** *B. korotnewi;* **E,** *B. costata;* **F,** *Acanthogammarus falvus;* **G,** *A. maximus.* [After Kozhov, 1963]

The faunas of the great lakes of Africa have frequently been compared with that of Baikal. Lakes Malawi (Nyasa) and Tanganyika occupy tectonic troughs and, like Baikal, are long, narrow, and deep. They are also ancient lakes, but exactly how old is unknown (according to one estimate Lake Malawi is between one and two million years old, Baikal 50 million years old). Lake Tanganyika, which is probably older than Malawi, has a maximum depth of 1,470 meters, while Malawi has a depth of 704 meters. A fundamental difference between these lakes and Baikal is that the latter has a twice-yearly circulation of its upper and lower layers of water, so that even the deepest layer is well-oxygenated and thus capable of supporting a rich abyssal fauna. In the tropical African lakes, however, the water stratification is permanent; the depths are completely anaerobic and even poisoned by hydrogen sulfide and methane, so that they are not inhabited by any higher organisms. Lake Victoria, which is much the largest of the great African lakes and the most recent in origin (area 68,800 square kilometers), is quite shallow, having a maximum depth of 80 meters.

There is abundant information concerning the very rich fish faunas of these lakes, but data on the invertebrate groups is much more fragmentary and for the most part based on rather old taxonomic studies that are unsatisfactory by modern standards. The species flocks of cichlid fishes in the African lakes are comparable to the gammarids of Lake Baikal. These cichlids show an amazing variety of size and form (Fryer and Isles, 1972); the smallest species is 3.8 cm long and the largest 80 cm.

The feeding habits of the cichlids are also extremely diverse. Fryer and Isles distinguish between:

Phytoplankton feeders
Epilithic algal feeders
Leaf choppers
Feeders on insects and other benthic Arthropods
Zooplankton feeders
Piscivorous species
Deposit feeders
Periphyton collectors
Mollusc feeders
Feeders on scales of other fishes
Fish egg, embryo, and larvae eaters

These specialized feeding habits are especially characteristic of the great genus *Haplochromis* and genera derived from it. Members of the genus *Tilapia,* by contrast, are ecological generalists.

Multiple colonization from different river systems has undoubtedly occurred in the case of all these lakes, e.g., the two species of *Tilapia* in Lake Victoria are derived from different stocks. But it has clearly been only a minor contributor to the species flocks of cichlids, which have originated, in the main, by intralacustrine speciation involving adaptive radiation to a great variety of ecological niches. As might be anticipated from the very different geography of Lake Victoria, it probably shows very different patterns of speciation from the "deep" lakes, Malawi and Tanganyika.

Although a number of aspects of the ecology and general biology of the African cichlids have been controversial (Fryer, 1959a, b, 1960a, b; Myers, 1960b) there seems to be a general consensus that these fishes exhibit an amazing diversity of trophic and other specializations, and that their speciation has, in general, been intralacustrine and allopatric, facilitated by a great diversity of habitats. However, Fryer (1960b) insisted that "the rocky and sandy shores of Lake Nyasa are very uniform habitats, and that, as these two major types dominate the littoral zone, the number of habitats is few." Later (1972) Fryer and Isles emphasized the fact that many species are confined to the littoral zone and consequently form a narrow band fringing the lake. Because of the frequent alternation of sandy and rocky stretches of coastline, each species is "broken up into innumerable populations of various sizes which are isolated to a greater or lesser extent." While this is undoubtedly the case, we have far too little information on the extent of migration between these demes or even the permanence of the shoreline ecology, which may have changed considerably in Lake Malawi, where the water level rose 5.8 meters between 1916 and 1963. Also, it is not really sufficient to describe the coastline as consisting of alternating stretches of rocks and sand. We need to know how steeply shelved the bottom is and to what extent rocky and sandy stretches of coast separated in shallow water by stretches of different character may be connected in deeper water.

One line of evidence Fryer and Isles rely on to a considerable extent to support their contention that cichlid intralacustrine speciation is essentially allopatric is the geographic localization of color morphs along the coastline, in species such as *Pseudotropheus zebra* in Malawi and *Tropheus moorei* in Tanganyika (Fryer, 1959b; Marlier, 1959; Matthes, 1962). They regard the various color forms of *T. moorei,* which inhabit different stretches of coastline (each about 50 kilometers long) at the northern end of Lake Tanganyika, as geographical subspecies and, by implication, potential species. There can be no doubt that the bright colors of these fishes ("black," "orange," "red and yellow," "green and yellow," etc.) are important in sexual recognition and serve as cues in complex premating behavior, which in some cichlids may involve the construction by the males of "sand scrape nests."

Whether the local populations of *T. moorei* can legitimately be regarded as different subspecies is a question that must await further studies, particularly of their allozymes. If they exhibit the sort of genetic differences that populations of *Mus m. musculus* and *Mus m. domesticus* do (see p. 206), then we would certainly be justified in regarding them as potential species; but there is no certainty that this is so. Moreover, the role of evolution of sexual recognition patterns in speciation is by no means clear. No populations of *T. moorei* seem to be polymorphic for these color patterns. Fryer and Isles have suggested that a population at Kashikezi, which is intermediate in color pattern between ones to the north and south of it (at Luhanga and Bemba, respectively), arose by hybridization between them. They also claim that a very closely related species, *T. duboisi,* which coexists with *T. moorei* at Bembra, apparently without interbreeding, arose from a population of *T. moorei* occupying a limited stretch of coastline, which was then reinvaded by another population of *T. moorei,* resulting in the present sympatry between the two. Their data, however, strongly suggest another interpretation. *T. moorei* adults live in shoals in very shallow water (0.5–1 meter), while *T. duboisi* occupy rock crevices at depths of 3 to 12 meters. If one is not committed to the universality of the allopatric model of speciation, one would certainly conclude that the splitting off of *T. duboisi* from the parental species took place in the two depth zones, in the manner that we have suggested was important for speciation of the Lake Baikal gammarids. A similar situation seems to exist in the case of two very closely related cichlids of Lake Malawi, *Pseudotropheus zebra* and *P. livingstonei,* which may have speciated in the same manner.

Geographical subspeciation clearly *has* occurred in a number of the Malawi and Tanganyika cichlids, however. For example, *Callochromis macrops* has a clearly marked subspecies at the northern end of Tanganyika (Poll, 1956). In the same lake, *Ophthalmochromis ventralis* has two subspecies, which differ markedly in dentition, although it is true they are connected by intermediates (Poll, 1956; Poll and Matthes, 1963).

Conditions for speciation seem to have been very different in shallow Lake Victoria from what they have been in the deep lakes (Malawi and Tanganyika). On geological grounds, it is probable that the Victorian basin contained several separate lakes at various times in the past. Thus, the huge flock of *Haplochromis* species may well have evolved in considerable part in these smaller lakes by classical allopatric speciation. However, Fryer and Isles (1972) go too far in claiming that no intralacustrine speciation has ever occurred in Lake Victoria. The likelihood that separate bodies of water were a factor in fish speciation in the Lake Victoria basin is strongly indicated by the evidence of the fauna of the small Lake Nabugabo (about 30 square kilometers), which at present is separated from Lake Victoria by a forested sandbar and a belt of swamp. Of the six known species of *Haplochromis* in Lake Nabugabo, five appear to be endemic but also closely related to the Lake Victoria species (and in some cases also to species endemic to Lake Edward or Lake Kivu).

Lake Ochrid, on the Albanian–Yugoslav frontier, is a much smaller lake than those previously discussed (270 square kilometers, 286 meters deep), and probably dates from the late Pliocene. It contains a number of species flocks that must have arisen by intralacustrine speciation (Polinski, 1932; Stanković, 1932, 1960; Brooks, 1950; Hubendick, 1960), including three ostracod, one turbellarian, several oligochaete, and two or three molluscan flocks. None is very rich in species, each having from three to five endemic species. Evaluation of the evolutionary situation is difficult in the absence of precise knowledge about the distribution of the various forms in the lake, their taxonomy, and (in the case of the turbellarians and oligochaetes) their reproductive mechanisms. There are only two gammarid species, which probably represent separate invasions.

The ancylid molluscs (freshwater limpets) of Lake Ochrid, studied by Hubendick (1960), comprise two genera. The genus *Ancylus* is represented by three closely related endemic species, *lapicidus, tapirulus,* and *scalariformis,* which presumably evolved from the widespread Palearctic *A. fluviatilis,* found in a spring draining into the lake. The genus *Acroloxus* is represented by two endemic species, *macedonicus* and *improvisus,* which must also have arisen from another widespread Palearctic ancestor, *A. lacustris,* of which two specimens were found on the shore of Lake Ochrid (presumably the endemic species effectively exclude their ancestors from the main body of the lake).

Hubendick has tried to explain intralacustrine speciation in these molluscs on the basis of allopatry, but prefers to speak of "spatial isolation" or "microgeographic isolation" on account of the very short distances involved. He recognizes four bathymetric zones in Lake Ochrid: (1) 0–7 meters, a littoral zone with a vegetation of *Phragmites, Potomageton,* etc.; (2) 7–18 meters, a zone of *Chara* meadows: (3) 18–35 meters, a zone with living and dead shells of the mollusc *Dreissena;* and (4) 35–286 meters, the profundal zone. *Ancylus lapicidus* and *Acroloxus macedonicus* inhabit the littoral zone, while *Ancylus tapirulus, Ancylus scalariformis,* and *Acroloxus improvisus* inhabit the shell zone. Ancylids are not found in the intermediate *Chara* zone. Consequently, Hubendick argues that the *Chara* zone could have functioned as a geographic isolating barrier, facilitating speciation (perhaps *A. tapirulus* and *A. scalariformis* represent the results of a "double invasion" of the shell zone by *A. lapicidus* from the littoral zone). Hubendick also points out that *Acroloxus macedonicus* is restricted to a special localized habitat (limestone boulders at a depth of 30–50 cm) and that its distribution is discontinuous. The shells of individuals from "site I" have a system of conspicuous radial ridges, whereas at "site II" the ridges are almost or totally absent. The two sites are probably about 10 km apart. Hubendick believes that the two populations diverged genetically as a result of geographic isolation, and argues from this that the intralacustrine speciation of the Lake Ochrid molluscs has been essentially allopatric.

Hubendick's conclusions in the case of *Acrolux macedonicus* seem sound, but his hypothesis that the *Chara* zone was a factor in isolating ancylid species must be regarded as more doubtful, in view of the much shorter distance involved. This zone is only about 11 meters wide and we have no information as to how continuous it is or whether it really forms an impenetrable barrier to the dispersal of ancylids at all stages of their life cycle. The occurrence of two species of *Ancylus* in the shell zone, living together on the shells of the family Dreisseniidae is peculiar, since we would have expected competitive exclusion to operate. Although Hubendick has presented *prima facie* evidence that spatial separation was a factor in the intralacustrine speciation of these molluscs, one may well ask where the line should be drawn between allopatric and sympatric speciation when the width of the isolating barriers is of the order of 10 to 100 meters.

Other molluscs of Lake Ochrid include three endemic species of the prosobranch genus *Pyrgula*. *P. macedonica* is said to occur at depths of 5–20 meters (littoral and *Chara* zones), *P. pavloviči* at 20–30 meters (shell zone), and *P. wagneri* below 50 meters (profundal zone). It would appear that this is another clear case of intralacustrine speciation due to a depth gradient, although in this instance the *Chara* zone has not had any isolating role.

An extremely interesting, but poorly known, instance of intralacustrine speciation is that of the cyprinid fishes of Lake Lanao in the island of Mindanao (Myers, 1960a). In contrast to the ancient lakes we have been discussing, Lanao is a young lake. Its age is not accurately known but has been estimated by one authority as only 10,000 years (it was clearly formed by the damming of its basin by one or more volcanic eruptions). Its area has been variously reported as 375 or 900 kilometers and its depth is somewhere between 180 and 280 meters.

There are 18 cyprinid species in Lake Lanao, all of which almost certainly evolved from *Barbus binotatus,* the most abundant and widespread cyprinid fish of the Sunda archipelago. Races of *B. binotatus* occur on Mindanao, but apparently not in Lake Lanao or elsewhere on the Lanao plateau. The 18 Lanao cyprinids include a species flock of 13 species that has been assigned to a genus called *Barbodes* (regarded by Myers as "not easily distinguished from . . . *Barbus*"). The other five species are assigned to four endemic genera characterized by peculiar jaw modifications but obviously derived from *Barbus*.

Any suggestions as to the exact mode of speciation of the Lake Lanao fishes would be mere speculation in the absence of solid evidence. Clearly, however, this is a case of "explosive" intralacustrine speciation, since at least 18 species (there may be some undiscovered ones) were produced from a single ancestral species in a remarkably short space of time.

The earliest interpretations of speciation in ancient lakes held that it was a special type of evolutionary phenomenon and, by implication, not allopatric (Woltereck, 1931; Herre, 1933; Worthington, 1940). Mayr (1942) suggested, however, that the species flocks of ancient lakes might owe their origin to multiple colonizations from different river systems—an explanation that has been generally rejected. In support of his "multiple colonization" model, Mayr (1942) claimed that "students of freshwater faunas have vastly underestimated the age of the species with which they are working." He thus rejected the notion that the speciation of species flocks in isolated lakes was explosive. In the case of the quite recent Lake Lanao, on an isolated plateau with no drainage from the outside, he was clearly wrong. The evolution of 18 or more species of fishes, including some highly divergent types, from one ancestral species in a period of the order of 10,000 years is explosive by any standard. The *Haplochromis* species flock of at least 170 species in Lake Victoria probably evolved from a single ancestor in about half a million years (Fryer and Isles, 1972). It is true that subdivision of Lake Victoria into a number of much smaller lakes in the past would have greatly facilitated allopatric speciation, but the time scale of speciation is almost as explosive as in the case of Lake Lanao.

Intralacustrine speciation in isolated lakes thus seems to be a reality; only the precise mechanism (or mechanisms) need further elucidation. In spite of the very extensive literature on this question, not a single case has been investigated in depth using modern techniques. The review of the literature by Brooks (1950), so heavily relied upon by Mayr (1963a, 1970), really does no more than demonstrate the diversity of habitats in lakes such as Baikal and Malawi and the fact that the various species of animals in these lakes are localized in certain regions, restricted depth zones, or specialized habitats.

It is extremely unfortunate that absolutely no cytogenetic studies have been carried out on any of the species flocks of ancient lakes. It is entirely possible that the problems of gammarid speciation in Lake Baikal or cichlid speciation in the African lakes might appear very differently if such investigations were performed. We simply do not know enough about these cases to predict whether karyotypic uniformity or diversity would be found.

It seems quite likely that intralacustrine speciation has followed a number of different courses. There is no *a priori* need to suppose that it must have been of the same type in the Baikal gammarids, the freshwater limpets of Lake Ochrid, and the cichlid fishes of Lake Victoria. That minor geographic barriers, such as alterations of sandy bays between rocky capes along the shoreline, may indeed in certain cases have played a role by causing isolation seems very probable. But to argue from this that all intralacustrine speciation is allopatric involves a number of assumptions. In particular, it implies that migration between local demes is negligible and that the demes retain their distinctness over periods of time long enough for them to attain species status. Those who argue in support of the universality of allopatric speciation seem to be using the term "geographic isolation" in quite different senses in the separate cases of terrestrial organisms and the faunas of ancient lakes. Almost all terrestrial organisms have their populations fragmented into local colonies, herds, or demes, but no one has claimed that their speciation must be allopatric because of this. For true allopatric speciation to occur there must be a geographic isolation of an entirely different order of magnitude. This point has been realized by Hubendick, who proposes that one should speak of "spatial isolation" rather than "geographic isolation" in the case of intralacustrine speciation. This seems a step in the right direction. But an entomologist dealing with a case of speciation that involved a switch in host plant from oak trees to pine trees (as seems to have happened in the case of the grasshopper *Dendrotettix australis,* see p. 239) would claim that this was a case of sympatric speciation if the oak and pine trees occurred in the same forested region. He would not argue that it must be regarded as allopatric because an oak tree and a pine tree do not occupy precisely the same location in space. Nevertheless, it must be admitted that what is important from the standpoint of speciation is not the actual distance in meters or kilometers between two populations of a species, but the "biological distance," i.e., the amount of migration between them. If because of different modes of life two populations are so isolated that no migration occurs between them in 10^5 or 10^6 years, they are effectively allopatric, whether situated 10 or 10,000,000 meters apart. Thus, it is our almost total ignorance of the vagility of freshwater species that is the root cause of our difficulties in understanding their modes of speciation.

One word of warning is necessary in connection with apparent intralacustrine speciation of some invertebrate groups, such as turbellarians, oligochaetes, and melaniid snails. Various forms of parthenogenetic reproduction are relatively widespread in these groups, and it is conceivable that some of the species flocks that have been reported are complexes of thelytokous biotypes, possibly complicated by polyploidy.

The work on the faunas of ancient lakes in the 25 years since Brooks's review has clearly shown that intralacustrine speciation has occurred on a large scale in the lakes that have been studied. Some degree of geographical separation of habitats ("allopatry") has probably been involved in every single case. But the separation in many cases has probably been of the order of only a few meters or tens of meters. Whether we should really call these instances allopatric or sympatric speciation is open to question. Certainly, speciation due to spatial distribution in depth gradients seems to have been extremely important in a great variety of cases (in gammarids, molluscs, and perhaps some fishes) and was insufficiently appreciated by earlier workers. Of the lake faunas discussed here, only that of Lake Victoria is comparable in a sense to the faunas of archipelagos such as the Hawaiian and Galápagos Islands (or would be if the islands of these archipelagos had become joined together). If there is a valid comparison between such lakes as Baikal, Malawi, and Tanganyika and oceanic islands it is with Saint Helena and Rapa, whose faunas will be discussed in Chapter 7. Nevertheless, there are many differences between the problems of speciation in ancient lakes and on oceanic islands (see p. 245), and they are best discussed separately.

Speciation mechanisms in marine organisms are, in general, poorly understood. Species of such wide-ranging fish as tunas would seem to be forms in which allopatric speciation due to complete geographic isolation could not be expected. Nevertheless, considerable genetic differentiation into geographic subpopulations (subspecies or semispecies?) seems to have occurred in a number of these oceanic fishes. For example, the albacore, *Thunnus alalunga,* has genetically distinct populations in the North Pacific, South Pacific, Indian Ocean, Atlantic, and Mediterranean (Fujino, 1970).

It is important, however, to distinguish between genuinely distinct populations, which preserve their differences throughout the life cycle (which is probably the case in the albacore), and a single Mendelian population that is widely dispersed by ocean currents during planktonic egg and larval stages and gives rise to a number of genetically different adult populations because of diverse selection pressures in different parts of its range. The latter kind of process, i.e., selection before the adult stage renewed in each generation, probably accounts for the geographic variation in hemoglobin allele frequencies in cod populations off the coast of Norway (Frydenberg *et al.,* 1965). A particularly significant case is that of *Anguilla rostrata* (American eel), whose populations, from Newfoundland to Florida, show considerable differences in allozyme frequencies (Williams *et al.,* 1973) in spite of the fact that all spawning occurs in a small area of the Atlantic northeast of the West Indies. The larvae disperse from this area and metamorphose into adults that move into estuarine or fresh waters, where they live for 10–20 years before returning to the breeding area, where they spawn and die. The selection pressures in this instance must be particularly intense, since it has been calculated that differentials of around 10 percent per generation exist for a number of gene loci. Yet the species shows no geographic variation in meristic characters, except that the number of vertebrae varies in individuals, from 104 to 111. The absence of geographic meristic variation is a strong argument against the existence of any true geographical races. Whether the situation in the European eel is essentially the same is controversial. It seems fairly certain that the two species are distinct, but they seem to breed in the same general area of the Atlantic, between 22–30° N and 48–65° W, and there is no adequate explanation for the fact that the American eel does not reach European coasts, or vice versa.

Some shallow-water marine organisms have planktonic larvae capable of wide dispersal. Mayr (1954) has discussed the situation in the tropical sea urchins. Many of the species are distributed over vast areas, such as the whole of the Indian Ocean together with the western Pacific, but a few are restricted to oceanic islands or archipelagos, e.g., *Eucidaris galapagensis,* from the Galápagos, and *Diadema paucispinum* from the Hawaiian Islands. As a general rule the species or geographic races replace one another allopatrically, a pattern that Mayr regards as strong evidence for classical allopatric speciation. For example, in the rock-boring echinoids of the genus *Echinometra, E. mathaei* inhabits the Indian Ocean and the western Pacific, while *E. vanbrunti* occurs on the Pacific coast of North and South America between northern California and Peru. *E. insularis,* a morphological intermediate between these two species, occurs in shallow waters surrounding the islands of the eastern Pacific (the Revilla Gigedo group, the Galápagos, and Easter Island). *E. lacunter* inhabits the warmer parts of the Atlantic (roughly between 30° N and 30° S). The only instance of sympatry in the genus is between *E. lacunter* and *E. viridis;* the latter is confined to the Caribbean area and has a very different ecology (it is not a rock borer). In the genus *Diadema, D. mexicanum* has a range that is almost exactly the sum of the ranges of *E. vanbrunti* and *E. insularis;* its distribution probably resembles one that existed in *Echinometra* before divergence into "continental" and "insular" species occurred.

Fragmentation of the range of a species—of a type liable to lead to allopatric speciation—may result from many causes. Elementary accounts tend to stress major geological phenomena: the separation of islands from nearby continents, e.g., Great Britain or Tasmania; the development of deserts in formerly fertile regions, e.g., the Sahara; and other such dramatic geological transformations. But competition between related species may also lead to separation of allopatric populations, which may subsequently diverge and become distinct species. Most species are capable of adapting to a considerable range of habitats. They may fail, however, to adapt to a habitat occupied by a better-adapted competing species. A good example is provided by the wingless grasshoppers of the genus *Psednura* (Key, 1972). The range of the coastal species *P. musgravei* in eastern Australia is divided into northern and southern areas by a stretch of coastline occupied by *P. pedestris*. The very narrow zone of overlap (11 meters) between *pedestris* and the northern race of *musgravei* has been studied by Key and Balderson (1972). Although the habitats occupied by the two species are somewhat different (*pedestris* is found in somewhat wetter situations), mixed populations do occur, apparently without hybridization.

Brown (1957) put forward a general model of geographic speciation ("centrifugal speciation") that is fundamentally different from the orthodox allopatric models. Although his attempt to provide an alternative model must now be regarded as only partially successful, certain features of it do have some merit. As an entomologist, Brown realized that strictly allopatric isolation by physiographic barriers cannot "account for all the overlapping complexities of many continental faunas, or the species swarms in ancient lakes, for instance." His model includes two main features: "emphasis on the center of a species' distribution as the principal source of evolutionary change leading to 'potent' new species and higher categories," and "the role of population density fluctuations in spreading characters and making and breaking the contacts between populations."

Instances in which all the geographically peripheral populations of a species are more similar to one another than they are to more centrally located populations were regarded by Brown as evidence for his model. Some of these are undoubtedly genuine and may possibly be interpreted in terms of the "stasipatric" model of speciation (White, Blackith, Blackith, and Cheney, 1967; White, 1968, 1974), inasmuch as the stasipatric model also envisages genetic changes (specifically, chromosomal rearrangements) establishing themselves well within the range of the species (i.e., in "central" locations) and giving rise to incipient species in nonperipheral (or at any rate "not necessarily peripheral") areas. However, some of Brown's examples seem rather forced, or are based on insufficiently studied cases. *Ceuthophilus uhleri* (camel cricket), very carefully studied by Hubbell (1956), is cited by Brown as evidence for his model. But Hubbell clearly showed that the central populations of this species, which differ from the peripheral populations in two characters of the external anatomy and two of the genitalia, inhabit the elevated terrain of the Cumberland Plateau and southern Appalachians. Thus, on physiographic grounds it is not surprising that they differ from the lowland populations. This pattern of variation is superimposed on a north-south size cline. A similar case occurs in *Ceuthophilus seclusus;* in the mountains of the Ozark region the males have a striking "deformation" of the pronotum, but as one moves away from the Ozarks the populations of the surrounding areas increasingly have "normal" pronota.

In Brown's model the distribution of a species is assumed to undergo successive expansions and contractions as a result of climatic or other changes. But the "largest continuous or near-continuous" area of occupation of a species is supposed to be the "chief fountainhead of its store of variation and general adaptive improvement." It is regarded by Brown as the "evolutionary center" of the species. The stasipatric model, while also claiming that new incipient species may arise within the main area of occupation of the original species, makes no assumptions about periodic fluctuations in population size, range, or density. It simply states that most chromosomal rearrangements leading to stasipatric speciation are likely to arise in nonperipheral populations because in fact only a minute fraction of all the individuals of the species are peripherally located.

The principle of character displacement, formulated by Lack (1947) and stressed especially by Brown and Wilson (1956), has sometimes been said to have originated in Darwin's more general concept of "divergence of character." In its modern form the principle simply states that in areas where closely related species overlap geographically they tend to be more different in morphology, ecology, or behavior than do individuals of the same two species from those areas where only one species occurs. There are, however, two very different types of character displacement. In the first—character displacement in the classical sense—genetic isolating mechanisms are not affected or involved. We are dealing simply with the fact that under conditions of sympatry of related species there is apt to be strong selection for specialized occupation of different ecological niches, tending to produce the condition of ecological exclusion that Mayr (1963a) has claimed is essential for producing full species. An example of such a case is the subterranean skink species *Typhlosaurus lineatus* and *T. gariepensis* in the Kalahari desert (Huey and Pianka, 1974). The range of *gariepensis* (the smaller of the two species) is entirely included within that of *lineatus,* whose linear measurements and prey size are larger in areas where the two species coexist than in areas where *gariepensis* is absent. This supports the hypothesis that, at least in the female, morphological and behavioral character displacement has occurred that reduces competition between the species for food.

This kind of character displacement is probably a very general phenomenon, and can be expected to occur regardless of whether the zone of overlap is primary or has arisen secondarily by range extensions of one or both taxa following a period of allopatry. Nursall (1974) has looked for instances of morphological and behavioral character displacement in 10 families of fishes. He found that it was frequently but not always detectable, and that it was relatively simple under conditions of r-selection, but complex, involving interactions between several species, under conditions of K-selection. A special case of character displacement is divergence of the breeding seasons of two related, sympatric species, which minimizes overlap of the periods when immature individuals of both species are present (Hutchinson, 1959). This may occur for ecological and trophic reasons alone, and have nothing to do with genetic isolation of the two species (which in any case may be completely isolated by other mechanisms).

The second type of character displacement involves the strengthening of genetic isolating mechanisms under conditions of sympatry, and is clearly different in principle from the other types we have been discussing. This is a somewhat controversial subject. The idea that genetic isolation may be greater between two sympatric species than between two allopatric species dates from the work of Dobzhansky and Koller (1938), in which it was shown that *Drosophila pseudoobscura* is more strongly isolated from *D. miranda* in areas where the two are sympatric than in areas where *miranda* does not occur (there are no areas where *miranda* occurs without *pseudoobscura,* so that the reciprocal situation does not exist). The explanation of this effect is that if interspecific hybrids are relatively inviable or sterile (which is the case in crosses between *miranda* and *pseudoobscura*), selection should operate to prevent such hybrids occurring, since their production decreases the reproductive potential of their parents.

Blair (1974) speaks of geneticic "reinforcement" of premating isolating mechanisms in areas of sympatry. He and his associates have described a number of examples of this phenomenon in their studies of amphibian mating calls (male calls, to which the females are attracted). The frog species *Gastrophryne olivacea* and *G. carolinensis* have an overlap zone about 200 kilometers wide in east Texas (Blair, 1955). In comparisons of the mating calls of the two species from the zone of overlap with those of *olivacea* from Arizona and *carolinensis* from Florida there is good evidence for character displacement. However, in comparisons with *olivacea* collected 80–500 kilometers west of the overlap zone and *carolinensis* from 8–100 kilometers to the east of it, the evidence for true "reinforcement" in the overlap zone looks much weaker.

A more convincing example is that of the frogs *Acris gryllus* and *A. crepitans*. Although mating calls of individuals of the former species from Florida and of the latter from Texas are remarkably similar (Blair, 1958), the calls of individuals of the two species from a mixed population in Mississippi are very different. On the other hand, Blair found no evidence of "reinforcement" of call differences in the species pairs *Bufo compactilis–B. cognatus, Hyla cinerea–H. gratiosa,* or *H. versicolor–H. phaeocrypta.*

Littlejohn (1965; and see Littlejohn and Loftus-Hills, 1968) has shown that *Litoria ewingi* and *L. verreauxi* in eastern Australia have almost indistinguishable mating calls except where they are sympatric. The area of overlap in this case is a great arc extending 750 kilometers east and north from Melbourne and 40–90 kilometers wide. This seems to be a very clear case of "reinforcement" of an ethological isolating mechanism in a zone of sympatry. It is interesting that the modification of the call in the overlap area is much more profound in *L. verreauxi* than in *L. ewingi*. More recently, Watson, Loftus-Hills, and Littlejohn (1971) have described a third species of this complex from northern Victoria. Its call is distinct from that of *L. ewingi,* but since females of the latter are attracted to both calls, the difference is probably ineffective as an isolating mechanism. The two species are genetically isolated, however, since there is a high degree of hybrid inviability.

Although species of anuran amphibians in general exhibit fairly uniform mating calls over their entire range, instances exist where several geographic races exhibit different mating calls. Littlejohn and Roberts (1975) have described the situation in *Limnodynastes tasmaniensis* of eastern Australia, in which "northern" and "southern" call races meet in a zone of intergradation 90–135 kilometers wide in central Victoria. The authors offer a hypothetical reconstruction of evolutionary events, according to which the southern race differentiated on the island of Tasmania and reinvaded the mainland 20,000 to 12,000 years ago, when Bass Strait was dry land; the zone of intergradation and hybridization would consequently be a secondary one, having followed a period of allopatry.

Those evolutionists who in the past have believed in the universality of the allopatric model of speciation have necessarily assumed that all zones of overlap between closely related species are secondary, having been preceded by a period of complete geographic isolation. According to this viewpoint, therefore, "reinforcement" of premating isolating mechanisms is a phenomenon that begins when sympatry due to range extension of one or other form is established. On the other hand, if other forms of speciation, not involving complete allopatry, are common in many groups (as now seems certain), many zones of overlap must be considered "primary." If "reinforcement" of genetic isolating mechanisms is occurring in these zones, it would be merely due to the continuing evolution of those mechanisms. Nevertheless, the known instances of "reinforcement" are still too few and too poorly understood for us to be certain whether it is a phenomenon of general importance. Its significance is that where it does occur it seems to represent the last stage of allopatric speciation, in which species that have not completely diverged (i.e., to the point where no hybridization can occur) in geographic isolation are finally severing their last genetic links. Walker (1974b) states that no convincing instances of character displacement of mating calls have been found in sound-producing insects (crickets, grasshoppers, cicadas).

All instances of character displacement for genetic isolating mechanisms that have been plausibly demonstrated seem to be ones in which the zone of overlap and potential hybridization is fairly broad, e.g., 50 kilometers or more in width). There seems to be a problem with regard to very narrow "parapatric" zones of overlap. On the one hand, it is believed these are kept narrow by strong selection against hybrids, and it might be thought that such selection would lead to "reinforcement" of premating isolating mechanisms. But if the zone of overlap is sufficiently narrow, any effects of selection in it are likely to be effectively swamped by gene flow from the zones on either side, where only one species exists. For example, the Australian frog species *Crinia laevis* and *C. victoriana* have a zone of overlap, in which hybridization occurs, that is only about two kilometers wide; no reinforcement of mating call differences was found in this instance (Littlejohn, Watson, and Loftus-Hills, 1971; Littlejohn and Watson, 1973).

5

Clinal and Area-Effect Speciation

Two models of speciation that are neither allopatric nor sympatric were briefly described in Chapter 1. Both involve the segregation of distinct species out of geographically continuous, genetically polymorphic, populations. Models of this type, for which Mayr (1942) used the somewhat ambiguous term "semi-geographic," seem to have been first discussed by Huxley (1939). However, they could hardly be seriously considered until the full extent of genetic polymorphism in the natural populations of most species of organisms had been revealed. The new picture of genetic variation, discussed in Chapter 2, has necessarily forced us to look on these models of speciation in a new light.

A reorientation of ideas on evolutionary questions seems now to have begun with the gradual realization of the extent of geographic heterogeneity of genetic polymorphism, i.e., the extent to which populations in different areas are characterized by different combinations of allelic frequencies. In one sense such genetic variability should not have occasioned surprise, because it has long been known in the case of the human blood group genes. With the growth of rapid and efficient biochemical techniques for studying the genetic structure of natural populations, the full extent of this type of variation must be taken into account by evolutionists. Certainly, it makes much more plausible the "clinal" and "area effect" models of speciation.

In contrast to the concept of biological species, which can be rigorously defined and which has been subjected to critical discussion (see Chapter 1), the notions of subspecies and geographic races lack precision and have been employed in varied ways by taxonomists working on different groups of organisms. In certain groups it is traditional to recognize the existence of geographic subspecies and even to give them formal status by the use of Latinized trinomials, while in the case of other groups this is seldom or never done. There is, however, a consensus that the concepts of geographical race and subspecies are equivalent. The very fact that these concepts have arisen and persisted in taxonomic and evolutionary biology implies that species distributed over wide geographic areas do show spatial variation in their genetic composition.

It has seldom been claimed even by the most ardent supporters of the universality of allopatric speciation that geographic variation within species requires absolute geographic isolation, in the sense of complete barriers to dispersal between the "subspeciating" populations. On the other hand, it is clear that where the range of a species has been fragmented by geographic features that ensure isolation of several allopatric populations, these populations are likely to become sufficiently distinct in genetic composition in the course of time to warrant recognition as geographic races, providing of course they have not gone a stage further and acquired genetic isolating mechanisms, which would compel us to recognize them as distinct biological species.

The question now arises whether geographic races that are spatially continuous, i.e., not separated by any insuperable geographic barrier, can proceed further along the path of genetic differentiation so that eventually they become full species. In Chapter 6 we shall examine instances in which chromosomal rearrangements appear to have played a part in differentiating geographically continuous species into two or more daughter species. But here we shall attempt to answer the question whether such differentiation is possible in the absence of major karyotypic changes.

In recent years genetic investigations of the geographic variation of wide ranging species with moderate or low vagility of the individuals have revealed the existence of so-called area effects, i.e., areas in which the genetic composition of the population is strikingly different from that in the surrounding contiguous area. Such area effects were first clearly recognized in the land snails *Cepaea nemoralis* and *C. hortensis* (Cain and Currey, 1963a, b). The natural populations of these species, which are broadly sympatric throughout western Europe, have been studied by a large number of workers on account of conspicuous genetic polymorphisms that involve the ground color of the shell and the presence or absence of dark bands on the shell (if bands are present there may be anywhere from one to five). In addition, the lip (peristome) of the shell may be pigmented or not, and the color of the mollusc's soft body is also variable. The genetic determination of some of these characters is known (Cain and Sheppard, 1957) but data on the composition of the natural populations are essentially based on phenotype frequencies rather than allele frequencies. The ground color of the shell is determined by a series of multiple alleles, with brown (C^B) dominant to dark pink (C^{DP}), which in turn is dominant to pale pink (C^{PP}), dark yellow (C^{DY}), and pale yellow (C^{PY}), the bottom recessive.

The locus for presence or absence of banding (B^O unbanded, B^B banded) is closely linked to C (Cain *et al.*, 1968) and so is the one for "spreading" of the bands (S^S spread bands, S separate bands), but the U^3 gene for the mid-banded phenotype (00300) and the gene for the three-banded one (phenotype 00345, genetical symbol T^{345}) are unlinked.

It was shown by Cain and Sheppard (1950, 1954), who worked primarily on *C. nemoralis* in southern England, that a certain degree of relationship existed between the frequencies of the color and pattern morphs and the nature of the habitat. Thus yellow, banded shells tend to occur in hedgerows and in rough herbage; non-yellow, unbanded shells are characteristic of beech and oak woods. Evidence was obtained that differential predation by birds (song thrushes) was responsible for these regularities of distribution. However, predation is not the only factor that determines the frequency of the morphs in the natural populations and it may be much less important in continental Europe than in southern England. It is now known that the various color morphs differ in temperate preference and tolerance (Sedlmair, 1956; Lamotte, 1951, 1954, 1959). Yellow snails and unbanded ones are more adapted to warm dry habitats than reddish or banded ones (Cain *et al.*, 1968), and Schnetter (1951) observed an increase of 16.2 percent in the frequency of yellow, unbanded individuals over the period 1942–1950, when the climate was warmer and drier than usual. It has also been shown by Wolda (1967a) that the color morphs differ in fecundity; yellow, unbanded snails produced more batches of eggs than yellow, banded or red, unbanded ones at both 20°C and 12°C. There were also complex effects on the number of eggs per batch. Unbanded snails also had a greater migratory tendency (Wolda, 1963).

It is therefore clear that the genetic equilibria in these polymorphic populations are extremely complicated and that the geographic distribution of the various color and banding types is only in part determined by the color and pattern of the natural background. In some areas, but not in others, there is a negative correlation between the frequency of the yellow color morph in mixed populations of *C. nemoralis* and *C. hortensis* (Clarke, 1969). This is interpreted as due to "apostatic selection" by predators that form learned associations between particular visual images and "good food."

Cain and Currey found that in certain regions of elevated grassland in southern England the populations of *C. nemoralis* are characterized by an overwhelming predominance of a few phenotypes, to the exclusion of others that are common elsewhere. For example, on Marlborough Downs an area of about 50 square kilometers was subdivided into four subareas (Figure 14). In area A, almost all the shells are unbanded or single-banded ("mid-banded"), with almost no five-banded ones. The unbanded condition is due to a dominant allele B^O, while the mid-banded phenotype is produced by the dominant U^3, which is

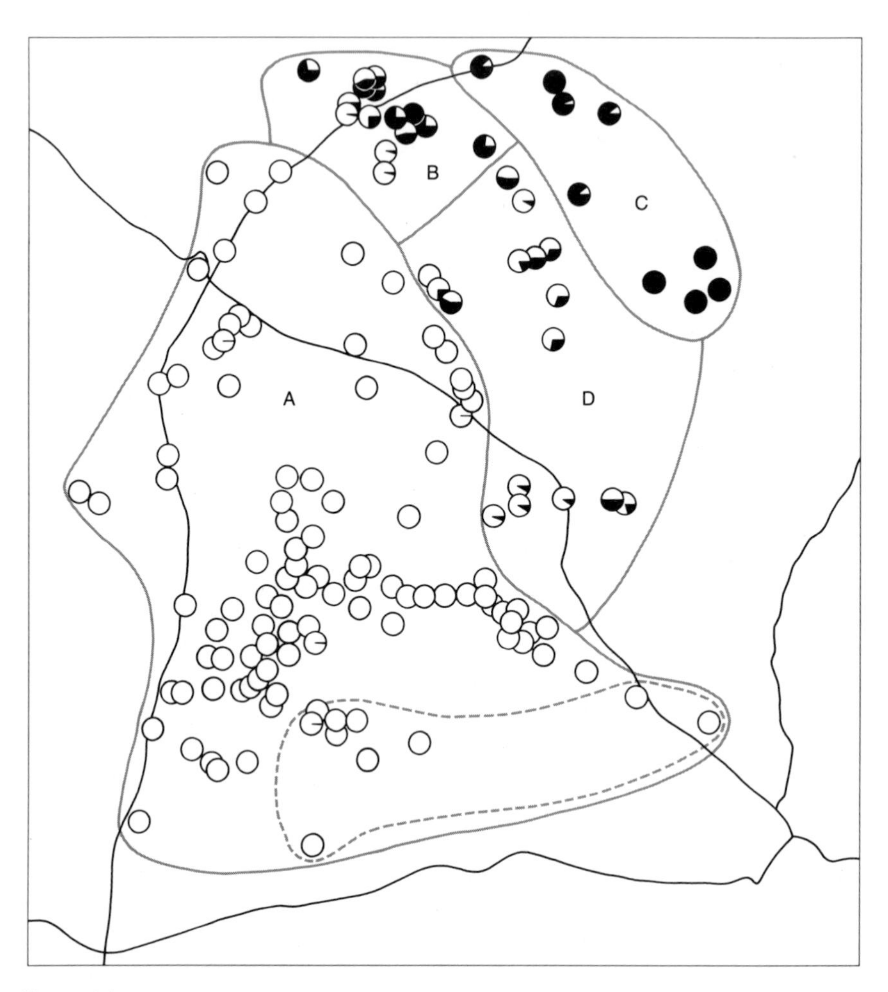

FIGURE 14

Area effects in *Cepaea nemoralis* on the Marlborough Downs, southern England. The proportions of five-banded shells in each sample is indicated by the black sectors. In Area A the populations consist almost entirely of unbanded and midbanded individuals. In area C these two dominant genes are virtually absent; area B is transitional. The dashed line encloses colonies with a high proportion of spread-banded shells. [After Cain and Currey, 1963]

only expressed in individuals homonygous for B^B. The gene B^O must have a frequency of about 0.4 in area A, while U^3 has a frequency of about 0.7. In area C, on the other hand, separated from A by the transitional areas B and D

TABLE 5-1
Phenotype frequencies of Cepaea nemoralis *populations on Marlborough Downs.*

	Yellow					Pink					Brown				
	00000	00300	sp-b	12345	00345	00000	00300	sp-b	12345	00345	00000	00300	sp-b	12345	00345
Area A	837	1,307	88	2	—	746	341	63	1	—	2,079	224	46	1	1
Area B	66	83	4	111	3	75	43	2	91	2	8	12	—	20	—
Area C	8	4	—	351	43	12	3	—	86	2	—	—	—	1	—
Area D	86	157	1	106	11	107	113	3	42	4	97	2	—	—	—
Total	997	1,551	93	570	57	940	500	68	220	8	2,184	238	46	22	1

00000: unbanded
00300: mid-banded
12345: 5-banded
00345: 3-banded
sp-b: "spread-banded"

(which constitute a corridor about two kilometers wide), five-banded shells predominate (Table 5-1). The same anomalies occur in the distribution of the color types. Brown shells are the commonest kind in area A (and almost the only type in part of it), but they are almost absent in C. In one area the frequency of brown shells changed from 73 percent to 12 percent over a transect of approximately 130 meters, the population being apparently continuous. These "area effects" do not seem to be related to elevation above sea level, type of soil or vegetation, or other obvious features of the habitat. There is some evidence (Cain and Currey, 1963a; Cain, 1971) that they have persisted since Neolithic times (4,500 to 2,500 years ago).

Similar area effects were also found in another area, Lambourn Downs. The areas involved are far too large for the observed distribution of phenotypes to be due to genetic drift or founder effects. In some instances it is definitely known that discontinuities in the distribution of allozymes coincide with area effects originally defined and described on the basis of visible shell characters (Johnson, 1976). Area effects are also known in *C. hortensis* (Cain *et al.*, 1968).

Broadly speaking, there seem to be two possible interpretations of these area effects. Cain and Currey (1963a, b) concluded that they must be due to selection for adaptation to cryptic differences in the environment. One factor could be temperature, since it is believed that the brown phenotype is cold-adapted.

Clarke (1966, 1968) has proposed a different interpretation of area effects, based on a mathematical model of gene interaction. According to this scheme (which could apply to either clinal or area effect differentiation) each area is characterized by a complex of interacting and coadapted genes, which would be adaptive to the "mean habitat" of the area. The process could start with a single polymorphic locus plus a modifier and might evolve until there was a sharp geographic discontinuity—a "phenotypic escarpment" in Mayr's terminology (1970)—between two adaptively incompatible gene complexes.

(incompatible in the sense that hybrids between them would be generally adaptively inferior). The process would not necessarily coincide with any sudden change in the environment.

This seems far the most likely explanation of area effects in *Cepaea*. Goodhart's interpretation (1963) contains elements of the founder principle. According to Goodhart, area effects are produced by populations of initially different genetic constitution in "small foundation stocks" that acquire different systems of coadapted polymorphisms in different areas and then spread out to meet. The separation of the initial founder populations, as proposed, is certainly unproven, seems improbable, and is unnecessary on Clarke's model.

Two questions may be asked regarding these area effects. First, are the populations characterized by unusual external phenotypes also carrying special combinations of alleles for allozymes and other characters not directly observable in the way that the shell color and pattern are? This question has been answered in part by Johnson (1976), who has shown that correlations between area effects for allozymes and for external phenotypes do in fact exist in some cases. More difficult is the general question whether area effects can be expected ever to lead to speciation. Arguing that they are unlikely to do so, Mayr (1970) claims that those organisms in which they have been observed, such as *Cepaea,* do not have large numbers of species (so-called species swarms) closely related to them. This is certainly the case for the genus *Cepaea,* in which area effects were first recognized as a distinct phenomenon. But they have also been observed in the land snail *Partula taeniata nucleola* (Clarke and Murray, 1969), which belongs to a species-rich genus in the Society Islands (see p. 131), and they probably exist also in the Caribbean genus *Cerion* (see p. 133). *P. taeniata nucleola* shows extensive genetic polymorphism, and the proportions of genotypes may differ significantly in populations as little as 10 meters apart. Thus, the range of this form is a mosaic of areas, each characterized by different frequencies of the various morphs. These areas do not seem to show any correlation with ecological variables and they traverse ridges and valleys, regardless of habitat differences. Presumably, they are occupied by different but coadapted complexes of genes, and the frequencies of the alleles in the individual populations depend more on the "genetic milieu" than on the external environment. The situation in *P. suturalis vexillum* (Clarke and Murray, 1971) seems to be broadly similar to that found earlier in *P. t. nucleola*.

In studies of *P. taeniata,* Clarke and Murray (1969; Clarke, 1968; Murray, 1972) examined in detail a 200-meter transect across an area transitional between one with approximately 37 percent purple shells to one with zero (Figure 15). The "frontier zone," about 100 meters wide, shows a number of peculiarities. In the first place, the population density is very low (the numbers

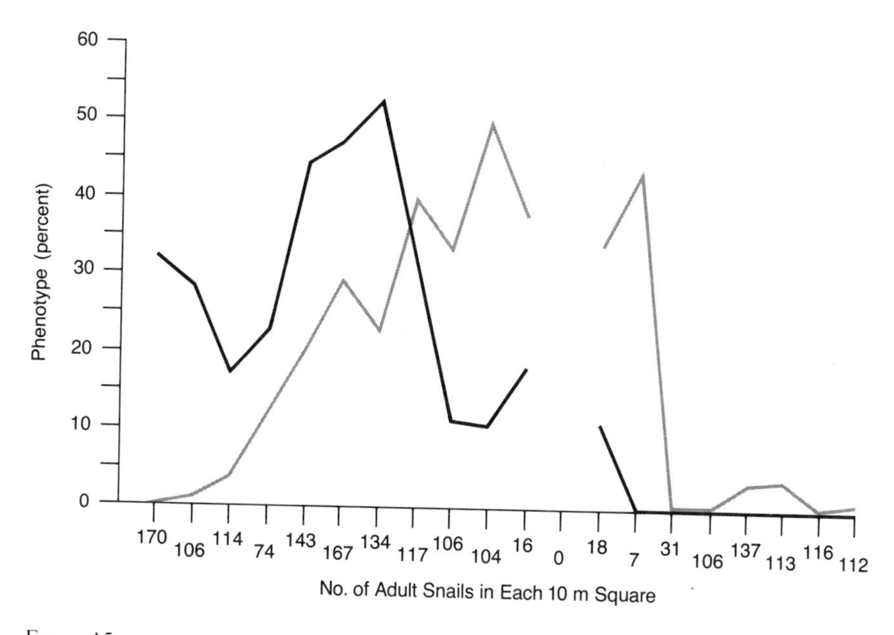

FIGURE 15

A transect of populations of *Partula taeniata* extending from an area of high frequency of "purple" shells to one of low frequency. The solid line shows the proportion of purple shells in a series of 10-meter squares, while the grey line shows the proportion of banded shells. The numbers along the abscissa are the total adult snails in each square. [After Clarke, 1968]

of adult snails in consecutive 10-meter squares across the zone were 170, 106, 114, 74, 143, 167, 134, 117, 106, 104, 16, 0, 18, 7, 31, 106, 137, 113, 116, and 112). This decline in population density in the frontier zone does not appear to be due to any ecological unsuitability of the habitat. Second, the frequency of banded shells suddenly increases in the frontier zone to 30–50 percent (elsewhere there are seldom more than 10 percent banded individuals). Third, body size changes across the zone; the mean body size in populations with a high frequency of purple shells is small (even though within samples there is no correlation between color and size). Fourth, the mean number of uterine eggs and embryos is higher in populations with a low frequency of purple shells. Fifth, the usual dominance of the gene for banding seems to break down in the frontier zone.

A quite different approach to the geographic variation of *Partula taeniata* was adopted by Bailey (1956), who was concerned with the biometry of shell shape. He found that the phenotypes of the main "varieties" recognized by Crampton *(propinqua, spadicea, nucleola, striolata, elongata,* and *simulans)* tend to become progressively more different the closer one gets to the boundaries between them (a form of character displacement leading to "reversed clines"). Unfortunately, Bailey's data were interpreted in terms of Crampton's postulated invasions of the territory of one form by another, which were purely hypothetical and are hardly likely to correspond to reality.

It seems entirely possible that in some instances area effects detected on the basis of external phenotypes may also be correlated with chromosomal rearrangements. Thus, populations like those of *C. nemoralis* on Marlborough Downs might be characterized by particular karyotypes, perhaps involving changes in heterochromatic chromosome segments. But nothing is known of this, and chromosomal races of grasshoppers, rodents, and annual plants (Chapters 3 and 6) do not seem to show special combinations of polymorphic alleles affecting the exophenotype.

Area effects are now well known in organisms having a much greater vagility (or at any rate greater potential vagility) than snails. The case of the butterfly *Maniola jurtina* in England (Creed, *et al.*, 1970) is well documented. Individuals of this species throughout most of southern England and in continental Europe from the Atlantic coast of France eastward to Finland and Romania show for the most part two spots on the hind wings of the males and no spots on those of the females. In an area of southwestern England (the county of Cornwall and part of Devonshire) the female population shows two modes for spot number, at 0 and 2. The interface between the two types of population is quite sharp, but moves from time to time, e.g., in 1967 it moved 70 kilometers to the west of its 1966 position). It is not clear whether this is a genuine area effect similar to those found in *Cepaea* and *Partula,* but there is certainly no reason to believe that it is likely to give rise to speciation.

Area effects, apparently similar to those that have been studied in molluscs, have also been recorded in a few plant species. Thus, on the basis of a large number of terpenoid characters, Adams (1975) has shown that *Juniperus ashei* is uniform over a large area of Texas, and that disjunct populations in Arkansas and the Ozarks, which have been isolated for thousands of years, are not significantly different from those in Texas. On the other hand, populations in south Texas and Coahuila are radically different in terpenoid constitution; there is a sharp transition in south-central and west Texas that has not been studied in detail. This area effect is apparently not due to introgression from related species, with which *J. ashei* probably does not hybridize. The distributions of *J. ashei* and *J. virginiana* are essentially parapatric in north-central Texas, suggesting an evolutionary origin of one from the other—the first stage being a sharp transition zone of the type that exists south and west of the Edwards Plateau in the former species at the present time.

No firm distinction can be drawn between area effects and geographic races or subspecies. The former concept has arisen out of genetic and biochemical work in which phenotypes have been counted and frequencies of alleles estimated (in the case of *Juniperus* it is the terpenoid molecules that have been identified rather than the alleles that determine them). Thus, area effects are defined in terms of the genetic composition of the population. On the other hand, the concepts of race and subspecies arose in classical taxonomy and are based on the phenotypic similarity of the individuals in each taxon and their differences from individuals comprising other subspecies or races (Mayr, 1969). Perhaps we might say that area effects are due to the existence of a special type of geographic race. All these types of populations are liable to intergrade in relatively narrow zones of contact.

Some authors (e.g., Lamotte, 1960) have claimed that in continental areas (as opposed to archipelagos) only species having vast areas of distribution or inhabiting mountainous regions, where migration is difficult, show local races. This is definitely not so. It seems likely that all species, except for relict ones occupying a minute fraction of their former territory, will show some degree of racial differentiation, whether expressed as area effects, cytological races, or geographic subspecies of the classical type.

A few evolutionists have gone beyond the claim that a complete geographic barrier is a *sine qua non* only of successful speciation; they have actually extended this principle to include the formation of well-marked geographic subspecies between which there are no genetic isolating mechanisms. Thus Turner (1971a, b, 1972; Turner and Crane, 1962) has done this in his most careful analysis of the Müllerian mimicry between the butterfly species *Heliconius erato* and *H. melpomene* in South America. There are 11 main subspecies of the former and 10 of the latter, distinguishable by striking color patterns on the wing. In general, the phenotypes of the two species in any one area are almost indistinguishable, i.e., they vary in parallel. Furthermore, a third species, *H. elevatus,* whose overall range is less extensive than that of any of the other species, has a series of six races that mimic those of the other two species.

In general, the races of these heliconids seem to be parapatric. Some of them have disjunct distributions, however, and have been given different names in different areas. Thus, if we consider the "red" races of *Heliconius erato* and *H. melpomene* we have the following situation:

	H. erato	*H. melpomene*
Central America	*e. petiverana*	*m. rosina*
Colombia	*e. demophoon*	*m. bellula*
Huallaga, Peru	*e. favorinus*	*m. amaryliss*

And in the case of the "pink" races:

	H. erato	*H. melpomene*
Peru	*e. amphitrite*	*m. euryades*
Northern Colombia, Venezuela, the Guianas, Trinidad, Brazil	*e. hydara*	*m. melpomene*

It is quite uncertain whether these disjunct distributions are the remains of formerly continuous ones or whether the same phenotype has evolved independently in different areas. Should we regard *petiverana–demophoon–favorinus* as a single race of *Heliconius erato* or as three virtually indistinguishable races? It is known that in different species the same phenotype may be produced by quite different genes; but whether *petiverana, demophoon,* and *favorinus* have the same genotype is unknown.

Brown, Sheppard, and Turner (1974), following on the earlier suggestions of Turner (1964), imply that the geographic races of these heliconids could not have crystalized out of widespread, uniform populations or have arisen by "budding off" peripheral populations that expanded into unoccupied territory. Instead, they believe that during dry periods in the Quaternary the now-continuous Hylaean forest of the Amazon basin was fragmented into a number of *refugia* separated by savannas unsuitable for heliconids. Such *refugia* have also been postulated to account for the distribution of bird and lizard species in the Amazonian region (Haffer, 1969; Vanzolini and Williams, 1970). The general theory of such neotropical *refugia* or centers of origin is expounded by P. Müller (1973).

In attempting to understand the raciation and speciation mechanisms of the heliconids it is important to realize that Müllerian mimicry does not involve color-pattern polymorphism in any one locality. In fact, the system ensures that in any one area, as a result of strong stabilizing selection, the phenotypes of each of the participating species is kept uniform and closely similar to that of the other (or others).

Vanzolini and Williams (1970) have made a very detailed morphological study of the south American forest lizard *Anolis chrysolepis*. They recognize four subspecies:

A. c. chrysolepis	The eastern Guianas ("core area I")
A. c. planiceps	Venezuela, Trinidad, and the northern half of Guyana ("core area II")
A. c. scypheus	Amazonian Colombia, Ecuador, and Peru ("core area III")
A. c. brasiliensis	Central Brazil to S. Paulo ("core area IV")

The "core areas" (defined as "areas of maximum overlap of distinctive distributions of a number of characters" and in which the subspecies exist in their most differentiated form) are all peripherally located and occupy only about a quarter of the total area inhabitated by the species. Between them the populations exhibit character intergradation to varying degrees. In the Amazonian area, centrally located between the four "core areas," patterns of intergradation are complex.

Vanzolini and Williams regard the core areas as having originated from forest *refugia* that are believed to have existed during periods of the Pleistocene when the great Hylaean forest was fragmented by the extensive development of savanna-like *cerrados* and frankly zerophytic *caatingas*.

Since such forest *refugia* have been postulated on the evidence provided by three very different groups of organisms (birds, heliconid butterflies, *Anolis*), it is worthwhile to consider whether the geographic pattern is essentially the same in all three cases. A fair degree of agreement exists between the *refugia* deduced from the distribution of forest bird species (Haffer, 1969) and the races of *Heliconius melpomene* and *H. erato* (compare Figures 3 and 4 in Brown, Sheppard, and Turner, 1974). But the agreement between these geographic patterns and the "core areas" of Vanzolini and Williams is very poor; core areas I and III possibly correspond in part to *refugia* postulated by Haffer, but Haffer has no *refugia* where core areas II and IV occur. Moreover, Haffer as well as Brown, Sheppard, and Turner postulate *refugia* in numerous areas where Vanzolini and Williams have found no core areas.

It is not possible to discuss here the extremely complicated evidence bearing on these problems of Pleistocene biogeography in tropical South America. Clearly, if the great neotropical forest was fragmented in the manner suggested, then the species of forest-dwelling organisms would most likely have developed geographical subspecies in each distinct forested *refugium*. But the converse proposition, that they would not have done so if the forest had remained continuous, seems open to the most serious doubt. It would obviously depend on the extent to which migration would disrupt local genetic races characterized by combinations of alleles and allelic frequencies not found elsewhere. The proposition is hence likely to be true for organisms whose vagility is extremely great, but untrue for those of limited vagility. We do not have any precise, objective data on the vagility of *Anolis chrysolepis* or *Heliconius* species. But it seems likely that vagility is fairly restricted in both cases.

A question that seems to have been inadequately discussed in the literature is the extent to which geographic races that have been recognized by taxonomists (to the extent of receiving formal subspecies status) are, under present conditions of geographic distribution, potentially incipient species. Of course, no one doubts that they would be so if they became isolated by a complete geographic barrier. But in a case like the North American song sparrow, with 34 recognized geographic subspecies ranging from Alaska and Newfoundland to central Mexico (Miller, 1956), it is unlikely that more than a very few of the geographic races are actually on the way to becoming species. However, it is necessary to recall here the case of *Peromyscus maniculatus* (deer mouse) (see p. 85), in which a great many and perhaps all the geographic races recognized by earlier taxonomists are in various stages of speciation (in this instance with the aid of chromosomal rearrangements).

There seems no fundamental difference between the geographic races of classical taxonomy and the area-effect populations of modern population genetics. As far as the snails *Cepaea* and *Partula* are concerned, the area effects extend over smaller areas of territory than is usual in the case of the subspecies of the taxonomists; but this is to be expected in view of the low vagility of these organisms. As with subspecies of the classical type, it seems likely that most area effects remain stable, as the *Cepaea* ones seem to have done, for thousands or tens of thousands of years, without proceeding all the way to complete speciation. Whether they will eventually do so is likely to depend on the degree of genetic incompatibility that is built up at the interface (the "tension zone" in Key's terminology) as a result of the spread of mutations from within a racial population or area population to the periphery of its range, where it meets another racial or area population in a parapatric contact. What happens then will be determined by the extent to which this incompatibility (in the sense of the production of ill-adapted and even inviable gene combinations in the zone of contact) becomes reinforced by genetic isolating mechanisms, whether of the premating or the postmating type and whether genic or chromosomal in origin.

In general, it can be plausibly argued that where two geographic races meet in a broad zone of intergradation genetic incompatibilities are unlikely to be occurring and therefore that speciation is an improbable outcome, at least in the near future. On the other hand, very narrow parapatric zones are clear evidence of genetic incompatibility of one kind or another. We should consequently expect to find that the parapatric zones of overlap or coexistence would be especially narrow in the case of semispecies, i.e., forms that are believed to be at a stage of evolutionary divergence close to the completion of speciation.

Numerous parapatric zones of contact between semispecies on continental land masses have now been described in the literature. But in most cases it is extremely difficult to determine whether these are primary or secondary (i.e., due to range extensions of populations that were formerly separated by geographic, climatic, or ecological gap). This question can sometimes be solved, however, if evidence from a number of different groups of organisms coincides. Thus, numerous vertebrate species have pairs of semispecies that meet in a parapatric zone of contact running in a fairly straight line from Denmark to the head of the Adriatic Sea. Populations that meet along this line include: *Corvus cornix* and *C. corone* (hooded and carrion crows) (Meise, 1928), the eastern and western taxa of *Mus musculus* (house mouse) (see p. 206) and those of the snake *Natrix natrix* (Thorpe, 1975); and the hedgehogs *Erinaceus europaeus* and *E. roumanicus* (Corbet, 1970; Gropp, Citoler, and Geisler, 1969). Some insects, e.g., the agromyzid fly *Phytomyza abdominalis* (Block, 1975) likewise have eastern and western semispecies that meet along the same line. The conclusion that this line corresponds to a southerly extension of the Pleistocene ice cap that covered northern Europe and that in each case the eastern and western populations developed in geographically isolated southeastern and southwestern "refuge areas" seems inescapable. But, in general, it is clear that upholders of the view that all speciation is allopatric have, in the past, greatly exaggerated the extent to which parapatric zones are secondary phenomena; the great majority of them are primary, especially in regions that were not glaciated during the Pleistocene.

The concept of clines, i.e., geographic gradients in phenotypic characters that are generally assumed to be adaptive to some geographical regularity of the environment, is a very old one in evolutionary biology. It was formalized and given a name by Huxley (1939) at a time when the extent of genetic polymorphism in natural populations was not appreciated. The geographic gradient may be in temperature, humidity, elevation above or below sea level, or the prevalence of other interacting organisms (competitors, predators, parasites) in the environment. An elementary mathematical consideration of genetic polymorphism in a cline (in terms of "beanbag genetics") was published by Haldane (1948). Mayr (1963a) defines clines as slopes from one extreme of a character to the other; when plotted on a map they are crossed at right angles by *isophenes,* contours connecting populations showing equal manifestation of the character.

Clines may be of many kinds. We may speak of a cline in the frequency of a single polymorphic gene or in a large number of variable characters each determined by multiple genes. Some clines are quite local phenomena, while others may extend over the whole range of a species (which may cover a continent). Some clines are smooth continuous gradients, while some are steepened locally, giving rise to a "stepped cline." Huxley (1942) cited the bird *Sitta caesia,* which shows an east-west cline in the plumage color of the underside (from chestnut to white) several hundred kilometers long across north-central Europe; the cline terminates at each end in uniform populations that extend over a considerable area.

The clinal model of speciation is based on the idea that "steps" in a cline become gradually steeper until an adaptive discontinuity is reached. The mathematical theory of a morph-ratio cline, based on interacting gene loci, has been worked out by Clarke (1966). There is, of course, no sharp distinction between area-effect and clinal models of speciation. Nevertheless, the latter essentially involves a single gradient, such as we might find in a species with a considerable north-south range, whereas the former is based on different adaptive complexes of genes occupying different areas in some kind of mosaic pattern.

Although clines generally involve biometrical (continuously variable) characters or ratios of sharply contrasted morphs, some involve chromosomal characters, such as extent of heterochromatic segments or number of chromosome fusions present (p. 200).

Special regularities of geographic variation may develop in organisms that exhibit the phenomenon of mimicry. A species may be mimetic in one part of its range but not in another where the model species is rare or absent; alternatively, a species may mimic one particular model in one part of its range and another in a second area. An instructive example was studied by Platt and Brower (1968). The North American butterflies *Limenitis astyanax* and *L. arthemis* were for over a hundred years regarded as either separate species or semispecies. *Arthemis* is a northern form, ranging from Nova Scotia westward across the Great Lakes region to Alaska and southward to Wisconsin, Pennsylvania, and northern Massachusetts; *astyanax* has a more southerly distribution, from Virginia westward to Arizona. The two overlap in a belt about 200 kilometers wide in which intergrading occurs. *Arthemis* has a broad white band across the fore and hind wings that is absent in *astyanax;* the latter has a diffuse blue-green iridescent area on the upper surface of the hind wings while *arthemis* usually has a row of orange or dull red spots between the white band and the margin of the hind wing. It is generally believed that *astyanax* is a Batesian mimic of the highly unpalatable *Battus philenor* (swallowtail butterfly), which it closely resembles in a superficial manner, and whose geographic distribution generally coincides with its own. The coloration of *arthemis,* with its conspicuous white wing band, is nonmimetic and appears to be an example of a disruptive pattern, i.e., one that helps the animal avoid the attention of predators.

Although earlier authors (e.g., Fisher, 1930, 1958) claimed that hybrids were rare in the zone of coexistence of *arthemis* and *astyanax* because of an ethological barrier to mating between the two, Platt and Brower have shown that mating is in fact random and that heterozygotes for the various differences between the two forms occur in Hardy-Weinberg proportions. Laboratory experiments revealed no heterozygote inviability and the male genitalia are not significantly different. This is then a case of a single species with two adaptive modes: a mimetic one where its distribution overlaps that of *Battus philenor,* and a disruptive one where the model is rare or absent. In the area where *B. philenor* occurs, another swallowtail butterfly, *Papilio glaucus,* is dimorphic in the female sex; one kind is black and mimetic to *B. philenor* while the other is yellow like the males and nonmimetic. The frequency of the black females is correlated with the abundance of the model species in the southeastern United States (Brower and Brower, 1962). Further north, where *B. philenor* is absent altogether, the black morph of *P. glaucus* is generally absent. Numerous other examples are known of butterflies that are mimetic in only one part of their range, where a model occurs, e.g., *Papilio dardanus* (Clarke and Sheppard, 1960) and *Hypolimnas bolina* (Clarke and Sheppard, 1975). The latter case is complicated by the highly migratory habits of the species. There does not seem to be any case of mimetic and nonmimetic races of a species undergoing differentiation into separate species, although it would seem on general principles that this kind of speciation should occur from time to time in organisms in which mimicry is a form of adaptation.

The area effect and clinal models of speciation have been grouped together by Murray (1972) as *parapatric speciation.* This term is best avoided for several reasons. In the first place, it is not clear that area-effect and clinal modes of speciation have enough in common to be included under a single heading of this kind. Second, although parapatric distributions of races or species, where they occur, are an objective fact, it is by no means clear that they have all arisen in the same way; that is, some of them are certainly primary, but others have probably arisen as secondary contacts between previously allopatric populations. Finally, not all parapatric situations are likely to give rise to speciation. The term "stasipatric speciation" was introduced by White and co-workers (1967) for one type of speciation that Murray would call parapatric, namely, one in which a major chromosomal rearrangement gives rise to a postmating isolating mechanism. It has been used by a number of workers concerned with speciation mechanisms in vertebrate groups (e.g., Wilson *et al.,* 1975). It therefore seems best to use the term "parapatric" to describe actual patterns of geographic distribution in nature but without any implication as to the mechanism of speciation or whether speciation is occurring at all.

Mayr, arguing against area-effect and clinal modes of speciation (called "semigeographic" by him), states that "everything we know about the cohesion of genetic systems" denies the possibility of their occurrence (Mayr, 1970). For the present, it seems desirable to avoid overestimating the strength and complexity of genetic interactions (see Haldane, 1964; Felsenstein, 1975). Numerous interactions undoubtedly exist, but in the area effects discussed in this chapter they are essentially properties of the populations inhabiting the several areas—it is not the genetic system of an entire species that is "cohesive" but the systems of individual populations. Parapatric transition zones a few tens or hundred meters wide (in organisms with low or moderate vagility) are thus essentially (in the terminology of Key, 1974) "tension zones" where "cohesion" breaks down, resulting in poorly adapted genotypes and (in the case of some *Partula* populations, at any rate) a marked reduction of population density. Such situations may remain essentially static for thousands of years, as they have probably done in *Cepaea*. But it seems impossible to avoid the conclusion that they are potentially species-generating, insofar as the genetic composition of the populations on either side of the "tension zone" may change in the course of time.

It is obvious that if some of the intermediate populations in a cline were to become extinct, the terminal populations might evolve into distinct species. This, however, would be allopatric speciation. The clinal model involves the fragmentation of an originally continuous cline into two or more species as a result of the development of "steps."

It is extremely difficult at present to arrive at an unbiased estimate of the extent to which area-effect and clinal modes of speciation are occurring relative to other modes (allopatric, sympatric, stasipatric). In view of the prevalence of karyotypic differences between closely related species, which have been repeatedly stressed in this book, perhaps the most favorable circumstances for area-effect or clinal speciation would be those in which one or more chromosomal rearrangements became superimposed on a preexisting area effect or cline. This would be essentially a combination of either the area-effect or the clinal model with the stasipatric one (see Chapter 6), and may be the mode occurring in *Peromyscus maniculatus,* for example.

6

Chromosomal Models of Speciation

In this chapter those models of speciation in which chromosomal rearrangements have a primary role in initiating divergence will be examined. Obviously, such models are inapplicable in cases of "homosequential" species, in which no chromosomal rearrangements have occurred. However, the term "homosequential" in the cytogenetics of organisms with polytene chromosomes refers only to the banding in the euchromatic segments. It is entirely possible that some pairs of "homosequential" species differ with respect to chromosomal rearrangements in the heterochromatic segments, which consist of satellite DNAs; in fact, Ward and Heed (1970) have described such a case.

The extent to which chromosomal rearrangements have occurred in different groups of organisms is not easy to quantify. Maxson, Wilson, and Sarich (1974) have estimated that the mean rate of evolutionary change in chromosome number has been about 20 times faster in mammals than in frogs and that "on the average it has taken about four million years for a pair of mammalian species to develop a difference in chromosome number, while the corresponding time for frogs is 75 million years. The rapid rate of chromosome change (and concomitant gene arrangement) in mammals parallels both their rapid organismal evolution and their rapid evolutionary loss of the potential for interspecific hybridization."

While this type of approach is interesting, it must be pointed out that there are many types of chromosomal rearrangements that do not change the chromosome number, and that some of these may have occurred very frequently in anurans, compensating in a sense for the rarity of chromosomal fusions and dissociations. Some evidence supporting this view comes from recent delta karyology studies on species of *Xenopus*. In both *X. laevis* and *X. mulleri* (morphologically almost indistinguishable species that can be crossed and give fertile hybrids) multiple genes for 5s ribosomal RNA are located very close to the telomeres of the long arms of most (probably all) the chromosomes (Pardue, Brown, and Birnstiel, 1973; Pardue, 1973; Brown and Sugimoto, 1973). But *X. mulleri* has less than half the total number of genes coding for 5s RNA that *X. laevis* has, and the "spacer" DNA sequences between them are quite different in the two species. Moreover, *X. mulleri* has an AT-rich satellite DNA that makes up four percent of the total DNA and is located in the short arm of each chromosome; this satellite, and another (of different buoyant density) are not present in *X. laevis* (Stern, 1972). It is not necessary to believe that the differences in the organization of the genomes of these two very similar species imply that chromosomal rearrangements of the classical type have occurred. They are, on the contrary, certainly indicative of the kind of changes listed under (3) on p. 8. In another anuran genus, *Bufo,* the amounts of DNA in the karyotype range from 8.9 to 14.6 pg (Bachmann, 1970), indicating that a considerable number of deletions and duplications of genetic material have occurred in the phylogeny of the genus.

Two other pieces of evidence should warn us against generalizing too widely along the lines suggested by Maxson and co-workers. In the first place the explosive tachytelic evolutionary pattern of the Hawaiian drosophilids (with very rapid organismal evolution) has been accompanied by an abnormally *low* rate of change in chromosome number (four such changes in 93 Hawaiian species compared with a minimum of 32 in 150 continental species, according to Carson *et al.,* 1970). Second, the grasshoppers of the superfamily Acridoidea, in spite of being very conservative in chromosome number, have diversified and speciated far more than the relatively archaic Eumastacoidea, in which there have been far more evolutionary changes in chromosome number (Blackith, 1973; Descamps, 1973; White, 1974).

The occurrence of spontaneous chromosomal rearrangements and the laws and principles that govern their survival are now fairly well known. Broadly speaking, the mechanisms and principles seem to be the same for higher plants (maize, wheat, *Tradescantia, Lilium, Vicia faba*) and animals (*Drosophila,* grasshoppers, mouse, chinese hamster, human). Nevertheless some differences between different groups undoubtedly exist. In particular, in certain groups of animals (e.g., bugs, butterflies and moths, earwigs) and a few plants (e.g., rushes of the genus *Luzula*) the chromosomes do not have a single localized centromere, but instead exhibit diffuse centromere activity (holocentric chromosomes). The laws and principles of chromosomal rearrangement in these organisms are not yet fully understood, but certainly they differ in some respects from those governing chromosomal rearrangement in species with the more usual monocentric chromosome.

In general, a chromosomal rearrangement has to be preceded by chromosome breakage, i.e., there is probably no possibility of two chromosomes simply fusing together, without loss of any material. Moreover, it seems that single chromosome breaks are usually and perhaps always incapable of giving rise to viable rearrangements, so that at least two breaks must occur (within a certain specified time) in the same nucleus. The great majority of spontaneous rearrangements, whether they result in an inversion of a chromosome segment, an interchange of material between two chromosomes (mutual translocation), or a deletion (loss of a segment) undoubtedly are the result of two breaks. Nevertheless, there are a number of recorded instances of 3- and 4-break rearrangements occurring spontaneously (e.g., White, 1963; White and Cheney, 1972). Such complex rearrangements may actually be a good deal more common than would be expected on the basis of a random occurrence of chromosome breaks. This is not altogether surprising since we know that abnormal physiological conditions in the cell, e.g., those due to virus infection, may increase the rate of chromosome breakage far above the normal level.

A full discussion of the mechanisms of chromosomal rearrangement is outside the scope of this book. For present purposes three points are especially important. (1) Something like one in 500 individuals, in populations of organisms as diverse as lilies, grasshoppers, and man, carries a newly-arisen chromosomal rearrangement. (2) Most rearrangements are deleterious, so that they are rapidly eliminated by natural selection. (3) For practical purposes, all chromosomal rearrangements must be regarded as unique, unlike gene mutations, which have a predictable rate of recurrence.

The third conclusion arises from the following considerations. If we imagine the *Drosophila* karyotype as having 5,000 bands (in the polytene chromosomes) and 5,000 interbands, then there will be 5,000 distinguishable breakage points.

Considering only the simplest type of chromosomal rearrangements, those that result from two breaks, the frequency of recurrence of the same rearrangement would be one in $5{,}000^2$ or 25,000,000. Thus, since a particular rearrangement arises spontaneously in one in 500 individuals, we would have to breed 125×10^8 individuals in order to have a reasonable chance of getting a second occurrence of the same rearrangement. After a very careful consideration of the evidence from *Drosophila,* Carson (1974) concludes: ". . . when both breaks of an inversion can be mapped in the euchromatic sections of the giant chromosomes, the probability of separate and independent origin of an identical two-break aberration becomes vanishingly small." Consequently, in a population of a species occupying a large geographic area and consisting of many millions of individuals, while a particular mutation of a single gene will occur again and again, at many points in the distribution of the species, a particular structural rearrangement will only occur *once,* at a single locality. Moreover, the gene mutations will be tried out repeatedly, in many different local environments, while each chromosome rearrangement only gets a single chance to establish itself.

The principles that determine whether particular chromosome rearrangements survive in natural populations are complex, and the following account of them is deliberately oversimplified. In the first place, many of them are cell-lethal, so that they have no chance of surviving even in the form of a patch of mutant tissue. Chromosomes that lack a centromere altogether as a result of rearrangement or have two centromeres situated far apart are usually incapable of survival, as are also ring chromosomes (which have no free ends). Even if a structurally altered chromosome is viable at the cellular level, it may still diminish the viability of the individual when present in either the heterozygous or the homozygous state. Thus, almost all deletions are lethal when homozygous, and in *Drosophila* deletions of over 50 bands (in the polytene chromosomes) are usually lethal even in the heterozygous state.

Even if they do not depress the viability of the individual, many rearrangements may be unable to pass through meiosis without giving rise to aneuploid gametes, that is, ones that either lack a chromosome or chromosome segment or carry one in duplicate (or gametes that have both abnormalities simultaneously). Production of aneuploid gametes is particularly likely to occur in individuals heterozygous for rearrangements. Broadly speaking, it results from two causes: nondisjunction and crossing-over (the latter may in some cases give rise to chromosomes that carry deletions of certain segments). It is necessary to point out, however, that some groups of organisms do possess various cytogenetic mechanisms that permit them to be heterozygous for such structural rearrangements as inversions and translocations without "paying the penalty" in the form of lowered fecundity (some of these mechanisms were described in Chapter 3).

In view of the multiple hurdles they have to overcome, it is not surprising that the great majority of spontaneous chromosomal rearrangements stand little or no chance of survival and that they are rapidly lost from the population by natural selection. The very few that do survive (perhaps one in 10^4 or 10^5 of those that occur) seem to be of two functional types. One type increases the fitness of the heterozygote above that of both homozygous genotypes. Such *heterotic* rearrangements are likely to become established in the population in a polymorphic condition. A genetic equilibrium will be established and both the "old" and "new" types of chromosome will persist indefinitely in the population. If, instead of heterosis, frequency-dependent selection occurs, the situation will be essentially the same, with the fitness of each homozygous genotype increasing when the genotype is rare and decreasing when it is common. The innumerable cases of polymorphism for chromosomal inversions in populations of *Drosophila, Chironomus,* and some species of grasshoppers undoubtedly owe their evolutionary success to one or the other principle. But, in general, chromosomal rearrangements existing in a balanced polymorphic state are more likely to be cohesive than divisive agents in natural populations, i.e., they are unlikely to lead to speciation.

The other functional type of chromosomal rearrangement leads to a decrease in fitness in the heterozygote (usually because it causes the production of a certain proportion of aneuploid gametes) but produces a high degree of fitness in the homozygous state. Theoretically, such rearrangements should be eliminated by natural selection, because initially a population would have only heterozygotes and it is not likely that homozygotes would occur until the rearrangement had reached a relatively high frequency (which it should never do if it is being constantly eliminated from the population by natural selection).

However, the innumerable instances of chromosomal differences between related species discussed in Chapter 3 clearly imply that chromosomal rearrangements can undergo fixation, in the homozygous state, in association with a speciational event. The question we have to answer is whether all karyotypic differences between related species were preceded by a situation in which the types of chromosomes that later underwent fixation in the two lineages coexisted in a state of balanced polymorphism in an ancestral population. No doubt this has sometimes been the case. A parallelism between the types of chromosomal rearrangements that occur in a floating (polymorphic) condition in certain natural populations and those that exist in a fixed condition as cytotaxonomic differences between related species is frequently observed. For example, in the genus *Drosophila* paracentric inversions are the most common type of both floating and fixed rearrangements, while in one section of the North American grasshopper genus *Trimerotropis* pericentric inversions are frequent in both roles. Such parallelism, where it occurs, is strongly suggestive of an evolutionary progression, from balanced polymorphism to fixation of cytotaxonomic differences.

Numerous exceptions to this parallelism are found, however. In the phylogeny of the Australian grasshoppers of the subfamily Morabinae (a total of about 250 species) 39 chromosomal fusions and 22 dissociations or fissions are known to have occurred (White, 1974). Yet not a single change of either type is known to exist in a balanced polymorphic condition in a natural population; the few heterozygotes for such rearrangements that have been found in nature have all been in zones of very narrow overlap and hybridization between different races or species. In this group the great majority of the species have visibly distinct karyotypes even by the criteria of beta karyology. Such a situation very definitely indicates that the cytotaxonomic differences have *not* passed through a state of balanced polymorphism, and this in turn suggests that they may perhaps have played a primary role in initiating evolutionary divergence, that is, in speciation. The only alternative to this conclusion—although it is a very improbable one—would be to suppose that the initial stage of speciation, when genetic isolation is established, is not associated with karyotypic changes, but that these occur during the subsequent phyletic changes. If this were so, we would expect to frequently encounter the situation where a "young" species or semispecies had different karyotypes in different parts of its range, without these being associated with any genetic isolation (and this is definitely not the case).

The opposition of some evolutionary geneticists to the idea that chromosomal rearrangements play a primary role in speciation has been based on the tacit assumption that these obey the ordinary algebraic rules of population genetics, and that they would consequently be eliminated from the population by natural selection whenever they diminished the fitness of the heterozygote. But since the chromosomal rearrangements that establish themselves in speciation are only a minute fraction of the total number that occur, it would not be surprising if some or even many of them exhibited anomalous properties. One such property might be meiotic drive, i.e., a transmission rate greater than 0.5 at meiosis. Such a phenomenon would be especially likely to occur in the female sex; the chromosome with the rearrangement would pass into the egg nucleus more frequently than into a polar body nucleus. It is now well known that a great many (probably most) so-called supernumerary chromosomes (the "B chromosomes" of some authors) possess accumulation mechanisms of this general type (Hewitt, 1973). Moreover, the number and variety of meiotic drive systems now known in *Drosophila* is considerable (Zimmering, Sandler, and Nicoletti, 1970). We shall consider later (p. 186) a case in which there is some evidence that chromosomal rearrangements that have been important in speciation do exhibit meiotic drive.

Although chromosomal rearrangements do seem to have played a role in speciation in many cases, if they diminish the fitness of the heterozygote it is probable that (except in self-fertilizing species of plants, considered later) the diminution must not be too severe, otherwise the rearrangement would simply be eliminated from the population by natural selection. This is especially so in organisms with considerable vagility, in which there is little opportunity for the production of homozygotes for a newly arisen rearrangement. For example, in the highly vagile genus *Drosophila* pericentric inversions always lower the fecundity of heterozygous females fairly severely (by 7–10 percent in the experiments of Alexander, 1952) and have probably played only a minor part in speciation. (One possible case of a pericentric inversion in *Drosophila* that may have been associated with a speciation event is the one that converted the originally acrocentric second chromosome of the *D. virilis* group into a metacentric in the *D. montana* subgroup.) Even very short pericentric inversions (which would be expected to lower fecundity only slightly because of the low frequency of chiasmata) do not seem to have played a significant role in *Drosophila* speciation. Stone (1955) listed a total of 32 pericentric inversions in the phylogeny of the *Drosophila* species that had been investigated up to that time, as compared with an estimated 35,000 paracentric ones that he believed had become established in the evolutionary history of the genus.

It follows from the above that when we speak of a chromosomal rearrangement giving rise to a genetic isolating mechanism, we are not necessarily implying that the fecundity of the heterozygote is reduced to zero, or even to 50 percent of that of the homozygotes. In certain cases a reduction in fecundity by only 5 or 10 percent might be quite enough to initiate divergence. Failure to appreciate this point has led more than one evolutionary biologist to deny the importance of chromosomal rearrangements in evolution.

Some chromosomal rearrangements, in heterozygous state, diminish the viability rather than the fecundity of the individual (or do both simultaneously). It is unlikely, however, that such types can play any role in speciation, because if they diminish viability when heterozygous they almost invariably do so much more when homozygous. Thus, when one speaks of a rearrangement initiating divergence between an incipient species and the parental population because it diminishes the fitness of the heterozygote, it is the fecundity component of fitness that is being referred to. This component depends on normality of meiosis, which may be disturbed in the heterozygote for a new rearrangement but which will generally be completely restored in the homozygote.

There have been three attempts to formulate more or less precise models of speciation in which chromosomal rearrangements play a direct role. These are the *triad* hypothesis of Wallace (1953, 1959), the *stasipatric* model of White, Blackith, Blackith, and Cheney (1967), and the *saltatory* model of Lewis (1962, 1966). Since these were based in the first instance on studies carried out on very different types of organisms (*Drosophila*, orthopteroid insects, and plants of the genus *Clarkia*, respectively) they should not be regarded as alternative hypotheses. In other words, each may be true or close to the truth for the particular group in question or for only some species within the group. There are some common features to these models, but also some fundamental differences.

Wallace (1953) pointed out that whereas a polymorphism involving only two mutually inverted sequences permits the building up of coadapted gene complexes in each sequence, because crossing-over is suppressed, it is possible to have populations containing three sequences ("triads") between which serial genetic recombination will occur by crossing-over. This will be the case (theoretically) whenever sequence A gives rise to B by a single paracentric inversion and B to C by an overlapping one. Assuming that coadaptation of mutually inverted sequences genetically isolated from one another because of the suppression of crossing-over between them is favored by natural selection, one might conclude that the existence of triads would break down this coadaptation or prevent it from developing in the first place. Wallace therefore argued that there should be an "avoidance" of triads in natural populations—only two sequences (A and B, B and C, or A and C) should be present in the populations, or if all three are present, one of them should be rare. Although the data on the distribution of inversions in natural populations of *Drosophila pseudoobscura* did seem to support the hypothesis, later data for *D. robusta* (Levitan, Carson, and Stalker, 1954) did not. However, Wallace (1959) points out that the data from *D. subobscura* (E. Goldschmidt, 1956) and *Chironomus tentans* (Acton, 1958) do support the triad hypothesis, so it is probably valid (although its validity is likely to be limited by the actual frequency of recombination in various chromosome segments).

An interpretation of the evolutionary separation of *Drosophila pseudoobscura* and *D. persimilis* on the basis of triads was attempted by Wallace (1953). He points out that as far as *pseudoobscura* is concerned there are at present four main associations of third chromosome inversions (see Figure 2, p. 35) in the western United States (others may occur in Mexico). One of these, characterized by the sequences Standard, Arrowhead, and Chiricahua (which do not form a triad), occurs from British Columbia to Baja California. It encloses a smaller area of California, in which the predominant sequences are Standard, Arrowhead, and Santa Cruz (again not a triad). In the Rocky Mountains and most of Texas the populations have high frequencies of the Arrowhead and Pikes Peak sequences, while in south Texas the Tree Line sequence is added to these. Thus, *pseudoobscura* is clearly a species that has developed different coadapted sets of inversions in different parts of its overall range. In the area between the "Pacific coadaptation" (ST/AR/CH) and the "Rocky Mountain coadaptation" (AR/PP) the populations are essentially monomorphic, cytologically, being homozygous for the Arrowhead sequence. Wallace assumes that there were originally two coadaptations, Santa Cruz/Hypothetical and Standard/Arrowhead, and that these merged into a SC/HY/ST/AR population in which coadaptation broke down because of the triad phenomenon. Extinction of the Hypothetical sequence left three populations: the ST/AR/CH coadaptation of the west coast, the ST/AR/SC complex, and a population carrying only ST, which gave rise to *persimilis*.

This particular scheme for the speciation event that gave rise to *persimilis* does not seem very likely. But it is very probable that there were a number of different coadapted inversion complexes in the common ancestor of *pseudo-*

obscura and *persimilis* and that *persimilis* evolved from one based on Standard and Klamath.

Wallace later developed the triad hypothesis in relation to the notion of a species extending its range into previously unoccupied territory (Wallace, 1959). He took as his starting point the hypothesis of Mayr (1954) that the periphery of a species' range is a place where gene combinations that are adaptive to the local environment are constantly being disrupted by gene flow from the center of the species' distribution, and that the precise position of the boundary represents a dynamic equilibrium between the building up of adaptive combinations of genes and their disruption. If this is so, a third member of an inversion triad that arises spontaneously in a peripheral population of a species may provide the basis for an adaptive combination of two sequences that is able to invade territory previously unoccupied by the species, while resisting genetic contamination from the territory previously occupied.

The triad model of chromosomal speciation does not seem to have found much favor since it was first put forward and it is difficult to imagine that it would be at all generally applicable. Nevertheless, the more general idea that in species with considerable inversion polymorphism peripheral populations with particular adaptive combinations of inversions might become segregated seems to have much merit.

The stasipatric model of chromosomal speciation was put forward by White, Blackith, Blackith, and Cheney (1967) and White (1968) in order to explain (1) the enormous number of cases in which closely related species of animals of limited vagility have visibly different karyotypes, and (2) the large number of instances in which the assumption of strictly allopatric speciation seems unreasonable, mainly because it would force us to assume that the parent species had a geographic range too small to be plausible. Essentially, the stasipatric model envisages a widespread species generating within its range daughter species characterized by chromosomal rearrangements that play a primary role in speciation because of the diminished fecundity or viability of the heterozygotes. The daughter species are assumed to gradually extend their range at the expense of the parent species, maintaining a narrow parapatric zone of overlap at the periphery of their distribution within which hybridization leads to the production of genetically inferior individuals (usually inferior because of irregularities at meiosis).

In order to illustrate the stasipatric model of speciation we shall examine in detail three groups of organisms that were interpreted in terms of this model by the original investigators. Three other cases that also seem to be examples of stasipatric speciation but which were not interpreted in terms of this model by the original investigators will then be considered.

The Genus *Vandiemenella*

The Australian morabine grasshoppers of the genus *Vandiemenella* (Key, 1976) were referred to as the "*viatica* species group" in the earlier literature (White, Carson, and Cheney, 1964; White, Blackith, Blackith, and Cheney, 1967; White, 1968; White, Key, André, and Cheney, 1969; Key, 1974; White, 1974). These are wingless insects of low vagility; they are largely feeders on shrubs of the family Compositae *(Olearia* spp., *Helichrysum* spp.) but they are not restricted to these food plants. Seven species will be recognized here (three with two cytological races and one with three races), although Key regards five of these as races of a single species. The genus has a relatively restricted distribution from Tasmania to the Eyre Peninsula (about 400,000 square kilometers altogether). All the taxa have parapatric distributions, i.e., no two are sympatric over any extensive area; in a number of instances they meet in zones of hybridization 200–300 meters wide.

Since only two species of the genus have been described, the other five will be referred to here by numbers (preceded by the letter P for "provisional species"). The complete list of taxa is as follows:

Species	Races
V. viatica	19-chromosome
	17-chromosome
P24	XO
	XY
	Translocation
P25	XO
	XY
P45b	XO
	XY
P45c	
P50	
V. pichirichi	

The distribution of these forms is shown in Figure 16. In this figure the lines separating the ranges are carried out to sea, since it is considered that speciation occurred during the Pleistocene glaciation when Tasmania and Kangaroo Island were part of the mainland and when Spencer's and St. Vincent's Gulfs were dry land.

We may regard the karyotype of $viatica_{19}$ as the approximate ancestral one for the group. With only minor modifications, this same karyotype occurs in the geographically peripheral species P45c and P50. It consists of two pairs of long acrocentric autosomes (A and B), an unequal-armed metacentric auto-

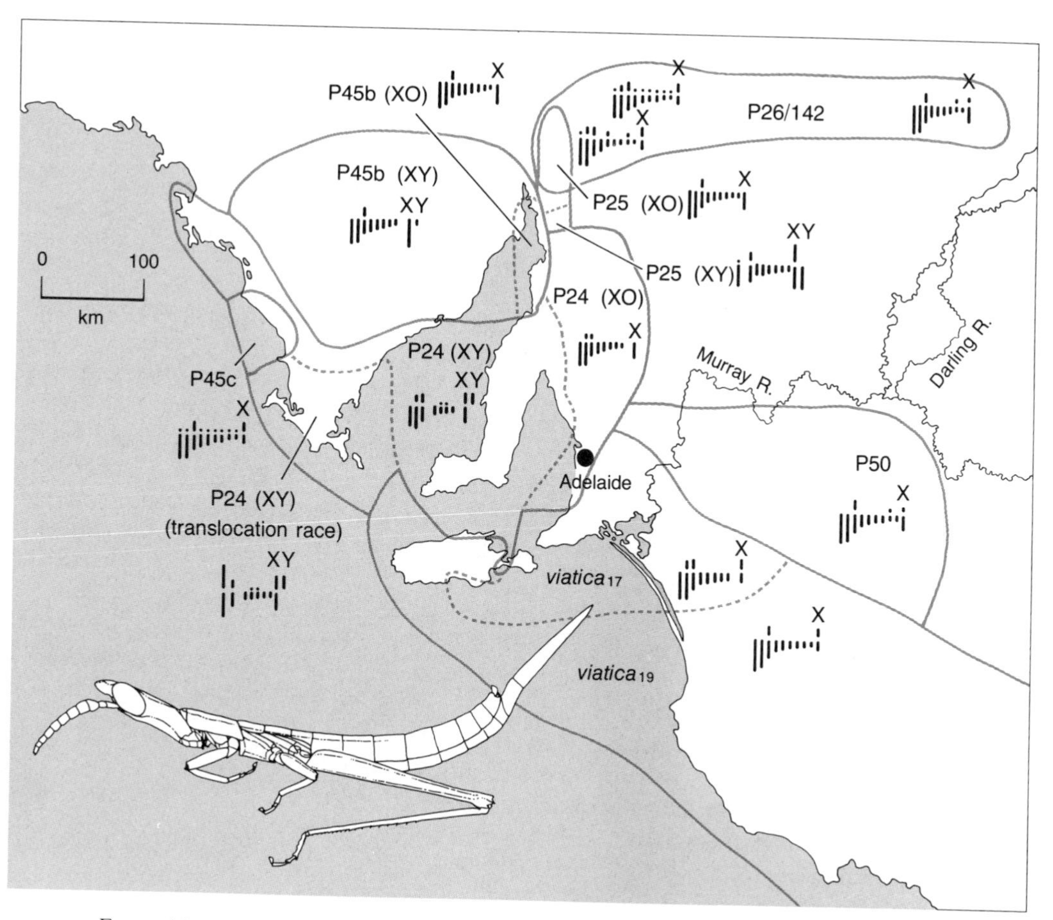

FIGURE 16

Distribution of the species and races of *Vandiemenella* in southern Australia. Ranges of taxa regarded as species are delimited by dashed lines, those of races by dotted lines. A diagram of the haploid karyotype of each taxon is given (except for taxa with XY males, both sex chromosomes are shown). "P26/142" = *V. pichirichi,* a chromosomally polymorphic species. The most easterly population assigned to it on external phenotype is karyotypically like "P50" and may belong to that species. The ranges are shown extending into areas of the Southern Ocean that were dry land during the Pleistocene. Insert: a male of *V. pichirichi*. [After Key, 1976]

some pair (CD), six pairs of small pairs of acrocentric autosomes (1 to 6), and a metacentric X chromosome. In populations of *viatica*$_{19}$ from near Melbourne the short arms of all the acrocentric chromosomes are exceedingly minute, but in *viatica*$_{17}$ from near Adelaide they are distinctly larger, and in P45c from the west coast of the Eyre Peninsula they are quite large. This seems to represent the remnants of an ancient east–west cline for the length of these short limbs. Since the material of the short arms is heterochromatic (revealed as late-replicating by autoradiographic techniques) it is clear that numerous duplications occurred in the short-arm material of the ancestral populations at the west end of the range of *Vandiemenella* (or that deletions of such material have taken place in the eastern populations).

Viatica$_{17}$ differs from *viatica*$_{19}$ in possessing a fusion of the long B autosome with one of the small acrocentric autosomes rather arbitrarily designated No. 6. It is thus homozygous for a J-shaped "B + 6" chromosome. The No. 6 autosome is late-replicating and heterochromatic in all members of the genus (Webb and White, 1975).

Species P24 also has the B + 6 fusion, except in the Translocation race, in which there has been a translocation between the B + 6 and the A chromosome, with both breakage points very close to the centromeres, so that the "new" chromosomes are effectively an AB metacentric and a free No. 6. The most primitive race of species P24 is the eastern XO one, which differs cytologically from *viatica*$_{17}$ only in that the X chromosome is acrocentric instead of metacentric, having presumably undergone a pericentric inversion. The XY race of P24 differs from the XO one in that this acrocentric X has undergone a centric fusion to the No. 1 autosome, thereby giving rise to a neo-XY system (the unfused No. 1 chromosome, now confined to the male line, becomes a neo-Y). This neo-XY mechanism exists also in the Translocation race.

Species P25 has an XO race that possesses the same acrocentric X as the XO race of P24. But it lacks the B + 6 fusion. The XY race of P25, known only from a very small area of the southern Flinders Ranges, has a centric fusion between the X and the B autosome.

Species P45b has an XO race that occurs on both sides of Spencer's Gulf. Its karyotype does not seem to differ from that of the XO race of P25 (but the two forms differ in the morphology of the male cercus). The XY race of P45b inhabits the northern part of the Eyre Peninsula and has a tandem, i.e., not a centric, fusion between the X and autosome No. 6.

Species P45c occupies a relatively small area on the west coast of the Eyre Peninsula. Its karyotype resembles that of *viatica*$_{19}$ except that the short arms of all the acrocentric chromosomes are considerably longer. These short limbs are composed of late-replicating DNA (Webb and White, 1975), so it is clear that there have been increases in the amount of satellite DNA in the evolution of this species.

There are two "inland" species of *Vandiemenella*. One, P50, occurs in areas of "mallee" vegetation in areas of western Victoria and the eastern edge of the state of South Australia. It has 2n ♂ = 19, but differs from *viatica* in having a submetacentric X chromosome and in the fact that one of the small autosomes is a metacentric. The other, *V. pichirichi,* consists of a series of populations in the Flinders Ranges. As in species P45c, the acrocentric chromosomes (all of them in northern populations, only some in southern demes) have much longer short arms than *viatica* and P50. The reason is quite different, however, since there has been no significant increase in the amount of late-replicating DNA in *pichirichi*. In consequence, we must conclude that pericentric inversions have been responsible for the difference between the acrocentric chromosomes of *pichirichi* and *viatica*.

Discussions of speciation in the genus *Vandiemenella* have been based mainly on the "coastal" species. Zones of overlap have been found in the following cases: (1) $viatica_{19}$ and $viatica_{17}$, on the mainland and on Kangaroo Island; (2) P24 (XY race) and $viatica_{17}$ on Kangaroo Island; (3) P24 (XY race) and $viatica_{19}$ on Kangaroo Island; and (4) P24 (XO and XY races) on the mainland. Other zones of overlap, e.g., between the XY races of P24 and P25, between the XO and XY races of P45b, and between the XO race of P24 and $viatica_{17}$, must have existed until recently, having disappeared because of the spread of agriculture and industry.

There is no reason to believe that any major geographic barriers ever separated the species of *Vandiemenella*. Their distribution was probably always continuous or essentially so, especially along the coast. An interpretation of their speciation along allopatric lines would have to postulate a gradual westward spread from the region of Bass Straits, with the invading populations acquiring the various chromosomal rearrangements *seriatim* (the B + 6 fusion, a pericentric inversion giving an acrocentric X, and the various fusions leading to XY mechanisms). On climatologic and biogeographic grounds this does not make much sense. The center of distribution of *Vandiemenella* is clearly the region of Spencer's Gulf and the Flinders Ranges, where the majority of the taxa now live, and during much of the Pleistocene the Bass Straits region was certainly too moist and cool for *Vandiemenella*. If there has been a substantial change in the total area occupied by the genus during the past 15,000 years, the change was probably due to an easterly migration of $viatica_{19}$ to coastal Victoria and Tasmania.

An interpretation of speciation in the coastal forms of *Vandiemenella* according to a chromosomal model, called *stasipatric,* was put forward by White, Blackith, Blackith, and Cheney (1967) and subsequently elaborated by White (1968, 1974). Briefly, the model (Figure 17) assumes that the area now occupied by the genus *Vandiemenella* (or most of it, anyhow) was originally occupied by a species we may call *proto-viatica*. This had a karyotype essentially identical to that of $viatica_{19}$, which may be regarded as its direct descendant,

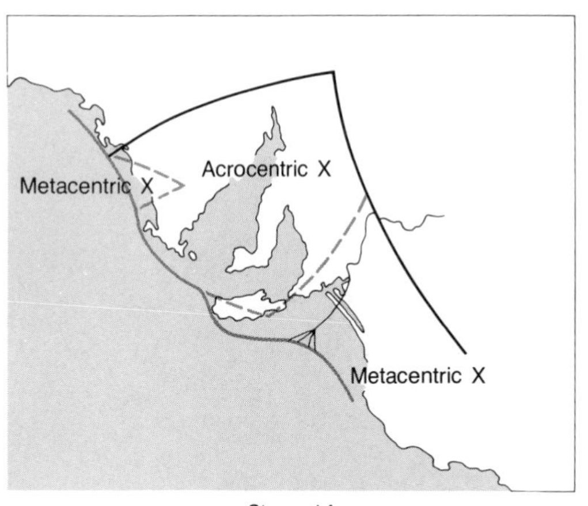

Stage 1A

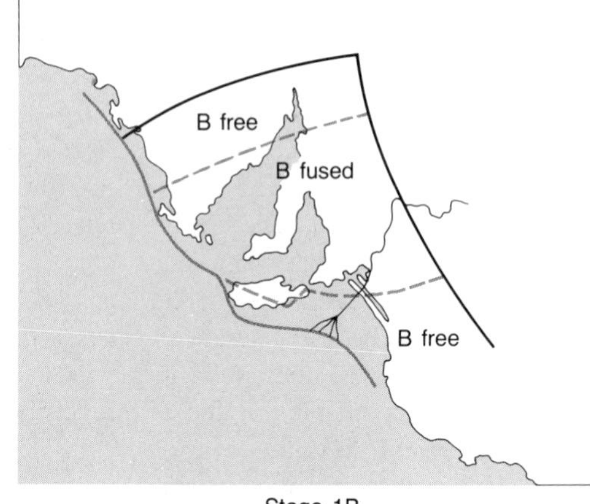

Stage 1B

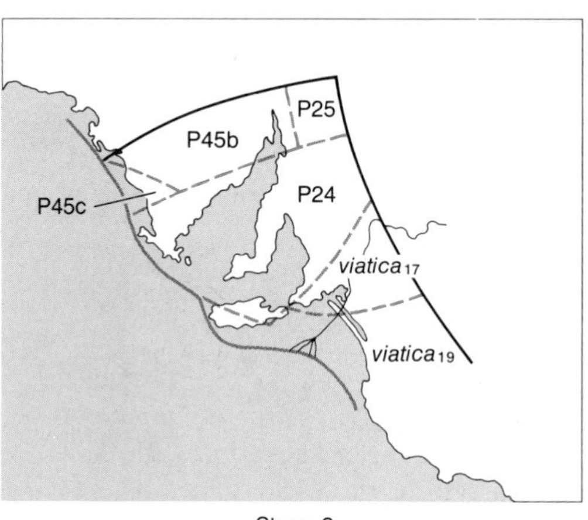

Stage 2

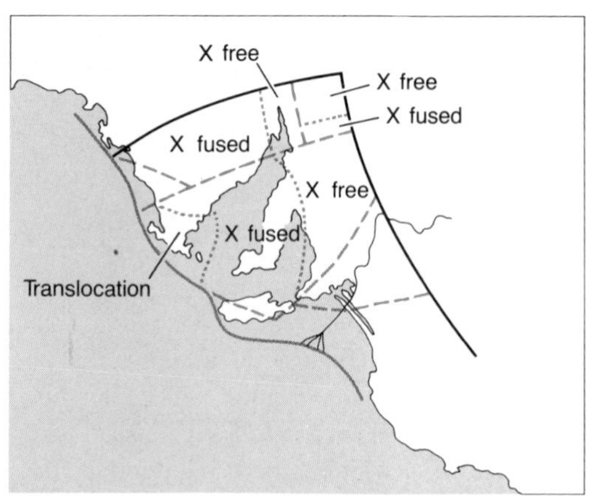

Stage 3

having undergone no major karyotypic changes. Two "basic" structural rearrangements are assumed to have occurred at points near the center of distribution of *proto-viatica:* the B + 6 fusion and a pericentric inversion giving rise to the acrocentric X chromosome of the XO races of species P24, P25, and P45b (no evidence exists as to which of these two rearrangements occurred first). Having established themselves in the populations in which they occurred, these rearrangements then spread out from their point of origin. This spreading process was *in spite of* the fact that heterozygotes for the rearrangements were disadvantaged by a segregational genetic load. It was presumably due to a combination of homozygote advantage (i.e., the homozygote for the new arrangement was adaptively superior to the homozygote for the original karyotype), genetic drift in a large number of small isolated or semi-isolated demes, and (perhaps) a segregational advantage at meiosis. The spread of the two arrangements took place northward and southward; populations that received neither produced the species $viatica_{19}$ (southeastern) and P45c (northwestern), those that received both produced P24, those that got the B + 6 fusion alone produced $viatica_{17}$, and those that got the X chromosome inversion but not the B + 6 fusion gave rise to P25 and P45b. In order to explain the distribution of the cytological races of species P24, P25, and P45b, it is postulated that a second round of chromosomal rearrangements (three different X autosome fusions and a translocation between the A chromosome and the B + 6) occurred later and that they spread in the same kind of manner. The entire process involved the fragmentation of the original territory of occupation into the territories now occupied by the various taxa, with minimal, or no, expansion of the total area inhabited by the genus. In each of the narrow zones of overlap, heterozygotes with reduced fecundity (due to meiotic accidents that lead to aneuploid gametes) would continue to be produced. It is assumed that premating isolating mechanisms did not play any part in the process.

The first investigations on the situation, in the zone of overlap between $viatica_{17}$ and $viatica_{19}$ on the mainland near Keith, showed that hybrids were in fact present in nature (White, Carson, and Cheney, 1964). A study of meiosis in the male hybrids (Figure 18) showed that they usually had a "lopsided"

FIGURE 17

Stasipatric model of speciation in coastal members of the genus *Vandiemenella* in southern Australia. It is assumed that two events occurred in an original *proto-viatica* population: a pericentric inversion that converted the originally metacentric X chromosome into an acrocentric (stage 1A) and a fusion between the number 6 and B chromosomes (stage 1B). The combination of these events (it is not known which occurred first) produced stage 2. The further addition of three different X–autosome fusions and a translocation in the southern part of the Eyre Peninsula produced stage 3, which is virtually the situation at the present day. [After White, 1968]

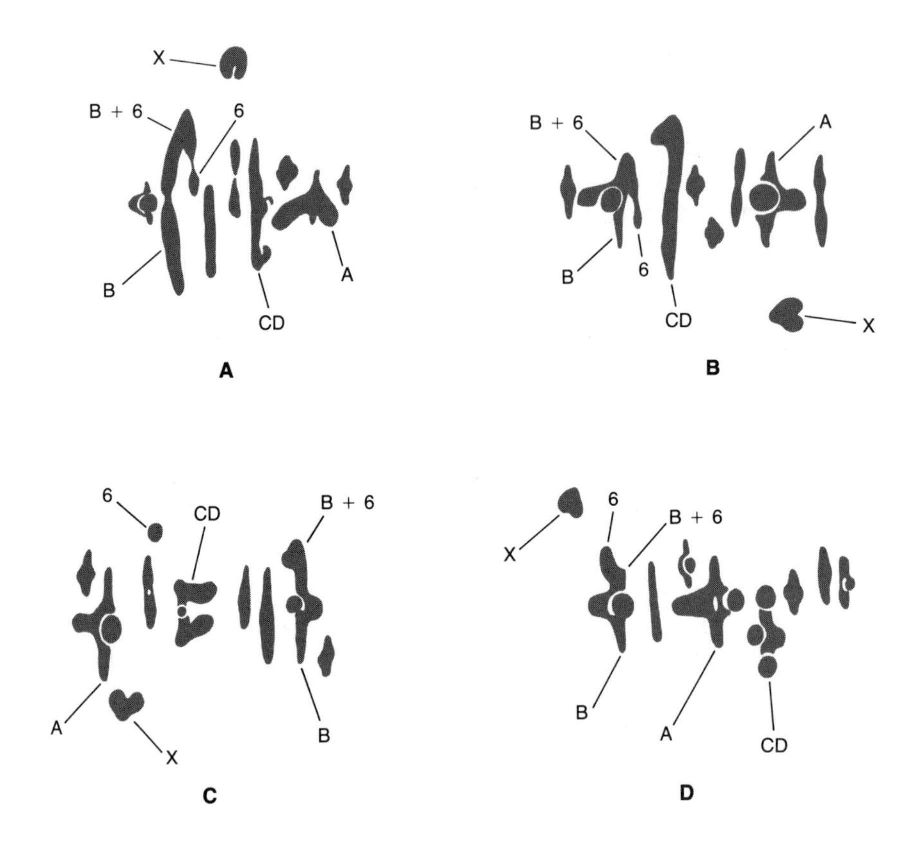

FIGURE 18

Male first metaphases (side views) of natural hybrids between the 19-chromosome and 17-chromosome races of *Vandiemenella viatica*. In figures *A*, *B*, and *D* the number 6 chromosome is synapsed with the B+6 chromosome from the 17-chromosome race, so that a trivalent is present, which appears maloriented in figure *D*. In figure *C* the number 6 chromosome is a univalent. Figures *A* and *B* are from an individual heterozygous for a pericentric inversion in the CD chromosome. [From White, Carson, and Cheney, 1964]

trivalent made up of the B + 6 chromosome from the *viatica*$_{17}$ parent and the long B and short No. 6 chromosomes from the *viatica*$_{19}$ parent. In a small percentage of spermatocytes, however, the No. 6 chromosome was a univalent. Such cells would be expected to give rise to aneuploid sperms in 50 percent of all cases. Another way in which aneuploid gametes could arise is through malorientation of the trivalent, with the B + 6 and the small No. 6 chromosomes passing to the same pole at first anaphase, but it is uncertain how significantly this contributes to aneuploidy.

The situation on Kangaroo Island, which has been cut off from the mainland for about 12,000 years, has been investigated in detail by White, Key, André, and Cheney (1969) and especially by Mrongovius (nee André) (1975). Three members of the genus *Vandiemenella* occur on the island: *viatica*$_{19}$, *viatica*$_{17}$, and the XY race of P24 (Figure 19). The relationships of these taxa in the three zones of overlap (*viatica*$_{17}$–*viatica*$_{19}$, *viatica*$_{17}$–P24(XY), and *viatica*$_{19}$–P24(XY)) have been studied in detail. Only the first of these also occurs on the mainland, because there P24(XY) is not in territorial contact with either of the races of *viatica* (see Figure 16).

The zone of overlap on Kangaroo Island between *viatica*$_{17}$ and P24(XY) is about 200–300 meters wide and does not correspond to any particular ecological discontinuity. Numerous hybrid females were found in this zone; hybrid males are presumed to be present as well, but they are undetectable cytologically as well as morphologically. Some biometrical studies, in particular on a hook-like structure on the male cerci, suggest that a substantial amount of introgression is occurring in this hybrid zone and that there has been a spread of genes from each taxon into the other well beyond the zone within which chromosomal hybrids occur. Nevertheless, the narrowness of this zone argues

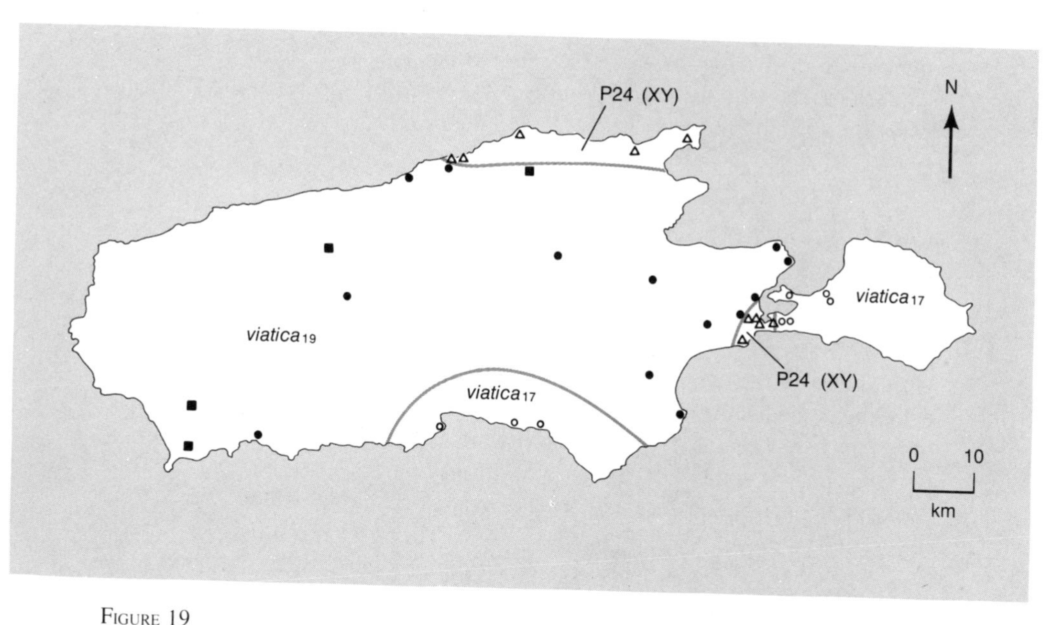

FIGURE 19

Distribution of three taxa of *Vandiemenella* on Kangaroo Island, Australia, showing the zones of narrow overlap. [After White, Key, André, and Cheney, 1969]

strongly for the view that in *viatica*$_{17}$ territory there is strong selection against the P24 X chromosome and vice-versa. The meiosis of the hybrids (including male hybrids reared in the laboratory) is effectively normal, but some female hybrids (not all) have a marked tendency to transmit the *viatica*$_{17}$ X chromosome to their offspring more frequently than the P24 one.

The zone of overlap between P24(XY) and *viatica*$_{19}$, about six kilometers further west, is quite different. It is at least as narrow, and perhaps narrower. However, the single hybrid that was collected in it was a young nymph and there is some evidence that these do not survive to the adult stage.

Morphometric differentiation in the genus *Vandiemenella* has been studied by Atchley (1974; and see Atchley and Hensleigh, 1974; Atchley and Cheney, 1974). Rather marked morphometric divergence between the chromosomal races was found to have occurred in general body shape and in the structure of the female "egg guide." In respect of the egg guide, 70–90 percent of the specimens of each race were discrete from those of other races of the same species, the divergence being that seen over and above interpopulation variation within the races. The results of these morphometric analyses were generally consistent with the chromosomal evolution that has occurred. Thus, among the species, *viatica* is most similar in its egg guide to P24 and most distinct from P45b and P45c among the coastal forms. The differences between the species are greater than those between the chromosomal races of each species, and similarities between the more primitive XO races of species P24, P25, and P45b are greater than those between the "derived" XY races of those species.

Didymuria violescens

The Australian stick insects belonging to the species (or species complex) *Didymuria violescens* occur in the eucalypt forests of predominantly montane environments in a great arc from the vicinity of Brisbane in southeastern Queensland to the Mount Lofty Ranges near Adelaide, South Australia. The cytogenetic studies of Craddock (1970, 1974a, 1975) are extremely complete, but there have been as yet no biometrical or isozyme studies on this most interesting case. A total of 10 cytological "races" have been discovered by Craddock; their distributions are essentially parapatric and the widest known zone of overlap is about six kilometers across. The most primitive karyotype, 2n♂ = 39(XO), is found in race I, which occurs, somewhat discontinuously, on both sides of the Great Dividing Range from near Brisbane southward to the vicinity of Sydney; it also occurs near Adelaide. Race I is thus by far the most widespread and has a distribution that largely encloses the "derived" races (this type of distribution is similar to that of *Vandiemenella,* in which the 19-

chromosome forms are peripheral and the "derived" 18-, 17-, and 16-chromosome taxa more central).

Race II of *Didymuria* seems to differ from race I only with respect to a pericentric inversion in the X chromosome. The other races all have one or more chromosomal fusions: race III has 2n♂ = 37 (one autosomal fusion), race IV has 2n♂ = 35 (two autosomal fusions), and race V has two additional autosomal fusions and some pericentric inversions (2n♂ = 31). These three forms all have XO males; races VI to X all have XY males, as a result of one or more X–autosome fusions (Figure 20).

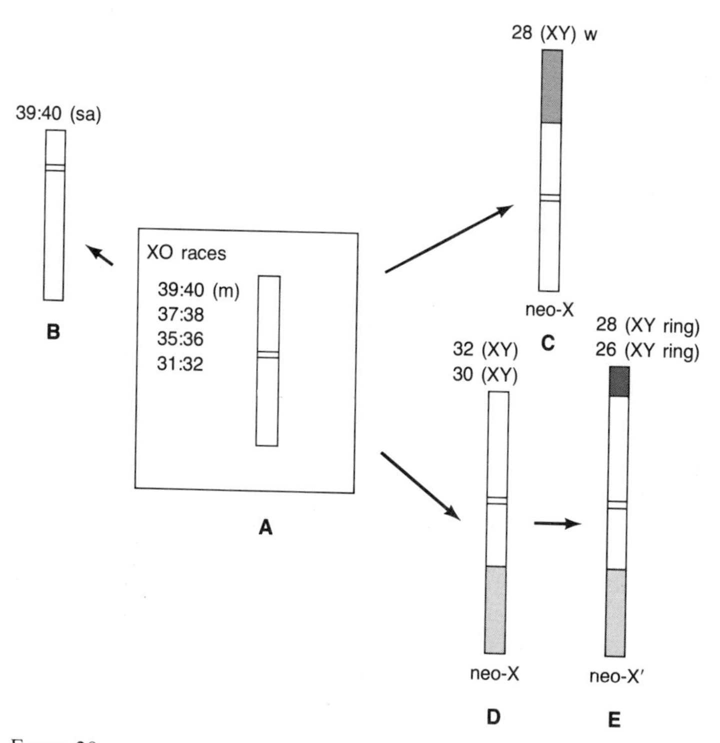

FIGURE 20

Postulated sequence of X chromosome rearrangements in *Didymuria violescens*, showing the interrelationships between the sex chromosome mechanisms of the 10 races. *A* to *B*, pericentric inversion; *A* to *C*, translocation of an autosome onto the short arm of primitive X; *A* to *D*, a different X–autosome translocation with pericentric inversion; *D* to *E*, translocation of another autosome onto the short arm of the neo-X. [After Craddock, 1975]

In race VI (2n♂♀ = 32) there has been a tandem fusion between the tip of the shorter arm of the original acrocentric X and one of the subacrocentric autosomes (the centromere of the autosome being lost in the process). This fusion, which gave rise to the primary neo-XY system, must have arisen in a now-extinct population with 2n♂ = 33, i.e., one intermediate between races IV and V. Race VII (2n♂♀ = 30) differs from race VI only in having an extra autosomal fusion and perhaps a pericentric inversion as well.

Race VIII (2n♂♀ = 28) has one more fusion than race VII. It has a neo-XY mechanism that differs in the relative lengths of the chromosome arms from the primary neo-XY of races VI and VII; this probably had an independent origin involving a tandem fusion of the X chromosome with a shorter autosome. Race VIII is only known from a single rather isolated peripheral locality; its origin is unclear and Craddock (1974a) implies that it may have evolved from the original ancestor (with the karyotype of race I) independently of all the other races.

Races IX and X (2n♂♀ = 28 and 26, respectively) have a neo-XY system in which both ends of the Y chromosome are associated with the two ends of the X at meiosis to form a ring XY bivalent. This type of neo-XY mechanism has presumably arisen from the primary one by two tandem fusions in the course of which the members of a pair of autosomes were attached to the "free" ends of the X and Y. Depending on which fusion occurred first, an X_1X_2Y or an XY_1Y_2 mechanism would have existed as an intermediate evolutionary stage between the primary XY and the ring XY mechanisms.

The distribution of these 10 taxa is shown in Figure 21. The primitive race I is peripheral in three senses: it occupies the two ends of the long arc of the overall distribution of *D. violescens,* the coastal areas near Sydney, and the west (the arid) side of the Great Dividing Range. Whether the populations known to occur along the River Murray in Victoria belong to this race or to the southern race IV is not known. Race II, which is perhaps rather closely related to race I, is also coastal (i.e., peripheral) in northern New South Wales. The highly derived races IX and X inhabit small areas that are more or less in the center of the main distribution of *D. violescens.*

Natural hybrids between some of the races have been found, and other hybrid combinations have been reared in the laboratory. There seem to be no ethological barriers to mating, and F_1 hybrids are viable. On the other hand, their meiosis is somewhat irregular, as a result of rather frequent failure of synapsis, failure of multiple chromosome associations to orientate disjunctionally at the first meiotic division, and occasional synapsis between nonhomologous chromosomes. Although the data have not been published in full, the fecundity of many of the hybrids is rather severely reduced (Craddock, 1974a). In addition to segregational sterility, due more or less directly to the chromosomal rearrangements, some hybrids also show evidence of genic sterility. There can be no

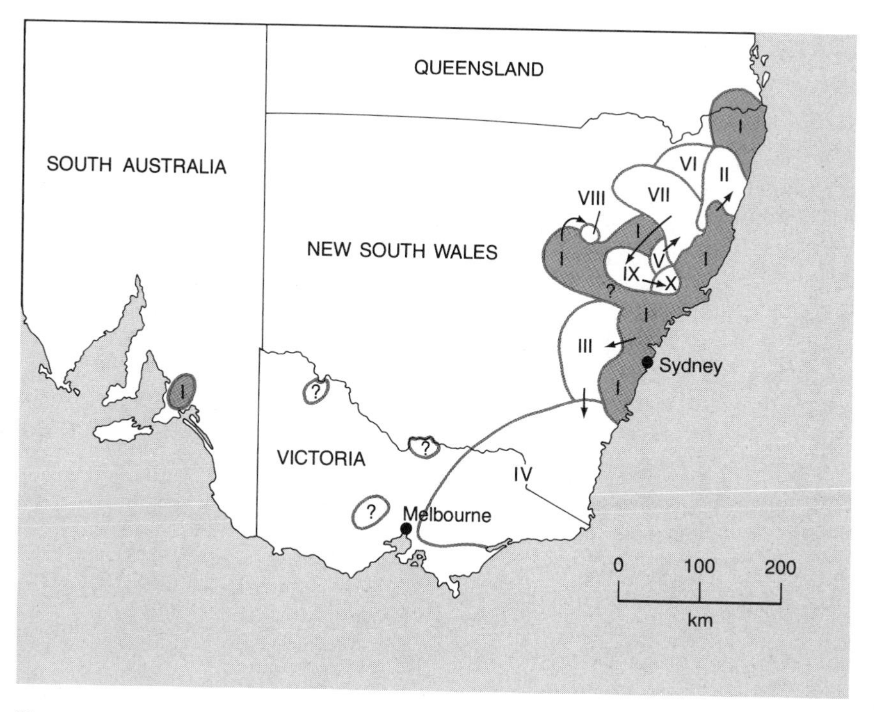

FIGURE 21

Distribution in eastern Australia of the 10 races of *Didymuria violescens*. [After White, 1976]

constitute fairly strong postmating isolating mechanisms and that premating isolating mechanisms are either absent or too weak to be detectable. As in the case of *Vandiemenella,* it is abundantly clear that chromosomal rearrangements have played a major role in the speciation of this group of insects. Craddock refers to all 10 forms as "races." They are, however, rather strongly isolated in some instances and might well be regarded as semispecies. There has been no biometrical study of these forms, similar to the ones carried out by Atchley on the taxa of the genus *Vandiemenella,* and no investigations on their genic differentiation, using gel electrophoresis have been performed.

Sceloporus grammicus

The North American lizard genus *Sceloporus* includes 64 described species and, according to Hall (1973), the total number is more than 70. The karyotypes of the 56 species that have been studied exhibit a range of chromosome numbers from 2n = 22–46, due to "Robertsonian" variation. However, only the *Sceloporus grammicus* species complex shows significant interpopulation chromosomal variation (Hall and Selander, 1973). The ancestral karyotype of this complex was undoubtedly 2n♂ = 31 (X_1X_2Y): 2n♀ = 32 ($X_1X_1X_2X_2$). This "Standard" karyotype (S), with six pairs of metacentric large autosomes and eight pairs of micro-autosomes, is rather widespread in Mexico from Coahuila in the north to Oaxaca in the south. The "polymorphic-1" karyotype (P1) differs from Standard in being polymorphic for a dissociation* of chromosome 1 into two acrocentrics; males have 2n = 31–33 and females 2n = 32–34. P1 populations occur in an area of less than 500 square kilometers above 3,200 meters elevation in the chain that includes the mountains Tlaloc, Ixtaccihuatl, and Popocatepetl. The dissociation is fairly rare, having an overall frequency of 0.1. In the "Fission-6" population (F6) a dissociation of the metacentric chromosome 6 has become fixed in the homozygous condition. This karyotype occurs across the Mexican central plateau, from the state of Tlaxcala to Michoacán; some apparently isolated colonies occur in northern Mexico. In Chihuahua a few populations (called F5) are homozygous for a dissociation of chromosome 5, and over a wide area of northern Mexico and south Texas there are populations (F5 + 6) homozygous for dissociations of chromosomes 5 and 6; they consequently have 2n = 35 (♂): 36 (♀). Finally, "multiple fission" forms of *S. grammicus* (FM) occur in an area of approximately 10,000 square kilometers north of Mexico City. Two subpopulations were recognized by Hall. Subpopulation FM1 is homozygous for dissociations of chromosomes 2, 3, 5, and 6 and polymorphic for dissociations of chromosomes 1 and 4, so that it has 2n = 37–41(♂): 38–42(♀). Subpopulation FM2 is also homozygous for a dissociation of one of the microchromosomes; it may be homozygous for the dissociation of chromosome 1, but this is not certain.

Some form of *S. grammicus* probably exists everywhere in Mexico north of the Isthmus of Tehuantepec and above 1,000 meters elevation, i.e., excluding the coastal *tierra caliente;* on the central plateau they are only absent from the hottest deserts. It is thus extremely unlikely that isolation by extrinsic geographic barriers has played any significant part in the origin and spread of the various chromosomal forms of the *grammicus* complex. It is probable that all these are esentially parapatric, with narrow zones of overlap in which hybridization occurs. This is certainly the case in the Valley of Mexico, where zones of overlap less than 500 meters wide have been found between populations P1 and F6, between S and FM2, and between S and F6.

Natural hybridization between P1 and F6 was studied by Hall and Selander

---

*Hall and Selander refer to such rearrangements as "fissions," believing them to be simple breaks through the centromere.

(1973) in their zone of overlap, using the karyotypic difference and two isozyme differences that are fixed in these taxa. Thirteen F_1 hybrids and 56 backcross hybrids were found (27 to F6 and 29 to P1), but no F_2 individuals. There was no indication of introgression into the F6 population, and if there was any introgression into P1 it was very slight. It was therefore concluded that despite little or no ethological isolation between P1 and F6, the survival rate of recombination products is poor and backcross hybrids leave few if any descendants. The two populations thus behave in nature as good biological species.

Similar results were obtained in the overlap zone between populations S and FM2. A few F_1 and backcross hybrid individuals were identified on the basis of their cytology, but no evidence of introgression was found. Consequently, these populations also seem to be behaving as good biological species. Unfortunately, the overlap zone between the two has not been studied in detail.

Although much more information on *S. grammicus* is obviously desirable, there is good reason to believe that the "populations" S, P1, F6, F5 + 6, and FM are actually biological species; F5 may also be a distinct species. Whether FM1 and FM2 are connected by a series of intermediate populations is uncertain. The actual phylogenic history of the members of the complex is somewhat uncertain. It seems clear, however, that the mechanism of speciation in the complex has been a chromosomal one. And the fact that the most primitive form (S) has a generally peripheral distribution, while the most highly derived karyotypes (FM1 and FM2) are central, is strongly reminiscent of the distributions of *Vandiemenella* and *Didymuria,* and suggests that the *Sceloporus grammicus* complex is the result of fragmentation of an ancestral population that extended from the Rio Grande Valley to the Isthmus of Tehuantepec, as a result of a series of chromosomal dissociations ("fissions" in Hall's terminology) that acted as isolating mechanisms. It seems unlikely that there is any ethological isolation between the species of the complex, but it is unknown whether F_1 hybrids are as fertile as the parental species. Minor geographic isolation may have facilitated the establishment of the various fissions, but the ancestral *S. grammicus* exhibits extreme ecological tolerance. An apparently continuous population, it occupies all habitats where tree trunks or fallen logs with crevices occur, from rain forests at high elevation (3,600 meters) to the most xeric *Yucca* and *Agave* deserts; all the "derived" taxa are also found within this ecological range.

Hall (1973) has considered three models of chromosomal speciation in *Sceloporus:*

1. The "peripheral version," which involves the establishment of chromosomal rearrangements only on the periphery of the species distribution (this is the one favored by Key in the case of *Vandiemenella*).

2. The "interstitial version," which assumes that establishment of the rearrangements leading to speciation can occur anywhere within the range of a species whose population is subdivided into a very large number of small, partially and temporarily isolated demes (this is the stasipatric model of White).
3. An intermediate model, which assumes that the rearrangements can establish themselves on "internal peripheries" in the case of a species that has a mosaic distribution (due to exclusion by ecological factors from numerous island areas within its overall range).

Hall is inclined to favor the second and third alternatives because of the continuous range and wide ecological tolerance of *grammicus,* although he admits that it is impossible to tell how far the distribution of the ancestral population may have been affected by the climatic fluctuations of the Pleistocene.

The members of the *S. grammicus* complex seem to have reached about the same stage of speciation as the taxa of *Didymuria violescens.* The fact that Craddock refers to the *Didymuria* taxa as "races" while Hall regards the *S. grammicus* taxa as "good biological species" must be regarded more as a reflection of the subjective judgments of the investigators than as a genuine difference between the two cases.

It is noteworthy that in all three of the relatively clear cases of stasipatric speciation considered above, premating isolating mechanisms have not been demonstrated; if they exist at all they must be very weakly developed, in view of the frequency of natural F_1 hybrids in the zones of overlap. There is absolutely no evidence in *Vandiemenella, Didymuria,* or the *Sceloporus grammicus* complex for speciation by any mechanism other than a chromosomal one.

Hall has made the point that iguanid lizards are in general karyotypically conservative; the great majority of the 29 genera that have been investigated show a karyotype of either 2n = 36 (six pairs of metacentric large chromosomes and 12 pairs of microchromosomes), which is believed to be primitive in the order Sauria, or 2n = 34 (with one less pair of microchromosomes). It is only in four species-rich genera that one finds significant deviations from these karyotypes:

Sceloporus (~ 70 species)	2n = 22–46
Anolis (> 200 species)	2n = 26–48
Liolaemus (> 50 species)	2n = 32–40
Polychrus (? species)	2n = 20–30

These facts in themselves suggest that these genera possess an ability to generate new species by processes of karyotypic change that are rare or lacking in the remaining genera of Iguanidae. Most probably, both the large number of species in *Sceloporus* and their karyotypic diversity indicate that earlier radiations of chromosomal speciation occurred in the phylogeny of the genus. Hall suggests that because this kind of chromosomal speciation is a consequence of particular types of genetic systems and population structure, and because the neospecies generated by it will themselves possess these properties and will occur in areas favorable for this type of speciation, there is likely to be a cascading process, the neospecies themselves generating yet more incipient species in a further round of speciation. This certainly seems to be the case in *Vandiemenella,* where the three XY races and the Translocation race of species P24 clearly represent a later round of speciational events than the primary one (the fact that they have not led to full speciation as yet is hardly relevant here).

One obvious difference between the *Sceloporus grammicus* case and the *Vandiemenella* and *Didymuria* ones is that in the *S. grammicus* complex a few of the dissociations do not seem to have led to genetic isolation but, on the contrary, have attained genetic equilibrium and established themselves as polymorphisms (in the P1 and FM species), whereas in *Vandiemenella* and *Didymuria* no fusions have established themselves in a balanced polymorphic condition.

The *Jaera albifrons* Species Group

It has been demonstrated by Bocquet (1953; and see Staiger and Bocquet, 1954, 1956; Lécher and Prunus, 1972; and Lécher, 1964, 1967, 1968) that the marine isopod species complex *Jaera albifrons,* inhabiting the Atlantic coast of Europe, exhibits a considerable range of karyotypes. The species karyotypes are:

Species	2n♀
J. albifrons	19,20,21 (polymorphic for a chromosome fusion)
J. forsmani	19
J. ischiosetosa	27,28,29,30,30,31,32,33 (polymorphic for fusions)
J. praehirsuta	27
J. syei	19,21,23,25,27,29 (in different populations)

Females of these species are the heterogametic sex and show an XY_1Y_2 trivalent at meiosis, the X being metacentric and the two Y chromosomes acrocentric. Males consequently are XX and have one chromosome less than the corresponding females. On the other hand, *J. posthirsuta,* a member of this species group from the east coast of the U.S., has 2n = 20 in both sexes and is presumably XY in the female (Lécher, 1962). The number of major chromosome arms *(nombre fondamental* or NF of Matthey) is always 36 in this species complex, apart from supernumerary chromosomes, which may occur in certain individuals. If one assumes with Bocquet (1969) that these crustaceans have a northern origin, it seems likely that a series of chromosomal fusions have occurred in the more southern populations.

Where several species of *Jaera* are sympatric, they tend to occupy different zones on the shore. Thus, in the region of Roscoff *ischiosetosa* occupies the subterrestrial and strongly euryhaline zone; *albifrons* and *syei,* more definitely marine, live in the *Sphaeroma serratum* zone; *praehirsuta* populations live lower down, in the *Fucus* zone; while *forsmani* lives partly together with *praehirsuta,* but mainly between the preferred zones of *praehirsuta* and *albifrons* (Bocquet, 1969). It is thus possible that the primary "geographic" factor in speciation has been the zonation of the seashore. Interspecific hybrids occur, but usually constitute less than one percent of the population, in the zones of contact. However, at Luc-sur-Mer (Calvados) there is extensive hybridization between *J. praehirsuta* and *J. syei,* 15–25 percent of the population being clearly F_1,F_2, or backcross hybrids (Solignac, 1969a, b). A similar situation exists near Kiel, Germany, where *J. praehirsuta* and *J. ischiosetosa* hybridize.

J. syei is a particularly complex species. At Wilhelmshaven, Germany it shows 2n♀ = 29. Further south, the chromosome number (2n♀) is 27 at den helder (Holland), 25 at Luc-sur-Mer, 23 at Roscoff, 21 at Royan, and 19 at Biarritz. But this is not a simple case of a north–south cline in chromosome number, since 2n♀ = 25 in southern Sweden and 23 in northern Sweden and at Oslo, Norway.

Along extensive stretches of coastline, the populations of *J. syei* are chromosomally uniform. Lécher (1967) studied a zone of hybridization between a 2n♀ = 25 and a 2n♀ = 23 population on the north coast of Brittany; the zone extends over about 20 kilometers of coast. There are two peculiar features about his data. In the first place, the frequency of the metacentric chromosome responsible for the 23-chromosome karyotype does not vary in a clinal manner, but very irregularly. Thus, eight populations sampled (from east to west) showed the following frequencies for this chromosome: 0.00, 0.40, 0.80, 0.65, 0.44, 0.05, 0.56, and 1.00. Second, all the polymorphic populations showed a deficiency of heterozygotes by comparison with the expectation based on the binomial square rule. The explanation of this situation is not obvious; it is possible that the "populations" are in fact mosaics of "micropopulations" of the two forms, between which there is only limited migration (such a situation could be due to the zonation of the seashore). Alternatively, we may simply be dealing with a situation where the F_1 hybrids between two incipient species are inferior in viability to the parental forms. A similar deficiency of structural heterozygotes for fusions or dissociations has been found in populations of *J.*

ischiosetosa on the coast of Iceland (Lécher and Solignac, 1972); this species shows a similar polymorphism on the coasts of continental Europe.

Just what role chromosomal rearrangements have played in speciation in this group is not really clear. Bocquet (1969) seems to favor a modified form of the stasipatric model, with the rearrangements arising at the periphery of the species distribution, as suggested by Key (1968, 1974). One thing is clear, however. The populations of certain species are regularly polymorphic for chromosome number, and in these the orientation of the trivalents is always disjunctional. There is thus not much reason to suppose that chromosomal fusions (or dissociations) have played a primary role in *Jaera* speciation because of a lowered fecundity of structural heterozygotes, although they may well have done so because of a lower viability of the heterozygotes. In all discussions of these species it is necessary to bear in mind the precise zonation of the seashore, a type of stratification of the environment not encountered by morabine grasshoppers, stick insects, or *Sceloporus* lizards, whose habitats may be variously diversified, but generally in a mosaic fashion. Also, there is a strong possibility that ethological premating isolating mechanisms may have played an important part in the speciation of the *Jaera albifrons* complex (Solignac, 1972).

The Genus *Chilocorus*

Speciation in North American species of the ladybird beetle genus *Chilocorus* (Coccinellidae) has been studied by S. G. Smith (1966) and in some Palaearctic species by Zaslavsky (1963, 1967a, b). While neither author interpreted the facts in terms of the stasipatric model—which had not been put forward at that time—the facts in both cases suggest that it is applicable.

Species of *Chilocorus* studied by Smith generally differ in karyotype. However a group of cytologically primitive species *(fraternus, cacti, bipustulatus, orbus)* have the same karyotype, when studied by beta karyology (Smith, 1962).

Chilocorus tricyclus (2n = 20) and *C. hexacyclus* (2n = 14) are very closely related species; by comparison with the karyotype that is primitive for the genus (2n = 22), the first has one chromosomal fusion while the second has four. *Tricyclus* occurs west of the Rocky Mountains in British Columbia, and the state of Washington, while *hexacyclus* inhabits Saskatchewan and southern Alberta. A hybrid population occurs on the east side of Crowsnest Pass in the Rocky Mountains. Of 389 specimens examined cytologically, a maximum of 202 were *hexacyclus* and 12 *tricyclus;* the remaining 175 all showed some evidence of hybrid ancestry (Table 6-1).

TABLE 6-1
Composition of the Crowsnest Valley population of Chilocorus.* *(S. G. Smith, 1966.)*

♂♂ with *hexacyclus* chromosome no. (2n = 14) and *hexacyclus* Y	132
hexacyclus " " (2n = 14) " *tricyclus* Y	24
tricyclus " " (2n = 20) " *tricyclus* Y	8
tricyclus " " (2n = 20) " *hexacyclus* Y	0
2n = 17 (possible F_1 hybrids) and *hexacyclus* Y	16
" (" " ") and *tricyclus* Y	30
2n = 15, 16, 18 or 19 (possible backcross and F_2 hybrid individuals)	53
♀♀ with *hexacyclus* chromosome no. (2n = 14)	70
tricyclus " " (2n = 20)	4
2n = 17 (possible F_1 hybrids)	29
2n = 15, 16, 18 or 19 (possible backcross and F_2 hybrid individuals)	23
Total	389

*The proportion of males in the sample reflects a preference of the investigator for analyzing males, and not a real excess of males in nature.

It is fairly clear that *hexacyclus* arose from a *tricyclus*-like ancestor, with three chromosomal fusions occurring during the process. Synapsis is virtually complete in the hybrids. Thus F_1 individuals show an XY bivalent, three autosomal bivalents, and three trivalents. Each of the trivalents is composed of a *hexacyclus* metacentric synapsed with two *tricyclus* chromosomes (the latter are actually metacentrics with one limb euchromatic and the other heterochromatic; it is characteristic of chromosomal fusions in *Chilocorus* that they occur between the euchromatic arms of such "diphasic" chromosomes, the heterochromatic limbs being lost in the process). In the hybrids these trivalents frequently orientate in a linear (nondisjunctional) manner, thus giving rise to aneuploid gametes and a considerable degree of sterility. In spite of this, a large amount of backcrossing obviously goes on, with consequent introgression of genetic material from one species into the other. It is fairly certain, from the frequency of occurrence of natural hybrids, that there is no ethological isolation between the species, which Smith refers to as "externally indistinguishable."

Clearly, the Crowsnest Valley population consists mainly of *hexacyclus;* it is apparently invaded yearly by individuals of *tricyclus* carried through Crowsnest Pass by westerly winds. Were it not for this the two species would be separated by the continental divide. As in the case of *Vandiemenella,* there must be strong selection against "unfused" *(tricyclus)* chromosomes in "fused" *(hexacyclus)* territory. But the hybrid zone is considerably wider than in *Vandiemenella* because these are winged insects, liable to be transported by air currents.

The precise mode of origin of *hexacyclus* is not clear, but it presumably arose from *tricyclus* in three stages, each characterized by a new fusion (it is, of course, conceivable that the three fusions arose in a single individual). Assuming, however, that the fusions occurred *seriatim,* the intermediate stages ("tetracyclus" and "pentacyclus") must have become extinct. Whether this case conformed to the stasipatric model in the strict sense is somewhat uncertain; perhaps it is more probable that the fusions arose in peripheral populations. It is clear, however, that they could never have existed in a balanced polymorphic state because of the strong tendency of the trivalents to orientate linearly in the heterozygotes.

A somewhat similar case has been reported by Zaslavsky (1963, 1967a, b) in Palearctic species of *Chilocorus*. *C. bipustulatus* is a very widespread species. In parts of Sinkiang, Kazakhstan, and Turkmenia it is replaced by *C. geminus*. Apparently both species have the same chromosome number (2n = 22). Some chromosomal rearrangements must have occurred, however, since trivalents were seen in the meiosis of F_1 hybrids. Unfortunately, however, the cytology of the hybrids was not studied in detail, so that the interpretation is less complete than is the case with the *tricyclus–hexacyclus* hybrids. Zaslavsky studied a zone of overlap, within which hybrids occur, along the course of the Syr-Darya river. It is stated to be 20–30 kilometers wide, which is probably comparable to the zone of overlap in the Crowsnest Valley. However, Zaslavsky states that the F_1 hybrids are completely sterile, due to extensive failure of synapsis at meiosis and formation of chromosomal bridges at the second division. It seems unlikely that all these meiotic abnormalities are the direct consequence of structural rearrangements; they probably indicate a more general incompatibility between the karyotypes of the two species. As in the case of *C. tricyclus–C. hexacyclus,* there seems to be absolutely no ethological isolation to prevent interbreeding.

Both these examples strongly suggest that speciation in *Chilocorus* can proceed as a result of structural rearrangements of the chromosomes, without any premating isolating mechanisms being developed, and that the old and the new species can then coexist for a long while, occupying distinct territories with only a narrow zone of overlap. These beetles are predaceous on scale insects and it is unlikely that speciation involves adaptation to any distinctly new ecological niche, although subtle ecological differences between sibling species may exist.

Several sections of the genus *Chilocorus* seem to be evolving in different ways. *Chilocorus similis* is polymorphic for a number of chromosomal fusions, and the trivalents regularly orientate disjunctionally, so that no aneuploid gametes are produced. In the *tricyclus–hexacyclus* section of the genus orientation of such trivalents is very frequently nondisjunctional, so that fusions act as genetic isolating mechanisms. In the *bipustulatus–geminus* section some kind of chromosomal speciation seems to have occurred, but fusions seem not to be involved.

The Genera *Spalax, Thomomys,* and *Ctenomys*

There are genera of burrowing rodents in many areas of the world that must be presumed to have a very low level of vagility, perhaps comparable with the morabine grasshoppers and stick insects considered earlier. A number of burrowing rodent species differing in karyotype have parapatric distributions that are strongly reminiscent of *Vandiemenella* and *Didymuria*.

Extremely interesting studies have been carried out in Israel on the superspecies "*Spalax ehrenbergi*" by Wahrman, Goitein, and Nevo (1969a, b; and see Nevo 1969; Nevo and Shkolnik, 1974; Nevo and Sarich, 1973; and Nevo, Naftali, and Guttman, 1975). There are four chromosomal forms (races or species) with the chromosome numbers 2n = 52, 54, 58, and 60. The distributions of these seem to be essentially parapatric (Figure 22), with very narrow zones of overlap in which some natural hybrids have been found (between the 52- and 54-chromosome populations and between the 58- and 60-chromosome ones). The karyotypes differ with respect to both chromosomal fusions (or dissociations) and pericentric inversions. Nevo has shown that there is some degree of ethological isolation (aggressive behavior, lower frequency of copulation) between these taxa. Aggression is stronger between geographically contiguous taxa (2n = 58–60 and 2n = 52–58) than between noncontiguous ones (2n = 52–60). Levels of genic heterozygosity are low in populations of these subterranean rodents—from H = .018 to .056, according to Nevo and Shaw (1972). Likewise, the genetic similarity between the different chromosomal taxa is very high (mean S = .96, range from .926 to .999). It would appear, therefore, that speciation has occurred in these forms without the "genetic revolution" some evolutionists have supposed is an essential part of speciation.

The morphospecies *Spalax leucodon* of southeastern Europe is probably also a complex of biological species characterized by different karyotypes (Soldatović *et al.*, 1967; Raicu, Duma, and Torcea, 1973). The situation has not been analyzed in detail, but populations with 2n = 48, 50, 54, and 56 from different regions of Yugoslavia, Romania, and Bulgaria have been recorded by a number of workers. Apparently both "Robertsonian" rearrangements and pericentric inversions have been involved in the karyotypic differentiation of these forms; thus the 2n = 54 form in Romania has an NF of 80 while material with the same chromosome number from Bulgaria has NF = 98 (Walknowska, 1963).

Members of the genus *Thomomys* (pocket gophers) of the western United States resemble *Spalax* species in their general mode of life. The chromosomal evolution of *Thomomys* has been studied by Thaeler (1968a, b, c, 1974), Patton and Dingman (1968, 1970; and see Patton, 1972, 1973), and by Nevo and co-workers (1974). In the case of *T. bottae,* taxonomists have described over 150 races or subspecies on the basis of minor morphological characters. In

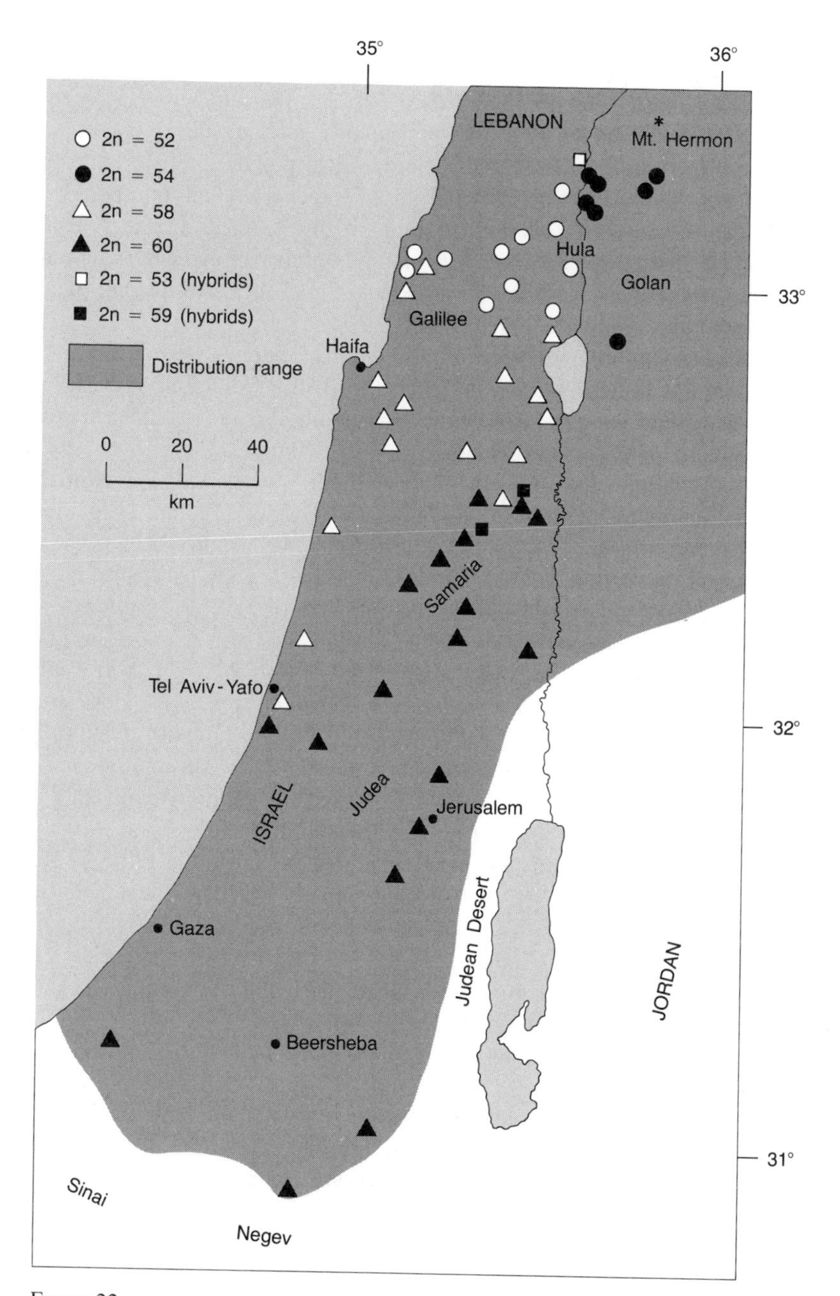

FIGURE 22

Distribution in Israel of the four forms of *Spalax ehrenbergi*. [After Nevo, Naftali, and Guttman, 1975]

a study of 24 local populations belonging to 16 of these, Patton and Dingman (1970) showed that all but one have 2n = 76. But the number of acrocentric chromosomes varies from 0 to 24, presumably as a result of at least 12 pericentric inversions (probably many more). Most populations are karyotypically uniform, i.e., inversion polymorphism is generally an interpopulation rather than an intrapopulation phenomenon in this species. In the *T. bottae* form from the Graham Mountains, Arizona, the number of acrocentric chromosomes increases progressively as one goes from 2,195 to 3,048 meters; populations at lower elevations are polymorphic for inversions.

The *Thomomys talpoides* complex, whose range extends from the southern Rocky Mountains in New Mexico northward to British Columbia, Alberta, and Saskatchewan, shows a rather different picture. Only the more southern populations, in Wyoming, Colorado, New Mexico, and Arizona, have been studied cytogenetically. Originally, a single highly variable species with approximately 60 subspecies was recognized by taxonomists. Thaeler (1968a, 1974, 1975) found a total of 15 different karyotypes with 2n = 40, 44, 46, 48, 56, 58, and 60. In general, populations with different karyotypes occupy parapatric ranges. But in several instances karyotypically distinct taxa (e.g., *T. t. pygmaeus* with 2n = 58 and *T. t. ocius* with 2n = 56; *T. t. pygmaeus* and *T. t. bridgeri* with 2n = 40; and unnamed populations with 2n = 48 and 2n = 56 near Gothic, Colorado) coexist at a locality, apparently without interbreeding. However, at two localities in western Colorado there is a detectable amount of hybridization; at South Fork this is between populations with 2n = 48 and 54, respectively, while along the Laramie River it is between populations with 2n = 46 and 48. In the South Fork zone of overlap the two taxa differ by at least three "Robertsonian" changes and four pericentric inversions; in the Laramie River zone the two forms differ by one "Robertsonian" rearrangement and three pericentric inversions. In both cases the zone of overlap and hybridization appears to be only a few kilometers wide and the number of hybrids small (some of which in Thaeler's study were not F_1 hybrids but probably backcross or F_2 hybrids). A different situation seems to exist west of Denver, where there is a zone perhaps 35–40 kilometers wide in which relatively free hybridization occurs between taxa having 2n = 48 and 54 (these karyotypes differ with respect to three "Robertsonian" rearrangements).

Obviously, several stages of speciation are represented in the *Thomomys talpoides* complex. Some of the forms presently included would appear to have reached the level of biological species, while others are still chromosomal races or incipient species. But regardless of whether raciation or speciation has occurred in a particular instance, karyotypic changes have always taken place and seem to have played a primary role in initiating and promoting divergence. As far as chromosome number is concerned, the individual populations seem to be monomorphic, but there may be a small amount of polymorphism for chromosome shape, probably due to pericentric inversions. More extensive chromosomal polymorphism exists in one form, *T. t. agrestis* (Thaeler, 1974).

Variation in the proteins coded for by 31 gene loci was studied by Nevo, Kim, Shaw and Thaeler (1974). In general, the levels of heterozygosity were low, i.e., in the same range as those found in the populations of *Spalax ehrenbergi*. So, likewise, were the numbers of loci polymorphic per population. On the other hand, the coefficients of genetic similarity (Table 6-2), although high, are slightly lower than those found in *S. ehrenbergi*. Nevo and co-workers conclude that speciation in the *T. talpoides* complex has been going on since the late Pleistocene without any major "genetic revolution," i.e. with rather few genic changes, and with chromosomal rearrangements undoubtedly playing a primary role. The low levels of genic heterozygosity and polymorphism in the individual populations are ascribed to the constancy of the subterranean niche occupied by these rodents. The exact pathways of karyotype evolution in the *T. talpoides* complex have not been worked out, and it is not clear whether the lower chromosome numbers are due to chromosomal fusions or the higher ones to dissociations (conceivably "fissions"). Some pericentric inversions have undoubtedly taken place as well, but the overall pattern is clearly different to that which has prevailed in *T. bottae,* where karyotypic differentiation has been almost entirely on the basis of pericentric inversions, with almost no changes in chromosome number.

The rodents of the genus *Ctenomys* seem to be the South American equivalents of *Thomomys*. Reig and Kiblisky (1969) studied the karyotypes of 11 species and found them to be very diverse; the chromosome number ranges from 2n = 22 to 68 and the NF from 44 to 122. Two of the species showed intrapopulational polymorphism for structural rearrangements (probably pericentric inversions). Reig and Kiblisky compared the karyotypes of the closely related species *C. talarum* and *C. porteousi,* which are allopatric forms

TABLE 6-2

Coefficients of genetic similarity between populations of the* Thomomys talpoides *complex having different karyotypes. (From Nevo* et al.*, 1974.)*

2n =	46	44	48(Wyoming)	40	60	48(New Mexico)
46	1.000	.967	.911	.871	.902	.888
44		1.000	.911	.874	.905	.886
48(Wyoming)			1.000	.849	.851	.856
40				1.000	.831	.795
60					1.000	.807

**S* = 1 − *D*, where *D* is the genetic distance (Rogers, 1972).

living in different areas of Pampas country in Buenos Aires Province; although they both have 2n = 48, there have been structural changes in 15 autosomes, and it is estimated conservatively that at least 30 chromosomal rearrangements have occurred during the evolution of these species from their common ancestor. There appear to be about 55 living species of *Ctenomys*, undoubtedly the result of fairly recent explosive speciation in which chromosomal rearrangements have rather clearly played a major part, although the published information is inadequate for establishing a precise speciation model. The species of *Ctenomys* that have been studied cytogenetically all seem to be quite easily separable on the basis of external morphology, size, ecological requirements, and serology; they are hence rather more differentiated than the "cryptospecies" of the *Thomomys talpoides* complex.

The situation in another neotropical rodent genus, *Proechimys*, is fundamentally of the same type (Reig and Useche, unpublished). *P. guyannensis* is a complex of seven species differing considerably in karyotype and distributed allopatrically or parapatrically. They show minimal morphological differences but may differ in aggressive behavior and escape reaction. Reig and Useche interpret speciation in this genus according to the stasipatric model. Species of *Proechimys* have restricted vagility but are not fossorial.

All the species of *Spalax*, *Thomomys*, and *Ctenomys* are both fossorial and sedentary. But low vagility does not invariably go with a fossorial mode of life. For example, the Chilean burrowing rodent *Spalacopus* is subterranean but migratory; rather significantly, however, animals from several localities up to 200 kilometers apart do not show any differences in karyotype (Reig *et al.*, 1972) and there is only one morphological species in the genus, which does not seem to show any tendency to fragment into chromosomal races. Chromosomal raciation, although most highly developed in small fossorial mammals, has occurred also in mammals having very different modes of life. Thus the central American bat species *Uroderma bilobatum* has three karyotypic races: one widespread race with 2n = 38, one in Colombia and Trinidad with 2n = 42, and one in Chiapas and El Salvador with 2n = 44 (Baker *et al.*, 1972).

Many groups of small mammals, particularly rodents and insectivores, show karyotypic diversity that is only slightly less striking than that of the three fossorial genera just considered. The vagility of small mice and shrews is probably in many cases no greater than that of more strictly subterranean members of those groups. It is thus to be expected that chromosomal rearrangements should have played a large part in their speciation patterns, although only in a few groups is the evidence sufficiently detailed to permit reconstruction of past events on the basis of a precise model, stasipatric or otherwise. The complicated case of the small African mice of the genus *Leggada* (sometimes regarded as a subgenus of *Mus*) will not be discussed here because data on precise distribution of the karyotypes, hybrid zones, genetic isolating mechanisms, and degree of genic differentiation are almost entirely lacking—although much cytogenetic information has been published by Matthey (1964, 1966, 1968a, 1970; and see Matthey and Jotterand, 1970; Jotterand, 1972, 1975). Nevertheless, it is clear that a process of explosive speciation is occurring in *Leggada,* with chromosomal rearrangements playing a major role. Similar patterns of chromosomal evolution have also occurred in the shrews of the holarctic *Sorex araneus–S. arcticus* complex, in which five species must now be recognized (Meylan and Hausser, 1973; Halkka *et al.,* 1974). As in the case of the *Leggada* mice, the amount of intra- and interpopulational variation in karyotype is so great that precise reconstruction of the past history is impossible at the present stage.

The case of the small Mexican pocket mouse *Perognathus goldmani* is peculiar in several respects. Patton (1969) has described six races of this species that differ in karyotype as a result of four different centric fusions and two pericentric inversions (one of which is in the X chromosome). The unique feature of this case is that although the ranges of the six races seem to be essentially parapatric they are separated by the riparian vegetation fringing the rivers of Sonora, which is inhabited by an entirely different species of *Perognathus, P. artus,* which appears to exclude *P. goldmani* from this habitat. Only three individual hybrids between the races α and δ were found by Patton, and none between any of the other races (presumably because, at least in part, they are not in direct contact). Patton does not accept a stasipatric interpretation of this case and regards it as an example of allopatric speciation. According to him the ancestral proto-*goldmani* population had 2n = 56 (NF 56). It occupied a relatively small area of the states of Sonora and Sinaloa and gave rise to *P. artus,* which possesses a chromosomal fusion (A6) so that 2n = 54 (NF 56). *P. artus* is believed to have extended its range to the coast of Sonora in prepluvial (early Wisconsin) times, displacing *P. goldmani,* which was then split into a northern population (which became race ϵ, with a different fusion, A2) and a southern population (which gave rise to race β, with a pericentric inversion of one chromosome, A1). Race ϵ extended its range in a northerly direction and gave rise, allopatrically, to race θ, which has an additional fusion (A3), so that 2n = 52 (NF still 56). By a different fusion, race ϵ gave rise to race γ, which in turn gave rise (by a pericentric inversion of the X chromosome) to race α, and this gave rise, by an additional fusion between autosomes, to race δ.

One of the outstanding features of *P. goldmani* is that, assuming Patton's chromosome phylogeny is accurate, the most "derived" races α and δ occupy a central position, while the more primitive races β and ϵ, which differ only by single rearrangements from the hypothetical ancestral karyotype, are peripheral. This is the type of geographic regularity that seems to be characteristic of stasipatric speciation (cf. *Vandiemenella* and *Didymuria*). It may well be, as Patton claims, that race θ extended its range, allopatrically, into previously unoccupied territory. But despite any complications produced by interaction with *P. artus,* the general pattern of differentiation of *P. goldmani* seems to fit well into the general framework of the stasipatric model.

All of the instances of stasipatric speciation we have discussed are multiple—that is to say, they are cases where an original species has given rise, by the stasipatric process, to half a dozen or more species. It is clear that this is to some extent inherent in the mechanism itself. Stasipatric speciation does not depend on isolation of a subpopulation by some fortuitous geographic barrier, neither does it necessarily involve specific vacant ecological niches, as sympatric speciation does (see Chapter 7). It is sufficient that the species have the cytogenetic properties and the type of population structure favorable to stasipatric speciation. If it possesses both, repeated production of new species is likely to occur in the course of time. Conceivably, however, there may well be cases where a species has given rise stasipatrically to only one descendant species, although such cases might be rather hard to recognize and distinguish from other types of speciation.

It seems probable that there have been many cases of stasipatric speciation in flightless orthopteroid insects, like those described in morabine grasshoppers and stick insects earlier in this chapter. Only one other group will be considered here, a species complex of fossorial insects whose vagility is certainly restricted. The mole crickets of Europe and the near East, traditionally assigned to the Linnean species *Gryllotalpa gryllotalpa,* are actually a complex of at least seven forms that differ in karyotype and may well have arisen stasipatrically. None of them are known to be sympatric and it seems probable that their distribution was originally (in Neolithic times) a "parapatric mosaic." However, the distribution of these insects has undoubtedly been greatly modified by the development of agriculture and urbanization, so that it may never be possible to interpret their past evolution in detail. At least in one respect they do not conform to the typical stasipatric pattern: both the karyotypically most derived form (northwest Europe) and the most primitive form (Israel) seem to be peripheral, instead of only the primitive form being peripheral, as in *Vandiemenella* and *Didymuria*. However, it is possible that if the distribution of all forms of *Gryllotalpa* in Europe were more fully known, the situation might look somewhat different.

Since a number of extra-European species of *Gryllotalpa* and related genera have 2n♂ = 23, we may take this as the primitive karyotype of the *G. gryllotalpa* complex. In Israel it is present in a population known only from a single locality, on the shores of the Dead Sea (Kushnir, 1948, 1952), but it may be much more widespread in the Arab countries. The other forms have 2n♂ = 19, 18 (XY), 17, 15, 14 (XY), and 12 (XY). The XY mechanisms may well have arisen from XO ones several times in the course of the evolution of this complex, as in *Vandiemenella* and *Didymuria,* but we cannot be certain. The views of some earlier authorities—that the 12-chromosome race was karyotypically primitive, the 19-chromosomes race has the small chromosomes reduplicated (polysomy), and even that polyploidy has occurred in the genus *Gryllotalpa*—must be rejected now that we know more about the principles of chromosomal evolution in general.

The 19-chromosome *Gryllotalpa* (Kushnir, 1948, 1952; Krimbas, 1960) presumably arose from the 23-chromosome form by two chromosomal fusions, but since the chromosomes in this genus seem to be invariably metacentric or submetacentric, it is unlikely that these were single centric fusions of the usual type. This race is widespread in the countries of the eastern Mediterranean (Israel, Rhodes, and Greece, including the Peloponnesus and western Macedonia, but not Turkey, Crete, Thrace, or eastern Macedonia).

The 18-chromosome race with XY males may have arisen from the 19-chromosome one by a single X–autosome fusion (Barigozzi, 1942; Tosi, 1959). It is known only in northern Italy. The form of mole cricket in Albania and Yugoslavia is presumably either the 19- or 18-chromosome one, although just what form occupies this area is unknown.

The 17-chromosome race is known from southern Spain and was named *G. 17-chromosomica* by Ortiz (1951, 1958). It is presumably derived from the 19-chromosome form by a chromosomal fusion, but where this event may have occurred is quite uncertain. It is most unlikely, but perhaps not quite impossible, that it arose from the 18-chromosome form by loss of the Y chromosome. If Senna's old account (1911) can be believed, 17-chromosome mole crickets also occur in central Italy. This seems reasonable, since a 15-chromosome form is certainly present in southern Italy (Barigozzi, 1933a, b).

A 14-chromosome form with XY males occurs in Turkey, the islands of Mytilene and Samos, Thrace, eastern Macedonia, and Romania (Steopoe, 1939; Kushnir, 1952; Krimbas, 1960). Some individuals in Romania and Greece, and perhaps Bulgaria, carry supernumerary chromosomes. The origin of this form is not clear, but it could have arisen from the 18-chromosome form by two further autosomal fusions; its distribution makes it unlikely that it arose from the 15-chromosome form that inhabits southern Italy.

Finally, a 12-chromosome form with XY males has a wide distribution in northern and western Europe (France, Belgium, Germany, northern Spain, northern Italy). It is presumably the form to which Linnaeus gave the specific name *gryllotalpa*. Bennet-Clark (1970) has described a second species from France *(G. vineae)* distinguishable morphologically and in song from *G. gryllotalpa*. He did not determine its karyotype, but it seems likely that it was the 18-chromosome form known from northern Italy.

It is probable that all these forms have reached the level of full species.

Kushnir demonstrated morphological differences between the 23-chromosome and 19-chromosome forms and Nevo and Blondheim (1972) showed that their calling songs are different. The karyotypic evolution of these insects will only be understood when hybrids (either natural or laboratory) between some of the taxa have been studied. If their speciation has been stasipatric, there is one important difference from the *Vandiemenella* and *Didymuria* cases, namely, that premating isolation (in the form of the calling song) has almost certainly played an important role; stridulation does not occur in *Vandiemenella* and *Didymuria* and the different taxa in these genera do not seem to be isolated by premating mechanisms of any kind. It is possible that allochronic isolation (see p. 249) may have played some part in the speciation of the European mole crickets. Morales (1940) has shown that the life cycle of one member of the complex (in Spain) extends over two years, so no gene exchange between populations born in alternate years would take place.

In general, we have assumed that any particular species would manifest a single mode of speciation, i.e., if it was "budding off" a whole series of descendant species the genetic mechanism would be the same in each case. However, the mouse *Mus musculus* seems to manifest two very different modes of speciation. Although the literature on this species is very extensive, there are still major gaps in our knowledge of its evolutionary biology. The exact number of subspecies that should be recognized, as well as their status, relationships, and precise geographic distributions, is still somewhat uncertain. In Europe there exist *Mus m. musculus, M. m. domesticus, M. m. brevirostris, M. m. spicilegus,* and *M. m. spretus* (Figure 23); in Asia there are *M. m. bactrianus, M. m. molossinus, M. m. wagneri,* and perhaps others. No major karyotypic differences between these races have been described and they all seem to have 2n = 40 acrocentrics. However, *M. m. musculus* and *M. m. molossinus* differ in the distribution of C-banding material in their karyotypes (Dev *et al.,* 1975). The European races are parapatric; using morphological criteria, Ursin (1952) defined a narrow zone of hybridization between the light-bellied *M. m. musculus* of eastern Europe and the dark-bellied *M. m. domesticus* of western Europe, which crosses the Jutland Peninsula of Denmark. The more recent electrophoretic studies of Selander, Hunt, and Yang (1969; and see Hunt and Selander, 1973) indicate that the position of this zone has not changed appreciably in almost 20 years. The two forms are best regarded as evolutionary semispecies, i.e., as having diverged genetically more than ordinary geographic subspecies. This conclusion is based on morphological, ethological, and allozymic differences. The coefficients of genetic similarity between *M. musculus* and *M. domesticus* are quite low (Table 6-3) by comparison with sibling species pairs in such genera as *Sigmodon* and *Peromyscus* (Johnson *et al.,* 1972; Selander and Johnson, 1973) and with the

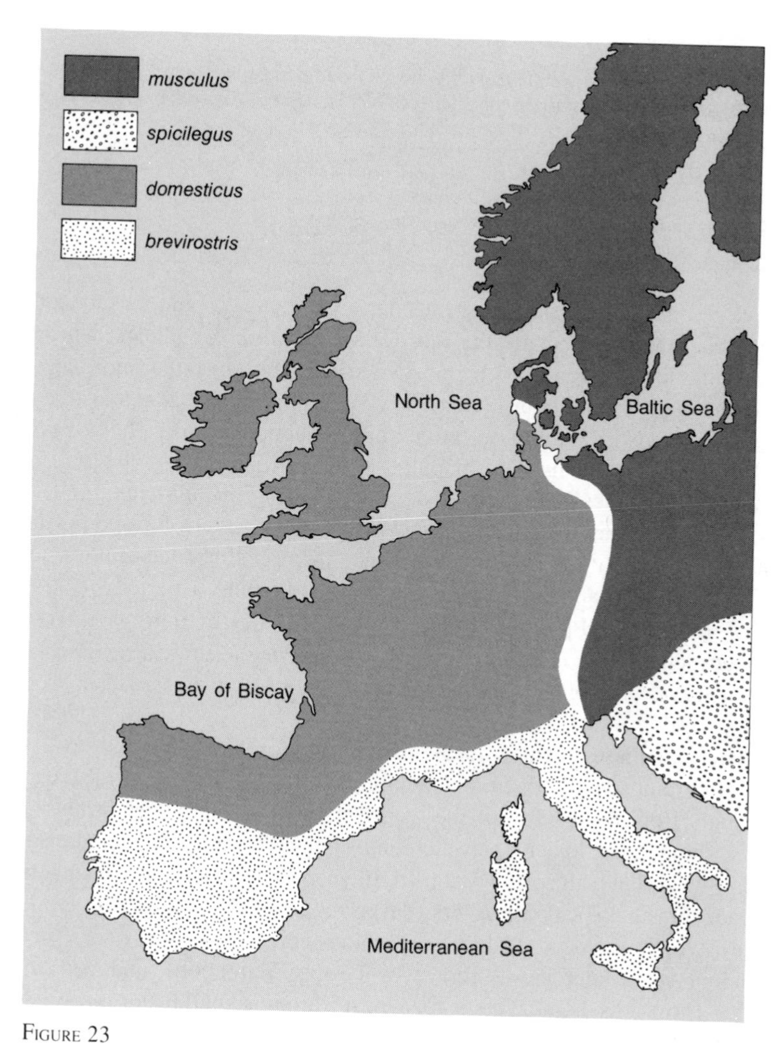

FIGURE 23

Distribution in Europe of four subspecies of *Mus musculus (M. m. spretus* from Portugal is not shown). The white strip indicates a zone of hybridization between *musculus* and *domesticus*. [After Hunt and Selander, 1973]

karyotype races, semispecies, or species in *Spalax* and *Thomomys*. Interbreeding between the two semispecies in the Jutland hybrid zone is apparently occurring freely, but the zone remains quite narrow; at one point 90 percent of the genetic change occurs along a transect of 20 kilometers. There seems to be

TABLE 6-3
Coefficients of genetic similarity between certain subspecies of Mus musculus *(based on 17 polymorphic enzyme loci). (From Selander, Hunt, and Yang, 1969.)*

Different populations of *M. m. musculus* in Denmark	.71–.76
" " " *M. m. domesticus* " "	.83
M. m. musculus and *M. m. domesticus* in Denmark	.43–.45
M. m. domesticus (Denmark) and *M. m. brevirostris* (California)	.71

little effective introgression of genes into the main populations outside the hybrid zone; this is especially the case for *M. m. musculus* alleles, which seem to be unable to introgress into *M. m. domesticus* (there is rather more introgression in the opposite direction). Gene flow across the hybrid zone must be rather strongly opposed by natural selection, i.e., selection against *M. m. musculus* genes in the *M. m. domesticus* population and vice versa.

Hunt and Selander (1973) have suggested that *M. m. musculus* and *M. m. domesticus* have been in contact and hybridizing in northern Europe for the past 5,000 years. The parapatric contact between them is almost certainly a secondary one that came about by range extension following a period of allopatric isolation during the Pleistocene (see p. 165). Studies of other contact zones, e.g., between *M. m. musculus* and *M. m. spicilegus* in eastern Europe and between *M. m. domesticus* and the mediterranean *M. m. brevirostris*, would be highly desirable.

Apart from these "genic" races, *Mus musculus* has given rise to a whole series of chromosomally distinct populations in the Alps and central Italy. Wild individuals from most parts of the world, as well as laboratory strains of this species, show 2n = 40 acrocentric chromosomes. However, in a single valley in eastern Switzerland, the Val Poschiavo, there exists a population distinguishable on minor morphological characters (darker color, body proportions, scaling of the tail), which was described as a separate species, *M. poschiavinus*, in 1869. A hundred years later it was shown by Gropp, Tettenborn, and von Lehman that this form was homozygous for seven chromosomal fusions, i.e., it had 2n = 26 (14 metacentric chromosomes, 12 acrocentrics). F_1 laboratory hybrids between *M. poschiavinus* and *M. musculus* show seven trivalents at meiosis. However, these frequently undergo linear orientation on the spindle of the first meiotic division, so that a large proportion of the gametes are aneuploid, and the hybrids are semisterile (Gropp, Tettenborn, and von Lehman, 1970; Tettenborn and Gropp, 1970).

This seems certainly to be a case of chromosomal speciation according to the stasipatric model. As Ford and Evans (1973) point out, centric fusions that lead to 20–30 percent infertility in the heterozygote cannot possibly have ever existed in a condition of balanced polymorphism in the population. They must have undergone fixation "by chance survival and matings in a very small isolate, *despite* the infertility of heterozygotes, followed by subsequent expansion with minimal hybridization." It seems likely that there has been selection for ethological isolation between *M. poschiavinus* and *M. musculus,* so that natural hybrids are not produced or are extremely rare.

The question arises as to whether the seven centric fusions occurred at the same time, as a result of some kind of cytogenetic revolution (conceivably as a result of viral infection), or *seriatim*. The latter alternative seems much more likely. The mice of Val Poschiavo are not the only population in which chromosomal fusions have established themselves. Gropp and co-workers (1972) have found populations with 2n = 38 (homozygous for one fusion) and 2n = 34 (homozygous for three fusions) in other areas of eastern Switzerland (Figure 24). Capanna (1973) has found a population with 2n = 32 in the

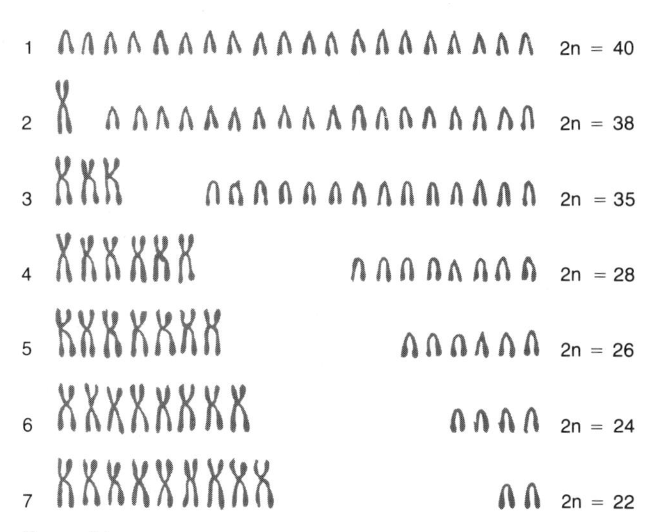

FIGURE 24

Haploid karyotypes from different *Mus musculus* populations in the Alps and the Apennines. (1) the standard karyotype for the species; (2) population from Albula, polymorphic for one fusion; (3) population from Chiavenna, with three fusions (polymorphic); (4) population from Val Mesolecina, with six fusions fixed; (5) population from Val Poschiavo, with seven fusions fixed; (6) population from the Apennines (Laga Mountains), with eight fusions fixed; (7) three different populations, two from the Apennines (Abruzzi and Molise) and one from the Alpi Orobie, all of which have nine fusions, but mostly different ones. [Courtesy of Professor Ernesto Capanna] western Dolomites (Capanna, Civitelli, and Cristaldi, 1973) and two different ones with 2n = 22 (homozygous for nine fusions) in the Apennines (Capanna, Civitelli, and Cristaldi, 1974; Capanna *et al.*, 1976). The evidence of Gropp and co-workers indicates that the fusions present in other areas of eastern Switzerland are, with a single exception, different from those in *Mus poschiavinus*. Capanna and co-workers (1975, 1976) have shown that only one of the Apennine fusions is identical with a Poschiavo one and that, with one exception, the fusions of the two Apennine populations are also different (Table 6-4). The two Apennine populations must come into contact east of Rome, but the situation in this area has not been fully studied. It would seem that there has been a general

TABLE 6-4
Chromosomal fusions known in European populations of Mus musculus. *(From Capanna* et al., *1976.)*

Fusion	Involving chromosomes	Known from
1	1-3	Poschiavo, Val Mesolecina
2	4-6	
3	5-15	Poschiavo, Apennines (Molise)
4	11-13	Poschiavo
5	8-12	
6	9-14	
7	16-17	Poschiavo, Chiavenna
8	10-11	Val Mesolecina, Val Bregaglia, Chiavenna, Apennines (Abruzzi)
9	4-12	Albula Valley and Basel area, Chiavenna
10	1-10	Val Bregaglia
11	2-14	Val Mesolecina
12	7-8	
13	13-16	Val Mesolecina, Apennines (Abruzzi, Molise)
14	1-6	Apennines (Abruzzi)
15	3-8	
16	7-9	
17	4-15	
18	2-18	
19	5-17	
20	12-14	
21	1-18	Apennines (Molise)
22	2-17	
23	4-11	
24	6-7	
25	3-9	
26	8-14	
27	10-12	

tendency for fusions to establish themselves in the Alpine–Apennine area and that different ones may have initiated the process in different localities. So far there is little information about the fertility of heterozygotes and the extent to which they occur in nature, except in the case of *M. poschiavinus* and the more northern of the Apennine populations, which hybridizes with 2n = 40 mice where the two are in contact.

The case of *M. poschiavinus* and the other populations homozygous for fusions follows the principle of the stasipatric model—that the "derived" forms are central and the ancestral karyotype occupies the periphery of the distribution. It is quite possible, however, that geographic isolation (by glaciers, during the Pleistocene) played a larger part in the case of *M. poschiavinus* than in the other examples of stasipatric speciation we have considered. However, this cannot have been so in the Apennine mice. The accumulation of fusions in these evolutionary lineages is a good example of the principle of "karyotypic orthoselection" (see p. 149). There seems no reason, from the data in Table 6-4, to assume that the involvement of the particular chromosomes in these fusions has been other than random.

In 1963 Mayr was not willing to admit that chromosomal rearrangements might play a special or primary role in speciation. They were regarded by him as merely a part of the general spectrum of mutations, and the possibility that they might generate reproductive isolating mechanisms based on heterozygote sterility was discounted. By 1969, however, this position was no longer tenable, and he discussed three different situations (Mayr, 1969a):

1. Occurrence of reproductive isolation without chromosomal rearrangements.
2. Occurrence of chromosomal rearrangements without production of reproductive isolation.
3. Occurrence of chromosomal rearrangements leading to reproductive isolation.

He believes, however, that the third can only occur under rather special circumstances: "Homozygotes for new arrangements would occur far too rarely in large populations to have any chance to become established and to acquire the other two prerequisites for successful speciation, perfected isolating mechanisms, and ecological exclusion. The evolutionist, therefore, believes that such chromosomal shifts can occur [he clearly means *can establish themselves*] only in very small peripherally isolated populations where they are exposed to selection pressures that differ from those in the main portion of the range." This is also the position adopted by Key (1968, 1974).

Now, in the first place, the distribution patterns of chromosomal races, semispecies, or recently evolved species in such thoroughly investigated cases as the grasshoppers of the genus *Vandiemenella,* the stick insects of the *Didymuria violescens* complex, the lizards of the *Sceloporus grammicus* superspecies, and the alpine and Italian populations of *Mus musculus* do not in the least support this concept of the peripheral establishment of rearrangements conducive to speciation. On the contrary, in each of these cases the karyotypically primitive form or forms have a peripheral distribution, with the derived forms occupying more centrally located areas, surrounded on most or all sides by the original population. This kind of distribution pattern is extremely characteristic and may be regarded as an important indicator of the stasipatric mode of speciation.

In the second place, both Mayr's and Key's arguments fail to take account of one certain and one probable factor. The certain factor is that environmental heterogeneity is such that even "large" populations are usually discontinuous and patchy, with many "internal boundaries" at the edges of regions where, for some reason, the species does not occur. Especially where vagility is low an entire species population may consist of a large number of small demes within which the probability of a few homozygotes for a newly arisen rearrangement will not be negligible, as Mayr and Key seem to imply. Of course we do not mean that establishment of a newly arisen rearrangement causing negative heterosis is impossible in a peripheral colony of a species, merely that it is statistically likely that the great majority of such rearrangements will establish themselves in "central" (i.e., nonperipheral) demes, which may be just as small, just as isolated geographically, and just as liable to periodic inbreeding as the demes situated on the periphery.

The probable factor that Mayr and Key have failed to take into account is that because the chromosomal rearrangements that manage to establish themselves are only a minute fraction (1 in 10^4 or 10^5) of all those that occur spontaneously, they may not be a random sample but may include a significant proportion that enjoy the benefit of some degree of meiotic drive. We have already referred to the general probability of such non-Mendelian genetic mechanisms being involved in speciation (p. 174). Thus far the only direct evidence for meiotic drive in a case of stasipatric speciation comes from the work of Mrongovius (1975) on female hybrids between the 17-chromosome race of *Vandiemenella viatica* and P24(XY). These two taxa differ with respect to two rearrangements in the X chromosome: the X of the latter form has been derived from that of the former by a pericentric inversion and a fusion with a small autosome. Some hybrid females, which occur naturally in a narrow zone of overlap on Kangaroo Island (about a fifth of those studied), transmit the $viatica_{17}$ X chromosome about twice as often as the P24(XY) one. This effect is not due to any zygotic mortality and must be assumed to be due to the $viatica_{17}$ X chromosome passing more frequently into the egg nucleus than into the polar body, i.e., it is a case of true meiotic drive. Its relevance to the stasipatric speciation process is not entirely clear, however, because the difference between the two chromosomes is a double one (due to two rearrangements), and because the chromosome that is driven (i.e., that has the advantage at meiosis) is the original rather than the derived type. Nevertheless, the fact that meiotic drive, even if of an unexpected kind, was found in the very first case of stasipatric speciation investigated from this standpoint seems highly significant.

Even if the chromosomal rearrangements initiating genetic isolation establish themselves on the periphery of the species distribution (the view of Mayr and Key) their subsequent fate may be imagined to follow one of two pathways, or a combination of both. On the one hand, homozygotes for the new chromosome arrangement may spread out into territory previously unoccupied by the species, being either preadapted to the new environment or rapidly acquiring alleles that are adapted to it. In this way, by an essentially allopatric process, we might have the development of two populations (incipient species) with a primary zone of contact between them, within which some hybridization would undoubtedly persist for a time. It is probably this model that Mayr and Key have in mind.

The alternative model envisages populations with the new chromosome arrangement invading territory already occupied by the species, displacing the original chromosome arrangement as they do so. This would be an essentially "stasipatric" process not essentially different from that envisaged by White (1968, 1974) except that the new arrangement would have a peripheral rather than a central point of origin. This should not be considered a sympatric model because it does not envisage that the two chromosome arrangements would coexist over any extensive area; the heterozygote disadvantage would render them incompatible except in a very narrow "tension zone."

Particularly puzzling at first sight, from the standpoint of speciation, are those cases in which two very closely related species have such different karyotypes that numerous chromosomal rearrangements must have occurred during their divergence. A good example would be the cotton rats *Sigmodon hispidus* and *S. arizonae* (Zimmerman, 1970; Zimmerman and Lee, 1968). The former, with 2n = 52, extends from Tennessee to Venezuela; an isolated population occurs at Yuma, Arizona. The latter, with 2n = 22, has a much more restricted distribution and is presumably the derived species of the pair. It occurs from central Arizona southward down the Pacific coast of Mexico as far as 21° latitude, where it is replaced by a third member of the complex, *S. mascotensis,* with 2n = 28. The electrophoretic studies of Johnson and coworkers (1972) on 23 loci for enzymatic and nonenzymatic proteins showed that *hispidus* and *arizonae* were very similar, with a coefficient of similarity of 0.76, as compared with an average value of 0.5 for sibling species of *Drosophila* (Rogers, 1972).

It is not necessary to suppose that in this or similar cases there was a sudden catastrophic repatterning of the karyotype, involving some 11 to 13 simultaneous chromosome fusions or dissociations. It is much more probable there was a stepwise series of changes. *S. mascotensis* may represent a surviving population of the numerous intermediate steps in the evolution of the *arizonae* karyotype from the *hispidus* one (or vice versa). The other intermediate karyotypes may be extinct or may yet be discovered in the *Sigmodon* populations of western or southern Mexico.

A similar case of chromosomal speciation is found in the lemmings of the *Dicrostonyx torquatus* complex in the Arctic, where three different populations that have been regarded as subspecies have the chromosome numbers 2n = 28, 30, and 57–60 (Kozlovsky, 1974). These mammals have an extraordinary sex chromosome system, with XY as well as XX females (Fredga *et al.*, 1976; Gileva, 1973, 1975).

An extremely interesting case of chromosomal speciation, but one that is very difficult to interpret in detail, was described by Staiger (1954, 1955) in the marine mollusc *Nucella* (also known as *Purpura* or *Thais) lapillus* on the coast of Brittany. In this genus there are no pelagic larvae; the young snails hatch as crawling miniatures of the adult, thus leading to a population structure favorable to local differentiation (Ahmed, 1974). Three species of *Nucella* are known to show n = 30 and three have n = 33. There are two forms of *N. lapillus* on the Brittany coast, one with n = 18 and one with n = 13. Apparently, only the 13-chromosome form occurs on the Atlantic coast of the U.S. (Mayr, 1963a). It has five large metacentric chromosomes and eight subacrocentric (J-shaped) chromosomes, whereas the 18-chromosome form has 18 acrocentric and subacrocentric chromosomes but no large metacentrics. It thus seems clear that the 13-chromosome form is homozygous for five chromosomal fusions of acrocentrics, or subacrocentrics, in the course of each of which the material of the short arms, or most of it, was lost. This 13-chromosome form lives in localities exposed to strong wave action where there is an abundance of mussels *(Mytilus)*, which constitute an important food for *Nucella*. The 18-chromosome form, on the other hand, inhabits and appears to be adapted to bays and other areas sheltered from wave action and characterized by the presence of the alga *Ascophyllum nodosum* and an absence of mussels. The zone between these two extreme habitats is inhabited by chromosomally polymorphic populations in which most of the individuals have chromosome numbers between 2n = 26 and 2n = 36 and show one or more trivalents at meiosis. Apparently the same two 26- and 36-chromosome forms and the hybrids between them occur also on the south coast of England (Bantock and Cockayne, 1975).

Staiger believed that the polymorphic populations showed heterosis in the intermediate zone, pointing to their increased shell thickness and slightly smaller average shell size. They have a high population density, but there may be some loss of fecundity, due to nondisjunction of some of the trivalents and formation of some aneuploid gametes.

One conclusion of major importance that clearly emerges from a tantalizingly incomplete investigation is that five different chromosomal fusions all seem to have the same adaptive properties—they all help render their carriers physiologically suited to life in the exposed zone. Although the coastline is extremely irregular in the area studied, the average transect from a typical 13-chromosome locality to a typical 10-chromosome one is probably one or two kilometers. With the intermediate zone occupied by polymorphic colonies, it is clear that there is plenty of opportunity for migration of individuals and hence of chromosomes. If fused chromosomes are absent from exposed capes and reefs and unfused chromosomes are absent from sheltered bays, this clearly indicates strong selection against these types of chromosomes when they migrate into the wrong environment. It seems probable that the heterosis in the intermediate zone, claimed by Staiger, is more a type of niche adaptation or frequency-dependent selection than simple hybrid vigor.

How could this situation have developed? Is it a stage in a speciation event, or are we confronted with a partial fusion of two originally separate species? Mayr (1963a) stated that "the different chromosome numbers and habitat preferences had apparently developed during a previous isolation . . . without, however, leading to reproductive isolation." But there is really no evidence for the polymorphic zone being due to a secondary contact of this kind (except that the 13-chromosome form occurs by itself on the other side of the Atlantic). If it were shown that, for example, the populations on the coast of Cornwall and Ireland were composed entirely of 13-chromosome individuals and ones in the Bay of Biscay solely of 18-chromosome snails, there might be some reason to accept Mayr's hypothesis of a secondary hybrid zone, although in the absence of any evidence for geographic isolation at some time in the past such a distribution would not prove conclusively that it was secondary. The one thing that seems fairly certain is that chromosomal fusions (rather than dissociations) have occurred, i.e., that the 18-chromosome race is more primitive than the 13-chromosome one. Two facts point strongly to this conclusion—the other species of *Nucella* have chromosome numbers even higher than n = 18, and the fusions seem to have involved losses of substantial chromosome segments. (On the alternative hypothesis that dissociations or centromeric "fissions" had occurred there would have been a *gain* of these segments, with no plausible explanation as to where they had come from.)

The suggestion that all five fusions have had similar effects in adapting these molluscs to a particular type of habitat appears at first sight far fetched. But if each fusion involved the loss of heterochromatic material in the short arms of two chromosomes, and if all this heterochromatin consisted of highly repetitive satellite DNA, the similarity of adaptive effects would begin to make sense. But the precise relation of this most interesting case to speciation theory remains unclear and will continue so until more information on *N. lapillus* in other geographic areas is available.

A model of speciation in higher plants that bears some resemblance to the stasipatric one was put forward by H. Lewis (1953a, b, 1962, 1966, 1973; see also Lewis and Raven, 1958; Lewis and Roberts, 1956; Vasek, 1958, 1968) primarily on the basis of his work on the genus *Clarkia* (Onagraceae). It should be recalled that the Onagraceae also include the genus *Oenothera,* well known for the structural heterozygosity of some of its species. It seems likely that a tendency to translocation heterozygosity and to the establishment of translocations in natural populations is rather generally present in this family of plants.

Three cases of speciation in *Clarkia* have been studied in detail by Lewis and his collaborators. All of these concern diploid species. They are (1) *C. franciscana* and *C. rubicunda* in the San Francisco area of California; (2) *C. biloba* and *C. lingulata* in central California; and (3) *C. unguiculata* and *C. exilis.* Other cases, not so intensively studied, include the origin of *C. springvillensis, C. tembloriensis,* and an undescribed species "Caliente" from *C. unguiculata* (Vasek, 1964, 1968); the origin of *C. stellata* from *C. milrediae* (Mosquin, 1962); and the origin of *C. virgata* and *C. australis* from *C. mosquinii* (Small, 1971). In all of these instances Lewis (1973) believes it is possible to identify with certainty the parent species and the derived one (neospecies). In general, the neospecies occupy very small areas of territory on the periphery of the range of the parental species. For example, *C. franciscana* grows as a single population just south of the Golden Gate Bridge in the city of San Francisco; *C. lingulata* is known only from two localities about three kilometers apart in the canyon of the Merced River, California; while *C. exilis* occurs in a small area of the Sierra Nevada foothills in central California; by contrast, the parent species mostly have extensive ranges (Figures 25 and 26). In the Myxocarpa section of the genus, however, all the diploid species occupy very small areas of territory (Small, 1971).

C. franciscana and its ancestor *C. rubicunda* both have n = 7 and are chromosomally monomorphic species. *Rubicunda,* which itself probably originated as a segregate from the more widespread and karyotypically variable *C. amoena,* grows in an area around San Francisco Bay and along the coast to the south of it. Artificially produced F_1 hybrids between *franciscana* and *rubicunda* (none occur in nature) are highly sterile; their meiosis indicates that the two species differ in respect of at least three translocations and four paracentric inversions. Thus, a very radical "repatterning" of the karyotype occurred during the speciation process that gave rise to *franciscana* (Lewis and Raven, 1958). There are minor differences in growth habit and floral morphology between *franciscana* and *rubicunda*. Whereas *rubicunda* and *amoena* are normally outcrossed by insect pollination, *franciscana* is regularly self-pollinated.

Clarkia lingulata resembles *C. franciscana* in being a neospecies that is virtually confined to a single locality, the Merced Canyon. It differs from its putative parent, *C. biloba,* in having n = 9 instead of n = 8, and is in fact a tertiary tetrasomic of *biloba,* having a chromosome in duplicate that consists of segments also present in two other chromosome pairs. However, this is not the only chromosomal difference between the two species; there is at least one other translocation and a paracentric inversion (Lewis and Roberts, 1956). Hybrids between *lingulata* and all three geographic subspecies of *biloba* have been obtained in experimental crosses, but they had greatly reduced fecundity, both when self-pollinated and when backcrossed to *biloba*.

Bartholomew and co-workers (1973) have made a very detailed study of the pattern of variation in *C. rubicunda*. They found rather marked differences between local populations, which they ascribed to periodical drastic reductions in population size, limited seed dispersal, and lack of seed dormancy—the same factors that are believed to underlie saltatory chromosomal speciation. All the populations studied showed seven bivalents, as did the interpopulation hybrids obtained by crossing.

Clarkia exilis, a neospecies that appears to have been derived from the more widespread *C. unguiculata,* has a wider distribution range than either *C. franciscana* or *C. lingulata* (Vasek, 1958, 1960). *Unguiculata* is a diploid species with n = 9, but many colonies contain translocation heterozygotes in high frequency (Mooring, 1958).

A series of six diploid species of *Clarkia* belonging to the section Myxocarpa was described by Small (1971). All of these occupy relatively minute montane areas in northern California. A more northerly triad of species, *C. borealis* (n = 7), *C. stellata* (n = 7), and *C. milrediae* (n =7), form one evolutionary cluster; *C. mosquinii* (n = 6), *C. virgata* (n = 5), and *C. australis* (n = 5) form another. Each species differs from the others by at least two translocations. There has apparently been a stepwise evolutionary reduction in chromosome number, culminating in the 5-chromosome species. All the species are reproductively isolated, due to the irregular meiosis of F_1 hybrids.

The general explanation favored by Lewis (1966, 1973) for the origin of neospecies in *Clarkia* involves the following steps (condensed and somewhat modified from his 1973 discussion):

1. An exceptional drought reduces a normally outcrossing population to very few plants or eliminates it.

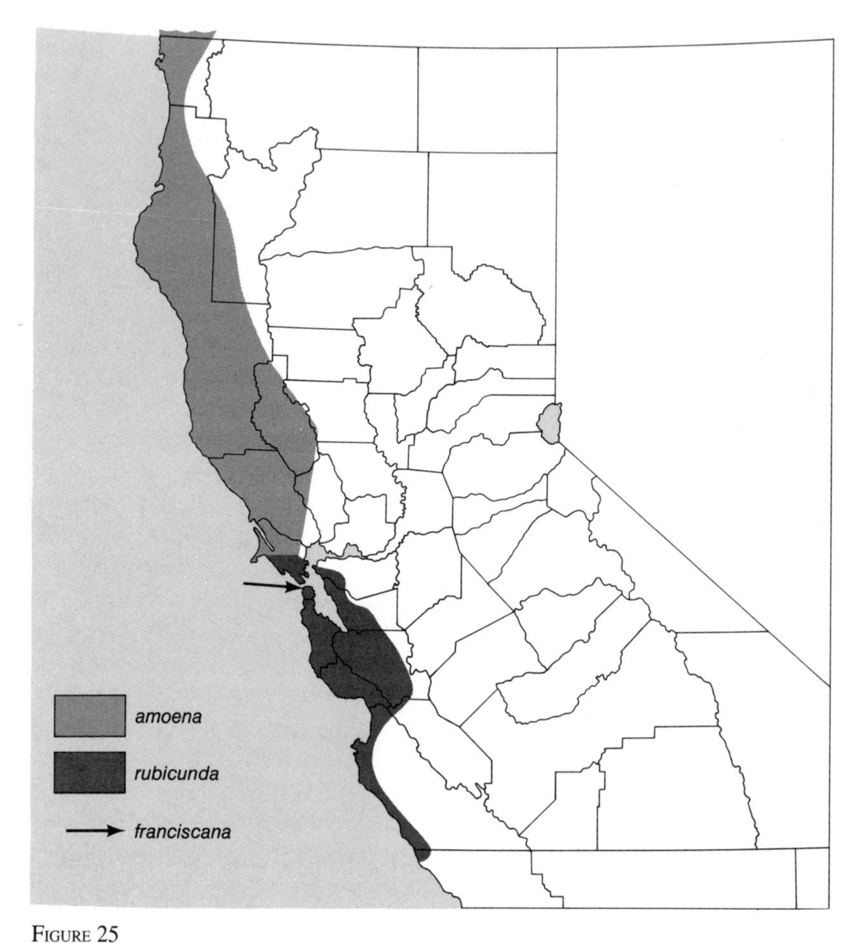

FIGURE 25

Distribution in California of *Clarkia rubicunda, C. franciscana,* and *C. amoena* (the latter extends northward as far as Vancouver Island). [After Lewis, and Raven, 1958]

2. The few survivors of founders that reestablish the population undergo self-pollination.
3. Extensive chromosomal rearrangements occur; structural heterozygotes are partly sterile.
4. Chance formation of a chromosomally monomorphic population from homozygous combinations of rearranged chromosomes.
5. The neospecies is genetically isolated from the parent species by hybrid sterility and because it is self-fertilizing.

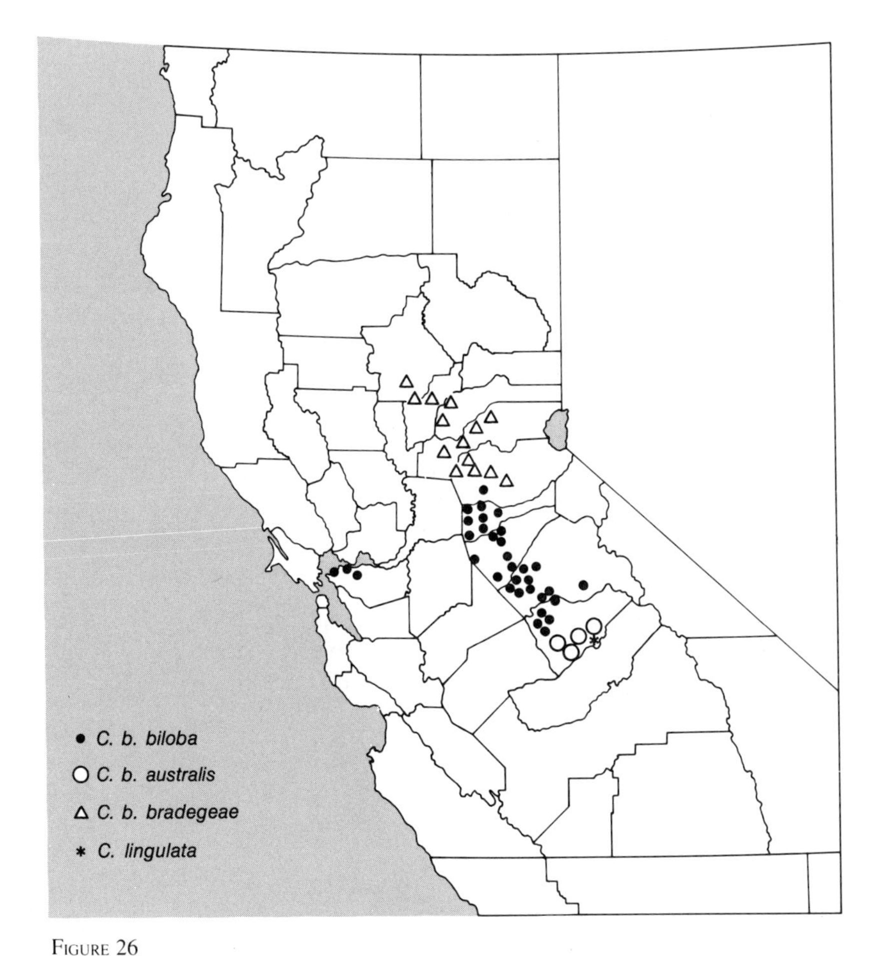

FIGURE 26

Distribution in the western United States of three geographic subspecies of *Clarkia biloba* and the location of the single colony of *C. lingulata*. [After Lewis and Roberts, 1956]

This process is envisaged as a rapid one; it has been referred to as "saltational speciation" due to "catastrophic selection" because of the role of drought or other exceptional circumstances in the first four stages. In general, the neospecies seems to grow in more xeric environments than the parental species, or at any rate in ones that are ecologically marginal. However, some populations of the parental species may grow in areas that are just as marginal as, or even more extreme than, the one occupied by the neospecies. In the case of *C.*

lingulata, extensive field, garden, and laboratory experiments suggest that it is no better adapted to the environment it occupies than the surrounding population of *biloba.* Hence, Lewis believes that the origin of *lingulata* was largely accidental; the neospecies, having no adaptive superiority to the parental one, is unlikely to spread and its long-term evolutionary future appears precarious.

Gottlieb (1974) has shown that there is a close similarity between the allozymes of *lingulata* and *biloba,* and this finding is consistent with a recent origin for the former. On the other hand, he found that *franciscana* was homozygous at loci controlling six out of eight enzyme systems, the alleles in each case being ones that were not present in either *rubicunda* or *amoena* (Gottlieb, 1973b). He therefore throws some doubt on the hypothesis of the very recent origin of *franciscana.*

Populations of some *Clarkia* species can be polymorphic for one or two translocations without this leading to any loss of fertility in the heterozygotes, because rings of four or chains of six chromosomes are regularly formed at meiosis and undergo alternate segregation at first anaphase, so that no aneuploid spores are produced. This is the case in some populations of *C. amoena, C. unguiculata,* and also in the neospecies *C. exilis.* When the number of translocations is increased, however, as it apparently has been in the origin of each of the neospecies, the meiosis of the heterozygotes becomes very irregular and virtually all the spores receive duplications and deficiencies of chromosome segments. It is thus possible that the critical event in the origin of a *Clarkia* neospecies is the addition of one more translocation to a karyotype that is already carrying one or two.

Stebbins (1971) believes that "saltational speciation . . . through the simultaneous occurrence of numerous translocations" explains these cases of speciation in the diploid species of *Clarkia* better than any other hypothesis, but states that similar situations are not known in other genera, and hence questions whether this mode of speciation is general. However, karyotype evolution of a similar type seems to have occurred in the Australian genus *Brachycome* (Compositae). But this is a large genus of at least 62 taxonomic "species" (many of which are undoubtedly complexes of several biological species) and there can be no doubt that polyploidy has also played a part in the overall pattern of evolution (Smith-White, Carter, and Stace, (1970). The species of the subgenus *Eubrachycome,* which have invaded desert and semidesert regions, are mostly ephemeral annual diploids. Among these (approximately 18 species altogether) the species complex "*B. lineariloba*" has been studied in detail. It includes five species referred to as A (2n = 4), B (2n = 23), C (2n = 16), D (2n = 8), and E (2n = 10). However, species A has three chromosomal races, two of which occur in the region of Spencer's Gulf and the Flinders Ranges, while the third grows in a small area about 500 kilometers to the east (Smith-White, 1968; Smith-White and Carter, 1970). All three races show supernumerary chromosomes in some individuals (Carter and Smith-White, 1972).

Race A_1, from the vicinity of Port Augusta, and race A_2, which grows on the western slopes of the Flinders Ranges, appear to differ by a translocation between the two chromosome elements. Some F_1 hybrids between these races were found in a narrow zone of geographic overlap. No backcross or F_2 hybrids were found, so the F_1 hybrids are probably largely sterile. Race A_3, known from a small area of New South Wales, appears to differ from A_1 by a translocation or a pericentric inversion; it may be a form that had a wider distribution in the past, linking it up geographically with the other two forms of A. It seems probable that these three "chromosomal races," confined to small areas, are in fact biological neospecies, analogous to such species of *Clarkia* as *C. franciscana* and *C. lingulata*.

The ancestral karyotype of the *B. lineariloba* complex seems to have been a $2n = 8$ one, as found in species D from the Eyre Peninsula; it is also seen in the related *B. campylocarpa*. The common ancestor of A_1, A_2, and A_3 must have evolved from a $2n = 8$ karyotype by two successive reductions in chromosome number. The other species (B, C, and E) have all *added* chromosomes to the basic karyotype. The origin of these additional chromosomes is obscure, but there are two of them, which may be called I_1 and I_2. Species B has $2n = 8 + 2I_1 + 2I_2$, species C has $2n = 8 + 2I_1 + 2I_2 + 2I_1' + 2I_2'$, where I_1' and I_2' are structurally modified variants of I_1 and I_2. Species E (Carter, Smith-White, and Kyhos, 1974), whose range exends from the Eyre Peninsula to the Nullarbor Plain, is cytogenetically most peculiar. Its karyotype consists of $2n = 8 + 1I_1 + 1I_2$. The two "extra" chromosomes I_1 and I_2 are univalents at meiosis and are inherited solely through the pollen.

Forms A_1 and A_2 hybridize with the more widespread species B and C at a number of localities and produce "amphihaploid" F_1 plants. These appear to be effectively sterile. A_3 and species C also produce sterile hybrids in nature in the very small area where the former exists.

Smith-White and his collaborators have suggested that "Many species [of *Eubrachycome*] are chromosomally unstable and may have been subject to catastrophic selection." This seems likely, but many details remain obscure. The ranges of the various species of the *B. lineariloba* complex are very much wider than those of most neospecies of *Clarkia* (only form A_3 appears to be geographically restricted, and this is probably due to insufficient collecting in the area where it occurs). The frequency of hybridization between the taxa of this complex indicates that, in general, they are outbreeding forms. If we envisage them as having arisen at a single point of origin, by "catastrophic selection," they would then have spread from there to occupy relatively large areas, without excluding other members of the complex. But there are many important differences between the mechanisms of speciation in the *B. lineariloba* complex and *Clarkia*. In the former, starting from a 2n = 8 karyotype, evolution seems to have proceeded in two quite different directions—a reduction in chromosome number in species A and an increase, by addition of I_1 and I_2 chromosomes, in species B, C, and E. Of these species, B and C are, broadly speaking, peripheral in distribution (especially C) and may be genotypes that were able to adapt to marginal environments, too arid for the ancestral form. Species E, on the other hand, which is found in southern Australia from the Eyre Peninsula to the Nullarbor Plain and the Hampton Tableland, is essentially coastal in distribution. It is referred to by Carter, Smith-White, and Kyhos (1974) as a "quasidiploid." Clearly, the whole evolution of this complex will not be understood until the nature and origin of the I_1 and I_2 chromosomes has been worked out. It seems probable that they were initially brought into the karyotype by hybridization and were later reduplicated and underwent differentiation (in species C) into I_1, I_2, and I_1', I_2'.

It is possible that the North American genus *Haplopappus* (Compositae) exhibits some of the same evolutionary changes as *Brachycome* (Jackson, 1957, 1962, 1965). The section Blepharodon includes a polyploid series based on n = 4 (species with 2n = 8, 12, and 16); but it also includes a widespread species *H. gracilis,* with 2n = 4, which may have arisen from *H. ravenii* (2n = 8), a species known only from a limited area of Arizona and southern Utah (Jackson, 1962). However, in view of the evidence from *Clarkia* and *Brachycome*—which suggests that diploid neospecies of arid habitats that have arisen by chromosomal repatterning usually have restricted ranges by comparison with the ancestral species—it seems more probable that *gracilis* gave rise to *ravenii,* particularly since it has apparently given rise to a form with 2n = 6 in a small area of south-central Arizona (Jackson, 1965). There are apparently two chromosome races of *H. gracilis,* which differ by a centromere transposition in the larger chromosome, and probably by a paracentric inversion as well (Jackson, 1973). Nevertheless, the published work on *Haplopappus* is inadequate for establishing any definite sequence of speciation events, although a general similarity to the presumed situation in *Clarkia* and *Brachycome* is evident.

There seem to be a number of basic differences between the modes of chromosomal speciation in *Clarkia* and such insect genera as *Vandiemenella* and *Didymuria.* The models of speciation by "catastrophic selection" (Lewis, 1962) or "quantum speciation" (Grant, 1963, 1971) in diploid plants are hence not equivalent to the stasipatric model of White (1968), although both do attribute a primary role to chromosome rearrangements.

In the first place, the formation of neospecies in *Clarkia* seems to occur on the periphery of the parent species' ranges. Second, although the neospecies remain confined to relatively minute areas (a single locality in some instances), they are strongly isolated genetically from the parent species, so that there is no hybrid zone surrounding their area of occurrence. Third, in some, but not all, cases the neospecies are self-fertilizing (which would not be possible in an insect).

Grant (1971) believes that "quantum speciation" is essentially restricted to annual herbs, especially in arid climates. This restriction seems unlikely. The kind of chromosomal repatterning that has occurred in the neospecies of *Clarkia* and *Brachycome lineariloba* is only likely to be detected and analyzed in diploid species with relatively low chromosome numbers. It is now well known that ephemeral annuals of arid habitats are especially liable to have low chromosome numbers (Stebbins, 1971). It seems likely that a significant number of speciation events similar to those in *Clarkia* postulated by Lewis and his collaborators have occurred in plant species of other growth forms (perennial herbs, shrubs, trees, etc.) of more mesic environments, but have been masked by the higher chromosome numbers, due in part to polyploidy.

The genus *Rumex* (Polygonaceae) consists mostly of perennial herbs of mesic or wet habitats. It includes several sections that are cytogenetically very different. Members of the section *Lapathum* are monoecious (hermaphroditic) and exhibit an impressive polyploid series. The section *Acetosella* likewise shows a polyploid series (n = 7, 14, 21, and 28), but these are dioecious (bisexual) forms, with an XY(♂):XX(♀) sex-determining mechanism. The section *Acetosa* likewise consists of dioecious species and races, most of which are XY_1Y_2(♂):XX(♀). The Eurasian species *R. acetosa, R. thyrsiflorus,* and *R. arifolius* have been studied by numerous workers, from Kihara and Ono (1925) to Świetlińska (1963). *R. hastatulus* is an annual that occurs in the eastern U.S. where its range extends from North Carolina southward to Florida and westward to Illinois, Missouri, and central Texas. There are three chromosome races of this species, all of which have fairly extensive ranges (B. W. Smith, 1969). The eastern race has 2n♂ = 9 (XY_1Y_2), 2n♀ = 8 (XX); the Texas–Oklahoma race has 2n♂ = 10 (XY), 2n♀ = 10 (XX); while the Illinois–Missouri race has 2n♂ = 9, 2n♀ = 8, like the eastern race, but differs from the latter in the shape of the smallest autosomal pairs and of the Y_1 chromosome. Numerous hybrids between these races have been found in nature; but their distribution at the present time is certainly not entirely natural because they are aggressive weedy colonizers of areas disturbed by agriculture and highway construction. It is not possible to put forward a precise scheme of the evolutionary differentiation of these chromosomal races, but obviously *R. hastatulus* is (or was, before human interference with the environment) undergoing a process of splitting into races or semispecies as a result of chromosomal rearrangements in both its autosomes and its sex chromosomes.

Most of the evolutionists who have put forward or discussed models of chromosomal speciation have assumed the neospecies carry certain chromosomal rearrangements that increase the fitness of the individuals carrying them in the homozygous condition, at least in certain environments, above the level prevailing in the original population (i.e., in homozygotes for the original chromosomal condition). However, this is not the view of Lewis (1973), who has put forward some evidence that the neospecies *Clarkia lingulata* is adaptively inferior in almost all respects, and in all environments, to its parent species *C. biloba*. If this were generally true, however, one would expect all neospecies to be very short-lived failures, destined to rapid extinction. It seems desirable for the time being to suspend judgment on this matter; *C. lingulata* may be an atypical neospecies, or it may have some adaptive advantage that was not shown in Lewis's experimental studies. Certainly the very numerous "chromosomal races" that seem to have given rise to new species of grasshoppers, lizards, and rodents can hardly be imagined to have established themselves by pure chance, without their having any adaptive superiority.

This advantage must almost always be in the viability rather than the fecundity component of fitness. There have been several attempts to explain the basis of this enhanced viability, two of which have been discussed by Grant (1971). The first, which Grant calls the *pattern effect theory,* harks back to R. Goldschmidt's (1940, 1955) view that speciation results from a radical repatterning of the karyotype. This was based on the so-called "position effect," according to which the phenotypic expression of genes is frequently modified when they are transferred to new locations in the karyotype, with new neighboring loci.

Goldschmidt's views did not find favor at the time they were put forward and have been generally rejected by evolutionists. It may in fact have been the radical and dogmatic nature of Goldschmidt's viewpoint that caused a whole generation of evolutionists to deny or undervalue the significance of chromosome "repatterning" in speciation. Goldschmidt's ideas met with strong opposition as soon as it became clear that most very closely related species, especially of *Drosophila,* did not differ by dozens or hundreds of karyotypic rearrangements (a complete "repatterning") but in only one or a few, and sometimes actually had indistinguishable karyotypes (the homosequential species). At the same time it was realized that a radical repatterning of the karyotype would never be able to pass through the "heterozygosity barrier," that any such repatterning would merely lead to the complete sterility of the individuals in which it occurred.

The more modern models of chromosomal speciation avoid these objections to Goldschmidt's earlier views. They envisage a less radical repatterning, one that involves one rearrangement at a time. Where two obviously closely related species differ in many rearrangements (e.g., *Sigmodon hispidus* and *S. arizonae,* considered earlier) it is assumed either that there was an intervening stage when the population was polymorphic for many rearrangements (the situation in the populations of *Nucella lapillus* occupying ecologically intermediate habitats) or (more usually) that there were a whole series of stepwise intermediate populations that have become extinct.

Instead of the pattern effect explanation of chromosomal speciation, Grant favors a *gene linkage theory,* according to which the rearrangements involved in speciation events bring together adaptive combinations of genes and prevent them from being disrupted by crossing-over. This interpretation has the advantage of explaining the association of "saltational" chromosomal speciation in plants with low chromosome numbers and low chiasma frequencies, i.e., with low levels of genetic recombination. The situation may, however, be somewhat more complex than this. It has been suggested (p. 174) that chromosomal rearrangements, in order to play a role in speciation, must not reduce the fecundity of the heterozygotes too much. This is certainly likely to be so in animals, and even in facultatively self-fertilizing plants it is probably true to a considerable extent. It is clear that most types of translocations, fusions, and other types of rearrangements will reduce the fecundity of the heterozygotes much less if the chiasma frequency is low and the chiasmata distally localized. It has long been recognized that this kind of chiasma distribution is a prerequisite for the evolution of complex translocation heterozygosity in the ring-forming Oenotheras and other plants having this type of genetic system. It has, however, also permitted the type of chromosomal speciation that has been so prominent in the genus *Clarkia*.

A third possible explanation of chromosomal speciation, not considered by Grant, may be important in the large number of cases in which changes in chromosome number have occurred. Such changes almost inevitably involve losses or gains of heterochromatic segments adjacent to the centromeres. Although the particular functions of the satellite DNAs of centromeric heterochromatin are still entirely mysterious, we may be fairly certain that this DNA has some adaptive significance. This explanation would not apply to cases of chromosomal speciation in which the only rearrangements that have occurred are simple two-break mutual translocations. Such cases seem to be extremely rare in animals, but may have been rather more common in plants.

Although the overwhelming importance of chromosomal rearrangements in speciation now seems abundantly clear, two main aspects are still obscure. One is their role in the diametrically opposite cases of frankly allopatric and entirely sympatric speciation (as opposed to the sort of intermediate geographic situations envisaged earlier in the cases of *Vandiemenella, Sceleporus,* and *Clarkia*). The second is the relative importance of the triad, stasipatric, and saltational models of chromosomal speciation. And are there perhaps other models that should be formulated and considered? What can be done to test the existing models more rigorously? It is difficult to formulate precise answers to these questions, but one might say in reply to the last question that the accumulation of electrophoretic data on critical cases should have high priority.

It seems unlikely that the triad model of speciation is of more than limited applicability, even in *Drosophila*. Nevertheless, the more general idea that *Drosophila* species have evolved from geographically localized coadaptations of inversions seems to be worthy of close consideration.

The stasipatric model appears to be applicable in a great many groups of insects of limited vagility. Although it was first put forward to explain the genus *Vandiemenella,* it probably also applies to other morabine grasshopper genera, including *Culmacris, Carnarvonella* (White and Cheney, 1972), *Warramunga,* and *Achuraba,* and a large proportion of the species of the subfamily. In lizards and small mammals the evidence for stasipatric speciation based on chromosomal rearrangements seems very strong. It does not necessarily follow that *all* speciation in these groups has been on this basis, but certainly the data suggest that stasipatric speciation has been a main (if not *the* main) mechanism of speciation. On the other hand, the evidence from anurans suggests quite different patterns of speciation in that group, and there seems no reason to question the view of Mayr that allopatric speciation has been universal or nearly so among birds.

The saltational model of chromosomal speciation put forward by H. Lewis seems to apply especially in annual diploids of arid lands. How far it applies in diploid plant species of other habitats, such as alpine meadows, saltmarshes, and tropical forests, must remain a fairly open question for the present.

Hanging over all discussions of chromosomal speciation for the time being is the admonition of Robertson (1975): "The evolutionists have hardly faced up to the evidence now accumulating from the multiple-copy sequences, such as the satellite DNA or those sequences which code for ribosomal RNA. I suspect that the mechanisms that we shall have to involve for the evolution of this part of the DNA will lie quite outside our present philosophy. . . ." Some reference has been made to this question in connection with the anurans *Xenopus laevis* and *X. mulleri;* it is at least a twofold problem, and perhaps a threefold one (see p. 101). Although conceivable, it is surely unlikely that changes in moderately and highly repetitive DNAs, which certainly occur frequently in phylogeny, do not play any part in speciation.

7

Sympatric Models of Speciation

The concept of sympatry is that of two (or more) genetic populations occupying the same geographic area. If we speak of "genetic populations" we are implying that they are at least partially isolated, reproductively. Thus the concept of the "same geographic area" has given rise to some semantic difficulties. In the strictest logical sense no two populations can be said to occupy precisely the same area, in the sense of being identically distributed in space. Such rigor of definition would obviously reduce the concept of sympatry to absurdity. Nevertheless, in practice questions do arise. Do oak and pine trees in a mixed forest constitute "different areas" for the insect species inhabiting the forest? A more practical question, perhaps, is whether stands of oak and pine of several hectares in a mosaic-type forest are to be regarded as different areas. And what about water at different depths in a lake or sea?

The answers to these questions must come from a consideration of what it is that we are really concerned with, rather than from formal logic or topology. We shall regard two populations as sympatric if their ranges overlap in such a manner that intermating could occur with a genetically significant frequency, unless prevented by genetic isolating mechanisms of some kind. Rivas (1964) proposed that the term "sympatric" should be used for species that have "the same or overlapping geographic distributions, regardless of whether they occupy the same macrohabitat (whether or not these species occur together in the same locality)" and suggested the term *syntopic* to refer to species that occupy the same macrohabitat, i.e., those that are "observable in close proximity and could possibly interbreed." From a geneticist's standpoint there seems little justification for using the term "sympatric" for cases in which separation of macrohabitats precludes any possibility of interbreeding. Rivas's terms "syntopic" and "allotopic" do not seem to have gained general acceptance.

Although nineteenth-century travelers and naturalists, such as Wagner (1889), seem to have generally favored the allopatric model of speciation (although unable to express it in genetic terms), the sympatric model was later supported by some entomologists, such as Brues (1924), Thorpe (1930, 1945), and H. S. Smith (1941), who claimed that, at least in groups of phytophagous insects with close obligate relationships to particular species of host plants, a species population may give rise to a race adapted to a new plant species and that two sympatric host races may diverge in evolution so as to become distinct species. This sympatric model was strongly criticized by Mayr (1942), who concluded that "bona fide evidence for sympatric speciation is very scanty indeed." Later (1963a) he stated that "the same old arguments are cited again and again in favor of sympatric speciation, no matter how decisively they have been disposed previously. . . . Sympatric speciation is like the Lernaean Hydra which grew two new heads whenever one of its old heads was cut off . . . ; the hypothesis is neither necessary nor supported by irrefutable facts."

It must certainly be admitted that a good deal of the evidence relied on by such workers as Thorpe (1930, 1940) and Emerson (in Allee *et al.*, 1949) was faulty. In particular, many supposed instances of sympatric "biological races" have turned out, when studied in depth, to be cases of entirely distinct sibling species that may well have been allopatric at an earlier period. One example that is now well understood is that of the sibling species of mosquitoes of the *Anopheles maculipennis* group. The European forms of this group (some, but not all, of which are vectors of human malaria) were originally regarded as biological races of a single species, and, since their distributions broadly overlap, were considered as evidence for sympatric speciation. These forms were originally separated taxonomically on the basis of the morphology of their eggs, and later shown to differ in respect of minor larval and adult characters; on the basis of the banding patterns of their polytene chromosomes (see p. 61), they must now be regarded as, for the most part, distinct species.

In retrospect, it is clear that Mayr's criticism of Thorpe's concept of "biological races," based largely on cases of sibling species whose genetic systems have diverged quite profoundly in spite of the rather minimal morphological differences between them, was well justified. But can we be sure that all apparent cases of biological races are of this type? In phytophagous insects the answer is certainly no (pp. 233–237). Unless host-range races of this type always arise allopatrically (a view against which there is considerable evidence), or unless they are incapable of proceeding to the status of full species (which seems improbable on general grounds), they are convincing evidence for the occurrence of sympatric speciation. Writing before Mayr's publications on speciation, the helminthologist Baylis (1938) cited evidence of "host races" in nematodes and stated that these may be looked upon as incipient species.

Nearly 40 years later the evidence he relied upon does not appear to have been disproved, but there is not much new that can be said in support of his conclusion as far as nematodes and other internal parasites are concerned.

While the main protagonists of sympatric speciation have been entomologists (or geneticists working with insects), the strongest opponents have been vertebrate zoologists. It is probably significant in this connection that vertebrates are, in general, larger than insects by one or two orders of magnitude. Hutchinson (1959) has spoken of "the impossibility of having many sympatric allied species of large animals," and stated "If I am right that it is easier to have a greater diversity of small than of large organisms, then the evolutionary process in small organisms will differ somewhat from that of large ones." There seems to be much truth in this rather simple distinction.

Evolutionists who have worked on plants have been, on the whole, unwilling to accept sympatric speciation (except in the case of allopolyploidy—see Chapter 8) as a significant evolutionary process. Thus, Grant (1971) recognizes three main modes of speciation, which he calls "geographical" (i.e., allopatric), "quantum" (see p. 222), and "sympatric." He regards the occurrence of sympatric speciation as an "unsettled question" and concludes: "Biogeographic evidence for primary sympatric speciation, unlike that for geographical and quantum speciation, is lacking. And the theoretical difficulties of speciation by disruptive selection within a cross-fertilizing population are very great. . . . Primary sympatric speciation is a possibility which has neither been confirmed nor ruled out. . . ." Stebbins (1950) states that "The available factual evidence points towards the rarity of speciation without previous geographic isolation . . . in plants just as in animals." But Jain and Bradshaw (1966), who have been concerned with plant populations that show sharp discontinuities in genetic composition corresponding to different soil types (polluted with heavy metals and unpolluted), state that "This [differentiation between geographically contiguous populations] is perhaps what has occurred in many cases in which sympatric speciation seems to have occurred. . . . Detailed analysis may reveal parapatric differentiation, with differential selection supported by geographical restriction of gene flow. . . ."

Two slightly different models of sympatric speciation in flowering plants, based on the extreme flower constancy of such pollinating insects as bees, have been put forward. Grant (1949) suggested that a conspicuous mutation affecting flower color or shape might be visited by a particular species of pollinator that avoided the original type. This would lead to homogamy, i.e., a tendency for like individuals in a population to be mated together. Mayr (1947) stated that "the evidence in favor of homogamy is virtually nonexistent . . . ; where it exists [it] is not of the type that would lead to the establishment of discontinuities within populations." But he had animal populations in mind, and some plant genera with highly specific pollinators may constitute exceptions to his generalization. Straw (1955), on the basis of his work on *Penstemon*, suggested a modified form of Grant's model. According to Straw, a diploid hybrid between two species might eventually be stabilized and established as a separate species if it were pollinated by a specific pollinator that avoided the parental species. Both models appear reasonable, but they would only apply to a very small number of plant species and can hardly be regarded as of general importance.

A conviction that the principles of population genetics exclude the possibility of sympatric speciation seems to be held by a good many non-geneticists, such as Fryer and Isles (1972), who in their book *Cichlid Fishes of Africa* state that "Perhaps the main argument that we can put forward against the concept of sympatric speciation . . . is that in the light of currently accepted genetical knowledge it seems to us to be theoretically impossible."

Geneticists, however, have been much less dogmatic on this point. Maynard Smith (1966), after a mathematical analysis of the sympatric model, stated that "the crucial step in sympatric speciation is the establishment of a stable polymorphism in a heterogeneous environment. Whether this paper is regarded as an argument for or against sympatric speciation will depend on how likely such a polymorphism is thought to be, and this in turn depends on whether a single gene difference can produce selective coefficients large enough to satisfy the necessary conditions. . . ." And Thoday (1972), whose laboratory experiments on disruptive selection we shall discuss later (p. 257), concludes that "Purely allopatric speciation is unlikely to be the only mode by which the origin of two species from one has occurred in evolution." The most detailed mathematical model of disruptive selection thus far developed is that of Dickinson and Antonovics (1973), who conclude: "There has been much circumstantial evidence that sympatric speciation may be possible . . . ; this model provides an unequivocal demonstration that it is feasible."

The main reason for the widespread belief that the algebra of population genetics rules out the possibility of sympatric speciation has been, of course, the conviction that mutations that render their carriers less likely to mate successfully with, and produce offspring by, other members of the Mendelian population of which they form a part will be eliminated by natural selection. Put in this somewhat simplistic manner the proposition is undeniable. It ignores two important considerations, however. The first is that it is somewhat unrealistic to think in terms of mutations whose sole effect is to give rise to genetic isolation between their carriers and the rest of the population. Such mutations will almost always have profound and diverse effects on the mode of adaptation and way of life of the individual, and it is entirely possible that the benefits conferred by some of these may outweigh the dysgenic effects resulting from partial genetic isolation from the rest of the population.

The second reason for being doubtful of the universality of the population genetics argument against sympatric speciation is, of course, that many natural colonies of organisms are not quite perfect random-mating populations inhabiting a uniform environment. Both because of low vagility and heterogeneity of the environment, which is fragmented into multiple habitats, such colonies undoubtedly provide considerable opportunity for the establishment of mutations that simultaneously affect the range of adaptation and the probability of successful interbreeding with other genotypes.

Maynard Smith (1966), having established mathematically the proposition that in a heterogeneous environment which includes different ecological niches a stable genetic polymorphism can be maintained by disruptive selection, then considers the conditions under which reproductive isolation might evolve between the two morphs. He discusses four ways in which this might occur. The first he calls "habitat selection"; individuals may have an inherent tendency to return to the habitat in which they developed for purposes of mating, and if such a tendency does not exist in the first place it will tend to be built up by natural selection. The second is "pleiotropism," in which the pair of genes that adapt individuals to different niches may themselves cause assortative mating. Maynard Smith regards this as very unlikely, and undoubtedly this is so if one thinks only in terms of single gene mutations; but it may be slightly more plausible in the case of chromosomal rearrangements affecting whole blocks of genes. The third way in which reproductive isolation might evolve is through modifier genes, i.e., the occurrence of mutations that modify the effect of the initial polymorphism so as to produce assortative mating. The fourth is through genes causing assortative mating *per se,* i.e., regardless of the genotype at the original polymorphic locus. Maynard Smith concludes: "If a stable polymorphism arises, this is likely to be followed by the evolution of reproductive isolation between the populations in the two niches . . . ; if a population is polymorphic for an allele pair $A,a,$ each conferring an advantage in one ecological niche, and if a second pair B,b causes assortative mating . . . , then, provided that there is some degree of habitat selection by egg-laying females, two reproductively isolated populations, *AABB* and *aabb* (or *AAbb* and *aaBB*) will evolve."

A special variant of the sympatric model of speciation is *allochronic* speciation, in which it is supposed that two populations develop genetic isolation without geographic separation through the acquisition of different breeding cycles, so that intermating is restricted or prevented. The evidence for this mode of speciation will be examined later in this chapter (p. 249).

Broadly speaking, there are three lines of evidence that have been or can be considered favorable to the hypothesis of sympatric speciation. These are: (1) the existence of "biological races," "host races," "ecological races," or "allochronic races" of a species within the same territory; (2) the occurrence of speciation on very small oceanic islands, where geographic isolation could not occur; and (3) the results of laboratory experiments on "disruptive selection," some of which seem to provide a plausible genetic basis for sympatric speciation. The first and the third of these lines of evidence have come under especially severe criticism from those who, like Mayr, regard allopatric speciation as universal, or nearly so. The second line of evidence has not been stressed much by either the supporters or the opponents of sympatric speciation, and has certainly been unduly neglected.

The problems of speciation in ancient lakes were dealt with in Chapter 4 because the evidence from these has traditionally been regarded as supporting the allopatric model. As interpreted in this book, however, most of the data on intralacustrine speciation in animals favors clinal, microallopatric, or frankly sympatric modes of speciation. Contrary to prevailing opinion, there is probably not much resemblance between the patterns and modes of speciation in ancient lakes and on oceanic islands. The degree of isolation of an island such as Saint Helena or Rapa (pp. 245–249) or an archipelago such as the Hawaiian Islands (considered as a whole) is much greater than that of any lake such as Baikal or Malawi connected with a number of different river systems. And the environmental uniformity that must prevail in the abyssal depths of these lakes is quite different from the ecological diversity that exists on the mountain chains of volcanic oceanic islands.

Much of the following discussion will be devoted to insects because the evidence is much more extensive than for other groups. There seems good reason to believe, however, that many of the conclusions would apply equally to other, less-studied invertebrate groups (i.e., the "small organisms" of Hutchinson).

It has been stressed by Janzen (1968) that, from a theoretical standpoint, species of host plants may be viewed as "islands" in their relationship to the insects that feed upon them and that colonization of a new host plant is analogous to immigration to an unoccupied island. He envisages the characteristics of a plant species considered as a vacant ecological niche as consisting of (1) its biomass and the way it is distributed in space; (2) distance between the species and other species of plants; (3) taxonomic and physiological similarities and differences between the species and its neighbors; (4) the insect faunas on these other plant species; (5) how anatomically and chemically varied the plant species is; and (6) how permanent or seasonal the plant organs (leaves, flowers, etc.) are.

The most detailed, and hence the most significant, studies on host races of insects have been carried out on the fruitflies of the family Tephritidae (Bush, 1966, 1969a, b, 1974, 1975a, b). Although the larvae of some species of Tephritidae are relatively polyphagous, feeding on a wide variety of fruits, many species are oligophagous or even monophagous (i.e., restricted to the fruits of a single plant species). Since many tephritid species are of considerable economic importance, as pests of cultivated fruit, they have been intensively studied in many countries. The adult insects are often brightly colored, with conspicuous body and wing patterns that serve as visual releasers in the courtship and mating rituals and, incidentally, facilitate the taxonomy of the group.

Bush has studied especially the genus *Rhagoletis* in North America. He has considered three species groups in detail. The *suavis* group includes five species (eastern United States to Arizona and Mexico) that feed on a total of eight species of walnuts (*Juglans* spp.) without showing any particular specificity. They may well have speciated allopatrically. The *pomonella* group includes four sibling species, each of which infests fruits of a different plant family. Although difficult to separate on morphological grounds, these siblings seem to be completely isolated reproductively and are quite different biologically. *R. pomonella* originally infested only Hawthorn *(Crataegus)* throughout the eastern United States, but a host race on introduced apples appeared in the Hudson River Valley about 1864 and rapidly spread until by 1916 it extended across the Great Lakes region to the vicinity of Winnipeg. In the southern United States the species is still only represented by the hawthorn race.

At the present time there are minor differences in size, number of postorbital bristles, and ovipositor length between some sympatric populations of the two races. There is also a striking difference in their seasonal cycles. Both have a single annual generation, but the emergence period of the apple race is from June 15 to the end of August (with a peak at July 25), i.e., about a month before the maturation of the apples, whereas the hawthorn race emerges between August 5 to October 15 (with a peak about September 12), i.e., approximately a month before the maturation of the hawthorn fruits. Whether this allochronic isolation completely prevents gene exchange between the two races at the present time is not known. The native crab apples are not infested by *R. pomonella* and were hence not the source of the apple race that must have arisen directly from the hawthorn race.

Another allochronically isolated race of *pomonella* infests native and cultivated plums in the eastern United States. It emerges considerably earlier than the apple race, in accordance with the earlier maturation of the plums. And in the 1960s a cherry race made its appearance for the first time in Wisconsin. The cherry race must have arisen from the apple race, since the gap between the time of maturation of cherries and haws is too great for a direct shift to have taken place.

There is thus reason to believe that host races that are effectively isolated genetically can arise in this species from time to time when new food plants become available. A most important fact is that the insects are attracted to their host plants by visual and olfactory cues and that mating takes place on the fruits, being then followed by oviposition. Bush (1974) points out that the biology of oligophagous and monophagous tephritids makes an allopatric origin of new host races highly unlikely. Migration of a few individuals to an area where a new species of fruit was present would have no effect unless these had the right genetic constitution for recognizing, ovipositing, and surviving on it. Most dispersal in *Rhagoletis* takes place before mating, so that both males and females would have to be attracted to the same fruits.

In place of such improbably allopatric models, Bush puts forward the following postulates in a sympatric model of *Rhagoletis* speciation (slightly modified and shortened from his original text):

1. The old and the new host plants must occur in the same area.
2. The maturation times of the two fruit species must overlap.
3. Diapause and emergence times must be under genetic control.
4. Orientation to and selection of a host plant due to a chemical cue.
5. Host selection is due to one main genetic locus.
6. h_1h_1 individuals are attracted to the original host; h_1h_2 individuals are attracted to both hosts, but because of modifiers (polygenes) will move preferentially to the original host; while h_2h_2 individuals are only attracted to the new host.
7. Survival on the original and the new host is controlled by another locus. s_1s_1 individuals survive on the original host; s_2s_2 individuals survive on the new one; s_1s_2 individuals survive on either.
8. Since mating occurs on the host plant, a high degree of homogamy is expected.

Under these circumstances one would expect that disruptive selection would lead to the rapid evolution of two host races, one homozygous for $h_1h_1s_1s_1$, the other for $h_2h_2s_2s_2$. If the time of maturation of the two kinds of fruits is different, disruptive selection would also be expected to occur for the genes controlling the time of emergence.

More complicated models involving two or more different loci for host recognition and survival are also possible. Although such models lessen the probability of successful colonization, they increase the likelihood that a shift to a new host will lead to a greatly reduced gene flow between the races, and hence the probability of complete speciation occurring.

The other species of the *Rhagoletis pomonella* group are *R. mendax,* whose host plants are *Vaccinium* spp. and *Gaylussacia* spp. (Ericaceae); *R. cornivora,* living on *Cornus* spp. (Cornaceae); and *R. zephyria,* whose host plants are species of *Symphoricarpos* (Caprifoliaceae). The first two species are broadly sympatric with *pomonella* in the eastern United States, while the fourth is a midwestern and western species whose range only overlaps that of *pomonella* in Minnesota and Manitoba.

The *Rhagoletis cingulata* species group includes an eastern *(R. cingulata)* and a western *(R. indifferens)* species infesting native and cultivated cherries. It also includes two Florida species, *R. chionanthi* and *R. osmanthi,* which feed, respectively, on the native "olives" *Chionanthus virginicus* and *Osmanthus americanus*. There is strong allochronic isolation between the Florida species; the first emerges in summer, the second in winter, when the *Osmanthus* fruits mature.

Bush (1966) briefly considered two other species groups. The *Rhagoletis tabellaria* species group includes two species whose host plants are unknown; *R. juniperina,* which develops on species of *Juniperus* (section *Sabina*); and *R. tabellaria,* which probably consists of two sibling species (an eastern one that attacks species of *Cornus* and is unable to attack *Vaccinium,* and a western one that lives on *Vaccinium* species and is not known to develop on *Cornus*). The *Rhagoletis ribicola* group includes two western species: *ribicola,* which attacks *Ribes* spp., and *berberis,* which attacks *Mahonia* spp. (Berberidaceae).

The sympatric origin of host races in *Rhagoletis*—and hence of species, because there can be no reasonable doubt that these host races are potentially incipient species—seems to be as firmly established as any case of allopatric speciation has been. Oligophagy and monophagy are very widespread in several families of Diptera, especially Cecidomyidae and Agromyzidae; they are also common in Lepidoptera, many families of Heteroptera and Homoptera, sawflies, and some families of Coleoptera (especially Cerambycidae and Curculionidae). These groups include perhaps a third of all insect species. Oligophagy or monophagy also occur sporadically in most other groups of phytophagous insects, and the enormous group of parasitic hymenopterans exhibit dietary specializations that in many instances are quite rigid. There is thus reason to believe that the formation of host races analogous to those that have arisen in *Rhagoletis pomonella* in the past century may be an extremely frequent phenomenon in many groups of insects. Clearly, in dealing with speciation in such organisms, we are in a different world from that with which the ornithologist or herpetologist is concerned.

Those evolutionists who have claimed that all (or *almost* all) speciation is allopatric have also emphasized that speciation involves a major "genetic revolution," i.e., the replacement of a large proportion of the alleles in the original genotype. Believers in the reality of sympatric speciation, on the other hand, have tended to think that a few changes at key gene loci would be all that is required to initiate a new host race.

It seems probable that this is more a misunderstanding than a disagreement in principle. Bush and other workers on host races in Tephritidae have built up a strong case for the view that a few changes at key loci can indeed lead to a switch in host plant. But it is clear that once two sympatric host races are in existence there will begin a long process of disruptive selection, involving a change in the season cycle of the new race and changes in those body and wing patterns (and perhaps behavior patterns as well) that serve as cues in the mating ritual. Huettel and Bush (1972) showed that two species of the tephritid genus *Procecidochares* (*P. australis* and an undescribed species "A"), which are sympatric on the Gulf coast of Texas, differ with respect to a gene locus that controls their ability to discriminate between different species of Compositae. *P. australis* infests *Heterotheca* spp. and cannot survive on *Macroanthera phyllocephala,* while the reverse is true of species A. When homozygous for one host-recognition allele, only one species of host plant is selected. F_1 hybrids can be reared on either host plant but normally, when offered the choice, F_1 adult flies deposit almost all their eggs on the host plant on which they themselves were reared. A study of backcross hybrids showed that although there is a primary locus with alternate alleles determining host plant recognition, there is also a separate genetic determination of the ability to be induced to oviposit on one host or the other. Other genes must be responsible for the ability to survive on a particular host species. Thus, Bush believes adaptation to a new host plant involves changes at gene loci concerned with the three separate processes of host recognition, induction, and larval survival.

Can such changes occur without a major "genetic revolution"? Bush (1974) has attempted to answer this question by studies of the European *Rhagoletis cerasi.* There are two races of this species, which live on cherry and honeysuckle *(Lonicera),* respectively. The evolutionary divergence of these species seems to be much older than that between the recently evolved apple and cherry races of *R. pomonella,* but there are no detectable morphological or karyotypic differences between them. Boller and Bush (1973) have shown by hybridization experiments that there is a considerable degree of genetic incompatibility (as measured by egg viability) between populations of the cherry race of *R. cerasi* from eastern and western Europe; in crosses between western males and eastern females almost no viable eggs were laid. However, a study of five polymorphic enzyme loci and seven monomorphic ones revealed no unique alleles or striking frequency differences between material collected in Spain, Turkey, Holland, and southern Italy. Thus the eastern and western populations seem to have achieved a considerable degree of genetic isolation without any general revolution in allele frequencies. It is not known whether they are connected stepwise by a series of intermediate populations across central Europe, each interfertile with the adjacent ones.

Sympatric populations of the cherry and honeysuckle races are not so strongly isolated (on the basis of laboratory experiments) as the eastern and western populations of the former. On the other hand, allozyme studies have demonstrated some differences between the cherry and honeysuckle races at some localities, but not at others.

In summary, the evidence, while not really extensive, suggests that in trypetids both allopatric differentiation into genetically isolated populations living on the same host plant and sympatric differentiation into races dependent on different host plants can occur without a genetic revolution of such magnitude as to affect a large number of enzyme loci.

We have referred above to the need for genetic changes at loci governing host recognition, induction, and larval survival. It is probable on general grounds that single mutations can affect the range of sensitivity of insect chemoreceptors, thereby leading to changes in host recognition and selection. Some trypetids, such as *Dacus* spp., are not highly specialized for development in a single species of fruit. But in *Rhagoletis,* although the adults may be induced to lay on strange kinds of fruit (e.g., *R. pomonella* will lay on tomatoes in laboratory experiments), larval mortality is very high if they do and few individuals survive to the adult stage. The actual causes of this larval mortality are not known, and although it is clear on general principles that genes for larval survival in particular types of fruit must exist, the details of this genetic determination have not been worked out.

As mentioned above, some *Dacus* species are generalized feeders that can develop in a large variety of species of fruits. The sibling species *D. tryoni* and *D. neohumeralis* are sympatric throughout a large part of eastern Australia, although *tryoni* extends further south than *neohumeralis,* which is restricted to coastal Queensland and northern New South Wales. Both species have been recorded as breeding in a large number of species of native fruits of tropical rain-forest trees and shrubs; they became adapted to a wide range of cultivated species of fruit after 1850 and spread southward as orchard pests (Birch, 1965). The two species are morphologically very similar, but the "shoulder patches" (humeral calli) are typically yellow in *tryoni* and brown in *neohumeralis.* Supposedly intermediate individuals occurring in nature were interpreted by Birch (1961, 1965) as due to the occurrence of hybridization. This conclusion seemed to be supported by the fact that when kept together in a laboratory population the two species hybridize readily and stabilize as a hybrid swarm. It was even suggested (Lewontin, in Birch, 1965) that the southward range extension of *D. tryoni,* which seems to have occurred in the past 60 years, had been facilitated by the introgression of genetic material from *D. neohumeralis.* However, Wolda (1967b), as a result of a most careful study, concluded that the intermediates are not natural hybrids but simply extreme phenotypes of one or other species. His analysis suggests that little or no natural hybridization is in fact occurring. One factor that undoubtedly operates to prevent hybridization in nature is the difference in mating time; *D. neohumeralis* mates from 9 A.M. to 4 P.M., while mating of *D. tryoni* is restricted to the period from 5 to 6 P.M. It seems likely that there has been disruptive selection for mating time; the mode of speciation that operated to produce these two extensively studied species in the first place is still entirely obscure. Birch and Vogt (1970) believe that the two species are "extreme variants of a continuum of quantitative characters." But field populations are strongly concentrated around two modes towards the extremes of the five morphological characters they studied. Taken in conjunction with the well-developed sexual isolation between them, this strongly suggests that we are dealing with two species of flies rather than one polymorphic one. To refer to them as "clusters," as Birch and Vogt do, rather than species, merely obscures the situation.

There are relatively few groups of animals in which host races equivalent to those of Tephritidae are well-known and fully documented. Monophagy, however, is equally shown in such groups as the gall-making cecidomyids and Cynipids, the agromyzids (Nowakowski, 1962), most groups of parasitic hymenopterans, some aphids and coccids and certain groups of lepidopterans and sawflies. In Coleoptera there are a number of phytophagous and wood-boring families in which the species are closely adapted to a single host plant or a small range of related species. All of these are groups in which the occurrence of sympatric speciation seems probable, on the same kind of basis as in the tephritids. Leafhoppers (Cicadellidae) are another group in which it is very likely that sympatric speciation is occurring, although critical evidence is lacking. Evans (1962) suggested that sympatric speciation haa probably taken place in the subfamily Typhlocybinae (about one hundred genera, some containing hundreds of species) because many of the species are restricted to a single species of host plant or a limited range of related species, and because the wide dispersal of these minute insects by wind currents is unfavorable to geographic isolation. There are about 700 species of bees belonging to the genus *Perdita* in North America, each of which visits the flowers of a single plant species or a small number of closely related species (Pimentel, Smith, and Soans, 1967); it is certainly difficult to imagine that their speciation could have been entirely allopatric.

Since complete geographic isolation, if continued long enough, will inevitably lead to speciation, we must expect that even among groups whose ecology, vagility, and population structure is most favorable to sympatric speciation there will be some cases of strictly allopatric speciation. The converse, however, does not hold; there are almost certainly groups whose general mode of life makes it impossible for reproductive isolation to develop without prior geographic isolation.

The gall midges (Diptera, family Cecidomyidae) include a great many species that make galls on a single species of host plant and in which speciation must frequently have involved adaptation to a new host plant. Numerous examples are to be found in the old works of Felt (1920, 1925) and Barnes (1946–1951). Here, however, in some genera we meet a different type of phenomenon, namely, that many closely related species make galls of widely different appearance on the *same* host plant. A conspicuous example would be the numerous species of *Caryomyia* that make leaf galls on hickory *(Carya)* species in the eastern United States (Felt, 1920). The relationship of the insects to different species of *Carya* has not been studied, but galls of three or four species may be found on the same tree and even on the same leaf.

Monophagy has been developed in some species of grasshoppers, although this is a group in which most species are moderately polyphagous. For example, in Texas the grasshopper *Chloroplus cactocaetes* (the only species of the genus) feeds exclusively on the cactus *Opuntia leptocaulis,* living among the dense tangle of spiny branches of the plant. In the eastern United States two species of tree-inhabiting grasshoppers, *Dendrotettix quercus* and *D. zimmermanni,* feed on oaks, while *D. australis* has adapted to a diet of *Pinus virginiana* (Friauf, 1957). Polyphagous, oligophagous, and monophagous species may be closely related, i.e., they may occur in the same genus. Thus, certain locusts of the genus *Schistocerca* in both the Old and New Worlds are notoriously polyphagous; but in central Florida there is a species that lives a concealed existence inside thickets of the shrub *Ceratiola ericoides,* which is its sole food plant (Hubbell and Walker, 1928).

The gall wasps of the genus *Cynips* (family Cynipidae) develop entirely in galls on the numerous species of oaks *(Quercus)* in North America and Europe. Kinsey (1930, 1936) studied the species of the United States and Mexico. He recognized 165 species, but it is certain that there are many yet unknown. Even his very extensive material (over 35,000 insects and 124,000 galls) came from a limited number of localities, and large areas of Mexico and the southwestern United States have not been explored for *Cynips.*

Some species of *Cynips* live on several, usually closely related, species of *Quercus.* Others inhabit galls on only one or two species. Since the genus *Quercus* is very rich in species in the mountain chains of Mexico and the southwestern United States, it is obvious that the speciation pattern of *Cynips* has been closely related to the multiple ecological niches provided by the diversity of the oaks. Kinsey, whose views on evolution are slightly peculiar by the standards prevalent today, but who had studied the genus in the field and the laboratory for many years, seems to have believed that speciation in *Cynips* had been partly on an allopatric and partly on a sympatric basis: ". . . these insects are restricted to limited areas that are effectively cut apart by deserts as though they were islands in an ocean. More than that, the strict host relationships of these insect species make the oaks the counterparts of so many more islands between which there are no means of migration." And with that, until someone undertakes another major research program on *Cynips,* the matter must rest.

Speciation in parasitic hymenopterans belonging to the superfamily Chalcidoidea has been discussed by Askew (1968), who laid particular emphasis on the widespread occurrence of inbreeding in this group, due to matings between siblings occurring immediately after emergence from the pupa. Many species of chalcids have very wide distribution areas and, due to their small size, are liable to be carried great distances by air currents. This group, then, is one in which enormous numbers of species exist—there are published estimates of 1,000 described species in the British Isles, 2,000 in the nearctic region, and 2,791 in Australia, but in the latter areas (in which opportunities for geographic isolation are rather limited) the number of undescribed species must be even greater.

A particularly interesting evolutionary situation, involving extremely close host specificity, exists in the small chalcidoid wasps of the family Agaonidae, which are absolutely necessary for the pollination of figs (*Ficus* spp.) and can only develop inside the "gall flowers" (Baker, 1961; Grandi, 1963a, b; Wiebes, 1965, 1966; Ramírez, 1970). There are many hundreds of species of *Ficus* in the Old and New World tropics and, with only very rare and slightly doubtful exceptions, each has a close obligate relationship to a single species of wasp (the exceptions being a few species of fig alleged to harbor two species of agaonids). Where a fig species occurs in several different regions it seems to have the same agaonid throughout. There is a relationship between the subgenera of *Ficus* and the genera of agaonids, e.g., the subgenus *Pharmacosycea* is the host for the genus *Tetrapus,* while *Urostigma* is the host of *Blastophaga.* Species of *Ficus* belonging to the section *Americana* harbor agaonids of the genera *Julianella* and *Valentinella,* which do not occur on any other species of figs. The genus *Ficus* has about 900 species and in most tropical areas several or many species are sympatric (e.g., there are 83 species in Java, which suggests much sympatry).

There are clearly two interrelated problems here—the speciation of *Ficus* and that of the wasps. The other genera of Moraceae (the family to which *Ficus* belongs) are wind-pollinated and have not speciated to anything like the same extent. It seems impossible to avoid the conclusion that the speciation of *Ficus* and of the *Agaonids* has been concomitant, i.e., that each incipient species of *Ficus* has evolved in parallel with an incipient species of wasp. But there is really no evidence as to whether this has usually been an allopatric or a sympatric phenomenon. Hybridization and polyploidy seem to have played a very minor part in the speciation pattern of *Ficus,* the overwhelming majority of the species being diploids with 2n = 26.

The above account, emphasizing the one-to-one relationship between species of figs and wasps, is based mainly on the careful recent work of Ramírez on the situation in central America. Baker (1961) was inclined to doubt that there was such a close correspondence; but he relied mainly on very old (nineteenth century) authorities for his evidence, and the opinion of such specialists as Ramírez and Wiebes is strongly in favor of absolute specificity.

The essential difference between this peculiar situation and that of *Cynips* is clearly the dependence of *Ficus* speciation on wasp speciation. The oaks are not in any way dependent on cynipids, and the various species can consequently be regarded as so many niches available for cynipid speciation, regardless of whether they already harbor a cynipid species or not. In the case of the figs and fig wasps there can never be such a thing as a vacant niche—the niches only exist as a result of being occupied.

Speciation in internal parasites, such as digenetic trematodes, cestodes, and many nematodes, is regarded by Mayr (1963a) as "strictly allopatric in most cases. The same species of parasite may have different hosts or intermediate hosts in different regions, and if the isolation is sufficient there will be the development of a geographical race of parasite in due time which will eventually reach species level." Mayr does not believe that the common situation of a parasite species having several alternative hosts in the *same* area leads to speciation.

It is extremely difficult to determine from the literature what the degree of specificity of any particular group of parasites really is. Certain rare species may have been recorded only from a single host species, thereby giving a false impression of extreme specificity. On the other hand, common parasite species may have been recorded from several or many host species, some of which they only parasitize occasionally or accidentally, so that they are in reality far more specific than appears from the published records. The latter appears to be the case in the monogenetic trematodes, of which Rogers (1962) states: "Parasites of this group are usually highly specific; indeed many can infect one species of host only." Digenetic Trematodes usually show fairly high specificity in the early stages of their life cycle, when they are parasitic in molluscs; but the adults, which parasitize vertebrates, exhibit a rather low degree of specificity (Baylis, 1938). In cestodes (tapeworms) the reverse is the case—it is the adults that show high specificity. The nematodes include, in addition to the numerous free-living species, some parasitic forms with poorly developed host specificity and others that are highly specific. The situation in the family Camallanidae has been described in detail by Stromberg and Crites (1974). This is a group of about 130 species that are endoparasites of fishes, anurans, and reptiles, mainly in tropical countries. Some have a considerable range of host species. Thus, the Nile river species *Procamallanus laeviconchus* has been reported in 20 species of fishes belonging to eight different families. *Procamallanus* includes 23 species that parasitize freshwater fishes in Asia, Africa, and Europe; five species that occur in marine fishes; and two in frogs of the genus *Xenopus*. Some of the marine species are host specific. The 54 species of *Camallanus* include a large number of host-specific forms, and are parasitic especially in frogs.

It must be obvious that in many groups such as this adaptation to new host species may have led to sympatric speciation, since the different host species would constitute so many sharply isolated ecological niches in the same region. Unfortunately, there have been hardly any modern studies of a critical nature on speciation in these groups. The position is slightly better with regard to insect parasites of vertebrates (Askew, 1971). The lice include four main groups: Amblycera, Ischnocera, Rhynchophthicina, and Anoplura. Hopkins (1949; and see Hopkins and Clay, 1952) has commented on the very marked host specificity of the mammal-infesting Amblycera and Ischnocera and on the equally developed specificity of the bird parasites. Thus, each of the European crow species—*Corvus frugilegus, C. corax, C. corone* (with two subspecies; see p. 165), and *C. movedula*—harbors a different species of *Philopterus;* only the two subspecies of *corone* have the same *Philopterus* species. The ultimate in parasite specificity seems to be reached in the case of the South African hyraxes *Procavia capensis* and *Heterohyrax syriacus*. Each geographic subspecies of these small mammals has its own species of *Procavicola* (Ischnocera). It is probable that the speciation of *Procavicola* has been allopatric, following the subspeciation of the host species.

Speciation in bird lice (Amblycera and Ischnocera) is discussed in detail by Clay (1949). Usually, each genus of louse is restricted to a single order of birds, but exceptions occur. Within genera, each species is, in general, restricted to a single species or subspecies or to a group of closely related species of birds (the latter situation seems to be uncommon and to be found mainly among parasites of grouse and terns). The head and wings of many bird species are occupied by lice belonging to different genera, but there do not seem to be instances of closely related species of lice occupying these different sites on the same species of bird.

Since the entire life history of all species of lice is spent on the vertebrate host, and since they can only survive off the host for a very limited time, migration from one individual host to another, whether of the same or a different species, is usually only possible when the two are in actual contact, e.g., in the case of avian hosts, when they are copulating or roosting or perching together or using common dust baths, or when one is a predator of the other. The one clear exception to the above generalization is "phoresy" (the transportation of ischnocerans from one bird to another by hippoboscid flies, which are themselves parasitic on birds).

Clay discusses allopatric speciation of a louse species following allopatric speciation of its host, i.e., the bird species' distribution becomes subdivided by a geographic barrier, whereupon it splits into two species by classical allopatric change, thus isolating the two louse populations, even if the two new bird species later become secondarily sympatric. There can be little doubt that this model of speciation is realistic.

Clay has some reservations about speciation in lice resulting from "isolation by the development of host specificity," and uses the example of *Cuculus canorus* (European cuckoo) to illustrate the difficulties in the establishment of lice on a new species of host. *C. canorus* is a brood parasite whose young are reared in the nests of a great variety of species of passerine birds. In spite of this it has never acquired any lice from any passerines. The main reason is probably that it already harbors three genera of lice found on cuculids throughout the world, i.e., the niche is already filled. Just how these specific cuckoo parasites get from one individual to another except during copulation is something of a mystery, as cuckoos are not gregarious. However, the numerous instances of populations of *Lipeurus caponis* (a common parasite of chickens) establishing themselves on pheasants, partridges, and guinea fowl under domestic conditions show clearly that transfer to and establishment on a new host species are quite possible, given the necessary close proximity.

A third mechanism of speciation considered by Clay is one resulting from isolation and later reunion of host populations. The host species acquires a disjunct range that persists long enough for speciation of the parasite species but not for the avian host. Later, due to range extension, the two bird populations merge geographically; at this stage they will harbor two sympatric species of lice. No actual examples of such a process are cited, however.

A fourth model of speciation considered by Clay is also a strictly allopatric one and depends on the fact that a host species with a continuous distribution may be parasitized by a species whose distribution is discontinuous, i.e., confined to two or more geographic areas because it has become extinct in the intermediate territory. Under such conditions (if maintained long enough) the parasite may be expected to undergo speciation.

Two possible models of sympatric speciation are discussed by Clay. The first is the possibility of strictly sympatric speciation on a single host species, the two species coming to occupy different areas of the plumage, e.g., head and wings. This is dismissed as inherently improbable, and examples that might appear to illustrate it are interpreted as due to speciation through geographical isolation (the third model discussed above).

The second is due to "secondary interspecific infestation," i.e., migration of a louse species to a new host species with subsequent divergence. Clay admits this occurs but considers the process essentially allopatric; host isolation is considered as equivalent to geographic isolation and colonization of a new host species is regarded as "analogous to the transoceanic colonization of oceanic islands by free-living animals. . . ." This kind of interpretation seems inherently unsound. Different sympatric host species that come into sufficiently close association (by sharing roosting places, nesting areas, dust baths, or the like) to permit migration of the parasites are emphatically *not* analogous to oceanic islands—they are alternative ecological niches within the same geographic area, habitat, or *patria*. The closest analogy is to the different species of host plants growing alongside one another, which we discussed earlier in connection with the host races of trypetids and other phytophagous insects. Thus, insofar as secondary interspecific infestations can be proved to have occurred in bird lice, they must be regarded as evidence for true sympatric speciation.

There has been much argument about the relationship of the head and body lice of man. Regarded by de Geer in the eighteenth century as distinct species, *Pediculus capitis* and *P. corporis,* these are now generally regarded as ecotypes or "biological races" of a single species, *P. humanus*. Fertile hybrids between them can be obtained in the laboratory, although some intersexes have been found among their progeny (Buxton 1939). Mayr (1963a) has suggested that they developed originally on different human races, e.g., Melanesians (with much head hair) and Eskimos (heavily clothed). Another species of *Pediculus* is parasitic on the chimpanzee and one or more species occur on South American spider monkeys (*Ateles* spp.). It has been suggested that the latter are derived from human lice brought to South America when *Homo sapiens* invaded that continent, possibly 30,000 or 40,000 years ago.

It seems clear that, just as in the case of the trypetid flies and other host-specific phytophagous insects, at least some of the speciation events in these groups of parasites are likely to have been sympatric. But in the absence of really comprehensive modern studies on speciation in any of them, we cannot claim at this time that they provide any absolutely critical evidence for sympatric speciation.

Speciation in mosquitoes was discussed in Chapter 6, because of the important evidence on chromosomal mechanisms of speciation in this group. It is worthwhile pointing out here that many species of mosquitoes are highly specific with regard to the vertebrates from which they obtain blood. Thus, just as in the case of true parasites, there is a possibility of speciation occurring sympatrically if it involves adaptation to a new species of vertebrate host.

If major geographic barriers are a *sine qua non* for speciation, speciation should be impossible on small oceanic islands (as opposed to archipelagos, such as the Hawaiian and Galápagos Islands). A critical case would be the island of Saint Helena, in the South Atlantic (about 120 square kilometers, 1,760 kilometers from the nearest point of Africa and 2,880 kilometers from South America). Saint Helena is essentially a single volcanic cone and there is no likelihood that it ever consisted of more than one island. Much of the insect fauna consists of cosmopolitan species, some of which are clearly recent introductions. However, out of 256 species of beetles present on the island, 137 are classified as "ancient endemics" and 20 as "recent endemics" by Basilewsky and co-workers (1972); only 99 species are considered introduced. The coleopteran fauna provides impressive evidence of speciation having occurred on Saint Helena (Table 7-1). This is certainly so in 17 wholly endemic

TABLE 7-1
Speciation of Coleoptera on the Island of St. Helena. (Data from Basilewsky et al., *1972.)*

Family	Genus	Endemic (+) or not (−)	Species evolved on St. Helena
Carabidae	*Pseudophilochthus*	+	9
	Apteromimus	+	2
Scarabeidae	*Melissius*	+	3
Elateridae	*Anchastus*	−	2
Tenebrionidae	*Tarphiophasis*	+	5
Anthicidae	*Anthicodes*	+	2
Chrysomelidae	*Longitarsus* [a]	−	3
Anthribidae	*Notioxenus*	+	12
	Homoeodera	+	14
Curculionidae	*Nesiotes* [b]	+	13
	Tychiorhinus	+	6
	Xestophasis	+	2
	Lamprochrus	+	2
	Eucoptoderus	+	2
	Chalcotrogus	+	3
	Acanthinomerus	+	13
	Microxylobius	+	15
	Isotornus	+	4
	Pseudomesoxenus	+	4
	Pseudostenoscelis	−	6

[a] In the opinion of Basilewsky the three species of *Longitarsus* on St. Helena probably constitute a separate, endemic genus.

[b] The genus *Nesiotes* is here considered as endemic to St. Helena. But it is very closely related to *Paranesiotes*, from South Africa, and the two might well be regarded as subgenera (Basilewsky *et al.*, 1972).

genera that have several species and probably also in a few nonendemic genera that are represented on the island by several obviously related species. Table 7-1 does not include a number of endemic genera with only a single species, closely related to genera listed in the table; in most instances these must also be due to speciation having occurred on the island rather than to independent colonization from Africa or South America.

It might be argued that even such a tiny land mass as Saint Helena is large enough to provide opportunities for geographic isolation. But some of the endemic genera with several species are confined to single small areas on the island and, for ecological reasons, were probably equally restricted even before the arrival of man. Thus, the nine species of *Pseudophilochthus* and the two species of the closely related genus *Apteromimus* are confined to elevated regions of the High Central Ridge, seven of them being recorded from Diana's Peak; they are hence sympatric in the strictest sense. Of the five species of *Tarphiophasis,* four are recorded from the same locality (Prosperous Bay Plain). The three species of *Longitarsus* occur at high elevations; *L. mellissi* is closely associated with plants of the genus *Sium* (Umbelliferae), *L. janulus* lives on *Senecio prenanthiflora,* and *L. helenae* on *Lobelia scaevifolia* (Basilewsky *et al.,* 1972). On the basis of evolutionary changes in host plant it would be difficult to conceive of a more convincing case of sympatric speciation than this. The species of *Notioxenus* also inhabit the High Central Ridge, living in dead or rotting wood of *Commidendron, Senecio,* and tree ferns. The situation with regard to *Homoeodera* is much the same. These anthribids do not seem to be restricted to wood of any single species of genus of plant; a complex trophic relationship undoubtedly exists between the beetles, the dead wood, and the fungi associated with it. The endemic weevils of the genera *Nesiotes* and *Tychiorhinus* also occur on the High Central Ridge, in dead and rotting wood of "Cabbage Trees" *(Melanodendron)* and *Commidendron,* but have been able to adapt to life in the wood of introduced trees and shrubs such as *Erythrina* and *Ulex;* they are hence not highly specific feeders. The distribution and ecology of most of the species of *Acanthinomerus* and *Microxylobius* seem to be generally similar.

With the exception of the species of *Notioxenus,* all these coleopterans are wingless and undoubtedly insects of very restricted vagility. The simplest explanation would be to suppose that each of the genera listed in Table 7-1 arose from a single immigrant species accidentally transported to Saint Helena at some time in the remote past. Actually the number of ancestral immigrant species was probably fewer than this. None of the endemic genera is at all likely to have arisen from more than one immigrant species, and in a number of instances several endemic genera (including genera with a single species, not listed in Table 7-1) must have arisen from a single immigrant ancestral species.

The taxonomy of the Saint Helena coleopterans has been carefully carried out by modern workers, following on the pioneer studies of nineteenth-century entomologists who collected a number of species that have almost certainly become extinct. The numbers of species given in Table 7-1 are those given by the modern workers (Basilewsky *et al.*, 1972) and are certainly conservative; in some instances a "species" is stated to include two "forms," which may in reality be distinct biological species.

Some other instances of insect speciation on Saint Helena may be mentioned here. They are similar to, but less striking than, the coleopteran cases mentioned above. There is an endemic genus of tettigoniid grasshoppers, *Phaneracra,* with two species (Ragge, 1970); there can be no doubt that this represents an instance of speciation on Saint Helena. Another case that is almost certainly of the same type involves the truxaline grasshoppers *Tinaria calcarata* and *Primnia sanctaehelenae* (Dirsh, 1970). Although these have been put in different genera, they seem to be rather closely related and do not appear to have any close relatives in Africa or South America. It is thus probable that, like the species of *Phaneracra,* they arose from a common ancestor on Saint Helena. The orthopteran fauna of Saint Helena is so restricted (the two species of Acrididae and three species of Tettigoniidae) that independent colonizations by closely related species must be ruled out as highly improbable. The grasshopper species mentioned above are probably all flightless; the ancestral forms that reached Saint Helena in the past almost certainly had much longer wings.

The beetles and grasshoppers of Saint Helena seem to prove overwhelming evidence for nonallopatric speciation, in the former case on a fairly large scale. But was this speciation sympatric in the strict sense? We do not have sufficiently clear and complete information on the distribution of the species and the range of their diets to be quite certain on this point. It is possible that very minor geographic barriers, on an extremely local scale, may have provided some temporary isolation in some instances. The balance of probability certainly points to a sympatric interpretation, with adaptation to specific host plants playing a role. We have of course no information as to the part played by cytogenetic processes of various kinds, nothing being known about the karyotypes of any of the species. The possibility remains that some stasipatric speciation has taken place on Saint Helena.

Some speciation has undoubtedly taken place on the islands of the Tristan da Cunha group about 2500 kilometers south of Saint Helena (Baird *et al.*, 1965). The weevil genus *Tristanodes* has 11 species, all except one being confined to single islands; the dipteran genus *Scaptomyza* has nine species, of which five are confined to single islands and one is present on all four islands. However, Tristan da Cunha is a small archipelago and it seems likely that speciation in these genera, as well as a number of others, has resulted from interisland migration, i.e., it has been of a strictly allopatric type.

The volcanic island of Rapa, in the south Pacific, is much smaller than Saint Helena (about 8 kilometers long, 6.5 kilometers wide, and a total area of about 25 square kilometers). However, Zimmerman (1938) collected and described 41 endemic species of the weevil genus *Microcryptorhynchus* from this tiny island. It is clear from his taxonomic treatment that he was not a "splitter," i.e., that his species are genuine and not just polymorphs based on differences of color or shape. The remaining species of the genus occur on other islands of the south Pacific, from Australia to the Marquesas. Zimmerman estimated that the 41 described species were only about half the total number present on the island. It is probable that Rapa was colonized by several ancestral species of *Microcryptorhynchus* (probably transported by cyclone activity), but it is clear that a great deal of speciation has taken place on the island. These weevils vary considerably in their host ranges; thus, *M. setifer* was collected from nine different genera of plants but 16 species were recorded from a single host plant each, i.e., they appeared to be monophagous. Looking at the map of Rapa one can imagine that the mountain ridges might have provided some degree of geographic isolation between the organisms in the valleys between them. But, apart from the fact that these valleys are well connected around the coast, it is clear that the ridges were originally forested, and to a considerable extent remain so, in spite of clearing and terracing by the Polynesian population. Although detailed distributions of the species of *Microcryptorhynchus* on Rapa has not been worked out, most species seem to be rather widely distributed, and some specialized and monophagous species were collected at opposite ends of the island. In general, it seems likely that the distribution of each species corresponds to that of the host plants.

It appears quite impossible that these numerous species of flightless weevils could have undergone speciation according to the classical allopatric model. No doubt there were several (conceivably as many as 10) immigrant species from the Austral and Cook Islands, to the north. But a number of species groups are apparently endemic to Rapa ("On Rapa alone are found groups of shiny black species. The squamose group represented by *M. fasciatus* and *M. squamosus* is peculiar to Rapa. . ."). Thus a very strong case seems to exist for sympatric speciation, facilitated by the limited host range of many species. To designate the degree of isolation between the habitats provided by different species of trees, shrubs, and ferns in a dense mixed forest as microallopatry, and to argue from this that there is no real difference between what has happened in this case and the allopatric model, seems quite unwarranted. The essential feature of the sympatric model is that individuals of the diverging populations must have cruising ranges that overlap, and this must surely have been the case in the forests of Rapa. No evidence exists to suggest that the clinal, area-effect, or stasipatric models of speciation would apply to the *Microcryptorhynchus* weevils of Rapa.

We have discussed this case at some length partly because it has been passed over in silence by almost all recent writers on speciation, but also because there seems no reason why this same type of sympatric speciation could not have been important in the case of many other insects or invertebrates of restricted host plant range occurring in the great forested tropical and subtropical regions of the world, whether continental or insular. Archipelagos such as Hawaii and the Galápagos have provided some of the best examples of classical allopatric speciation; but it seems now that isolated oceanic islands such as Saint Helena and Rapa provide equally strong evidence of the sympatric process. Although in both cases the insular situation enables us to infer the events that are occurring more clearly, there seems no reason to suppose that these processes are occurring only on islands. If this argument is sound, sympatric speciation would not simply be a very exceptional mode of little general importance but one of the main processes of organic evolution, at least in those organisms that, like phytophagous insects and many parasitic forms, are specialized in diet and way of life.

It has been repeatedly suggested that two sympatric populations whose mating periods were nonoverlapping, or only slightly overlapping, might evolve into distinct species due to the restriction of gene transfer between them. There are two variants of this model. The first assumes separate spring and summer or fall generations of an animal in each year, the individuals of each spring generation being the offspring of the previous spring population and those of the fall generation the offspring of the previous fall population. An analogy to this in plants would be two populations, one of which blooms in early spring, the other in summer. The second variant assumes an organism whose life cycle extends over several or many years, as in certain cicadas, so that the adults of a particular year are the offspring of those of a year 13 or 17 years before. Numerous examples of closely related species of marine organisms that become sexually mature at different seasons of the year were cited by Lo Bianco (1909). While such cases do not necessarily prove that allochronic speciation has occurred, they certainly suggest that it has.

It will be appreciated that the allochronic model is really only a special case of sympatric speciation, in which the "niches" are temporal instead of spatial. Criticism of the allochronic model has been based on two arguments: (1) If two broods of a species are sympatric but genetically isolated by nonoverlapping mating periods, this condition was preceded by allopatric separation, during which the different life cycles evolved, i.e., that the sympatry of the genetically isolated biotypes is due to later range extensions of one or both forms. (2) The temporal separation of two broods is seldom complete or absolute, and the occurrence of some "stragglers" will lead to gene transfer between them on a scale that prevents speciation.

A simple case of allochronic isolation between two populations, but one that has not been claimed to have led to speciation, occurs in *Nemobius sylvestris* (European forest cricket) (Gabbutt, 1959). In southern England and probably also in Czechoslovakia, both marginal areas for this species, the life cycle lasts two years, the individuals of the even and uneven years forming two isolated populations. This is what Allee and co-workers (1949) referred to as *annual cyclic isolation*. In France and Germany, however, *N. sylvestris* apparently has an annual instead of a biennial cycle. Given sufficient time, and with continued isolation from the continental population, one might expect that *N. sylvestris* would give rise to two allochronic species in southern England. The cyclic isolation may have evolved when a *Nemobius* species with an annual life cycle extended its range northward into a region with a cooler climate and shorter summers; alternatively, it may have originated when the climate of southern England became cooler following the so-called postglacial thermal maximum. Mayr's statement (1963a) that "It would be altogether unlikely for a species to adapt itself simultaneously to two different breeding cycles. . . ." almost certainly does not apply to this case or others like it. Analogous to the case of *N. sylvestris* is that of certain grasshoppers, such as *Pardalophora apiculata* and *Xanthippus latifasciatus,* which have a two-year cycle in Saskatchewan but a one-year cycle further south (Pickford, 1953).

A case of speciation that was originally interpreted as due to allochronic isolation is that of the crickets *Acheta pennsylvanicus* (which overwinters in the egg stage) and *A. veletis* (which overwinters in the nymphal stage) (Alexander and Bigelow, 1960). These two species are sympatric over a vast area of eastern North America, from northern Georgia and Arkansas to southern Canada. However, *Pennsylvanicus* occurs in Nova Scotia, while *veletis* does not. The two species occur together and there is not the slightest evidence of habitat separation. Both species have a single annual generation, but the adults of *veletis* are present in early summer (mid-May to end of July in Quebec, end of April to end of July in North Carolina), while *pennsylvanicus* adults are present in late summer (end of July to mid-October in Quebec, beginning of July to end of October in North Carolina). There is consequently a very short period (less than a week in Quebec, less than a month in North Carolina) in which mating between the two species theoretically *might* occur (but we do not know whether very old individuals of *veletis* are capable of mating with very young ones of *pennsylvanicus*—the mating season of each species may be shorter than the period when adults of the species are present in the populations).

Alexander and Bigelow interpreted this as a case of speciation "without geographic separation through a seasonal isolation of breeding populations."

They emphasize the fact that the calls of the two species are indistinguishable and state that they know of no other example of sympatric species of Orthoptera where this is the case. Three steps in the speciational process are envisaged: (1) repeated elimination during the winters of all stages except late instar nymphs and eggs, leading to reduction or elimination of gene flow between the two populations; (2) development of an egg diapause in one population and a nymphal diapause in the other; and (3) development of reproductive incompatibility. Morphological differences between the two species are minimal—in any one locality the females of *pennsylvanicus* tend to have longer ovipositors. In laboratory crosses copulation and sperm transfer sometimes occurred, but no offspring were obtained.

The interpretation of Alexander and Bigelow, according to which speciation in this case occurred by a sympatric-allochronic mechanism, has been challenged by Mayr (1963a), who states: ". . . it would seem far more reasonable to assume that the range of the ancestral species was fractionated into several geographical isolates (during one of the glaciations?) in one of which fall breeding proved more adaptive, in the other spring breeding. Otherwise the two species remained very much the same, so that they could occupy essentially the same region after they embarked on their post-Pleistocene range expansion." However, he considers that "the difference in breeding cycle of *pennsylvanicus* and *veletis* was not yet complete, when they first met, after emerging from their geographic isolation. If so, competition eliminated any tendencies for spring breeding in *pennsylvanicus* and of fall breeding in *veletis*." Mayr (1963b) has also suggested that the split between the two incipient species was an east–west one (possibly on opposite sides of the Appalachians) and that one population became exclusively nymph-overwintering and the other solely egg-overwintering.

These views have been criticized by Alexander (1963, 1968), whose most pertinent argument seems to be that no related cricket species are known to show the combination of allopatry and allochrony postulated by Mayr, whereas four (*Acheta firmus, A. fultoni* in the eastern United States, *Gryllus campestris* in southern Europe, and *Scapsipedus aspersus* in southern Japan) exhibit sympatric allochronic populations to various degrees.

The *A. pennsylvanicus–A. veletis* case looks very different, however, when one takes into account the cytogenetic data of Lim, Vickery, and Kevan (1973), who have shown that the karyotypes of the two species are very different. Not only is there a difference in chromosome number (2n ♂ = 31 in *veletis,* 29 in *pennsylvanicus*), but the former species has 10 pairs of acrocentric chromosomes while the latter has only three pairs. Unfortunately, Lim and co-workers appear to have compared samples of the two species only from southwest Quebec, and it is consequently unknown whether there is any geographical variation of either karyotype. On the face of it, one chromosomal dissociation (or fusion) and at least seven pericentric inversions or other changes would be required to convert the karyotype of one species into that of the other; it is quite possible the karyotypic reorganization involved more changes than this.

The problem of the evolutionary divergence of these two species is hence a complex cytogenetic one and not merely one based on different life cycles, as imagined by Alexander and Bigelow. We may suspect that the primary factor in this case of speciation was a cytogenetic event (perhaps the fusion or dissociation responsible for the change in chromosome number) and that if the heterozygotes were handicapped (due to irregularities at meiosis) the two classes of homozygotes may then have evolved different seasonal cycles, thereby minimizing the production of heterozygotes. If so, the course of events would have been different from that imagined by Alexander and Bigelow, although allochronic isolation would still have played a role. The very wide geographic overlap between the two species at the present time and the fact that the life cycles are essentially the same from Quebec to North Carolina both seem to argue against Mayr's view that the life cycles of the two forms evolved during a period of geographic separation, as adaptations to different climatic conditions. If we assume a cytogenetic event was the primary initiator of speciation in this case, we might expect different calling songs to have evolved (as has doubtless occurred in many other cases). Instead, an effective isolating mechanism based on allochronic breeding periods was developed. *Veletis* does not occur in Nova Scotia, New Brunswick, or Prince Edward Island (Lim *et al.*, 1973), perhaps because spring comes too late in the area roughly north of 45°. The karyotype of *pennsylvanicus* seems to be more typical for the genus than that of *veletis,* which appears more aberrant; it is thus probable that a metacentric chromosome in the 29-chromosome stock underwent dissociation and gave rise to two acrocentrics in *veletis* (rather than the reverse).

South of the range of *A. pennsylvanicus* and *A. veletis,* in eastern North America, occurs a species, *A. firmus,* which seems to be in an early stage of splitting into two allochronically (Alexander, 1968). In southern Florida this species appears to breed continuously—all stages are present throughout the year. In northern Florida adults are present more or less throughout the year but with peaks in early and late summer. Further north, in coastal North and South Carolina and Georgia, there is a distinct seasonal separation between egg-overwintering fall adults and nymph-overwintering spring adults. Unfortunately, there is no information on the karyotypes of the various populations of *A. firmus*. Another plausible case of allochronic speciation in crickets has been made out for *Gryllus ovisopis* in Florida. This is a "taciturn species" lacking a mating call (Walker, 1974a).

Where a clear-cut difference in breeding season exists between closely related species it is of course always tempting to interpret the speciation process that has occurred as an allochronic one. This may be true in many such cases and to some extent even in all of them. But it is always well to look more deeply into the situation and determine what cytogenetic events have taken place. Speaking of two species of European lycaenid butterflies, Hogben (1940) stated: "Apart from the fact that the single brood of *Lysandra coridon* Poda emerges between the two broods of *Lysandra bellargus* Rott., there is no known obstacle to interbreeding between these two butterflies." He was speaking of the seasonal cycles in Britain. In some areas of southern Europe natural hybridization between these species *does* occur (de Lesse, 1960, 1969). Natural hybrids from the Abruzzi were given the name *L. italoglauca* while those in the Pyrenees were called *L. polonus*. But because of multiple cytogenetic differences between the parent species (*bellargus* has n = 45 while *coridon* has n = 88, except in peninsular Italy, where it has n = 87) the hybrids are probably quite sterile in both sexes. Speciation in this case has consequently involved a long and complex series of cytogenetic changes; it seems probable that the difference in the breeding cycles of the species has been acquired secondarily in northern Europe, being favored by selection as it eliminated the production of sterile hybrids (but even in Britain a few presumed individuals of *polonus* have been recorded). *L. coridon* and *L. bellargus* belong to a species group that also includes *L. syriaca, L. punctifera, L. hispana, L. albicans, L. coelestissima, L. caeruleossmar,* and *L. ossmar*. With the exception of *syriaca* and *punctifera* (with n = 24), which are karyotypically primitive members of the group, the others all have very high chromosome numbers (n = 45–90). The two species *coridon* and *bellargus* are probably not especially closely related; the nearest relative of *coridon* is *hispana,* with which it coexists in southern France. Laboratory hybrids between these two species are quite fertile, but it is uncertain how much hybridization occurs in nature (which in any case is not much, according to de Lesse).

The most complicated case of allochronic isolation that has been studied is that of periodical cicadas of the genus *Magicicada* in the eastern United States (Alexander and Moore, 1962). There are three pairs of sibling species: *septendecula* and *tredecula, cassini* and *tredecassini,* and *septendecim* and *tredecim*. The first-named species of each pair has a 17-year life cycle, the second a 13-year cycle. The siblings of each pair are extremely similar or indistinguishable morphologically and have indistinguishable songs; but both types of difference exist between the three sibling pairs.

In the case of the 17-year species, *septendecula, cassini,* and *septendecim,* the individuals that become adult and form the breeding population in any particular year are the offspring of individuals that were adult and died 17 years previously. Since "stragglers," i.e., individuals that become adult a year early or a year late, are extremely rare and may not occur at all in many instances, the individuals that become adult in a particular year are genetically isolated from individuals of the same species emerging the year before or the year after; thus each "brood" of each species is a genetic isolate. The situation is the same, in principle, for the 13-year species.

In theory we would expect there to be 17 different broods of 17-year cicadas (I to XVII) and 13 broods of 13-year cicadas (XVIII to XXX). However, certain broods seem not to exist. In general, each brood includes all three species, but a few of them lack one species (Table 7-2). The periodicity of the species of *Magicicada* has undoubtedly been established for a very long time. The earliest record is that of the 17-year brood XIV in the year 1634, 20 generations before the most recent brood XIV (1974).

Alexander and Moore postulate that an ancestral species of *Magicicada* underwent classical allopatric speciation to give three species: the *-decula* ancestor, the *-decim* ancestor, and the *-cassini* ancestor (perhaps, southeast, southcentral, and southwest, respectively). It is not clear whether these had a

TABLE 7-2
Latest emergence years for "broods" of the species of the genus Magicicada. *The genus is composed of three pairs of sibling species; in each pair one sibling has a 17-year life cycle, the other a 13-year cycle. In theory there should be 17 different broods of 17-year cicadas (I to XVII) and 13 of 13-year cicadas (XVIII to XXX). However, broods VII, XI, XV, XVI, XX, and XXI are very local and are regarded as doubtful by Alexander and Moore (1962).*

17-year Cycle		13-year Cycle
septendecula	Siblings	*tredecula*
cassini		*tredecassini*
septendecim		*tredecim*

Brood	17-year Cycle	Brood	13-year Cycle
I	1944, 1961, 1978 . . .	XVIII[c]	1958, 1971, 1984 . . .
II	1945, 1962, 1979 . . .	XIX	1959, 1972, 1985 . . .
III	1946, 1963, 1980 . . .	XX	1960, 1973, 1986 . . .
IV	1947, 1964, 1981 . . .	XXI	1961, 1974, 1987 . . .
V	1948, 1965, 1982 . . .	XXII[c]	1962, 1975, 1988 . . .
VI[a]	1949, 1966, 1983 . . .	XXIII	1963, 1976, 1989 . . .
VII	1950, 1967, 1984 . . .	XXIV	1964, 1977, 1990 . . .
VIII[a]	1951, 1968, 1985 . . .	No broods XXV–XXX	
IX	1952, 1969, 1986 . . .		
X	1953, 1970, 1987 . . .		
XI	1954, 1971, 1988 . . .		
XII[a]	1955, 1972, 1989 . . .		
XIII	1956, 1973, 1990 . . .		
XIV	1957, 1974, 1991 . . .		
XV	1958, 1975, 1992 . . .		
XVI	1959, 1976, 1993 . . .		
XVII[b]	1960, 1977, 1994 . . .		

[a]Lacks *septendecula* [b]Lacks *cassini* [c]Lacks *tredecassini*

17-year cycle, a 13-year one, or some other type. By range extension they are believed to have become secondarily sympatric, after which each underwent allochronic speciation, which resulted in the 17-year and 13-year siblings. The broods are assumed to have arisen later. It was postulated that major climatic fluctuations were responsible for brood formation and that there were at least two major periods of brood formation.

The factors responsible for the evolution of the extended life cycle have been discussed by Lloyd and Dybas (1966a). They conclude that this facilitated the buildup of very large populations because the various arthropod predators are age specific, i.e., some prey on the eggs, some on large nymphs, and others on the adults). Since the life cycles of the predators are in all cases shorter than those of the cicadas, the benefit to the predator populations is dissipated during the 16 (or 12) years between successive emergences.

The same authors (Lloyd and Dybas, 1966b) have discussed the evolutionary origin of the 17-year and 13-year life cycles and of the broods. They postulate that there was a long period of evolution of an ever-longer life cycle, under the influence of a hypothetical synchronized parasitoid, until a perfect 13-year cycle was achieved. Some of the populations then evolved a supplementary sixth nymphal instar (four years duration), which increased the life cycle to 17 years, thereby throwing the parasitoid out of synchrony and causing its extinction. Later, in the absence of the parasitoid, some of the southern populations reverted to the 13-year cycle, losing the sixth instar. The three ancestral species are believed to have diverged from one another after periodicity had been evolved, since it is difficult to imagine all the changes involved in the development of periodicity evolving simultaneously in three species. The three species pairs as they exist at present do show some habitat preferences. Most non-periodic cicadas have long life cycles (three to four years, seven years in two oriental species). These species, as far as is known, all have five nymphal instars, as do the 13-year-cycle cicadas. Whether the 17-year forms really have a sixth instar appears probable, but not quite certain.

For the most part, the various broods of periodic cicadas have their own exclusive territories. Broods that are one year out of phase are frequently parapatric, but two never coexist in zones of contact in the same woods. Periodic cicadas emerge in enormous numbers and are extremely vulnerable to predation by birds. In fact, the broods probably only survive by being in excess of the numbers necessary to satiate the bird populations. There is thus very strong selection against "stragglers," which emerge in small numbers one or two years early or late (a factor that makes it very difficult to understand how 17-year forms could have evolved from 13-year ones, or vice-versa).

No cytogenetic work has been carried out on the genus *Magicicada,* so it is quite uncertain whether any cytogenetic differences exist between the various species or (as is not impossible) between the different broods. Some allozyme studies on two of the 13-year species have been carried out by Krepp and Smith (1974). They showed a high degree of genic similarity. Seven out of 15 gene loci studied (47 percent) were polymorphic and the average genic heterozygosity was 0.15 to 0.175, which is about average for invertebrates with small body size. It would seem highly desirable to extend these investigations to the 17-year species and to different broods of both 13-year and 17-year cicadas.

We have described the situation in the periodic cicadas at some length because it is certainly the most remarkable example of biological differentiation on an allochronic basis that has been recorded. It is still far from clear, however, exactly how it should be interpreted from an evolutionary standpoint.

Brooks (1950), commenting on the presence of two endemic species of pelagic fishes belonging to the genus *Comephorus* in Lake Baikal, states that "these two similar species maintain their identity because of a difference in the time of breeding. *C. dybowski* reproduces in February–March, whereas *C. baicalensis* breeds during July–August. The only way in which these two species could have arisen is through separation of the lake into two basins, thus isolating two populations." Clearly, for Brooks the possibility of allochronic speciation was not even worth considering. Kozhov (1963) also notes that the mating seasons of the two species do not overlap. *Comephorus* species do not spawn but produce living young, so there is a considerable time interval between mating and the birth of the young (2,000 to 3,000 per female). *C. dybowski* mates in September–November and gives birth in February–April; *C. baicalensis* mates in April–June and gives birth in September–October (Koryakov, 1955, 1956).

It is possible that about the beginning of the Quaternary period Baikal consisted of two or even three lakes that subsequently merged. However, there seems to be absolutely no reason why *Comephorus* populations in separate lakes, which clearly must have originated from a single population, would have evolved different seasonal cycles when geographically isolated, since the climatology of the two lakes would necessarily have been virtually identical. On the other hand, a genetic polymorphism for time of breeding in a single population makes sense, since it enables the population to exploit additional sources of food for the young. Once such a polymorphism became established, the basis for allochronic speciation would have been laid. In fact, the evidence for allochronic speciation in this case seems considerably stronger than in the various insect cases considered earlier.

Numerous cases of seasonal isolation in plants are known, i.e., related species having different blooming times so that hybridization is avoided. However, Stebbins (1950) implies that such isolation arose after speciation had occurred and that it did not play a primary role in initiating divergence. In the absence of critical evidence no firm conclusions seem possible.

A number of geneticists have carried out laboratory experiments on so-called "disruptive selection" that seem to be very relevant to the problem of sympatric speciation in nature, especially if the latter can be the result of a relatively small number of "key" genetic changes (see p. 41). Most of these experiments have been performed with *Drosophila*.

Several experiments demonstrate that genetic diversity is increased by disruptive selection. Powell (1971) set up 13 population cages each with 500 flies of *Drosophila willistoni* from the same stock. Some cages were supplied with only one kind of food medium, one kind of yeast, and kept at a constant temperature. Other cages provided a choice of either different media or different yeasts, while still others had variable temperatures (alternate weeks at 19° and 25° C) in addition to these choices. After approximately one year the amount of heterozygosity at 22 different enzyme loci was determined. In the cages with constant environment heterozygosity averaged 7.8 percent, in cages with one variable (medium yeast or temperature) it was 9.6 percent, while in the ones with all three variables it was 13.4 percent. The result was presumably due to alleles that are rare in the constant environment cages, but which became more common in the variable environments.

The experiments that seem most relevant to the problem of sympatric speciation are those of Thoday and Boam (1959; and see Thoday and Gibson, 1962, 1970, 1971; Thoday, 1965, 1972) in which divergence for a polygenic character and ethological isolation were produced simultaneously by disruptive selection for a morphological character (sternopleural chaeta number). Two replicate experiments were carried out with a stock ("Southacre") of *Drosophila melanogaster* derived from four fertile wild females captured in the same garbage can. Four bottle cultures were set up, each containing four pairs of flies. Twenty males and females from the next generation in each bottle were scored for chaeta number while still virgins. The eight males and eight females with the highest chaeta number together with the eight males and eight females with the lowest number were selected to be the parents of the next generation. These 32 flies were placed in the same vial for 24 hours in the dark, i.e., they were given a free mating choice. The males were then discarded, the females were separated again into groups with high and low chaeta number; the "high" females (those with the high number) were put into two bottles to oviposit (four flies per bottle), the "low" females (those with the low number) being treated likewise. The process was repeated in each generation. One replicate experiment was maintained for 21 generations; the other failed at the 16th generation but was restarted, using residual flies from the previous generation, and then maintained until the 28th generation.

In both experiments there was rapid divergence for chaeta number. In one experiment the progeny—those from high and low females—formed two nonoverlapping groups with respect to chaeta number by the twelfth generation, while in the other experiment this result occurred as early as the seventh generation. In each generation tests were carried out to determine the types of progenies produced by forced high–high, high–low, and low–low matings. These tests confirmed that the absence of flies with intermediate chaeta numbers after the seventh or twelfth generations was due to reproductive isolation. Another type of test required rearing, scoring, selecting, and mating additional flies in each generation; after mating each female was put in a separate bottle. The numbers of chaetae in progenies from high–high, high–low, and low–low matings were sufficiently different to be able to determine whether a female had mated with a "high" or "low" male. The results were as follows:

	H × H	H♀ × L♂	L♀ × H♂	L × L	Failures
Experiment 1	42	5	6	36	39
Experiment 2	71	15	27	62	17

Since there were relatively few sterile cultures in these experiments, it was concluded that the disruptive selection for chaeta number had produced some type of ethological isolation.

Unfortunately, attempts to repeat these experiments in a number of different laboratories have led to substantially negative results. Using the same methods as Thoday and Gibson, Chabora (1968) failed to obtain reproductive isolation in a number of strains of *Drosophila melanogaster,* some of which had been recently derived from wild populations. She concludes that "disruptive selection for sternopleural chaeta number leads to reproductive isolation in the laboratory very rarely." Robertson (1970) worked with a laboratory strain; he failed to obtain an increasing difference in chaeta number between the progeny of high and low females after the fifth generation, and so the point at which reproductive isolation might have been expected to occur, on the basis of Thoday and Gibson's results, was never reached. His explanation is based on the mating behavior of the females, and he concludes that in this organism selection programs with populations of the size used are inherently unstable and will not lead to stable divergence in phenotype between the two progeny groups. Barker and Cummins (1969) obtained results essentially like those of Robertson and Sharloo, who in a series of experiments (Sharloo, 1964, 1971; Sharloo, den Boer, and Hoogmoed, 1967; Sharloo, Hoogmoed, and ter Kuile, 1967) had also failed to obtain ethological isolation. Clearly, the flies used to start Thoday and Gibson's experiments must have been somewhat unusual in some respect. Barker and Cummins suggest that, for some reason, there may have been a genetic correlation between bristle number and isolation tendency. Sharloo, den Boer, and Hoogmoed (1967) suggested that the four original wild females collected by Thoday and Gibson may have come from more than one geographic source (possibly imported with fruit from different regions of origin). One might suggest, very tentatively, that one of the naturally occurring chromosomal inversions of *Drosophila melanogaster* was present in Thoday and Gibson's experiments, and that this might account for the kind of correlation postulated by Barker and Cummins.

From the standpoint of speciation theory, however, the sort of result obtained by Thoday and Gibson, even if it occurs only once in ten or once in a hundred experiments, is highly significant. Speciation is a rare event in evolution. Disruptive selection, because of the multiplicity of ecological niches, is constantly occurring in most species of organisms. Even if it only occasionally leads to the development of reproductive isolation this could, on the evolutionary time scale, provide a basis for sympatric speciation as a process of major importance.

A number of disruptive selection experiments on organisms other than *Drosophila* have been carried out. Pimentel, Smith, and Soans (1967) carried out an experiment with houseflies *(Musca domestica)* in which they produced two subpopulations that had preferences as high as 90 percent for ovipositing on food media based on banana and fish meal, respectively. In these experiments there was 5–30 percent migration between the subpopulations; hybrids between the two strains showed no particular preference for one medium or the other. Unfortunately, no tests were carried out to determine if there was any reproductive isolation between the two subpopulations.

The relevance of the experiments of Thoday and Gibson to speciation has been contested by Mayr (1963a), who states that "the severity and absoluteness of the selection, the complete prevention of gene flow, and the elimination of mutual competition are, of course, a set of artificial conditions that could scarcely be expected ever to occur in nature." This criticism seems to have been effectively answered by Thoday (1972), who points out that the selection intensity used in his experiments was less than in the case of some natural situations, especially in plants (Barber and Jackson, 1957, for *Eucalyptus;* Antonovics and Bradshaw, 1970, for grasses). The "complete prevention of gene flow" was definitely not a condition of Thoday and Gibson's experiments; if it occurred, it was a consequence of the disruptive selection for chaeta number.

On the other hand, when Thoday (1972) states that "every hybrid zone which is not expanding" (and in which heterozygotes are at a disadvantage) is evidence for the role of disruptive selection in speciation, there seems to be a distinct flavor of the *post hoc ergo propter hoc* type of fallacy in the argument. The fact that disruptive selection can be observed in nature maintaining species differences in zones of overlap is hardly evidence that it led to speciation in the first place, or would be capable of doing so in the particular group of organisms under consideration.

Very striking data on disruptive natural selection have been obtained from plants growing on soils contaminated by salts of heavy metals, such as lead and zinc derived from mining operations (Bradshaw, 1952; Antonovics, Bradshaw, and Turner, 1971; Antonovics, 1971; Snaydon, in Bradshaw *et al.*, 1965). The boundary between contaminated and uncontaminated soil is often extremely sharp, and populations of pasture grasses, such as *Anthoxanthum odoratum*, on either side of it differ not only in tolerance to the toxic metals but in numerous other genetic characters. These differences occur over distances of 1–2 meters. Snaydon showed that plants growing immediately below a galvanized wire fence that was only 30 years old were zinc-tolerant as a result of contamination of the soil, whereas plants growing a few centimeters on either side of the fence were not. Snaydon (1970) also demonstrated genetic differences between *Anthoxanthum* populations growing on adjacent experimental plots with calcareous and acid soils.

This work clearly demonstrates that disruptive selection can occur in plant populations growing on adjacent soils of contrasting type, and that gene flow is insufficient to prevent the development of very marked differences in less than 50 generations. There is so far no evidence of the development of any genetic isolating mechanisms developing in such situations. Clearly, however, gene flow across the boundaries is disadvantageous, since it retards the development of well-adapted populations, so that the situation would be favorable for the development of such isolating mechanisms, i.e., for incipient speciation. The studies cited above have all been concerned with the effects of human interference with the environment, but there must be many entirely natural situations in which there are similar sharp transitions in soil type. The extent to which such heterogeneities of the environment really lead to full speciation in plants is, however, still a fairly open question.

A zoological equivalent of the work of Bradshaw and co-workers is the work of Blair (1947), who demonstrated sharp geographic discontinuities in coat color in the case of mice *(Peromyscus eremicus)* inhabiting light- and dark-colored soils (the latter due to lava flows across the desert sands in New Mexico).

It is extremely unfortunate that there have been no detailed cytogenetic studies on any of the organisms for which evidence for sympatric speciation is really strong. The species of *Rhagoletis,* which until now have provided the best evidence for this type of speciation, have polytene chromosomes that are unsuited for analysis, and the somatic karyotypes published by Bush (1966) have only established that the *pomonella* group of species differ from the rest of the genus with respect to two chromosome pairs (they are acrocentric rather than metacentric).

However, in spite of a rather woeful lack of evidence as to the exact nature of the genetic processes involved, it seems impossible today to deny the reality of sympatric speciation, at least in many groups of insects.

8

Speciation by Polyploidy

A special type of chromosomal speciation, one that has been recognized for the past 50 years as a distinct mode, is the origin of polyploid biotypes, which are necessarily isolated genetically from their diploid progenitors. Polyploidy arises when the original (usually diploid) chromosome set is doubled (giving rise to tetraploids, with four sets; hexaploids, with six; octoploids, with eight; etc.) or when a gamete (usually haploid) fuses with one of the opposite sex that is itself polyploid (giving rise to triploids, with three sets; pentaploids, with five; etc.). In sexually reproducing species only the even-numbered polyploids (tetraploids, hexaploids, etc.) can succeed, the odd-numbered types (triploids, pentaploids, etc.) being necessarily sterile because of meiotic irregularities.

Evolutionary polyploidy is much more widespread in the plant kingdom than among animals, so that most of the examples in the present chapter will be higher plants. Evolutionary polyploidy in animals is largely confined to hermaphroditic and parthenogenetic forms, although a few genuine cases are now known in some bisexual groups, particularly the anuran amphibians.

Two extreme types of polyploidy can be recognized: *autopolyploidy* (or autoploidy), in which all the chromosome sets have been derived from a single ancestral species, and *allopolyploidy* (or alloploidy), in which the sexual chromosome sets have originated from two or more ancestral species. In the latter case we have a combination of polyploidy and hybridity.

The distinction is clear-cut in theory, but in practice many intermediate types of polyploidy exist. Thus if we call the haploid chromosome sets of three species A, B, and C, we may have autotetraploids AAAA, BBBB, and CCCC, and allotetraploids AABB, AACC, etc. At the hexaploid level there would be hexaploids such as AAAAA and allohexaploids such as AABBCC; but we can also have autoallohexaploids of the type AAAABB. Also, many species of plants have considerable racial variation, so that a diploid AA may appropriately be designated A′A′ in another area, where A′ represents a set of chromosomes that is genetically fairly different from A. In that case we can have a tetraploid AAA′A′, which is in most respects intermediate between an autopolyploid and an allopolyploid.

The stability of polyploid biotypes depends on meiotic regularity. In general, this means that they must form bivalents rather than multivalents at meiosis. In theory, a tetraploid that always formed quadrivalents would show regular inheritance and hence genetic stability; in practice, however, a genotype that forms quadrivalents will not do so with complete regularity, but will sometimes form trivalents and univalents and will consequently show reduced fertility.

It has been universally believed by plant cytogeneticists that a high frequency of multivalent formation at meiosis (i.e., trivalents, quadrivalents, etc. at first metaphase) indicates autopolyploidy, while a low frequency points to allopolyploidy. In a general sense this is undoubtedly true. But other factors affect the situation. Many short chromosomes seldom or never have more than one chiasma and they are simply incapable of forming multivalents (if they do so at the zygotene stage of meiosis these will fall apart long before first metaphase). Thus the difference between a plant which forms 0–3 quadrivalents at meiosis and one that forms 10–12 may not be a difference between allopolyploidy and autopolyploidy; it may simply reflect a difference in the chiasma frequency of the diploid ancestors.

The other assumption that is generally made by plant cytogeneticists is that the failure of an allopolyploid to form multivalents is due to structural differences between the karyotypes of the ancestral species (due to translocations, inversions, etc.). In some instances this is probably the correct explanation, particularly if we stress the "minute" rearrangements (types 2 and 3, on p. 8) rather than the major ones (type 4). But in the hexaploid wheats, where meiotic synapsis is restricted to homologous chromosomes and prevented between homoeologous ones (i.e., those derived from different ancestral species), this is due to the action of a specific locus *(Ph)* in the 5B chromosome (Sears and Okamoto, 1958; Riley and Chapman, 1958; Riley, 1966, 1974). Waines (1976) has given reasons for believing that loci analogous to *Ph* occur in diploid species of wheats and related grass genera and do not have to arise by mutation after polyploid formation. To what extent similar factors govern the meiotic behavior of natural allopolyploids is not known, but the fact that another locus in the 3D chromosome of wheat has a similar but weaker effect (Mello-Sampayo, 1971) suggests that they may be of widespread occurrence and hence may have facilitated the adaptive success of many allopolyploids that would otherwise have shown severely reduced fertility as a result of irregular multivalent formation. If so, the conclusion of the earlier plant cytogeneticists, that suppression of multivalent-formation in polyploids is mainly due to structural differences (and hence phyletic divergence) between the ancestral species, may be seriously in error.

The morphological and physiological effects of polyploidy are complex. Some of them are due to the polyploid condition as such and are hence shown by autopolyploids as well as allopolyploids. Others are essentially caused by hybridity and are therefore only seen in allopolyploids; they depend on the precise genetic constitution of the parental species and are hence specific to each individual case.

Among the general effects of polyploidy as such is an increase in all the dimensions of the cell. Usually, this is not reflected in a proportionate increase in the size of the plant, since there is a reduction in the number of cell divisions in the course of development, so that the mature plant consists of fewer but larger cells (Stebbins, 1950, 1971). On the other hand, increases in size (so-called *gigas* effects) may be produced in such organs as sepals, petals, anthers, and seeds), which have a more or less strictly determinate pattern of growth. Thus, it is often possible to identify polyploid genotypes by comparing the sizes of their stomata or pollen grains with those of their diploid relatives, where a comparison of leaf size or overall height or biomass of the plants would be valueless. Among the various, often obscure, morphogenetic effects of polyploidy *per se* are a thickening of the leaves and petals and a general reduction in the amount of branching.

It is unlikely that any of these effects would be adaptive (if they were, it is likely that they would have already been evolved in the diploid ancestors), and many of them are likely to be actually deleterious. The rarity of autopolyploid species of plants is thus more likely to be due to their being poorly adapted whenever polyploid biotypes arose than to a reduction of fertility due to multivalent formation, as has been widely believed.

The positive, adaptive, effects of allopolyploidy may be ascribed essentially to their hybrid condition, i.e., to heterosis in the most general sense. In order for this to be effective it must be more powerful than the negative effects (referred to above) of polyploidy as such. In some cases allopolyploids may be adapted to a wider range of environments than their diploid progenitors, presumably because they carry the genes of both.

The significance of polyploidy as a mechanism of speciation is that even-numbered polyploids are necessarily isolated from their ancestors because any hybrids they produce will necessarily be odd-numbered polyploids, which are largely or entirely sterile as a result of irregularities of meiosis. Newly arisen even-numbered polyploids thus constitute distinct biological species from the moment of their origin. Some cases, however, may represent hybridization between diploids and tetraploids, giving triploids, and a further doubling of the chromosome number to produce hexaploids. This is the situation in the Australian grass *Themeda australis* (Hayman, 1960). Broadly speaking, this consists of a diploid biotype with 2n = 20 (perhaps the term "species" would be more strictly correct), which grows throughout the mesic eastern coastal and mountain region and in Tasmania, and a tetraploid one (2n = 40), which occurs throughout the arid central and western part of the continent and in the mesic southwestern corner. But at a number of localities triploid, hexaploid, and (in a few cases) pentaploid plants have been found. The triploids and pentaploids must be largely sterile, but at one locality Hayman found a population of plants with chromosome numbers such as 2n = 21, 22, 23, and 25, which may have arisen by hybridization between diploids and triploids or by occasional crossing of two triploids. The difference between the adaptive range and hence the geographic distribution of the diploid and tetraploid forms of *T. australis* suggests that the latter arose from a cross between two different ecotypes, which Hayman suggested may have originated more than once. In the absence of competition from the diploid form it has been able to invade quite mesic environments in southwestern Australia. The African and Asian *Themeda triandra* shows an even greater range of chromosome numbers from diploid to octoploid, but it reproduces at least in part by aposporous apomixis (Brown and Emery, 1957), which probably does not occur in *T. australis*.

In addition to the distinction between autopolyploidy and allopolyploidy, Stebbins (1950, 1970, 1971) has drawn one between inter-ecotypic (interracial) allopolyploids (of which the tetraploid form of *Themeda australis* would probably be an example), genomic allopolyploids, and segmental allopolyploids. These categories seem useful, provided one realizes that they are not hard and fast ones and that intermediates exist.

There are very few certain cases of strictly autopolyploid taxa known to exist in the wild; the evolutionary significance of autopolyploidy is hence rather minimal. However, the existence of a few species with diploid and polyploid forms in the absence of any close relatives proves that evolutionary autopolyploidy can occasionally be successful. Stebbins (1950) cites the case of the North American *Galax aphylla* (Diapensiaceae), a monotypic genus that includes diploid and tetraploid forms (Baldwin, 1941); the tetraploid, which has a slightly wider range, is a more stocky plant with thicker leaves. Another example is *Achlys triphylla* (Berberidaceae), which includes diploid and tetraploid forms on the Pacific coast of North America (Fukuda, 1967); the only other species of the genus, a diploid, occurs in Japan. The genera *Galax* and *Achlys* are undoubtedly ancient members of the Arcto-Tertiary flora (Stebbins, 1971) and the only reason for doubting that the tetraploids are strictly autopolyploids is that we do not know what species or races may have become extinct throughout the Tertiary period. The well-known *Larrea tridentata* (creosote bush) of the

North American deserts includes diploid, tetraploid, and hexaploid forms, the latter two biotypes being presumably autopolyploids, since the only other species of the genus occur in South America (Yang, 1967, 1968). Many earlier examples of supposed autopolyploids were shown to be actually allopolyploids in a critical review by Clausen, Keck, and Hiesey (1945). In more recent years a number of claims to have demonstrated the existence of strictly autopolyploid taxa have been published, but in no case is the evidence absolutely convincing. Thus, Mosquin (1967) argued that the 4n and 6n races of *Epilobium angustifolium* (Onagraceae) must be autotetraploid and autohexaploid, respectively; they show a high frequency of multivalents at meiosis. The species is very distinct from all others in the genus and is unlikely to have hybridized with them; but the diploid has a vast range throughout the Holarctic and the polyploids, which have a more southern distribution, may well be examples of interracial allopolyploids. Another tetraploid that has been claimed to be an autopolyploid is *Vaccinium uliginosum* (Ericaceae), investigated by Rousi (1967); but it, too, may be an instance of an interracial allopolyploid.

A more certain example of interracial allopolyploidy, since both ancestral diploids can be identified, is *Dactylis glomerata* (cocksfoot grass), a tetraploid that is widespread in the Palaearctic region (Stebbins, 1971). It apparently arose by doubling the chromosome number in hybrids between two diploid forms that were formerly given specific rank as *D. aschersoniana* (occupying the forests of central and northern Europe, and the mountains of the Balkans) and *D. woronowi* (from the steppes of southwestern Asia). The artificial diploid hybrid between these is quite fertile in the F_1 and F_2 generations, so that they should almost certainly be regarded as ecotypes of the same species, i.e., races adapted to different habitats. By doubling the karyotype of such hybrids one obtains plants that closely resemble typical *D. glomerata* and, like it, show many multivalents at meiosis (Stebbins and Zohary, 1959).

In contrast to these interracial allopolyploids, the genomic or typical allopolyploids of Stebbins's classification result from spontaneous doubling of the karyotype of a hybrid between two different species whose chromosomes are sufficiently similar that there is reasonably complete pairing in the meiosis of the diploid hybrid (they are hence said to have the same genome). The third category, segmental allopolyploids, have arisen by doubling the chromosome number of a hybrid between two species with different genomes, i.e., a hybrid in which pairing at meiosis is extremely incomplete or irregular. The term "segmental allopolyploidy" seems rather undesirable (although the category undoubtedly exists), since it is by no means clear that the failure of synapsis in species hybrids is always, or even usually, due to structural ("segmental") differences between their chromosomes. It may, in fact, be frequently due to genic factors similar to those that Sears and Okamoto (1958) and Riley have shown to be responsible for failure of pairing between homoeologous chromosomes in the allopolyploid wheats. Perhaps the term "uberrestitutive allopolyploidy" (Latin: *uber,* fertile + *restituere,* to restore) may be found preferable, since it merely indicates what actually happens without implying cause.

The prevalence of polyploidy has been studied in different systematic groups of plants, in plants of different growth forms (ephemeral herbs, perennials, shrubs and trees), and in the floras of different latitudes and different life zones (rainforests, deserts, grasslands, etc.). Significant regularities have emerged from these considerations.

Polyploidy is rare or absent in fungi, but has been recorded in some groups of algae and is well known in bryophytes (Vaarama, 1950, 1953, 1956). It is extremely prevalent in ferns and very high levels of polyploidy are known in some genera (Manton, 1950, 1951, 1953; Manton and Sledge, 1954; Ninan, 1958; Mehra, 1961; Löve and Kapoor, 1966, 1967). Nonsexual reproduction (apogamy) is common in certain groups of ferns (Manton, 1950; Walker, 1962) and has permitted the establishment of many triploid and hybrid biotypes whose karyotypes could not undergo normal meiosis.

High degrees of polyploidy are also found in three other classes of spore-bearing plants, which are clearly survivals from Palaeozoic times: Psilotales, Equisetales, and Lycopodiales. In Psilotales, *Tmesipteris tannensis* has $2n > 400$, and two species of *Psilotum* have $2n \approx 52$, although races with twice this number occur in one of them. In Equisetales, a class consisting of only a single living genus, *Equisetum,* all the species seem to have $2n \approx 216$ (Manton, 1950). They might perhaps be ancient 24-ploids with a basic number of $n = 9$. In Lycopodiales, high levels of polyploidy occur in *Lycopodium, Phylloglossum,* and *Isoetes,* but species of *Selaginella* are diploid with $n = 9$ (some species of *Isoetes* are also diploid). *P. drummondii,* the only species of *Phylloglossum,* has $2n \approx 502$ (Blackwood 1953); it shows some univalents at meiosis, suggesting an ancient hybrid origin or an aneuploid constitution.

The highest recorded level of polyploidy in ferns (and indeed in both the plant and animal kingdoms) is reached in *Ophioglossum reticulatum,* which shows approximately 630 bivalents at meiosis i.e., $2n \approx 1{,}260$ (Figure 27). Other species of the genus have $2n = 240$, 480, 720, and 960; *O. reticulatum* is perhaps 84-ploid (assuming the basic haploid number is $n = 15$).

There have been no biochemical studies that would reveal the architecture of these fantastic genetic systems. Ferns are a very ancient group and many existing genera undoubtedly go back to Palaeozoic times. It is possible that the living species of *Ophioglossum* have gone through three to six rounds of allopolyploidy, at intervals of 50 to 100 million years since the Palaeozoic, undergoing a gradual "diploidization" (i.e., genetic divergence, by mutation between originally largely homologous chromosomes) in the intervals from one

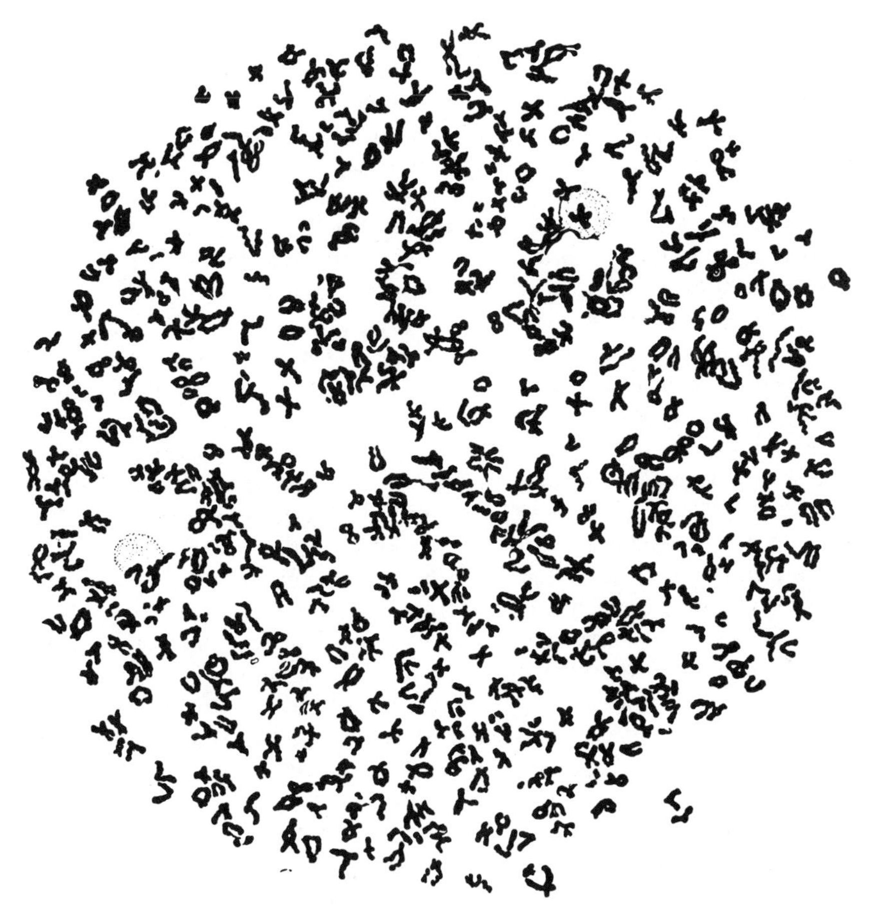

FIGURE 27

Meiotic prophase (diakinesis stage) in a sporophyte of the fern *Ophioglossum reticulatum,* showing approximately 630 bivalents. [After Ninan, 1958]

round to the next. That allopolyploidy rather than autopolyploidy is the basis for these systems is fairly clear from the existence of levels of polyploidy that are not powers of two, e.g., 48-ploidy and 84-ploidy. Unlike the situation in the high polyploids of the *Claytonia virginica* complex (see p. 274) aneuploidy is either absent or has played only a minor role in the evolution of *Ophioglossum.*

In contrast to the lower vascular plants, polyploidy in gymnosperms is rare, and high levels of polyploidy are never found. The cycads and the single species of *Gingko* are all diploids (Khoshoo, 1969) and the only known polyploids in the conifers are the hexaploid *Sequoia sempervirens,* two species of *Juniperus,* and the tetraploid *Fitzroya cupressoides.* The only gymnosperm genus in which polyploidy seems to have played a major evolutionary role is the peculiar desert *Ephedra,* in which eight out of 18 species that have been studied cytogenetically are tetraploids (probably allotetraploids). The reasons for the rarity of evolutionary polyploidy in this ancient group of plants are not clear, but have been discussed by Khoshoo (1959); their woody habit is undoubtedly a contributing cause (see p. 270). A large number of genera of conifers in the southern hemisphere, many of which contain only a single species, have not been studied cytogenetically (Li, 1953; Khoshoo, 1960). The remarkable *Welwitschia mirabilis,* with the high chromosome number $2n = 42$ (Khoshoo and Ahuja, 1963), might be an ancient hexaploid, but it has no close relatives for comparison.

The prevalence of polyploidy in flowering plants (angiosperms) varies greatly from family to family and even from genus to genus within the same family. Stebbins (1971) states that 30–35 percent of all species of flowering plants have gametic chromosome numbers that are multiples of the "basic" number of the genus (or of one of the basic numbers if there are several) and are hence "straight polyploids." Other species with high chromosome numbers that are not multiples may be secondary derivatives of polyploids. And there is a tendency among plant cytogeneticists to believe that a number of plant families (Magnoliaceae, Salicaceae, Hippocastanaceae, Ulmaceae, Juglandaceae, Winteraceae, and Bombacaceae) in which low chromosome numbers (below $n = 16$ or, according to some authorities, below $n = 11$) do not occur are of ancient polyploid origin, i.e., the living species are descended from ones that became polyploid a long while ago, perhaps in early Tertiary times. The idea that all truly basic, i.e. genuinely haploid, numbers in higher plants are lower than about 12, and that apparently basic numbers between 12 and 30 must be of ancient polyploid origin, is rather firmly established in plant cytogenetics. In many cases it may be true, but firm evidence is very difficult to obtain. Certainly, many groups of animals in which polyploidy has definitely not occurred show genuine haploid numbers much higher than $n = 12$, and there seems no *a priori* reason to suppose that the same might not be true for some genera and families of plants. Thus for the present one can only conclude that Stebbins's estimate of 30–35 percent polyploidy in flowering plants is a minimum, without attempting to arrive at a maximum figure. Grant (1971), working on the assumption that species with numbers higher than $n = 13$ are polyploids, calculates that 47 percent of all angiosperm species are polyploid (dicotyledons 43 percent, monocotyledons 58 percent). This compares with estimates (on the same basis) of 1.5 percent for conifers and 95 percent for ferns and their allies. Among angiosperm families, high frequencies of polyploidy occur in

Ranuculaceae, Polygonaceae, Crassulaceae, Rosaceae, Malvaceae, Araliaceae, Gramineae, and Iridaceae. Contrary to the statement of Stebbins (1950), polyploidy is not unknown or even particularly rare in Berberidaceae or Cucurbitaceae.

In the predominantly Australian and South African family Proteaceae, which shows absolutely no intrageneric polyploidy, Smith-White (1959) has postulated that a doubling of the chromosome number occurred independently in four lineages, leading from n = 7, assumed to be the primitive condition, to n = 14. This seems a doubtful assumption, and is based merely on a belief that haploid numbers above 11 or 12 cannot be primitive and must have been derived by polyploidy. In point of fact, n = 7 is only known in the genera *Persoonia* and *Placospermum,* where the relatively large size of the chromosomes suggests that fusions have occurred. In the lineages supposed by Smith-White to have undergone polyploidy, 17 genera show n = 14, three show n = 10, 11 show n = 11, five n = 12, six n = 13, and one n = 15 (Smith-White, 1959; Johnson and Briggs, 1963). It would seem more in accordance with the evidence, and the principle of parsimony, to suggest that polyploidy has not occurred at all in this undoubtedly ancient southern hemisphere family.

Evidence that the percentage of polyploid species in the total angiosperm flora increases with latitude has been repeatedly presented. Thus, in Eurasia and the Arctic the frequency of polyploids goes up from 37 percent in Sicily (latitude 36–38° N) to 51 percent in central Europe, 57 percent in Sweden and Finland, 71 percent in Iceland, and 86 percent in Peary Land (latitude 82–84° N) according to the compilations of Löve and Löve (1957) and Hanelt (1966). The reasons for this correlation have been repeatedly discussed and are summarized by Grant (1971), who concludes that several causes act together to produce this result. One may be the different contribution that certain families, such as the grasses (which show a particularly high percentage of polyploidy), make to the floras of cold, temperate, and arctic zones. The Löves have suggested that polyploids are better adapted to cold and other extreme conditions. But Stebbins (1950, 1970) has produced evidence that polyploids are better adapted to colonize newly exposed habitats, such as those made available by the retreat of the ice sheets at the end of the Pleistocene—a somewhat different concept. Favarger (1957) has shown that the percentage of polyploids in the flora at high elevations in the Alps is no higher than in that of the lowlands of central Europe. Stebbins (1970) points out that there is 53 percent polyploidy in the flora of the Queen Charlotte Islands (53° N, extensively glaciated in the Pleistocene, present climate mild) and only 56–59 percent in northwestern Alaska (68° N, largely unglaciated, present climate very severe). He concludes that the disappearance of the Pleistocene glaciation has had more to do with the increase in polyploidy (from an estimated 36 percent in the Coast Ranges of Northern California) than the severity of the climate at the present time.

An association between polyploidy and life form is present in many angiosperm families; perennial herbs have the highest percentage of polyploids, while annuals and woody plants show significantly lower percentages (Müntzing, 1936; Stebbins, 1938). A number of species-rich genera of trees, such as the oaks (*Quercus*, 2n = 24) figs (*Ficus*, 2n = 26), and *Eucalyptus* (2n = 22), consist entirely of diploids. Some other large genera of trees, such as *Populus*, *Salix*, *Acer*, and *Carya*, show a few polyploid species. On the other hand, the alders (*Alnus*), birches (*Betula*), and elms (*Ulmus*), in which the lowest chromosome number is 2n = 28, probably contain only tetraploids, together with a few hexaploids, octoploids, and (in *Betula*) higher polyploids.

Allopolyploidy (sometimes termed "amphiploidy") is based on two genetic events: hybridization and doubling of the chromosome number. The hybridization may be between two diploids, two polyploids, or between a diploid and a polyploid. It is generally assumed that the hybridization occurred first and the chromosome doubling later, and this is probably the usual sequence. But Clausen, Keck, and Hiesey have suggested that in some instances a cross between two autopolyploids may have produced an allopolyploid.

A number of artificially produced allopolyploids, such as the famous cases of *Primula kewensis*, *Galeopsis tetrahit*, and *Raphanobrassica*, have been studied experimentally and provide models for the kinds of processes that must have occurred on innumerable occasions in the evolution of the higher plants. The tetraploid *Primula kewensis* arose by somatic doubling of the chromosome number, leading to tetraploid branches on an originally diploid hybrid individual (Newton and Pellew, 1929; Upcott, 1939). Unlike *P. kewensis*, which does not exist in nature, the tetraploid *Galeopsis tetrahit* (2n = 32) is a wild species growing in the Palaearctic region. Müntzing (1932, 1938) was able to "synthesize" it by crossing the diploid species *G. pubescens* and *G. speciosa* (both with 2n = 16). The diploid hybrid had irregular meiosis and was fairly sterile. Nevertheless, some reproduction did occur and an F_2 was obtained that included one triploid plant that must have arisen from fertilization of a diploid (unreduced) gamete by a haploid one. This triploid was backcrossed to the *G. pubescens* diploid and yielded a seed that gave rise to a tetraploid plant. The artificial *G. tetrahit* thus arose as a result of hybridization followed by two successive meiotic accidents, one in the diploid hybrid and one in the triploid. Whereas the mechanism that gave rise to the tetraploid *P. kewensis* ensures that precisely two complete sets of chromosomes derived from each of the parent species are present, the processes involved in the origin of the artificial *Galeopsis tetrahit* would permit the tetraploid to have more *pubescens* than *speciosa* chromosomes (or the reverse, if the backcross had been to *speciosa*).

The tetraploid *Raphanobrassica* seems to have arisen in yet another manner, by union of two unreduced (diploid) gametes derived from a largely sterile diploid hybrid between raddish, *Raphanus sativus,* and cabbage, *Brassica oleracea* (Karpechenko, 1927).

It is simply not possible to state the extent to which these various kinds of processes (and others that are conceivable) have played a role in the origin of natural allopolyploids. The essential point is, however, that all of them are capable of immediately giving rise to new allopolyploid species, although it is undoubtedly true that such "instant species" will have to undergo a number of generations of natural selection before they are sufficiently adapted to successfully spread in nature.

The conditions that favor successful speciation by allopolyploidy have been discussed by Grant (1971). He lists three of these: (1) a long life cycle, usually combined with some method of vegetative reproduction; (2) the existence of "primary" speciation combined with, or due to, chromosomal rearrangements; and (3) frequent spontaneous interspecific hybridization. He regards the second and third of these as more important factors than the first. Plant genera that show the first and third, but whose species do not differ with respect to chromosomal rearrangements, are said by Grant to show the *Ceanothus* pattern of speciation. *Ceanothus* (Rhamnaceae) is a genus of North American woody shrubs that are predominantly outcrossing. All the species are diploids with $2n = 24$ (Nobs, 1963). They are kept apart in nature largely as a result of growing on different soil types, but hybridization occurs rather freely in zones of contact or overlap. Nobs found that within the subgenus *Cerastes* all the experimentally produced species hybrids were vigorous and fully fertile, with normal meiosis. On the other hand, crosses between members of the subgenera *Cerastes* and *Euceanothus* mostly produced no hybrids or ones that died as seedlings. The few intersubgeneric hybrids that were produced had extremely irregular meiosis and were entirely sterile.

It is easy to see that there would be no "incentive" to polyploidy in *Ceanothus*. Any intrasubgeneric hybrids that occur naturally and that possess adaptive qualities are successful as diploids, and in view of their regular meiosis and high fertility they are not adaptively improved by polyploidy. The hybrids between different subgenera might have their fertility restored by chromosome doubling but they are inherently weak, stunted, and probably ill-adapted to any available habitat.

Natural hybridization occurs rather freely in oaks *(Quercus)* and *Eucalyptus* (Muller, 1952; Tucker and Muller, 1956; Forde and Faris, 1962; Pryor, 1959) but there are no polyploid species in these genera, in which vegetative reproduction by suckering does not occur. On the other hand, polyploidy has been a major evolutionary mechanism in *Ulmus* and *Populus,* trees notorious for suckering. In both *Quercus* and *Eucalyptus* the diploid hybrids are usually fertile provided that the parental species are not too distantly related; a number of species seem to be of hybrid origin. These genera therefore have the *Ceanothus* pattern of speciation.

Polyploidy plays a highly variable role in the evolution of plant genera. We may distinguish several types of situations. On the one hand, in genera in which most of the species are diploid but a small number are tetraploid, the role of polyploidy is rather minimal. On the other hand, in genera in which all the species are polyploid the diploids have become extinct.

The large genus *Geum* (Rosaceae), which has been studied in general detail by Gajewski (1957, 1959), consists, as far as is known, entirely of polyploid species (Table 8-1). These are perennial herbs that outcross in nature; their overall distribution extends from the Arctic to the Auckland Islands in latitude 51° S, mainly in cold or mountainous areas. A few diploids, with 2n = 14, occur in the closely related genera *Coluria* and *Waldsteinia,* but the 33 species of *Geum* whose chromosome numbers are known are tetraploids (2n = 28), hexaploids (2n = 42), octoploids (2n = 56), decaploids (2n = 70), and dodecaploids (2n = 84). Allopolyploidy seems to have been favored by the low chiasma frequency of the chromosomes, which prevents multivalent formation in the meiosis of polyploids. Gajewski recognizes 11 subgenera of *Geum,* of which seven have been studied cytogenetically, to some extent. The large subgenus *Eugeum,* the best known, consists largely of hexaploids. Presumably, all the diploid ancestors and most of the tetraploid ancestors of the hexaploids and higher polyploids are extinct. It is possible, however, that some members of the subgenera *Acomystylis, Sieversia,* and *Neosieversia* (Arctic and high mountains of Asia and North America) and *Oncostylus* (southern hemisphere), which have not been studied cytologically, may turn out to be diploids.

TABLE 8-1
Levels of polyploidy in Geum *and two related genera. (From Gajewski, 1959.)*

	No. of polyploid species					
	Di	T	H	O	De	Do
Waldsteinia	1	—	1	—	—	—
Coluria	1	—	—	—	—	—
Subgenera of *Geum*						
Oreogeum	—	1	1	—	1	—
Acomastylis	—	—	2	2	—	—
Erythrocoma	—	—	3	—	—	—
Eugeum	—	—	15	—	1	4
Stylipus	—	—	1	—	—	—
Orthorus	—	1	—	—	—	—
Woronowia	—	—	—	—	1	—
Total	2	2	23	2	3	4

ABBREVIATIONS
Diploid, **T**etraploid, **H**exaploid, **O**ctoploid, **De**caploid, **Do**decaploid

The species of *Geum* are moderately distinct in appearance and hybridization in nature is rare, although it occurs throughout Europe between the very widespread hexaploid species *G. rivale* and *G. urbanum*. Gajewski was unable to obtain hybrids between *Geum* species and members of the related genera *Waldsteinia* and *Coluria* but did succeed in producing numerous experimental hybrids between different *Geum* species assigned to different subgenera. Many of these hybrids showed a relatively high degree of fertility. He has presented strong evidence that the subgenus *Eugeum* arose by doubling of the chromosome number in a triploid hybrid or hybrids between a tetraploid species close to *G. montanum* and one or more diploid species of arctotertiary *Waldsteinia*. Although a few dodecaploid species of *Eugeum* are known, the hybrid between the hexaploid *G. rivale* and *G. urbanum* does not seem to have given rise to an allopolyploid species anywhere in Europe, in spite of the fact that it is continually arising. There is therefore an upper limit to polyploidy in *Geum*, even if it is a somewhat elastic one.

Speciation in the genus *Galium* (Rubiaceae) has followed essentially the same plan as in *Geum*, except that the diploid species are not extinct, although most of them are relicts (Ehrendorfer, 1949, 1964, 1965). The *G. pumilum* group includes a single diploid species, *G. austriacum*, which has a discontinuous distribution as a glacial relict in the Alps, and a number of widespread polyploid species (tetraploid to octoploid). The situation is similar in the *G. anisophyllum* complex. The diploids occur in small areas of the Alps, Tatra Mountains, Transylvanian Alps, and Macedonia where glaciation was minimal, while tetraploids are widespread in central and southern Europe. Hexaploid and octoploid species occur mainly in areas of the Alps and Pyrenees that were covered by Pleistocene ice sheets, while a decaploid grows in the northern Apennines.

A very different type of polyploid complex occurs in the *Gilia inconspicua* group of species (Polemoniaceae), studied by Grant and Grant (1960), and by Grant (1964, 1971) and Day (1965). These are the so-called cobwebby gilias; there are 13 diploid species and 10 tetraploid ones in western North America. In the diploid species five different "genomes" can be recognized, i.e., chromosome sets that give good pairing (synapsis) when homozygous but poor pairing when two different genomes are present. For example, the diploid *G. aliquanta* has two A genomes (constitution AA), *G. minor* and *G interior* are both TT, *G. clokeyi* is CC, and *G. mexicana* is OO. *G. malior*, a tetraploid derived from *G. aliquanta* and *G. minor*, is AATT, *G. transmontana* and *G. ophthalmoides I* are both CCTT, while *G. ophthalmoides II* is CCOO.

It might be supposed that polyploids, and especially high polyploids, would not be particularly sensitive to loss or gain of one or even several chromosomes, and that aneuploid biotypes would consequently establish themselves fairly frequently in nature. There is, however, little evidence to support such a view. The vast majority of plant polyploids seem to carry complete genomes, with no chromosomes missing or in excess. It is true that many so-called "aneuploid series" have been described by plant cytogeneticists. But the concept of aneuploidy has frequently been used in a very loose general sense to designate any condition where a chromosome number is present that is not an exact multiple of a "basic" number. In some cases such "aneuploid" numbers may be due to structural rearrangements leading to changes in chromosome number, while in others they may be due to allopolyploids having arisen from crosses between species having different basic numbers.

One of the most remarkable polyploid complexes that have been studied is the North American genus *Claytonia* (Portulacaceae), investigated by Lewis, Oliver, and Suda (1967). *C. caroliniana,* which occurs from the Appalachian mountains northward to Nova Scotia and the Great Lakes region, shows 2n = 16 throughout most of its range, but polyploids occur in the southern Appalachians. *C. virginica,* whose range extends over most of the same area and southwest into the Mississippi valley and northeast Texas, includes diploids with 2n = 12, 14, and 16, and polyploids with chromosome numbers 2n = 17 through 37, 40, 42, 44, 46, 48, 50, 72, 81, 85, 86, 87, 91, 93, 94, 96, 98, 102, 103, 104, 105, 110, 121, 173, 177, and 191. It is probable that allopolyploidy has occurred repeatedly in this group, involving both species, and in particular the three basic karyotypes of *C. virginica* (with n = 6, 7, and 8 respectively), which must differ from one another by structural rearrangements such as fusions and translocations. In this way tetraploids with every number between 2n = 24 and 32 and hexaploids showing every number between 2n = 36 and 54 could have arisen (Stebbins, 1971). The assumption that the vast range of biotypes in *C. virginica* (which are morphologically rather uniform and restricted ecologically to deciduous woodlands with rich soil) form a true aneuploid series—in the sense of one consisting of biotypes showing actual loss or gain of chromosomes from a "balanced" karyotype—is hence likely to be true only in part.

Among the families of angiosperms, the grasses *(Gramineae)* show perhaps the highest frequency of polyploidy, about 70 percent of the species that have been studied being polyploids. The genus *Danthonia,* whose distribution includes North America, South Africa, Australia, and New Zealand, shows the basic numbers n = 7, 9, and 12, the latter being probably a secondary one, resulting from a doubling of n = 6, since some South African species have 2n = 12 (De Wet, 1954). The numerous Australian species have been studied by Brock and Brown (1961). Apart from *D. frigida,* which is a hexaploid (2n = 42), and more closely related to certain alpine New Zealand species, the Australian species are "diploids," tetraploids, hexaploids, octoploids, and a decaploid with 2n = 120. Several taxonomic species are complexes of forms showing different levels of ploidy. Thus, the widespread *D. caespitosa* includes diploid, tetraploid, and hexaploid biotypes that may all occur at the same locality; somewhat surprisingly, artificially obtained hybrids between the diploid and the tetraploid form proved inviable. The montane species *D. induta* includes a hexaploid, an octoploid, and a decaploid form, which may occur on different massifs. But high polyploidy is not necessarily associated with a montane habitat, since diploid forms of three species occur at an elevation of over 2,000 meters on Mt. Kosciusko. The only species to have successfully colonized the arid zone of central Australia is a hexaploid. A number of natural hybrids were found, which included triploids, tetraploids, and a pentaploid; although their meiosis was somewhat irregular, they did set some seed. In addition to these, some atypical tetraploids (2n = 48) were found in the wild, which could have been due to recent doubling of the karyotype in hybrids between diploid forms.

Much work has been carried out on the polyploid evolution of the species of *Aegilops* and *Triticum* on account of the economic importance of the wheat *T. aestivum,* a hexaploid with 2n = 42 (the basic number of the tribe Triticinae is n = 7). The latest general summary of the evolutionary situation is that of Kimber (1974), which differs somewhat from earlier accounts. Important additional information has been provided by Gill and Kimber (1974), Johnson (1975), and Dhaliwal and Johnson (1976).

Diploid, tetraploid, and hexaploid "wheats" occur. *T. aestivum* carries three different "genomes," and has been designated AABBDD. It arose, probably in neolithic times, by a cross between a tetraploid wheat, *T. turgidum* (formula AABB) and a diploid *T. tauschii* (= *Aegilops squarrosa* of earlier authors) with the formula DD. The history of *T. turgidum* is still uncertain, however. It obtained its A genome from the diploid *T. monococcum,* but the origin of the B genome, earlier believed to be derived from *T. speltoides,* is not known with certainty. It may have been derived from *T. urartu* (Johnson, 1975), but the possibility that it arose by modification of one A genome in an AAAA tetraploid, rather than by hybridization, now has to be considered. It is clear that in addition to allopolyploid evolution in this complex there has also been natural hybridization between tetraploids, and that this has probably modified the properties of the original "genomes" (Zohary and Feldman, 1962, Pazy and Zohary 1965). The wheats are normally self-fertilizing, but crossing apparently has occurred on rare occasions in the past.

The only species-rich genera of grasses that consist almost entirely of diploids are *Melica* and *Dichanthelium* (Brown, 1948). The reason why polyploids have not arisen or have not established themselves in the phylogeny of these genera is probably that the species are strictly self-fertilized. At least in

Dichanthelium the fertile flowers are cleistogamous and probably incapable of cross-fertilization.

The cases of polyploidy considered earlier all occur in biotypes that reproduce sexually. However, in many groups of higher plants there is an association between polyploidy and asexual reproduction (apomictic or vegetative). Grant (1971) has described such "agamic complexes" in a number of plant groups. They will be considered later (pp. 313–314) after the evolutionary mechanisms of the simpler, diploid parthenogenetic organisms have been discussed.

It was formerly believed that polyploidy was an irreversible process, but there is now evidence that it is not invariably so. Müntzing (1943) obtained two spontaneous diploid plants in some thousand progeny from the tetraploid *Dactylis glomerata,* and Raven and Thompson (1964) have suggested that the diploid populations that occur sporadically throughout Europe may have arisen as reversions of this kind, rather than being primitively diploid. De Wet (1968) has suggested that sexually reproducing diploid populations have arisen repeatedly from apomictic tetraploids in the grass genus *Dichanthium,* and Siew-Ngo (1973) has demonstrated a possible cytological mechanism for reductions of chromosome number to lower ploidy levels (due to multiple spindles at meiosis) in polyploid species of the Asiatic genus *Globba* (Zingiberaceae). It is unlikely that reversion from polyploidy to diploidy has been a major evolutionary mechanism in any group of plants, but Grant (1971) probably goes too far in suggesting that diploids that have arisen in this way must necessarily be sterile.

Polyploidy has been extremely infrequent in bisexually reproducing groups of animals. Muller (1925) long ago suggested that this was because in organisms with XY:XX or XO:XX sex chromosome mechanisms doubling the chromosome number would in effect abolish the heterogamy on which the production of two sexes depends (in an XXYY individual one would expect that in most cases an XX and a YY bivalent would be formed at meiosis, instead of two XY bivalents). If the XY gametes produced by XXYY individuals are fertilized by the XX gametes from an XXXX parent, the result would be XXXY tetraploids. In many groups of animals, with the *Drosophila* type of sex determination, these will be sterile intersexes. Muller's argument still carries considerable weight. However, it is now well known that there are groups of both animals and plants in which sex determination does not depend on the *Drosophila* type of genic balance but on a "dominant Y" principle (or "dominant W" in the case of female heterogamy). Thus, the tetraploid race of the dioecious plant *Melandrium album* has an XXXY (♂):XXXX (♀) system (Westergaard, 1958) and the artificially produced tetraploid stock of silkworms *(Bombyx)* a ZZZW (♀):ZZZZ (♂) one (Astaurov, 1972). Both mechanisms seem to function efficiently in producing a 1:1 sex ratio in the progeny without any intersexes or sterile genotypes.

Among plants, the docks (genus *Rumex,* in the broad sense) illustrate the relationship of sex determination to polyploidy extremely well. The subgenus *Lapathum* consists of monoecious species that exhibit a polyploid series from diploids, such as *R. alpinus* (2n = 20), tetraploids, hexaploids, octoploids, decaploids, 16-ploids, and 20-ploids, such as *R. hydrolapathum* (Kihara and Ono, 1926; Jaretsky, 1928; Jensen, 1936). The subgenus *Acetosa* includes three sections, *Acetosae, Insectivalves,* and *Americanae,* that consist of dioecious species. Except for the Texas race of *Rumex* (*Acetosa*) *hastatulus,* these are all XY_1Y_2 in the male plants, which show a trivalent at meiosis (the Texas race of *hastatulus* has XY males and is clearly derivative). The European and Asiatic acetosas have been studied by numerous workers (Kihara and Ono, 1923, 1925; Świetlińska, 1963; Źuk, 1963; Löve, 1943, 1957, 1969). Sex determination in all these species depends on a genic equilibrium between female-determining factors on the X chromosome and male- and female-determinants on the autosomes, as has been demonstrated by extensive studies on artificial polyploids; the Y chromosomes only affect male fertility, not maleness as such. The sex determination mechanism is consequently similar in principle to that of *Drosophila;* that is, it seems to have acted as an absolute barrier to polyploidy. Löve regards the Palearctic species as having a tetraploid set of autosomes (six pairs), by comparison with a diploid set (three pairs) in the North Carolina and southern Illinois–Missouri races of *R. hastatulus,* but there is no good reason to accept this interpretation. It is much more likely that the difference in number of autosomes is due to structural rearrangements, especially as the Texas race of *R. hastatulus* has four pairs of autosomes (Smith, 1969), and that the other races of this species are derived from it rather than vice versa (Stebbins, 1971).

One section of the subgenus *Acetosa* includes two species with XY male plants that inhabit the mountains of western North America: *Rumex* (*Acetosa*) *paucifolia* is a diploid with 2n = 14, while *R.* (*A.*) *gracilescens* is a tetraploid (2n = 28, sex chromosomes XXXY in the male). In this section the Y, which must certainly have had an evolutionary origin entirely different from the Y_1 and Y_2 of the other sections, is strongly male determining. But experiments with artificially produced octoploids showed that the Y was not strong enough to prevent the appearance of intersexuality in them (Löve and Sarkar, 1956; Löve and Evenson, 1967).

In the subgenus *Acetosella* there are four species: a diploid, a tetraploid, a hexaploid and an octoploid, all with XY males. Here the very strongly male-determining Y chromosome has permitted polyploidy to occur. Studies on artificial 12-ploids have shown that the Y is in fact strong enough to produce males rather than intersexes in the presence of 11 X chromosomes and 12 pairs of autosomes (Löve, 1957, 1969).

If Muller's explanation for the rarity of evolutionary polyploidy in animals—based on the genetic system of sex determination in *Drosophila,* the only one known to him in 1925—is largely unjustified, what is the correct explanation? Some authors have suggested that the more complex histological differentiation of animals might serve as a barrier to successful polyploid evolution. That this explanation is also unjustified is clear from the existence of numerous polyploid biotypes among parthenogenetic animals (see Chapter 9). This fact supplies the essential clue—polyploidy in animals is rare because most animals are bisexual and depend absolutely on cross-fertilization. Thus, if a tetraploid individual is produced it can only mate with diploids and give rise to sterile triploids.

If this were the whole story we would expect to find cases of evolutionary polyploidy in those groups of animals such as flatworms, earthworms, and pulmonate molluscs.

In Turbellaria there are a number of polyploid species and races. Unfortunately, it is not always easy to determine whether they are reproducing sexually or by parthenogenesis, especially since pseudogamy (parthenogenetic reproduction triggered off by a sperm that does not transmit genetic material to the offspring) is relatively widespread. Thus, *Dendrocoelum infernale* (2n = 32) is a tetraploid by comparison with its close relative *D. lacteum* (2n = 16) according to Aeppli (1952); but Benazzi (1957) believes that it may reproduce by pseudogamy rather than sexually. In the genus *Mesostoma* the chromosome numbers n = 2, 4, 5, and 8 occur (Valkanov, 1938; Ruebush, 1938; Husted and Ruebush, 1940); it seems likely that the species with n = 4 and n = 8 are tetraploid and octoploid. And in the genus *Macrostomum* one species has n = 6, while 10 others have n = 3, which is again suggestive. There is also a possible tetraploid species in *Phaenocora,* but in *Dalyellia* all of the 11 species that have been studied cytologically have n = 2 (Ruebush, 1937, 1938).

In the digenetic trematodes (Digenea), which are hermaphroditic with the single exception of the family Schistosomatidae, Jha (1975) has reviewed the karyotypes of 107 species, in which the haploid number ranges from 6 to 14. He concludes that there is no evidence for evolutionary polyploidy, since the two species with n = 12 and the one with n = 14 are not closely related to the species with n = 6 and n = 7. In tapeworms (Cestoda) Jones (1945) found that 15 species showed a range of chromosome numbers from n = 5 to n = 8, which obviously excludes polyploidy.

In oligochete worms and leeches (Hirudinea) there are a number of polyploid species or probable polyploids. In the oligochetes polyploidy and parthenogenetic reproduction are frequently combined, but there are also many polyploid species that reproduce sexually. Christensen (1961) found 35 out of 115 species of sexually reproducing Lumbricidae and Enchytraeidae that were polyploid.

No information exists that would enable us to decide whether these sexually reproducing invertebrate polyploids are allopolyploids or autopolyploids. By analogy with what is known of plant polyploids, we are probably justified in thinking that they would not have been adaptively superior and would hence not have established themselves in evolution unless they were allopolyploids. Obviously, a hermaphroditic organism capable of both cross- and self-fertilization is most likely to give rise to new polyploid biotypes—by cross-fertilization it can form hybrids, and these, after doubling the chromosome number, can give rise by self-fertilization to a local population of polyploids. Unfortunately, we still have only very fragmentary information as to the prevalence of cross- and self-fertilization in the hermaphroditic groups of invertebrates. In triclad turbellarians self-fertilization appears to be rare, but is known to occur in a few species. The same is true of the pulmonate and opisthobranch molluscs, in which hermaphroditism is universal. The difficulty is largely one of determining the extent to which self-fertilization can occur occasionally in species that are normally cross-fertilized.

Polyploidy is quite rare in the pulmonate molluscs. In the terrestrial Stylommatophora the chromosome numbers of about 80 species are now known; there appear to be no polyploids among them, perhaps because cross-fertilization is obligatory. In the aquatic Basommatophora there are a few polyploid forms. In fresh-water limpets (Ancylidae) Burch, Basch, and Bush (1960) recorded a few polyploid species (tetraploids and an octoploid). The situation has also been studied in the genus *Bulinus* (Planorbidae), which are important as vectors of schistosomiasis in Africa (Burch, 1964; Brown and Wright, 1972). Many of the species are diploid ($2n = 36$) but tetraploid, hexaploid, and octoploid populations occur in the *B. truncatus* complex. The tetraploids are widespread in central and north Africa, but the higher polyploids are only known from the highlands of Ethiopia, where some diploids also occur. Polyploidy is not known to occur in the three large genera of freshwater snails, *Lymnaea, Physa,* and *Planorbis*.

The occurrence of polyploidy in groups of bisexually reproducing animals has been a controversial question for many years. Claims to have established its existence as an evolutionary phenomenon in lepidopterans, beetles, earwigs, and even mammals were put forward in former times, but no credible evidence for its occurrence in these groups remains (White, 1973, 1976). Apart from mammals, amphibia, and some fishes, in which the dominant Y mechanism (or dominant W mechanism in the case of female heterogamy) seems to be general, and *Drosophila,* in which an X–autosome genic-balance mechanism of sex determination occurs, we have little evidence as to which mechanism operates in most bisexual animal groups. However, it is clear that the *Drosophila* type of mechanism must exist in groups in which all or most of the species have an

XO:XX sex chromosome constitution, a Y being simply absent (the possibility remains, however, that species that have become secondarily XY in such groups, as a result of X–autosome fusions, may have acquired an entirely different genetic mechanism of sex determination, dependent on a dominant neo-Y).

It is in the three vertebrate groups mentioned above, therefore, that we might expect to find cases of evolutionary polyploidy. We would certainly not expect to encounter them in such groups as dragonflies or grasshoppers, in which most of the species have XO males.

The evidence for the existence of a few polyploid species of frogs and toads is now incontrovertible, but in Urodela the only known cases of polyploidy are in some parthenogenetically reproducing species of *Ambystoma* (see p. 296). The known polyploid species of Anura are members of the South American family Ceratophrydidae (Saez and Brum, 1959, 1960; Saez and Brum-Zorrilla, 1966; Beçak and Beçak, 1973; Beçak, Beçak, and Rabello, 1966; Bogart, 1967; Beçak *et al.*, 1967); the family Hylidae (Wasserman, 1970; Beçak, Denaro, and Beçak, 1970; Batistic *et al.*, 1975); the family Leptodactylidae (Barrio and Rinaldi de Chieri, 1970a, b); or the family Pipidae (Tymowska and Fischberg, 1973). No cases are known in the families Ranidae and Bufonidae, a large number of whose species have been studied cytologically.

Diploid species of *Odontophrynus* with 2n = 22 and of *Ceratophrys* with 2n = 26 are known (Saez and Zorrilla, 1963; Beçak *et al.*, 1967). A tetraploid form of *O. americanus* with 2n = 44 is widespread in South America, but the ancestral diploid form is known from very few localities (Beçak, Beçak, and Vizotto, 1970). The tetraploid usually forms 11 quadrivalents at meiosis, but sometimes a few trivalents, bivalents, and univalents are present (Beçak, Beçak, and Rabello, 1966). It was shown by Beçak and Goissis (1971) that the amount of DNA per milligram of kidney tissue of the tetraploid was twice the amount found in the diploid; but there was no significant difference in the amount of RNA per milligram in the two forms. When the amount of lactate dehydrogenase per milligram of heart muscle was compared in the diploid *O. cultripes* and the tetraploid *O. americanus,* the former actually had the higher value, but the red cells of *americanus* are slightly larger than those of *cultripes* and contain a similarly larger amount of hemoglobin (Beçak and Pueyo, 1970). Probably there is some degree of repression of duplicated structural genes in the tetraploid (Beçak, 1969). It is likely that polyploid *O. americanus* with numbers other than 2n = 44 exist; Saez and Brum-Zorrilla (1966) reported 2n = 42 and 50 from localities in Argentina.

Ceratophrys dorsata and *C. ornata* (if, indeed, these are distinct species), with 2n = 104, are octoploid by comparison with *C. calcarata,* which has 2n = 26 (Saez and Brum-Zorrilla, 1966; Beçak, 1969; Bogart, 1967). The DNA values confirm octoploidy and multivalents are seen at meiosis; in view of the difficulty of counting such large numbers of chromosomes, the claim of Saez and Brum-Zorrilla that multivalents are also found in individuals of *C. dorsata* with 92, 96, and 108 chromosomes should be regarded with some reserve. Diploid populations of *C. ornata* have been found (Barrio and Rinaldi de Chieri, 1970b) but no tetraploid populations of either species are known.

In the tree frogs (Hylidae) there are two tetraploid species known: the tetraploid South American *Phyllomedusa burmeisteri,* with 2n = 52 (Beçak, Denaro, and Beçak, 1970), of which a diploid form is also known (Batistic *et al.*, 1975), and the North American *Hyla versicolor,* with 2n = 48 (Wasserman, 1970). The latter is closely related to the diploid *H. chrysoscelis;* the diploid and tetraploid are morphologically almost indistinguishable and occur sympatrically in nature, but can be distinguished on the basis of their calls (*chrysoscelis* is "fast-trilling," *versicolor* "slow-trilling"). *H. versicolor* forms quadrivalents at meiosis, but with what regularity is unknown. Multivalents are also formed in *Phylomedusa burmeisteri,* but apparently not in two tetraploid species of the genus *Pleurodema* (Leptodactylidae) studied by Barrio and Rinaldi de Chieri (1970a).

Two polyploid species of the African genus *Xenopus* have been reported by Tymowska and Fischberg (1973). Most diploid species of *Xenopus* have 2n = 36, but *X. bunyoniensis* with 2n = 72 is a tetraploid and *X. ruwenzoriensis* with 2n = 108 is a hexaploid. These polyploids usually show only bivalents at meiosis, but occasionally multivalents occur. The high frequency of multivalents in the polyploid ceratophrydids would be regarded by a plant cytogeneticist as evidence for autopolyploidy, but it is more probable that they are interracial allopolyploids, as the hexaploid *Xenopus ruwenzoriensis* almost certainly is. It is a mystery how the American species of polyploid anurans (except for *Pleurodema bibroni* and *P. kriegi,* which form bivalents) maintain their fertility in spite of a high but irregular frequency of multivalents at meiosis. Bogart (1967) suggested that these polyploid anurans had arisen by "chromatid autonomy," an ancient and now discredited concept. He and Wasserman (1972) later suggested that each of the polyploids had arisen directly from a single diploid ancestral species, with a triploid individual as an intermediary. By analogy with what is known of plant polyploids, this seems unlikely.

Although eight or nine species of polyploid anurans are now known, these are only a minute fraction of the total number of species whose karyotypes are known. As a mechanism of speciation, polyploidy has played a minimal role in amphibians. There are no anuran genera that seem to have had an "ancient polyploid origin."

The only other vertebrate group in which evolutionary polyploidy had been claimed by numerous workers is the teleost fishes. The evidence, however, is much less clear than in the case of the anurans previously discussed, and a few of the claims that have been made are demonstrably unjustified. The early suggestions of Svärdson (1945) and Kupka (1948) on possible polyploidy in coregonid fish have been criticized by Matthey (1949), Bogart (1967), and White (1973) and must now be regarded as insufficiently based. Svärdson argued for a basic haploid number of n = 10 in Salmonidae, so that *Salmo salar* (Atlantic salmon) and *S. irideus* (rainbow trout), both with 2n = 70, would be hexaploids, *S. trutta* and *S. alpinus,* with 2n = 80, would be octoploids, while *Thymallus thymallus* (grayling), with 2n = 102, might be a decaploid. However, Rees (1964) has found that the DNA values of *S. salar* and *S. trutta* are almost exactly the same, and a large number of chromosome numbers are now known in Salmonids that are not multiples of 10 and that fill in the gaps between 2n = 60, 80, and 100. A more probable case of polyploidy is that of the *Lucioperca sandra* (pikeperch), which has 2n = 48 in Finland and 2n = 24 in Sweden (Suomalainen, 1958). In some species of teleosts, however, although the chromosome numbers suggest polyploidy, detailed study of the karyotypes throws doubt on this interpretation. Thus, Raicu and Taisescu (1972) claimed that the European fish *Misgurnus fossilis,* with 2n = 100, is a tetraploid by comparison with the Japanese *M. anguillicaudatus,* which has 2n = 50 (Ojima and Hitotsumachi, 1969). But the European species has 36 clearly bi-armed chromosomes (136 major chromosomes altogether) while the Japanese one has only 14 bi-armed chromosomes (64 major chromosome arms).

In Cyprinidae, Ohno (1974) has claimed that the carp *(Cyprinus carpio)* and goldfish *(Carassius auratus),* both with 2n = 104, and *Barbus barbus,* with 2n = 100, are tetraploids by comparison with a number of related species, including two species of *Barbus* that have 2n = 50. In this case the DNA values support the polyploidy hypothesis, but the numbers of major chromosome arms are not in good agreement with it. Conversely, Ohno (1974) has claimed that the rainbow trout *Salmo irideus* (2n = 58–64, according to him, and 104 major chromosome arms) is a tetraploid by comparison with the Pacific anchovy *Engraulis mordax* (2n = 48, with 48 major chromosome arms). Here the case is obviously very weak and would hardly be worth considering were it not for the fact that Ohno found some quadrivalents in the meiosis of *S. irideus.* However, such quadrivalents (and higher multivalents) may very well be due to translocation heterozygosity rather than to polyploidy. There are at least two vertebrate species that are certainly not polyploids but which show such multiple associations of chromosomes at meiosis. One is the frog *Eleutherodactylus binotatus,* which shows the same chromosome number as related species, although somewhat surprisingly it has a much higher DNA value (Beçak and Beçak, 1974). The other is the fish *Sphaerichthys osphromonoides,* which has the lowest chromosome number (2n = 16) known in fishes (Calton and Denton, 1974).

Ohno (1974) has gone so far as to state his opinion that "polyploidy at the stage of fish and amphibians was the key to evolution by gene duplication of higher vertebrates." His view is based largely on the incontrovertible biochemical evidence for duplication of a number of gene loci. But similar instances of duplicated loci are well known in organisms that are certainly diploid, such as *Drosophila*. Utter and co-workers (1973) have commented that "While there is undisputable evidence of considerable gene duplication in salmonid fishes . . . the case for polyploidy as postulated by Ohno . . . has not been proven and the uncritical assumption that salmonids are tetraploid stifles creative thinking concerning alternative models."

One of the species of salmonids claimed as a tetraploid is *Oncorhynchus keta,* with 2n = 74 (Ohno *et al.*, 1967; Ohno, Wolf, and Atkin, 1968; Ohno, 1970). Levels of genetic polymorphism and heterozygosity in this species seem to be quite low (11–18 percent loci polymorphic, mean heterozygosity per locus 2–3 percent, according to the data of Altukhov *et al.*, 1972). This is hardly what we would expect on the polyploidy hypothesis.

The possibility that thelytokous populations might occasionally give rise, by reversion, to sexual ones, and that some polyploid bisexual species of animals might have arisen from diploid ancestors via a thelytokous intermediate stage has been advocated by Schultz (1969) and Astaurov (1972). The latter author and his colleagues were in fact successful in producing sexually reproducing allotetraploid silkworms in laboratory experiments involving artificial parthenogenesis as an intermediate stage. Their procedure involved the following steps: (1) production of diploid all-female clones of *Bombyx mori;* in each generation eggs extracted from the moths were heat-treated to induce apomictic thelytoky; (2) occasional production of tetraploid females in these clones; (3) establishment of tetraploid all-female clones by the use of heat treatment in each generation; (4) production of sterile allotriploid males and females as a result of mating these females with diploid males of the very closely related species *Bombyx mandarina;* (5) production of an all-female allotriploid clone by heat treatment of the eggs; many of these allopolyploids are in fact mosaics of triploid (2B + IM) and hexaploid (4B + 2M) tissue; and (6) mating of the mosaics with normal males of *mandarina*. Tetraploid male and female offspring were obtained from hexaploid oocytes that had undergone reduction and had been fertilized by haploid *mandarina* sperm, and these offspring could be propagated as an artificial allotetraploid "species," called *"Bombyx allotetraploidus"* by Astaurov. This "species" would of course be completely isolated from its diploid ancestors, since it could only produce sterile triploid offspring if backcrossed to them.

Interesting though this experiment is, there are two powerful reasons for believing that no polyploid bisexual species have arisen in nature by Astaurov's model. In the first place, the technique of heat treatment did not suppress or remove from the genotype the genes necessary for meiosis. In natural apomictic parthenogenesis, on the other hand, these genes are lost either by mutation or deletion from the karyotype and it is unlikely in the highest degree that they could be reacquired at a later stage.

There remains the possibility that the automictic type of thelytoky (in which meiosis still occurs, but is compensated for by some kind of pre- or post-meiotic doubling of the chromosome number) could in some instances have served as a step to bisexual polyploidy. However, for this to happen the chromosomal doubling mechanism would have to be lost again and a stable XXXY:XXXX mechanism based on a "dominant Y" sex-determining system established. While neither type of change is impossible (and the second must certainly have occurred in the tetraploid amphibia referred to above), they are sufficiently improbable events that their coincidence would be in the highest degree unlikely.

The second reason for believing that Astaurov's model for the origin of polyploidy is not a realistic one is simply that natural parthenogenesis is totally unknown in the one group of animals (Anura) in which there is incontrovertible evidence for polyploid bisexual species. Moreover, although a few cases of natural parthenogenesis are known in teleosts, they occur in families that are not particularly closely related to those in which instances of polyploidy have been claimed to occur.

The various earlier claims for evolutionary polyploidy in several insect orders characterized by well-developed sex chromosome mechanisms have been criticized by White (1973, 1976). The situation is clearly different in Hymenoptera, however, an order in which the males are haploid and develop from unfertilized eggs. In fact, a number of workers have claimed to have discovered cases of evolutionary polyploidy in this order.

Although the basic principles of the genetic mechanisms of sex determination in Hymenoptera are still not fully understood, it may well be that they do not, in themselves, constitute a barrier to the establishment of polyploidy, although the fact that mating of two different individuals is necessary for the perpetuation of the population (except in the rare cases of thelytokous hymenopterans) is probably a major barrier to evolutionary polyploidy.

We shall not discuss here whether the remote ancestor of all hymenopterans was a haplo-diploid species or a diploid-tetraploid one; it presumably existed in Permian times so we are never likely to know. The question is—granted that most present-day hymenopterans are haplo-diploid—are there nevertheless some diplo-tetraploid species that have recently become so? In the ants, which have been investigated fairly extensively, there are no probable cases (Crozier, 1975). The sawfly *Diprion similis,* with $n = 14$ as compared with $n = 7$ in six other species of the genus (S. G. Smith, 1941, 1960; Crozier, 1975) is certainly a possible case, but not definitely proven. In bees, Kerr and da Silveira (1972) have claimed five instances of evolutionary diplo-tetraploidy, but most of these are considered doubtful by Crozier. The most plausible is that of *Melipona quinquefasciata,* with n = 18 by comparison with eight other species of the genus with n = 9. But Mello and da Silveira (1970) showed that *M. quinquefasciata* and a related species with n = 9 have the same quantity of DNA in their Malpighian tubule nuclei. This may be evidence against the polyploidy interpretation, but it is also possible that the Malpighian tubule nuclei of *M. quinquefasciata* have undergone one less endoreduplication cycle (in all insects the Malpighian tubule nuclei are endopolyploid, having undergone several cycles of DNA replication without mitosis). Accordingly, we can say that there are only two species of Hymenoptera, out of a total of over 300 that have been studied cytogenetically in which a reasonably plausible case for evolutionary polyploidy has been made. Clearly, polyploidy has not been a major factor in the speciation mechanisms of this species-rich order of insects.

In summary, allopolyploidy must be regarded as a supplementary mode of speciation, one that has been extremely important in the evolution of certain groups of plants (ferns, grasses, etc.). In so far as the number of habitats available to these groups is limited and polyploid biotypes occupy many of them, polyploidy has probably had an inhibitory effect on speciation by other methods in those groups in which it has been especially prevalent. In some other groups of plants, and in animals as a whole, polyploidy has been so rare that its overall evolutionary role in them has been fairly insignificant.

9

Asexual Speciation

There are some evolutionists who would claim that the biological concept of the species implies that there can be no species of asexually reproducing organisms and that consequently "asexual speciation" is a contradiction in terms. Even if this is admitted, however, it is clear that whenever a population that reproduces by vegetative or parthenogenetic processes arises from a sexually reproducing one, an evolutionary event has occurred that involves reproductive isolation, the severance of genetic continuity. It seems better to extend the meaning of the term "speciation" to cover such events than to invent a new term.

A further problem exists, however, with regard to populations that reproduce *exclusively* by parthenogenesis, apomixis (in the botanical sense), or by some vegetative process. Can these be said to constitute species at all, or are they composed of an indefinitely large number of separate clones that exhibit a chaotic range of variation that defies analysis and classification? If speciation cannot be said to occur in such populations (apart from the initial speciation event by which they arose from a bisexual population) what processes of phylogenetic branching and diversification *do* occur in them? Or is the evolutionary life of asexual "species" so brief that this is not a problem?

A relatively large number of genetic systems are known in which sexual and asexual reproduction are combined in some way. In many aphids, cynipid wasps, and trematodes there is an obligate alternation of generations, each sexual generation being followed by one or more parthenogenetic ones. In many plant species, so-called apomictic reproduction (see next page) is combined to a greater or lesser extent with sexual or vegetative reproduction, and in some cases all three methods of reproduction may coexist. In general, those mainly asexual species in which genetic recombination occurs occasionally through sexual reproduction, even if only once in a number of generations, will have a genetic population structure closer to that of sexually reproducing organisms than to that of entirely asexual species. The present chapter will therefore be concerned with species in which the male sex is absent, or in which males are produced only as rare, nonfunctional anomalies. Such all-female species are said to reproduce by *thelytoky,* a term that is preferable to the more general one "parthenogenesis," which denotes, in addition to thelytoky, cases of rare, accidental, or artificial parthenogenesis in normally sexually reproducing species. Thelytoky is an alternative term to "arrhenotoky," the genetic system present in hymenopterous insects and a few other groups in which males develop from unfertilized eggs and females from fertilized ones.

The number of described "species" of animals whose reproduction is exclusively thelytokous is probably of the order of 1,000—possibly more but certainly not as many as 10,000. This means that, among animals, thelytokous genetic systems are approximately one in 1,000 of all those that exist. Species that exhibit vegetative reproduction (corals, bryozoans, some worms of various groups, some tunicates) are not included in this estimate because they are not parthenogenetic and most of them reproduce sexually as well as vegetatively. It does not seem possible to arrive at a similar estimate for the plant kingdom at the present time, although estimates for a few groups, such as the grasses, might be attempted.

The cytogenetic details of thelytokous reproduction, which are extremely varied and largely irrelevant to the issues discussed in this chapter, have been described for animals by Narbel-Hofstetter (1964) and White (1973a), and for plants by Gustafsson (1946–1947). Broadly speaking, we can distinguish two fundamentally different cytogenetic types of thelytoky. In the first of these, usually called *apomixis,* meiosis is totally lacking (thus it is also called ameiotic thelytoky); the divisions in the occyte are simple mitoses. Since this seems by far the more usual of the two types in plants, botanists are apt to use the term "apomixis" as equivalent to "thelytoky."

The second type, in which meiosis is preserved, is *automixis*. Since fertilization has been abolished, it must be compensated for in some manner by a doubling of the chromosome number. There are, broadly speaking, three ways in which this can happen. In the first, which is certainly the commonest, two of the four nuclei formed as a result of meiosis fuse, or (which comes to the same thing) fail to separate at either the first or the second anaphase or telophase. The second mode of compensation is for a doubling of the chromosome number to occur during the cleavage divisions of the embryo. This may involve a fusion of the cleavage nuclei in pairs, but more usually it seems to consist in an endomitotic reduplication of the chromosomes within the cleavage nuclei, i.e., an extra DNA replication without a corresponding mitosis. The third way in which meiosis can be compensated for is by a *premeiotic* doubling of the chromosome number; in this case meiosis takes place with double the number of chromosomes, which is then reduced to the original number in the course of meiosis.

The effects of these different mechanisms on the genetic system are somewhat diverse, and this is no doubt reflected in the evolutionary patterns of apomictic and automictic organisms. Apomixis removes all those restraints that meiosis normally imposes on evolutionary mechanisms. Apomictic forms can be diploid, polyploid, or aneuploid, and they can be heterozygous for all types of chromosomal rearrangements that are capable of surviving through ordinary somatic mitoses. If they have been in existence for a considerable length of time they must be expected to be highly heterozygous, in a genic sense, since whenever a mutation establishes itself in a phyletic line there is no mechanism—apart from the rare chance of an identical mutation in the homologous chromosome—whereby it can become homozygous. No doubt there are regularities of cytogenetic evolution in such forms, but by comparison with the evolution of bisexual species they appear anarchic and genetically unstructured.

Automictic forms that rely on a doubling of the chromosome number during the cleavage divisions represent a fairly complete genetic contrast to apomicts. Their karyotypes must be capable of passing through meiosis without leading to aneuploid nuclei, and this eliminates a considerable range of structural changes. Moreover, they must, if diploid, be completely homozygous, since their two haploid sets result from replication of a single one.

Organisms that are automictic and rely on fusion of products of meiosis are genetically intermediate between the above types. If it is the products of the first division that fuse, then crossing-over may give rise to new genotypes in the offspring by segregation, e.g., a double heterozygote aa′bb′ may give rise to single heterozygotes aa′bb and aa′b′b′ in the offspring. The result after a number of generations will be complete homozygosity except for gene loci very close to the centromere i.e., those that do not separate from it by crossing-over. If the nuclei resulting from the second meiotic division fuse, it will make a difference whether they are "second-division sister nuclei" or "second-division nonsisters." If the former, genes in distal regions of the chromosomes may be preserved in a heterozygous state from generation to generation; if the latter, proximally located genes may be maintained in a heterozygous state.

Finally, in those species with meiotic (automictic) thelytoky that compensate for meiosis by a premeiotic doubling of the karyotype, everything depends on whether pairing (synapsis) of the chromosomes is restricted to sister chromosomes that are molecularly identical or is unrestricted, so that genetically different chromosomes may pair and segregate to opposite poles. The former mechanism is genetically equivalent to apomixis, without any meiosis.

It will be obvious that the evolutionary significance of thelytoky will depend greatly on whether it promotes heterozygosity (apomixis and automixis with premeiotic doubling) or enforces homozygosity (automixis compensated by doubling in cleavage nuclei). Those forms of automixis that depend on a fusion of nuclei resulting from meiosis, while theoretically intermediate between the two extreme systems, will in general tend to promote homozygosity rather than heterozygosity. A reasonably large proportion of thelytokous species are also polyploids. Generally speaking, those cytogenetic types that favor heterozygosity will permit the development of polyploidy, while those leading to homozygosity will not.

The basic fact about thelytokous reproduction is of course that new genotypes are not being constantly generated by the union of gametes in fertilization. The result is bound to be relatively uniform populations of individuals. The few biometric studies that have been carried out on thelytokous organisms, e.g., those of Zweifel (1965) on the lizard *Cnemidophorus tesselatus,* do indicate considerably lower variances for most characters, compared with related bisexual species.

The number of cases of thelytoky that definitely promote homozygosity is strictly limited. Most of them are in insects of the order Homoptera. Many species of the family Lecaniidae (soft scale-insects) are parthenogenetic and have few or no males (Theim, 1933). *Pulvinaria hydrangeae,* studied by Nur (1963), is a diploid that forms eight bivalents at meiosis. After two meiotic divisions the haploid egg nucleus divides and the two daughter nuclei fuse again, thus restoring diploidy. A few embryos appeared to be male, since they showed heterochromatinization of one haploid set in their nuclei, a feature normally shown only by male Lecaniidae (Nur interprets this as due to cytoplasmic factors, since the two haploid sets must be genetically identical). Restoration of diploidy by fusion of cleavage nuclei is also known in the scale-insects *Eucalymantus tesselatus* (Nur, 1971) and *Gueriniella serratulae* (Hughes-Schrader and Tremblay, 1966). Several other species of scale-insects were recorded by Brown (1965) as having meiotic (automictic) thelytoky, but the mode of restoration of diploidy was not determined. Other thelytokous animal species known to restore diploidy by embryonic diploidization are the white fly *Trialeurodes vaporarium* (Thomsen, 1927) and the mite *Cheyletus eruditus* (Peacock and Weidman, 1961).

In these cases the related bisexual species have haploid males, or ones that are physiologically haploid due to inactivation of the paternal set of chromosomes, and one might imagine that this kind of genetic system could only evolve in species that, because of male haploidy, were not to any extent dependent on heterosis. However, the thelytokous race of the stick insect *Bacillus rossii* has a mechanism of diploidization that involves endomitotic doubling of the chromosome number in the nuclei of the blastoderm (Pijnacker, 1969), and the related bisexual race definitely has diploid males (Montalenti and Fratini, 1959). However, in certain areas of central Italy a few rare males of this species occur and may fertilize occasional females, which then give rise to bisexual progenies (Benazzi and Scali, 1964; Scali, 1968; Bullini, 1964, 1965). Thelytoky is thus incomplete in this case, and some heterozygosity may from time to time be generated by sexual reproduction.

It is also worth noting in connection with the adaptive significance of homozygous thelytoky that none of the thelytokous species of Hymenoptera (an order of insects in which the males are invariably haploid) has a mechanism of embryonic diploidization (Crozier, 1975). It therefore seems necessary to rely for an explanation of these homozygosity-generating thelytokous systems on the hypothesis of White (1970b), according to which the males of certain species with strong sexual dimorphism have become a liability for a variety of reasons—they are simply biologically unsatisfactory. It is well known that in scale-insects the adult males are very fragile, ephemeral, and without mouthparts. A minor environmental change might well create conditions intolerable for the males of a particular species and hence confer a high selective advantage on thelytokous reproduction, whether of the homozygosity- or heterozygosity-generating type. In fact, most cases of thelytoky in scale-insects seem to be apomictic, i.e., leading to heterozygosity.

Another case where there is a strong suggestion that males became a liability is the thelytokous "race" of the Mediterranean embiid ("web-spinner") *Haploembia solieri*–actually, the "race" is a distinct species, since it differs in chromosome number, egg-structure, and mode of reproduction. The parthenogenetic form is widespread in the countries of the western Mediterranean, coexisting with the bisexual one in the Balearic Islands, southern France, and parts of Italy (Stefani, 1956, 1959). In Sardinia it occurs alone, without the bisexual sibling species. Oogenesis is apomictic, so that we might expect the thelytokous species to be highly heterozygous. The main interest of this case, however, is in the observation that males of the bisexual species are often rendered sterile as a result of parasitization by a gregarine protozoan. When populations at a number of localities were sampled, a strong positive correlation was found between the proportion of females at each place belonging to the thelytokous species and the proportion of individuals of both sexes parasitized. It thus seems possible (since parasitized females are not sterile) that parthenogenetic reproduction proved adaptive in this case because it led to loss of dependence on males, which were being rendered largely useless to the species because of their vulnerability to the parasite.

It seems unlikely, but not totally impossible, that some of the homozygosity-enforcing thelytokous species of animals may have had a hybrid origin; it is far more likely that each of them arose from a single bisexual species. On the other hand, it is becoming increasingly clear that a great many, and probably the majority, of the heterozygosity-generating thelytokous species of animals arose in the first place as interspecific or interracial hybrids. Opportunistically taking advantage of their initial heterosis (which is due to their hybrid origin), they undoubtedly add to it in the course of time and modify it by further mutational change.

The chief evidence in favor of a hybrid origin of some thelytokous species is in those cases where the karyotypes of the parental species are strikingly different and the thelytokous form, a diploid, combines the haploid complements of both. However, complications may occur if the karyotype of the thelytokous form has undergone secondary changes as a result of chromosomal rearrangements. Thus, Fritts (1969) put forward convincing arguments for the view that the all-female Mexican lizards *Cnemidophorus cozumela, C. maslini,* and *C. rodecki* have arisen as stabilized thelytokous hybrids between the bisexual species *C. deppei* (2n = 50) and *C. angusticeps* (2n = 44), both of which are sympatric with *maslini* in different parts of its range (although the ranges of the two bisexual species are separated by a gap a few kilometers wide at the present time). The karyotype of *maslini* supports this interpretation in general, since it has the expected chromosome number 2n = 47; but most of the individuals examined lacked a large metacentric chromosome characteristic of *angusticeps*. The karyotypes of *cozumela* and *rodecki* have not been studied; they may be regarded as having also arisen by hybridization between *deppei* and *angusticeps,* the only bisexual species of the genus to occur in that area of Mexico.

A hybrid origin has been plausibly claimed for at least five other thelytokous species of *Cnemidophorus* in the southwestern United States, on the basis of either karyology, allozyme studies, or graft compatibility. *C. tesselatus* is a complex of at least six biotypes differing in external phenotype, which inhabit west Texas, New Mexico, and southern Colorado (Figures 28 and 29). Biotypes C, D, E, and F are all diploid and were regarded by Lowe and Wright (1967a, b) as having one haploid set of chromosomes from the bisexual species *C. tigris* and one from *C. septemvittatus*. The triploid biotypes A and B, which inhabit small areas of southern Colorado, were believed to be trihybrid, an additional haploid set being derived from *C. sexlineatus*. These conclusions were strongly confirmed by the allozyme studies of Neaves (1969), which are summarized in Table 9-1. They were also in agreement with the skin-graft experiments of Maslin (1967), in which it was shown that biotypes C–F were generally all graft-compatible and could successfully donate grafts to biotypes A and B, while A and B could not successfully act as donors to the diploid biotypes, presumably because they were carrying the genes of the third species, *C. sexlineatus*.

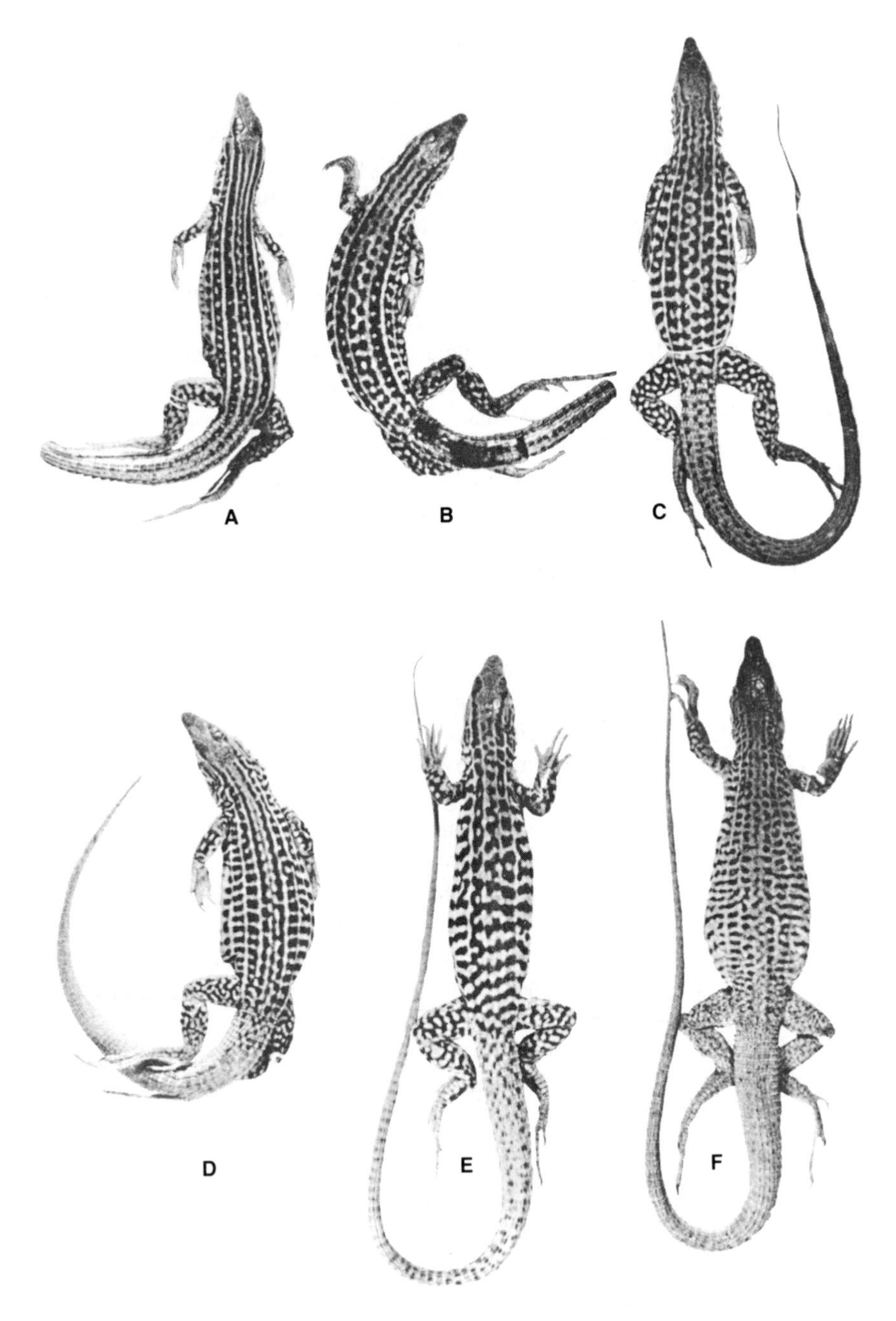

FIGURE 28

The six biotypes of *Cnemidophorus tesselatus*. [After Zweifel, 1965]

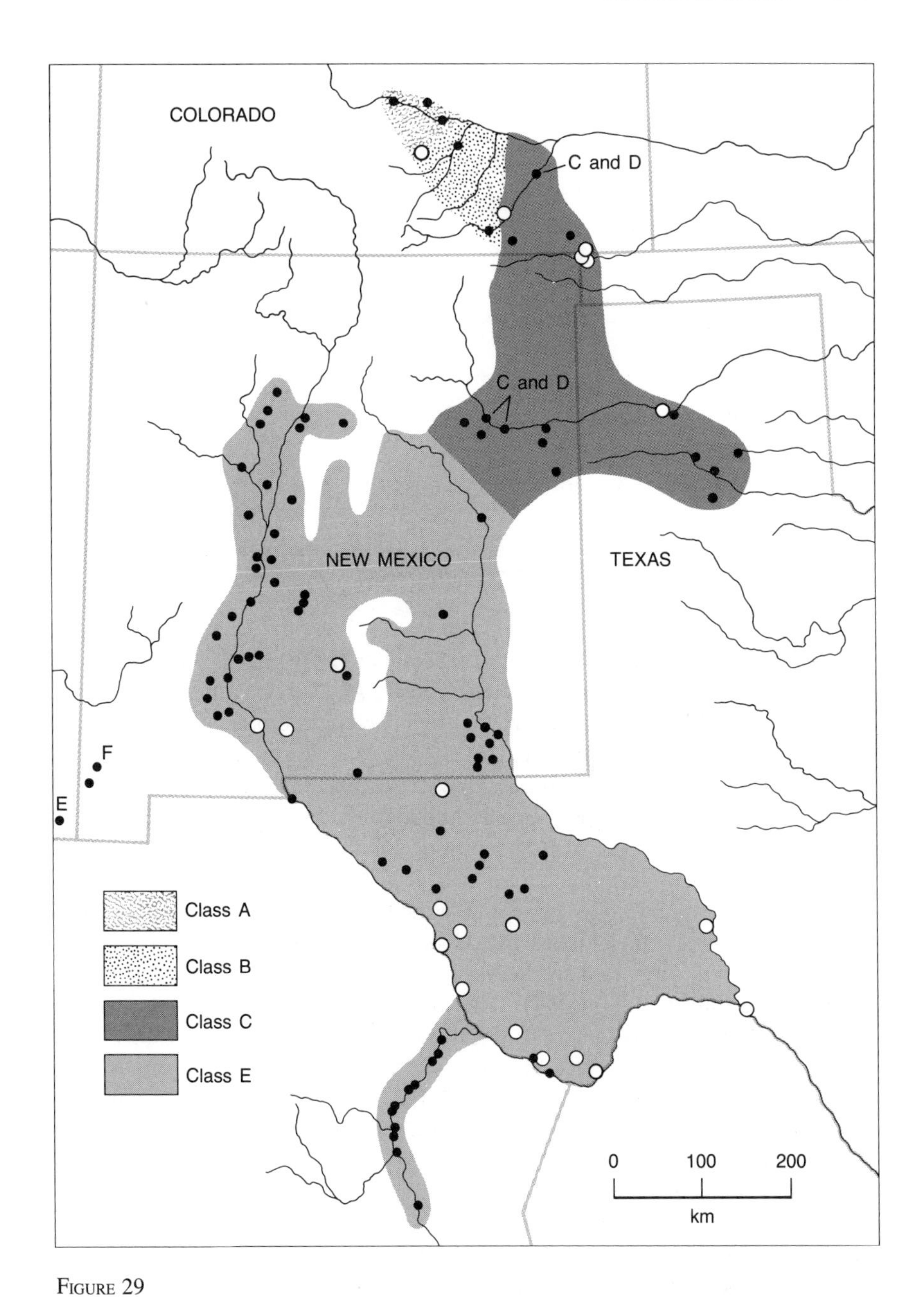

FIGURE 29

Distribution in the southwestern United States of the six biotypes of *Cnemidophorus tesselatus*. The letters refer to the individual forms pictured in Figure 28. [After Zweifel, 1965]

TABLE 9-1
Allozymes of bisexual and thelytokous Cnemidophorus *lizards from New Mexico, Colorado, and Arizona. (Data from Neaves, 1969.)*

Species	Reproduction	Karyotype	LDH	ADA	PGDH
tigris	B	TT	b′/b′	3/3	1/1
sexlineatus	B	SS	b/b	3/3	2/2
inornatus	B	SS	b/b	4/4	1/1
gularis	B	SS	b/b	2/2	1/1
septemvittatus	B	SS	b/b	1/1	1/1
neomexicanus	Th	ST	b/b′	4/3	1/1
tesselatus					
biotypes C–F	Th	ST	b/b′	1/3	1/1
biotypes A, B	Th	SST	b/b/b′	3/3/1	1/2/1
exsanguis	Th	SSS	b/b/b	2/3/4	1/1/1
uniparens	Th	SSS	b/b/b	3/4/4	1/1/1
velox	Th	SSS	b/b/b	3/4/4	1/1/1

ABBREVIATIONS:
B, bisexual; **Th,** thelytokous; **T,** *tigris*-type karyotype; **S,** *sexlineatus*-type karyotype; **LDH,** lactate dehydrogenase; **ADA,** adenosine deaminase; **PGDH,** 6-phosphogluconic dehydrogenase.

Later allozyme studies by Parker and Selander (1976) have greatly extended the earlier work of Neaves. The heterozygosity of *C. tesselatus,* based on an investigation of 21 gene loci, is very high: 56 percent in diploid clones and 71.4 percent in triploid ones, compared to 5, 5.8, and 7 percent in the bisexual species *C. tigris, C. septemvittatus,* and *C. sexlineatus,* respectively. Part of this high level of heterozygosity is clearly due to the hybrid origin of the clones of *C. tesselatus;* but some of it is due to accumulation of mutations and some may be the result of recombination.

On the basis of karyology, Lowe and Wright (1966) claimed *Cnemidophorus neomexicanus* was a thelytokous hybrid between *C. tigris* and *C. inornatus;* this conclusion is supported by Neaves's studies on lactate dehydrogenase and adenosine deaminase (Table 9-1). The allozyme work also suggests that the triploid all-female *C. exsanguis* is trihybrid *C. gularis* × *C. sexlineatus* × *C. inornatus.* Lowe and Wright (1966) thought the triploid *C. uniparens* was *C. inornatus* × *C. inornatus* × *C. gularis,* but the evidence of the deaminases makes it more likely that it is *C. inornatus* × *C. inornatus* × *C. sexlineatus.* The origin of the triploids *C. velox, C. flagellicauda, C. sonorae,* and *C. opatae* has not been worked out.

Although the hybrid origins of these thelytokous lizard species seem to be firmly supported by several independent lines of evidence, nevertheless, there remains a major theoretical difficulty in accepting hybrid origins of thelytokous species. Each evolutionary origin of thelytoky involves some major modification of the maturation divisions in the oocyte—a cytological *tour de force* of one kind or another. It is extremely difficult to conceive either that hybridization would automatically cause the new mechanism to appear or that, in each case, an act of hybridization and a cytological macromutation happened, fortuitously, to coincide in time. These difficulties were sufficient to cause Cuellar (1974) to be skeptical of hybrid origins of thelytoky in general and in *Cnemidophorus* lizards in particular. Unfortunately, there is only one thelytokous biotype in the genus *Cnemidophorus* whose maturation divisions have been studied; that is the triploid *C. uniparens,* in which there is a mechanism involving a premeiotic doubling of the chromosome number, followed by normal meiosis (Cuellar, 1971). Only bivalents were observed, but it is not certain that they were all formed by synapsis between sister chromosomes.

A hybrid origin has been denied by Peccinini-Seale and Frota-Pessoa (1974) in the case of diploid thelytokous populations of *Cnemidophorus lemniscatus* in the Amazon valley, since this is the only species in the region. Two kinds of bisexual populations occur on the upper Amazon, with karyotypes designated D and E. Both forms have 2n = 50 and are structurally homozygous; they differ in the arm-ratio of the largest chromosome pair. All-female populations were found on the lower Amazon; each local colony contains only one karyotype, but three different karyotypes (A, B, and C) are known. Each of these is structurally heterozygous for one or more chromosome pairs.

It is fairly clear that these Amazonian thelytokous lizards did not arise by interspecific hybridization, like those of the southwestern United States and Mexico. Nevertheless, they may well have originated by crossing between races of the bisexual *C. lemniscatus* that differed in karyotype. There are two other possible interpretations, however. The thelytokous biotypes may have arisen polyphyletically from one or more chromosomally polymorphic populations, in which case each of them may have trapped and fixed a particular heterotic combination of structurally different chromosomes. Alternatively, the structurally altered chromosomes may simply have arisen since the origin of thelytoky. The simplest interpretation would be that *all* the thelytokous biotypes in the genus *Cnemidophorus* have a genetic system involving a premeiotic doubling of the chromosomes; that all of them have arisen by hybridization (interspecific or interracial); and that hybridization automatically gives rise to the premeiotic doubling process. Occasional males have been reported in a number of thelytokous species of *Cnemidophorus;* they may be the offspring of crosses with related bisexual species.

Apart from the genus *Cnemidophorus,* there are a number of other lizard genera (in several families) in which all-female species have evolved. Unfortunately, little of the cytological details and mode of origin of thelytoky is known in these cases. Three of the species in question are geckos. *Hemidactylus garnotii* is a triploid with 3n = 70 (Kluge and Eckardt, 1969); it is presumably derived from one or more ancestral species having n = 23, an extra chromosome having been acquired somehow. Gorman (1973) has suggested that it cannot have arisen by hybridization because it has several metacentric and submetacentric chromosomes, whereas all the related members of the genus with n = 23 have exclusively acrocentric chromosomes. *Gehyra variegata,* with 3n = 63 is probably also a triploid, derived from n = 21 ancestors (Hall, 1970). On the other hand, *Lepidodactylus lugubris,* another parthenogenetic gecko, has 2n = 44 and is certainly a diploid (Cuellar and Kluge, 1972).

Thelytoky occurs also in four species of the Old World genus *Lacerta* in the Caucasus mountains. These forms, which belong to the *L. saxicola* group, are diploids with 2n = 38 (Darevski and Kulikova, 1961, 1964; Darevski, 1966). They are largely sympatric with one another in the area between Erevan and Tbilisi. The bisexual forms of the *saxicola* group seem to be excluded from this area, but show narrow parapatric zones of overlap with the thelytokous forms. Occasional triploid hybrids between thelytokous and bisexual species have been found in the zones of overlap; they include both males and females, but the latter are sterile.

The only known thelytokous Amphibia are two triploid species of the salamander genus *Ambystoma, A. tremblayi* and *A. platineum,* in the eastern United States. These all-female populations are closely related to two diploid bisexual species, *A. jeffersonianum* and *A. laterale* (Uzzell, 1963, 1964; Uzzell and Goldblatt, 1967). Electrophoretic studies of the serum proteins suggest that both *tremblayi* and *platineum* are of hybrid origin, the former having two *laterale* sets of chromosomes and one derived from *jeffersonianum* (constitution LLJ), while *platineum* has two *jeffersonianum* sets and one from *laterale* (JJL). The triploids have a premeiotic doubling of the chromosome number in the oocyte (from 42 to 84), which is followed by a normal meiosis (Macgregor and Uzzell, 1964). Synapsis is apparently not absolutely restricted to sister chromosomes, as it is in the thelytokous grasshopper *Warramaba virgo* (see p. 298), since some quadrivalents were seen in meiotic nuclei. There is thus the possibility of some segregation occurring.

The triploid eggs usually do not develop unless penetrated by a sperm from one of the bisexual species, but this "fertilization" serves merely to activate the egg to development and the sperm nucleus does not contribute any genetic material to the embryo. This phenomenon is variously known as pseudogamy or gynogenesis. It is also found in the thelytokous fish *Poecilia formosa* (see p. 297) and in a number of invertebrate groups, and an equivalent phenomenon is known in some plants. It leads to *Ambystoma tremblayi* and *A. platineum* being "reproductive parasites" on the corresponding diploid species. Thus, *tremblayi* (LLJ) normally coexists in the same localities as *laterale,* and *platineum* (JJL) lives with *jeffersonianum,* the triploids mating with males of the sympatric diploid species. However, in some localities one or other of the thelytokous species occurs alone, without the diploid form being present, so that gynogenesis is apparently not always essential for development. It is interesting that there seem to be no localities where *tremblayi* coexists with *jeffersonianum* or *platineum* with *laterale*.

It has been pointed out by Uzzell and Goldblatt that the origins of four aspects of the *Ambystoma* triploids need explanation: hybridity, the triploid state, the premeiotic doubling mechanism, and the mechanism by which the chromosomes of the sperm nucleus are rejected in the ovum. It seems probable that both the triploid species arose from diploid thelytokous populations, i.e., that the bisexual *A. jeffersonianum* gave rise to a thelytokous *jeffersonianum* in which the premeiotic doubling mechanism was established by mutation (which mutation was itself the initiator of thelytoky), and similarly in the case of *A. laterale*. Later, hybridization of these diploid thelytokous biotypes with males of the bisexual forms gave rise to triploids, which may have enjoyed some of the benefits of heterosis. The hypothetical diploid thelytokous biotypes may be extinct, if they proved adaptively inferior to the presumably heterotic triploids, or they may simply not have been discovered yet.

There is, however, another possibility, which is at least suggested by the known facts in the cases of the *Cnemidophorus* lizards and the grasshopper *Warramaba virgo*. Diploid hybrids may have been produced first, and these may have possessed the premeiotic doubling mechanism, spontaneously generated by the fact of hybridity. Only the discovery of a diploid thelytokous population or the rearing of hybrids between *laterale* and *jeffersonianum* in the laboratory can solve this question.

In fishes there are four genuinely thelytokous species, all of which are of hybrid origin. *Poecilia formosa,* native to south Texas and northern Mexico, is an all-female, gynogenetic species whose eggs have to be triggered to develop by sperm from any of three related bisexual species (Hubbs and Hubbs, 1932; Meyer, 1938; Haskins, Haskins, and Hewitt, 1960; Kallman, 1962). True fertilization occasionally does occur, in which case triploid offspring are produced (Rasch *et al.,* 1965); a natural triploid population occurs in one region of Mexico (Rasch, Prehn, and Rasch, 1970). *P. formosa* most likely arose by hybridization between *P. latipinna* and *P. mexicana,* since it is heterozygous for serum albumin alleles for which those species are homozygous. The oogenesis of *P. formosa* has not been studied, but its genetic system is obviously capable of maintaining heterozygosity. This, and the fact that a triploid population exists (probably due to backcrossing with one of the parental species) proves that thelytoky must either be apomictic or depend on premeiotic doubling in this case.

Three species of thelytokous triploids are known in the genus *Poeciliopsis*.

All Mexican, they are of hybrid origin and, exactly like the triploid *Ambystoma* discussed earlier, are reproductive parasites on the corresponding bisexual species, each of which contributed two haploid sets of chromosomes (Schultz 1967, 1969, 1971). Several additional all-female diploid biotypes of hybrid origin of *Poeciliopsis* are known, but these are not truly thelytokous, since their eggs have to be fertilized by sperm from related species and in this case a genuine fertilization occurs. However, the chromosomes brought in by the sperm are not transmitted to the next generation, but eliminated in oogenesis (Schultz, 1961, 1966, 1971; Vrijenhoek and Schultz, 1974). This type of reproduction, which clearly differs from true thelytoky, has been called "hybridogenesis" by Schultz; it apparently also exists in one species of European frog (see p. 333).

According to Uzzell (1970), as things stand at the present there is no counter-evidence against the possibly extreme hypothesis that all the thelytokous species of fishes, amphibia, and reptiles (diploids and triploids) are permanent hybrids, maintained by a premeiotic doubling mechanism in oogenesis.

The situation in the Australian parthenogenetic grasshopper *Warramaba virgo,* which has been studied by White, Cheney, and Key (1963; and see White, 1966; White and Webb, 1966; White and Webb, 1968; White, Webb, and Cheney, 1973; Webb and White, 1975; White, Contreras, Cheney, and Webb, 1977) is similar. It is a diploid, with a peculiar, structurally heterozygous karyotype, which is also heterozygous for a number of late-replicating DNA segments and for numerous C-banding sections. There is a premeiotic doubling of the chromosome number in the oocyte (from $2n = 15$ to $4n = 30$). This is then followed by synapsis, formation of 15 bivalents, and two meiotic divisions that reduce the chromosome number to $2n = 15$ again. Synapsis is known in this case to be absolutely restricted to sister chromosomes (which are molecular copies of one another). The mechanism consequently preserves all types of heterozygosity from one generation to the next and ensures that all the offspring of a particular female shall be genetically identical to one another, and to their mother (except for newly arisen mutations).

It is now clear that *W. virgo* arose, probably some tens of thousands of years ago, by hybridization between two related bisexual species, "P196" and "169," which occur in the same general area of Western Australia (Hewitt, 1975; White, Contreras, Cheney, and Webb, 1977). The first of these has an X_1X_2Y(♂):$X_1X_1X_2X_2$(♀) sex chromosome system, while the second is XY(♂):XX(♀). *W. virgo* thus carries the X of "P169" and the X_1 and X_2 of "P196." White and co-workers have reared "synthetic" *virgo* females by crossing the two parental species. These individuals were themselves capable of thelytokous reproduction, and embryos that developed from their eggs had the same hybrid karyotype as their mother. It is not yet known, however, whether the "synthetic" *virgo* females manifested the premeiotic doubling mechanism in their oocytes spontaneously. If kept virgin, females of the parental species lay eggs, but these have not undergone the premeiotic doubling process and give rise only to haplo-diploid mosaic embryos. Much of the structural and genic heterozygosity of *W. virgo* is clearly due to its hybrid origin, but some of it has been built up by subsequent mutational changes of various kinds, since the karyotype differs from locality to locality in the details of its C-banding pattern and sometimes in other ways.

The examples of thelytoky considered so far have arisen from bisexual ancestors. In earthworms and flatworms (and of course in higher plants), on the other hand, thelytoky has been derived from hermaphroditism. In such cases, instead of loss of the male sex, we are likely to find some degree of atrophy of the male organs in a basically hermaphroditic anatomy.

Numerous parthenogenetic species of Oligochaeta (Lumbricidae and Enchytraeidae) have been studied by Omodeo (1951, 1952, 1953, 1955) and Christensen (1961). Some of these are gynogenetic. Spermatogenesis seems to be generally abnormal and chaotic in the thelytokous oliogochetes. Most thelytokous earthworms (Lumbricidae) have an automictic mechanism that involves a premeiotic doubling of the chromosome number. Since the majority of them are polyploids, very large numbers of bivalents occur in the oocyte nucleus. It is probable, although it has never been proved, that synapsis is restricted to sister chromosomes. In the enchytraeids there is a different type of automictic thelytoky, in which restoration of the original chromosome number is achieved by an abortive second meiotic division. The "lumbricid mechanism" is compatible with odd-numbered polyploidy, and many of the thelytokous earthworms are in fact triploids and pentaploids. The "enchytraeid mechanism," on the other hand, does not permit odd-numbered polyploidy, so that only even-numbered polyploids occur. A few species in both families show apomictic rather than automictic thelytoky.

The taxonomy of these earthworms is not well understood and biochemical evidence is lacking. There seem to be a number of "superspecies," each of which may include a number of different morphological types. Thus, the species *Dendrobaena rubida* includes the morphological types *typica, subrubicunda, tenuis, constricta,* and *norvegica*. There are sexual diploids (2n = 34), tetraploids, hexaploids, and "suboctoploids" (with about 120 chromosomes), but these do not correspond to the morphological types, e.g., *subrubicunda* includes diploid, tetraploid, and hexaploid populations. In addition, there is a triploid thelytokous race with 51 chromosomes that has been recorded only from England; this is a good example of the premeiotic doubling system, and shows 51 bivalents at meiosis. Presumably it arose by interracial hybridization between diploid and tetraploid biotypes. The hexaploid sexual form may have arisen independently in the Arctic (Iceland, Greenland) and in Italy. Another species of the same genus, *E. octaedra,* found in Iceland, Greenland, and the Alps, is entirely apomictic and consists of high polyploids (hexaploids with 2n = 108 and clones with 2n = 99–124).

Allolobophora caliginosa includes a diploid sexual biotype in England and triploid and tetraploid thelytokous biotypes in continental Europe and North Africa; these latter show the premeiotic doubling mechanism. *A. rosea* has triploid, pentaploid, and hexaploid thelytokous populations in various parts of Italy; a tetraploid sexual population occurs at Naples and a decaploid or dodecaploid sexual one with about 200 chromosomes is found at Monte Pollino in Calabria. Other species in which diploids are not known are *Eiseniella tetraedra* and *Octolasium lacteum,* both of which have triploid and tetraploid thelytokous races. *E. tetraedra* is morphologically variable, with the male pores on the thirteenth segment *(E. t. typica)* or the fifteenth *(E. t. hercynia)* and the spermathecae more or less rudimentary.

Omodeo (1954) has argued, on the basis of geographic distribution, that the thelytokous forms of some earthworms, such as *Dendrobaena octaedra* and the triploid biotype of *Allolobophora caliginosa,* date from the Miocene, i.e., as much as 30 million years ago. While this may well be so, the evidence is not really clear and certain polyploid biotypes may well be polyphyletic, having arisen independently in different areas. The species referred to above are all European and, for the most part, widespread; some of them occur also in Greenland *(Dendrobaena rubida* and *E. octaedra)* and North America *(Allolobophora caliginosa).* Omodeo (1953) stresses the fact that the triploid biotype of *A. caliginosa trapezoides,* which is widespread in peninsular Italy, occurs also in Sardinia, and concludes that it must be at least 30–40 million years old. If this conclusion can be trusted, it implies that at least some of these thelytokous biotypes are very old and have shown considerable adaptability to changing conditions. It is interesting to note that the triploid *A. c. trapezoides* shows quite a high coefficient of variation for number of body segments (13.0 to 17.2 in several samples, compared to 7.2 to 11.9 in diploid sexual earthworms, according to Omodeo, 1955). This may indicate some genetic segregation, perhaps due to synapsis of nonsister chromosomes at meiosis. All the forms of thelytoky known in earthworms seem to be heterozygosity-promoting and almost all thelytokous biotypes are polyploid.

The forms of thelytokous reproduction in freshwater planarians have been studied in great detail by Benazzi and Benazzi-Lentati (Benazzi, 1960, 1963, 1967, 1968, 1969; Benazzi, Baguñá, and Ballester, 1970; Benazzi, Pulcinelli, and Del Papa, 1970; Benazzi-Lentati, 1966, 1970). Most of the work has been carried out on members of the *Dugesia gonocephala* species group, whose total area of distribution includes most of Europe and extends eastward to Persia and

Afghanistan and southward to central Africa. *D. gonocephala* itself, from central Europe, is a diploid sexually reproducing species with 2n = 16; so are *D. etrusca* (Tuscany), *D. ilvana* (island of Elba), *D. sicula* (Sicily and Elba), and *D. cretica* (eastern Mediterranean islands and Middle East). However, extra chromosomes are present in some individuals of most of these species—in the female germ line and somatic cells, the male germ line being always diploid.

The *Dugesia benazzii* complex (which also belongs to the *gonocephala* group) is most interesting. Biotype I, which occurs in the Mediterranean islands of Sardinia, Corsica, and Capraia, is diploid and reproduces sexually; it cearly represents the primitive condition. Biotype II, from various localities in Corsica and Sardinia, is basically triploid and reproduces by pseudogamy. In the female germ line the chromosome number is first doubled (from 24 to 48 chromosomes) and then reduced again, by normal meiosis, to 24. The eggs are hence triploid; they are activated by the haploid sperms but the sperm nuclei are rejected and degenerate in the egg. Biotype III, found in one area of Sardinia, is tetraploid and has apomictic thelytoky; reproduction is pseudogamous. A population of *D. benazzii* on the small island of Tavolara, which is basically triploid, reproduces exclusively by vegetative fission. Some peculiar populations from limited areas of Corsica and Sardinia are believed by Benazzi (1968) to have originated by crossing between biotypes I and either II or III. The *D. benazzii* complex probably includes a whole range of genetic systems from normal sexual diploidy with free recombination to triploid and tetraploid thelytokous or vegetative reproduction with limited recombination or none at all. Benazzi-Lentati has carried out crosses between the sexual and thelytokous biotypes of *D. benazzii*. Her main conclusion from the experiments was that all the characters—polyploidy, asynapsis versus synapsis, premeiotic doubling, meiosis, and pseudogamy, which, when correlated, constitute the genetic systems of the natural biotypes—were independently inherited in the hybrids in a complex polygenic manner. Two populations of *D. benazzii* from Corsica showed triploid and hexaploid oocytes in the same individual (Benazzi and Giannini, 1970); they may be the result of natural hybridization between sexual and thelytokous biotypes.

Dugesia lugubris is also a complex of different biotypes. There are four sibling species that are sexual diploids; designated A, F, E, and G by Benazzi, they all differ in karyotype. Form B is a pseudogamous triploid-hexaploid with meiotic thelytoky like biotype II of *D. benazzii,* although some populations are tetraploid-octaploid. Form C is an apomictic triploid (in the female germ line, the male one being diploid). Form D resembles C, but is tetraploid.

The impression one gets from these and other investigations on thelytokous planarians is that thelytoky is almost certainly combined with hybridity, but diploid thelytokous forms, if they ever existed, have become extinct. One may well ask why thelytoky should have been successful in a group with a widespread capacity for simple vegetative reproduction by fission. Perhaps we need to know the true significance of pseudogamy in order to answer that question. Clearly, in pseudogamic reproduction the sperm contributes something to the future individual, even if all its chromosomes are cast out. The thelytokous planarians are probably extensively heterozygous, assuming that synapsis is restricted to sister chromosomes.

The weevils (beetles of the family Curculionidae) include a large number of thelytokous forms in Europe, North America, and Japan. The cytology was reviewed by Suomalainen (1969). According to Takenouchi (1972), 46 different thelytokous forms in Europe, North America, and Japan. The cytology was reviewed by Suomalainen (1969). According to Takenouchi (1972), 46 different evolved in at least 11 genera belonging to the subfamilies Otiorrhynchinae and Brachyderinae. Most of them are wingless, or at any rate flightless, i.e., of restricted vagility. As far as is known, thelytoky in this group is always apomictic with a single mitotic maturation division in the oocytes. The taxonomic treatment of the group has followed conventional lines; in a large number of instances a so-called "species" includes a diploid bisexual "race" and one or more polyploid thelytokous "races." In a few cases, however, only the polyploid thelytokous biotypes exist (e.g., in *Otiorrhynchus subdentatus, Peritelus hirticornis, Barynotus moerens,* and *Catapionus gracilicornis),* the bisexual biotypes being unknown and presumably extinct. The widespread triploid thelytokous *Otiorrhynchus singularis* may be derived from a rare bisexual diploid, *O. carmagnolae,* restricted to a small area on the border of Switzerland and Italy.

In Europe the bisexual forms of these weevils usually occupy quite small areas in the mountain ranges, while the thelytokous biotypes are much more widespread. For example, in the genus *Otiorrhynchus,* with a total of 1,000 species, a number of bisexual forms inhabit small areas of the eastern Alps and Carpathians while related thelytokous "races" are much more widely distributed in central and northern Europe at lower elevations (Figure 30). Apparently, the bisexual forms are relict populations that have survived in *refugia* that were ice-free during the glacial period, while the thelytokous ones have colonized the area to the north of the mountains that was exposed by the retreat of the Würm glaciation. Some bisexual species of very limited distribution (e.g., *O. mulleri,* from the summit of Monte Baldo, and *O. kunnemanni* and *O. decipiens* from Monte Arera in the Italian Alps) never seem to have given rise to thelytokous races. The situation in the genus *Trachyphloeus* is broadly similar to that in *Otiorrhynchus;* there are at least six thelytokous biotypes in Poland, several of which are represented by bisexual forms in southern Europe (Petryszak, 1975).

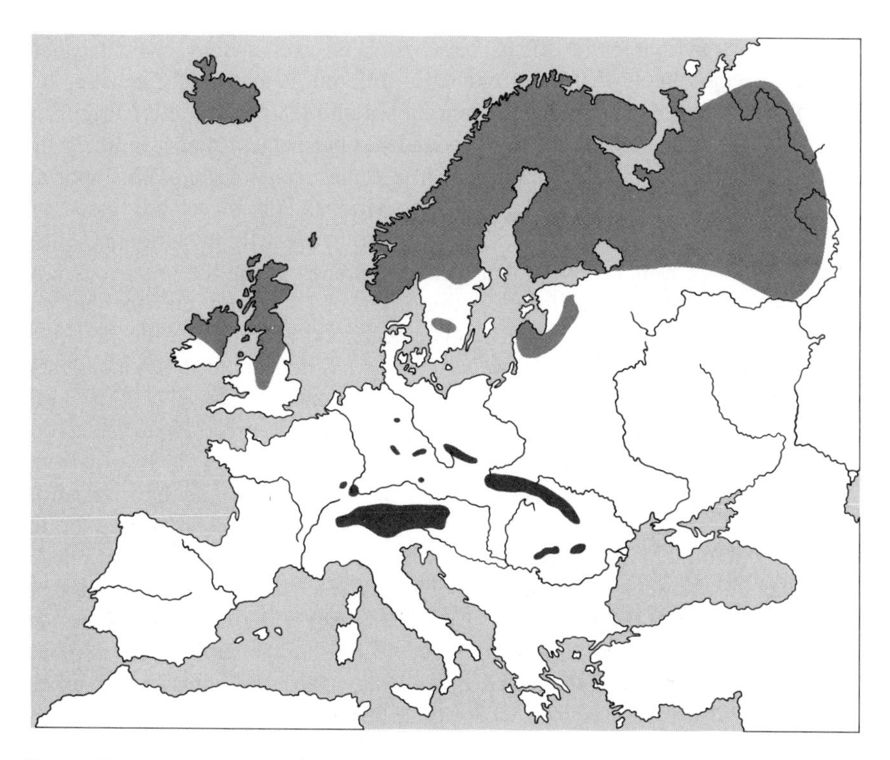

FIGURE 30

Distribution in Europe of the diploid bisexual biotype (black) and the parthenogenetic biotype (grey) of *Otiorrhynchus dubius*. [After Suomalainen, 1970]

In a number of instances the bisexual and thelytokous populations of a "species" of *Otiorrhynchus* are present at the same locality. For example, Suomalainen and Saura (1973) collected equal numbers of diploid bisexual and triploid thelytokous individuals of *O. scaber* at Plesch in southeastern Austria. However, at Lunz in central Austria only the triploid form was found; in Switzerland and Poland both triploids and tetraploids occur (sometimes at the same locality), while in Finland only the tetraploid exists. There is no published information on the precise ecology of these various populations, but the biometry of some of them has been studied by Suomalainen (1961). Significant differences in both size and shape exist between races of different ploidy and between populations of the same ploidy from different localities. In general, the tetraploids are somewhat larger than the triploids; in Poland the triploid and tetraploid biotypes of *O. scaber* can be distinguished by size, shape, convexity of the elytra, and color (Mikulska, 1960; Smreczyński, 1966).

Although environmental effects have not been excluded by critical experiments, it seems likely that these biometric differences are mainly genetic. It is not possible to rule out the occurrence of multiple (polyphyletic) origins of thelytoky in these weevil "species" (actually superspecies), but it is likely that Suomalainen has obtained evidence of true evolutionary changes in important biometric characters in the thelytokous populations. This is certainly incompatible with the view that thelytokous species are necessarily evolutionary "dead ends," although it may well be that their adaptability is limited in various ways not revealed by biometry. However, the origin of polyploidy and thelytoky in the *Otiorrhynchus* weevils is not really understood. That is, we do not know whether in a "species" like *O. scaber,* which has no diploid thelytokous race, apomixis and triploidy appeared simultaneously or consecutively. If the latter was the case, the triploidy probably, but not certainly, arose by hybridization between a now extinct diploid thelytokous race and the ancestral bisexual one. The origin of the tetraploid race is even more uncertain. It may have arisen by doubling the chromosome number in either the diploid bisexual race or the hypothetical diploid thelytokous one (which may never have existed), or by hybridization between the triploid thelytokous race and the diploid bisexual one. Suomalainen (1970) believes that in all cases a diploid thelytokous race preceded the appearance of polyploid races, i.e., that thelytoky and polyploidy occurred consecutively and in that order, and regards the diploid thelytokous "race" of *Polydrosus mollis* as evidence for this view. But even now, with many additional (mostly Japanese) thelytokous weevil biotypes known, only two diploid ones have been found; the others may be all extinct or confined to very small geographic areas, but it is also surely possible that they never existed.

Extremely interesting allozyme studies have been carried out on some of these weevils by Suomalainen and Saura (1973). In the case of *O. scaber* the diploid bisexual race (or at any rate the Plesch population of it) shows a heterozygosity per locus of 30.9 percent, a very high figure. There are technical obstacles to arriving at truly comparable figures for the polyploid races because of difficulties in determining the dosages of the various alleles; but one can say that the heterozygosity of the thelytokous populations is of the same order of magnitude (e.g., the two triploid samples from Plesch and Lunz give estimates of 35.6 and 23.8 percent heterozygous loci per individual, while a tetraploid one from Finland showed 44.4 percent heterozygous loci, if one does not distinguish between different allelic dosages). Clearly, the thelytokous populations are highly heterozygous (although less so than the clones of the lizard *Cnemidophorus tesselatus*—see p. 294), but heterozygosity has not increased in an unlimited manner, as might have been expected on theoretical grounds, since a considerable number of loci are monomorphic.

By comparison with the thelytokous earthworms, which Omodeo believes are 30–40 million years old in some cases, the thelytokous weevils of Europe are regarded by Suomalainen as of relatively recent origin. Certainly their distribution, in the area covered by the Würm ice sheet, suggests an origin no more than 20,000 years ago, and perhaps in some cases only 12,000 years ago. If, in the case of each bisexual species, thelytoky arose only once, a considerable amount of evolutionary change must have occurred under conditions of asexual reproduction to give rise to the existing biometric and allozymic differences between the geographic populations. Suomalainen is of the opinion that the polyploid biotypes arose by hybridization with males of the bisexual forms, usually of the same species, but in some instances of other, related species.

The thelytokous biotypes of the moths of the genus *Solenobia* parallel in many ways those of the *Otiorrhynchus* weevils. They were studied in great detail by Seiler over a period of 45 years. *S. triquetrella* is a complex of three main biotypes: a diploid bisexual, a diploid thelytokous, and a tetraploid thelytokous one; the bisexual populations may have either XY or XO females. All three forms occur in Switzerland, but their geographic distributions are different (Seiler, 1961). The bisexual race occurs in areas that were on the edge of the Würm ice sheet and in areas *(nunataks)* that were ice-free during the Würm glaciation. The diploid thelytokous form occupies, in the main, the area that was the first to be uncovered when the ice sheet retreated, while the Alpine zone, which was the last to become available for colonization, is inhabited solely by the tetraploid thelytokous race. At a few localities two races coexist, and near Chasseron, Switzerland and Linz, Austria all three occur together. Further north, in Germany, Poland, Czechoslovakia, Scandinavia, and England, the tetraploid race is widespread. Seiler believed that the diploid thelytokous race may have arisen about 20,000 years ago and passed through a stage when reproduction was inefficient, due to failure of many of the eggs to develop. As a result of natural selection the reproductive mechanism was gradually stabilized. The tetraploid race arose from the diploid thelytokous one and then, following the retreating ice sheet, spread out over the whole of central and northern Europe.

The thelytoky of *S. triquetrella* is automictic. Meiosis is normal but, as in the oogenesis of other lepidopterans, achiasmatic. The original chromosome number (diploid or tetraploid as the case may be) is restored by fusion of second-division nonsister nuclei; since female *Solenobias* are XO or XY, one of these nuclei carries an X chromosome and the other a Y or no sex chromosome at all, and the XO or XY condition is restored at the nuclear fusion.

The allozymes of *Solenobia* populations have been studied by Lokki and co-workers (1975). The mean heterozygosity per locus of the bisexual race was determined as 23 percent; the corresponding figures for the XY and XO diploid thelytokous populations were 25 and 20 percent, respectively. A number of alleles were found in diploid thelytokous populations that were not encountered in the bisexual race. A total of 22 genotypes (i.e., clones) were found in 116 individuals from 10 localities in Switzerland. A number of tetraploid populations from central Europe and Finland were also studied (Figure 31). Thirteen alleles that had not been encountered in either the bisexual or the diploid thelytokous races were found in these tetraploid populations. In central Europe 15 genotypes were found in 80 individuals from 10 localities, no genotype being found in more than one locality. In Finland there was one common "western" genotype and a common "eastern" one, both found in many localities. In addition, there were a number of deviant genotypes, most of which could be derived from the foregoing by a small number of mutational changes.

The one feature that seems to be common to the thelytokous earthworms, weevils, and *Solenobia* is the absence or limited geographic range of diploid thelytokous biotypes. In general, only the polyploid thelytokous forms seem to have been evolutionary successes. One reason for this is suggested by the work of Suomalainen and his colleagues on the allozymes of *Otiorrhynchus* and *Solenobia*. It seems clear that the heterozygosity of diploid thelytokous populations is, on average, no higher than their bisexual progenitors, presumably because at many loci any allelic substitution is deleterious, i.e., two doses of the original allele are required for a high adaptive value. It is only when more chromosome sets are added, in triploid and tetraploid strains, that the additional alleles become "available" for mutation. In all probability the triploid and tetraploid weevil biotypes are, for the most part, allopolyploid, while the tetraploid *Solenobia triquetrella* is autopolyploid.

The bisexual species that have given rise to thelytokous ones seem to be characterized by high levels of genic heterozygosity (*Otiorrhynchus scaber,* 31 percent; *Solenobia triquetrella,* 23 percent). Some of their close relatives that have not produced thelytokous biotypes have much lower levels (12.5 percent heterozygosity per locus in *Solenobia manni,* according to Lokki *et al.,* 1975).

FIGURE 31

Numbers of individuals of each overall genotype identified in central European tetraploid parthenogenetic populations of *Solenobia triquetrella*. Each vertical column represents a district clone. The enzymes are: adenylate kinase (*Adk-1* and *Adk-2*), esterase (*Est-2*), fumarase (*Fum*), α-glycerophosphate dehydrogenase (α-*Gpdh*), hexokinase (*Hk*), isocitrate dehydrogenase (*Idh-1* and *Idh-2*), malate dehydrogenase (*Mdh-1* and *Mdh-2*), malic enzyme (*Me*), phosphoglucomutase (*Pgm*), superoxide dismutase (*Su-1* and *Su-2*), and triosephosphate isomerase (*Tpi*). One, two, or three alleles at each locus were identified in each individual (dashes indicate alleles not studied). The localities sampled were: Schöfflisdorf (S), Netstal railway station (NR), Cozzo (C), Bellavista (Be), Netstal (N), Uzwil (U), Monte Generoso (M), Bodio (Bo), Gschaid (G), and Herrliberg (H). [After Lokki *et al.,* 1975]

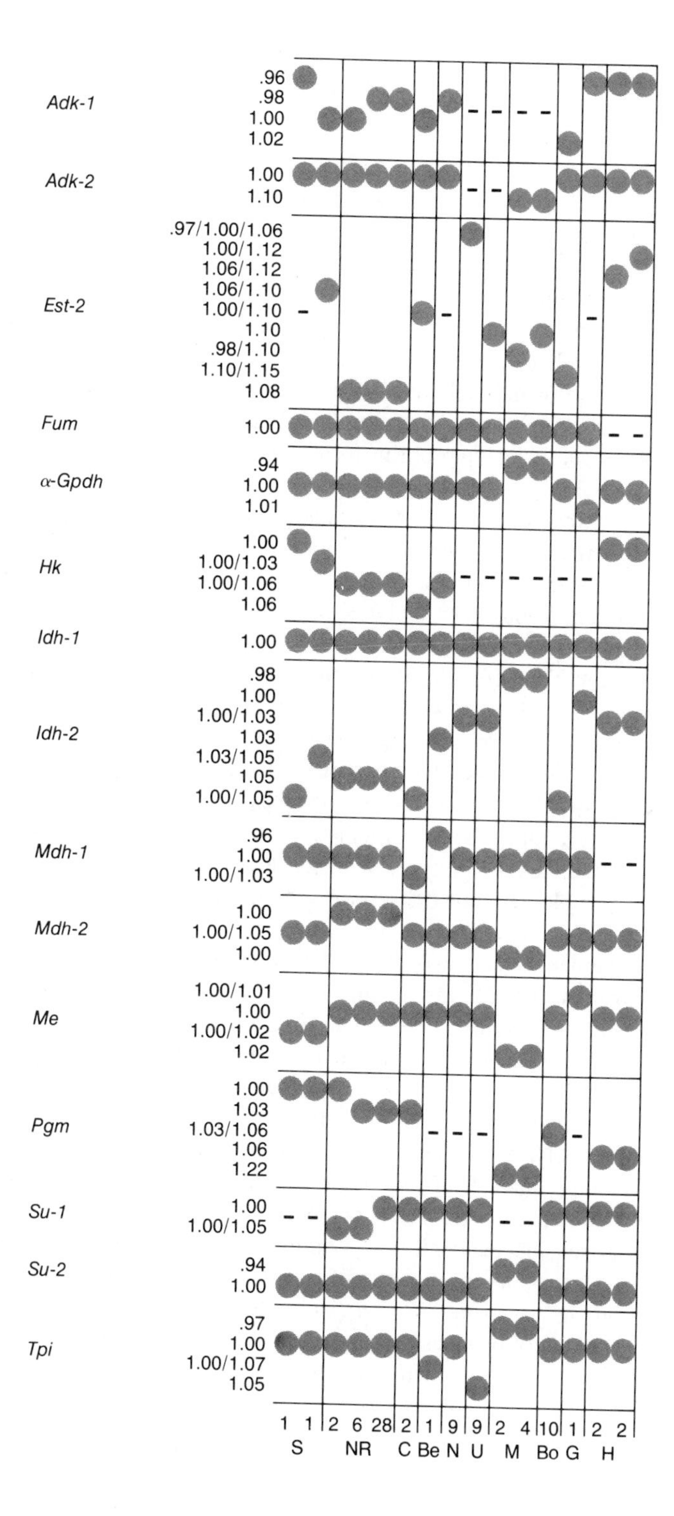

Adk-1
.96
.98
1.00
1.02
Adk-2
1.00
1.10
Est-2
.97/1.00/1.06
1.00/1.12
1.06/1.12
1.06/1.10
1.00/1.10
1.10
.98/1.10
1.10/1.15
1.08
Fum
1.00
α-Gpdh
.94
1.00
1.01
Hk
1.00
1.00/1.03
1.00/1.06
1.06
Idh-1
1.00
Idh-2
.98
1.00
1.00/1.03
1.03
1.03/1.05
1.05
1.00/1.05
Mdh-1
.96
1.00
1.00/1.03
Mdh-2
1.00
1.00/1.05
1.00
Me
1.00/1.01
1.00
1.00/1.02
1.02
Pgm
1.00
1.03
1.03/1.06
1.06
1.22
Su-1
1.00
1.00/1.05
Su-2
.94
1.00
Tpi
.97
1.00
1.00/1.07
1.05
1 1 2 6 28 2 1 9 9 2 4 10 1 2 2
S NR C Be N U M Bo G H

It is tempting to suggest that, as a general rule, sexually reproducing species that have given rise to heterozygosity-promoting thelytokous forms are ones that had already reached high levels of heterozygosity, and were in fact dependent on extensive genic heterozygosity for their continued existence. However, the thelytokous lizard *Cnemidophorus tesselatus,* although highly heterozygous, has arisen by hybridization between bisexual species that themselves show rather average levels (for vertebrates) of heterozygosity (see p. 24). One rather mysterious feature of the thelytokous weevils is a rather considerable variation in chromosome number (see especially Takenouchi, 1969). It is not at all clear whether this is due to polymorphism for chromosomal fusions or dissociations or to true aneuploidy.

The thelytokous cockroach *Pycnoscelus surinamensis* has an almost worldwide distribution in tropical and subtropical countries. It is a complex of at least eight biotypes, with chromosome numbers ranging from 2n = 34 to 54 (Roth and Cohen, 1968). The thelytoky of these insects is apomictic in type, with two maturation divisions in the oocyte probably preceded by an extra DNA replication. They are related to a bisexual species, *P. indicus,* which itself includes chromosomal races with 2n♂ = 33, 35, and 37. The history of this complex of thelytokous forms has not been worked out and can only be guessed at. The different thelytokous biotypes may have arisen, polyphyletically, from different bisexual races of the *indicus* complex. Alternatively, they may have arisen from one another by evolutionary processes that included karyotypic changes. The thelytokous biotypes with the lower chromosome numbers must be essentially diploid, while those with chromosome numbers above 50 are probably basically triploid.

There are numerous parthenogenetic species of black flies (Simuliidae) in the northern parts of the Palaearctic and Nearctic regions, almost all of them triploid (Basrur and Rothfels, 1959; Chubareva, 1968a, b). As in the weevils, diploid thelytokous forms have either not arisen at all or have become extinct. The case of the Canadian *Cnephia mutata* is unusual in several respects. There is a diploid bisexual race that is not polymorphic for chromosomal inversions and a triploid thelytokous one that shows extensive inversion heterozygosity. Apparently, a large number of genotypes exist in the natural populations, carrying different combinations of inversion sequences. Either the triploid *Cnephia mutata* has arisen independently on a number of different occasions from the bisexual ancestor, or its mechanism of parthenogenesis is such as to lead to segregation in the offspring of a particular female (White 1970b). Unfortunately, the cytology of oogenesis has not been studied. The simuliids are apparently yet another group in which a number of triploid thelytokous forms occur but in which diploid thelytokous biotypes are unknown. The situation is, however, somewhat different in other families of Diptera. The chironomid midge

Lundstroemia parthenogenetica is an apomictic triploid that is heterozygous for two inversions and a deletion of a chromosome segment (Porter, 1971). Other apomictic chironomids, some diploid others triploid, have been studied by Scholl (1956, 1960), who found most of them to be extensively heterozygous for inversions.

The thelytoky of the diploid fly *Lonchoptera dubia* in the United States is interesting on account of its four clones, which differ with respect to the inversions for which they are heterozygous (Stalker, 1956). In some localities three or even four clones coexist, while in others only one or two occur. There are four other species of the genus in North America, all bisexual. The maturation of the oocytes of *L. dubia* is meiotic, with fusion of two of the four postmeiotic nuclei occurring in such a manner as to preserve the inversion heterozygosity.

Thelytoky in species of *Drosophila* can be studied in the laboratory in a much deeper manner than is possible in most other organisms. However, out of almost 1,500 species of the genus now known, only one, the neotropical *D. mangabeirai,* is naturally thelytokous. This is a rare species of the *willistoni* group, whose range extends from Central America southward to Brazil (Carson, Wheeler, and Heed, 1957; Murdy and Carson, 1959; Carson, 1962). At least in the northern part of its distribution, all individuals are heterozygous for the same three inversions in chromosome 2. This is apparently a diploid automictic species in which meiosis is "compensated for" by fusion of nonsister nuclei following the second meiotic division. A few males were found in nature, but they are apparently sterile and nonfunctional; over 4,000 females were reared in the laboratory without the appearance of any males. The extreme rarity of this species, in spite of its wide distribution, suggests that it is biologically unsuccessful.

Drosophila parthenogenetica, in spite of its name, reproduces bisexually in nature. Stalker (1954, 1956) reared thelytokous laboratory stocks for 80 generations; after 17 generations 1.56 percent of the eggs underwent successful development. In addition to diploid females, triploid females and sterile XO males were produced. The maturation divisions in the eggs were meiotic in type, and were followed by fusion of second-division sister or nonsister nuclei; fusion of three of the products of meiosis gives rise to the triploid females.

Similar thelytokous laboratory stocks of *Drosophila mercatorum* were obtained by Carson (1967, 1973c). The initial rate of successful parthenogenetic reproduction was extremely low—about one adult female per 1,000 unfertilized eggs in strains from El Salvador (Carson, 1967) and 0.04 per 1,000 eggs in Hawaiian strains (Templeton, Carson, and Sing, 1976). By artificial selection, using alternating cycles of bisexual and parthenogenetic reproduction, it was possible to increase the rate of successful thelytokous development as much as a thousandfold. Thus, natural populations of *D. mercatorum* have been proved to contain alleles that favor thelytokous reproduction. Such populations show high levels of heterozygosity; individuals from the Hawaiian localities had about 18 percent of their allozyme loci heterozygous and the populations were polymorphic for up to 50 percent of their loci.

Thelytoky in the all-female laboratory stocks is of course automictic. The predominant mode of restoring diploidy is through duplication of haploid egg nuclei, but sometimes diploidization is through fusion of postmeiotic nuclei (Carson, Wei, and Niederkorn, 1969). However, the prevalence of the first method of diploidization leads to the thelytokous laboratory stocks being completely homozygous for all loci.

Natural thelytoky is known to occur in a large number of invertebrate groups, such as mites and ticks (Oliver, 1971) and crustaceans (Vandel, 1928). In all these the number of thelytokous species is very small by comparison with the sexually reproducing ones, and they occur sporadically. With a single exception (see p. 319) no genus of any size, let alone a family or higher taxonomic group, is composed entirely of thelytokous biotypes.

The factors that determine whether thelytokous species occur in a particular group of organisms or not are very obscure. Why should there be about 25 species of thelytokous lizards, in at least five different families, while there are no thelytokous species of birds or mammals? And why should there be no thelytokous species of spiders, when there are many thelytokous mites and ticks? As far as the human species is concerned, there is now considerable evidence (Linder and Power, 1970) that a form of diploid parthenogenesis occurs rather frequently, as a pathological phenomenon, but leads only to histologically disorganized dermoid ovarian cysts rather than to viable embryos.

When we consider the successful origins of thelytoky in relation to the habitats occupied by all-female populations, several regularities become apparent. On the one hand, a rather large number of thelytokous forms seem to have successfully colonized the area uncovered by the retreating Würm ice sheet (pp. 202–203). These include biotypes that are apomictic and others that are automictic, but in all cases they seem to be highly heterozygous and usually polyploid.

On the other hand, the isolation of small populations in spatially restricted habitats seems to be strongly conducive to adoption of the thelytokous mode of reproduction. Thus, of eight species of the insect order Orthoptera that have abandoned sexuality for thelytoky, three are cave-inhabiting and one (a cricket) lives as a commensal in ant nests (Baccetti, 1960; Baccetti and Capra, 1969; Lamb, 1975; Lamb and Willey, 1975). In the case of the three species of cave crickets (one in Italy, two in North America) it is clear that the origin of thelytoky has been fairly recent, since bisexual populations assigned to the same species occur in other caves in the same general area. The thelytokous populations of cave crickets are all diploid, but the type of oogenesis has not been determined. Of the three thelytokous species of long-horned grasshoppers (Tettigoniidae), two *(Poecilimon intermedius* in the U.S.S.R. and *Xiphidiopsis lita* on a number of Pacific islands) seem to be successful in the habitats they occupy; their cytology is unknown. The third, *Sago pedo,* is an apomictic tetraploid (Matthey, 1941). Although widespread in the countries north of the Mediterranean, it appears to be on the verge of extinction, surviving precariously in small colonies from Portugal and Sicily eastward to Siberia. It probably has a peculiar life cycle, extending over a number of years; there are several bisexual species of the genus in the eastern Mediterranean, but it is not known whether they have the same type of extended life cycle.

The idea that thelytokous organisms vary in some kind of disorganized anarchic manner, and hence cannot be regarded as constituting distinct biological species, has frequently been expressed. In all those (animal) cases that have been carefully studied, this is demonstrably incorrect. The various thelytokous lizards considered earlier (pp. 291–296) are either single discrete taxa or (in the cases of *Cnemidophorus tesselatus* and *C. lemniscatus)* complexes of a small number of biotypes. Such well-studied thelytokous insect forms as *Warramaba virgo, Saga pedo,* and *Drosophila mangabeirai* are individual taxa, easy to define. They may show geographic variation, but no more than many bisexual species. On the other hand, some thelytokous earthworms and the cockroach *Pycnoscelus surinamensis* (p. 308) include a large number of clones, differing in karyotype, which may have arisen polyphyletically from bisexual ancestors, or in some instances by hybridization between already existing thelytokous biotypes and a bisexual ancestor.

The evolutionary significance of asexual reproduction in higher plants is somewhat different in kind. Although many coelenterates, flatworms, oligochetes and other invertebrates exhibit purely vegetative reproduction by fission, either regularly or occasionally, this form of asexual multiplication is lacking in most of the larger phyla of the animal kingdom. In the plant kingdom, on the other hand, reproduction by suckers, runners, creeping rhizomes, bulbils, and other like means is extremely widespread and may be the main or exclusive method of reproduction in some species.

Some botanists (e.g., Stebbins, 1950) group all forms of vegetative reproduction, together with the various types of parthenogenesis, under the term "apomixis," a word that has been used in this book in a much more restricted sense. Gustafsson (1946–1947), Stebbins (1950), and Grant (1971) distinguish between purely vegetative reproduction, in which seeds are not produced, and *agamospermy,* which includes all forms of nonsexual reproduction in which seeds are nevertheless formed. Agamospermy is further subdivided into *adventitious embryony,* a form of reproduction in which no gametophyte generation occurs, the daughter sporophytes arising directly from somatic cells of the mother sporophyte; and *gametophytic apomixis,* in which the alteration of sporophyte and gametophyte generations is retained. A great variety of different types of gametophytic apomixis exist, and are discussed in detail in the works of Gustafsson and Stebbins cited above, and in the more recent ones of Nygren (1954) and Fryxell (1957).

It will be evident that, from the standpoint of population genetics and evolution, there is no fundamental distinction between vegetative reproduction by bulbils or runners and adventitious embryony, in which seeds are produced. Both are absolutely asexual in the sense that even occasional hybridization with sexually reproducing forms is excluded. On the other hand, such hybridization is at least potentially possible in gametophytic apomixis (which corresponds in a general way to thelytoky in animals).

Adventitious embryony, known in a few Rutaceae, Euphorbiaceae, Ochnaceae, Myrtaceae, Buxaceae, Cactaceae, Liliaceae, and Orchidaceae (details in Gustafsson, 1946–67), generally occurs in single species rather than in species groups. Such forms are not capable of giving rise to the "agamic complexes" that frequently arise as a result of hybridization between sexually reproducing forms and gametophytic apomicts. From an evolutionary standpoint, *vivipary* (the asexual formation of propagules other than seeds within the inflorescences) is yet another type of vegetative reproduction. All these reproductive mechanisms are equivalent to "fission," as it occurs for example in many planarians (Benazzi, 1974) and some oligochetes (Bell, 1959). Classic examples in plants are *Euphorbia dulcis* (Carano, 1926), *Ochna serrulata* (Chiarugi and Francini, 1930), several species of *Eugenia* (van der Pijl, 1934), and the European orchid *Nigritella nigra* (Afzelius, 1932). These types of reproduction do not seem to occur in Compositae, a family in which gametophytic apomixis is widespread. Most of the species reproducing by vivipary or adventitious embryony include a number of biotypes differing in chromosome number, but it is not easy to determine to what extent this is due to polyploidy, genuine aneuploidy (trisomy, monosomy, etc.), or to chromosomal rearrangements involving fusions or dissociations.

Botanical writers make a distinction on embryological grounds between two types of gametophytic apomixis, generally called *displospory* and *apospory.* In the former the embryo sac arises from an archesporial cell, after a more or less modified meiosis that does not involve a numerical reduction, while in the latter it is formed from a cell of the nucellus or inner integument, as a result of somatic mitoses. Apospory is hence always ameiotic, whereas in diplospory some aspects of meiosis may be retained. But in contrast to the situation in thelytokous species of animals—where the maturation divisions in the oocytes, whatever their precise type, are uniform and regular in each particular species or biotype—the situation in plants is frequently irregular, with several different types of cytological behavior coexisting in the same species. Thus, Bergman (1941) found that bivalents were frequently, but not always formed in the apomictic microspecies of the composite genus *Hieracium*. In the embryo mother cells of *H. rigidum* the divisions were always mitotic, but in those of *H. caesium* 10 percent were meiotic, while in *H. scandinaviorum* 56 percent were meiotic. In such cases restitution normally occurs by fusion of postmeiotic nuclei, thus permitting and promoting the perpetuation of heterozygosity. In *Rubus,* according to Thomas (1940), a complete meiosis sometimes leads to haploid embryo sacs that later undergo diploidization; as in analogous cases in animals, this must lead to completely homozygous progeny. Pseudogamy also occurs in some apomictic plants, where a nucleus derived from the sperm is required for the formation of the endosperm, althought no fertilization of the ovum occurs.

Although the cytological picture is highly variable and even chaotic in apomictic plants (by comparison with the regularity generally found in thelytokous animal species), its genetic consequences are probably in almost all cases heterozygosity-promoting and polyploidy-permitting. The cytological details may hence be largely irrelevant to the evolutionary consequences of apomixis in plants.

Much has been written about the vast agamic complexes that have developed in certain plant genera as a result of the combination of apomixis, hybridization, and polyploidy. It is therefore important to point out that only a relatively small number of such complexes are known, most of which inhabit the temperate and arctic zones of the northern hemisphere.

One of the best understood of these complexes occurs in the genus *Crepis* (Compositae) in western North America (Babcock and Stebbins, 1938; Stebbins and Babcock, 1939; Stebbins, 1950). This complex includes six diploid sexual species ($2n = 22$) of very restricted distribution in California, Nevada, Oregon, Washington, and British Columbia, and a seventh *(C. acuminata)* with a more widespread distribution extending across the Great Basin to Wyoming and Colorado. These diploid species do not coexist at the same localities and hence do not hybridize directly.

The apomicts derived from these bisexual species are far more widespread; they form a vast complex of polyploid forms that includes triploids, tetraploids, pentaploids, heptaploids, and octoploids. Some are facultative apomicts, i.e., they retain some capacity for sexual reproduction. Some of these biotypes closely resemble one or another of the sexual diploid species, while others are intermediate in morphology and ecological and climatic tolerance between two or more of the sexual forms. As Stebbins (1950) puts it: "The diploids can be thought of as seven pillars which by themselves are sharply distinct from each other. . . .They are connected by a much larger superstructure of polyploid apomicts which represent all sorts of intergradations and recombinations of their characteristics." The *Crepis* agamic complex is undoubtedly relatively recent, since the sexual species from which it was formed are all still extant.

The other main northern hemisphere agamic complexes that have been studied are in the genera *Rubus, Poa, Antennaria, Taraxacum,* and *Hieracium.* The three latter are all members of the family Compositae, which also includes agamic complexes in the genera *Chondrilla, Youngia, Ixeris,* and *Parthenium. Parthenium* is confined to the arid southwestern United States and consists of facultatively apomictic polyploids related to only two ancestral sexual species, *P. argentatum* (guayule) and *P. incanum* (Rollins, 1944, 1945, 1946).

The agamic complex in *Taraxacum* and two complexes in different subgenera of *Hieracium* are huge, both from the standpoint of the number of existing biotypes and their vast distribution. The *Taraxacum* complex is based on at least 12 original species. No really comprehensive study of these complexes has ever been undertaken, but Stebbins (1950) regards them as similar in character to the one in *Crepis,* only much larger.

The agamic complexes in *Rubus* and *Potentilla* are composed mostly of facultative apomicts. In both cases hybridization has clearly occurred on a large scale and some of the ancestral sexual species are probably extinct. In *Rubus* the situation is also complicated by the existence of some polyploid sexual species.

The grass genus *Poa* appears to contain about seven different agamic complexes, but Stebbins (1950) suggests that these are all parts of one vast and almost worldwide agamic complex, which would be by far the largest in the plant kingdom. As in *Rubus,* many of the ancestral sexual species no longer exist.

Clearly, the main evolutionary significance of these agamic complexes in plants is that they have permitted hybridization to occur to a far greater extent than in the case even of polyploid sexual complexes. An enormous range of microhabitats can hence be colonized by apomictic biotypes representing a great number of different combinations of genetic material from several or many ancestral species. But a really detailed study of the genetic architecture of such complexes, now possible through modern biochemical techniques, is a task for the future. The potentiality for further adaptive evolution in agamic complexes would seem to exist so long as some sexual relatives persist or, alternatively, if the apomixis is only facultative. Facultative apomixis may involve a combination of sexual and apomictic forms of reproduction in the same individual plants or even in the same inflorescence. For example, the hawkweed *Hieracium aurantiacum* in the Polish Tatra mountains has both sexual and apomictic ovules within the same flowering heads (Skalinska, 1967); the ratio of sexual to asexual achenes varies from plant to plant, being no doubt genetically controlled.

There seem to be few or no apomictic species of plants comparable to the grasshopper *Warramaba virgo,* the mantid *Brunneria borealis* (White, 1948), or the flies *Lonchoptera dubia* and *Drosophila mangabeirai,* discussed earlier. These are discrete diploid species that are quite distinct from, and do not hybridize with, their bisexual relatives. A possible example among plants might be *Houttuynia cordata* (Saururaceae), referred to by Stebbins (1960) as "an isolated apomictic species without living sexual relatives" and perhaps "the last relict of an ancient agamic complex." It is, however, a complex of clones with chromosome numbers ranging from 2n = 52 to 104 (details in Gustafsson, 1946–47). Most agamic complexes of plants consist of hundreds of microspecies, differing in morphology and often in chromosome number. Within each microspecies the individual plants are extremely alike, both morphologically and physiologically. No doubt some of the agamic complexes referred to above began to develop in Pliocene times; some individual microspecies are at least 1,000 years old and may be much older. Christiansen (1942) found that some Icelandic biotypes of apomictic *Taraxacum* introduced into Greenland by human agency about 1,000 years ago still resembled the Icelandic plants in every detail of appearance.

The evolutionary significance of asexual forms of reproduction may be considered at several different levels. In recent years there has been a revival of interest in the general question, What biological use is sex? The mathematical consideration of this question by Maynard Smith (1971) has been followed by a stimulating book by Williams (1975) and an attempt, by means of computer simulation, to determine whether sex (recombination) accelerates adaptation and evolution (Thompson, 1976).

Clearly, there are several different issues involved here. One is the question as to why sex first established itself in prokaryotes and the earliest eukaryotes. Another is why the great majority of modern eukaryotes retain sexual reproduction—only about one in a thousand genetic systems are asexual.

The main argument tending to establish the biological superiority of asexual reproduction may be called the demographic one; it is independent of any genetical considerations. If males and females consume equal amounts of all resources, a parthenogenetic population will be able to maintain twice as many females on a given area, and each of these will be able to regenerate a population in the event of local extinction or migration to new areas. If a sexual population is reduced by a catastrophe to a very small number of individuals the distances between them may be too great for mating to occur. Thus, for sexual species there is a minimum density necessary for survival, while this is not the case for thelytokous species.

The classic argument for the long-term advantages of sexual reproduction was put forward by Fisher (1930) and Muller (1932). Briefly, it is that if two favorable mutations, $a \rightarrow A$ and $b \rightarrow B$, occur in different individuals of a population they can only be incorporated in a single individual if reproduction is sexual. If reproduction is clonal they must occur sequentially, in the same lineage, and in most models this is not very likely. Crow and Kimura (1965) formulated this argument mathematically and showed that sexual reproduction is genetically advantageous where the total rate of occurrence of favorable mutations in the population is large and the selective advantage per locus is small. Maynard Smith (1971) calculated that sexual reproduction will accelerate evolution if $Nu/2 > \ln 20\ Ns$, where N is the population size, u the mutation rate per locus, and s the selective advantage per locus. In practice, this means that sexual reproduction gives a higher rate of evolution than clonal reproduction for populations of over about 10^6, the advantage being approximately fivefold over a wide range of population size. This calculation assumes that mutations are recurrent, i.e., not unique events. If a considerable number of them are chromosomal rearrangements, each of which is unique, the advantages of sexual reproduction would be greater; in view of modern work on the evolution of the blocks of highly repetitive satellite DNA, this may be an important point. The essential correctness of the Fisher–Muller argument was verified by Felsenstein (1974), using computer simulation.

Williams (1975) has put forward the view that sexual reproduction involves a 50 percent "cost of meiosis." The argument (essentially a more sophisticated version of the simple demographic one) is based on a model population containing two kinds of females: genotype A_1A_2 (which is thelytokous and produces A_1A_2 daughters) and genotype A_3A_4 (which is sexual and produces equal numbers of A_3 and A_4 gametes). The sexual offspring will consist of A_3A_m and A_4A_m individuals in equal numbers (where A_m is the A allele from the sperm). Comparing the parental generation with the offspring, Williams claims that the genes A_3 and A_4 for sexual reproduction will have been halved in frequency. The basic fallacy in this argument seems to be that alleles of the A_m type will necessarily also be ones for sexual reproduction, i.e., identical to or equivalent to A_3 and A_4.

A full discussion of these questions would be beyond the scope of this book, but we are here concerned with the problems of what causes thelytokous biotypes and species to arise, why they do not arise more often, and what their adaptive potentialities are. In general, it seems that the old Fisher–Muller argument, namely, that sex is highly advantageous for long-term adaptation, is still valid, although part of the advantage may depend on group selection. As Maynard Smith (1974) says: "it [the advantage of sex] relies on selection acting between groups or species and not between individuals. It is the species which evolves or goes extinct, not the individual. To the extent that species go extinct much less frequently than individuals die, group selection is less effective than individual selection." Thus, thelytokous populations are blind alleys in evolution in the long term, but may display considerable adaptive capacity and ability to diversify genetically in relatively brief periods, as the evidence of Suomalainen and his colleagues on weevils and *Solenobia* moths shows. The great majority of thelytokous biotypes seem to be of hybrid origin. There are, basically, only three ways in which organisms can maintain a permanently hybrid constitution without an associated heavy genetic load of inferior homozygotes (as distinct from a high level of heterozygosity associated with genetic polymorphism). The first, structural heterozygosity, is confined to the few species of plants (mostly members of Onagraceae), and perhaps a few scorpions and cockroaches (whose genetic systems are not properly understood), that are multiple translocation heterozygotes. The second is allopolyploidy. The third includes those types of thelytoky that are heterozygosity-generating, especially where these are initiated by interspecific or interracial hybridization.

These three types of genetic systems are not simply equivalent; the first two allow some recombination, which thelytoky does not. Adaptation to new environments in thelytokous species is thus extremely dependent on mutation. Clearly, some thelytokous species, such as *Warramaba virgo,* have been in existence long enough to permit a good deal of diversification by mutation; and the fact that a fairly large number of structural chromosomal changes have occurred in *W. virgo* since the origin of thelytoky implies that a large number of allelic substitutions have also taken place in this species. But there is very little evidence that diversification and divergence of thelytokous clones by mutation is likely to proceed to the point where they would be regarded as different species by a museum taxonomist guided solely by phenetic criteria. In all cases of thelytokous clones as different as that, it is probable that they have arisen independently from their bisexual ancestors, whether by different acts of hybridization or otherwise. This indicates clearly the limitation of thelytokous reproduction; its adaptive potentiality is restricted to an extent that renders really long-range progressive evolutionary change out of the question.

It is assumed by Williams (1975) in his model, referred to earlier, that the difference between asexual and sexual reproduction may depend on a single gene. This may be so in prokaryotes and unicellular eukaryotes but in all higher organisms the situation is likely to be more complicated. In view of the fact that a capacity for rare, accidental parthenogenesis (so-called tychoparthenogenesis) is extremely widespread, and that workers like Carson and Stalker have succeeded by artificial selection in greatly increasing the rate of thelytokous reproduction in such normally bisexual species as *Drosophila mercatorum* and *D. parthenogenetica,* many evolutionists have assumed that the natural origin of thelytokous biotypes has followed the same course, i.e., that natural selection gradually increases the number of alleles, at a number of loci, favoring a mechanism such as fusion of meiotic products or endoreduplication in cleavage nuclei. While this may be so in some cases, there are certain types of thelytokous mechanisms that seem to have an all-or-none type and can only have arisen by some kind of macromutation that radically alters essential cytogenetic processes. In animals, apomixis, which involves the complete suppression of the whole sequence of events involved in meiosis and their replacement by a simple mitosis, is one such system and so are mechanisms that include an extra premeiotic chromosomal replication. No intermediate stages between normal meiosis and a mitotic type of maturation divisions in the oocyte are easily conceivable, and coexistence of two kinds of oocytes in the same individual, one kind undergoing meiosis, the other kind apomictic, seems generally improbable in animals, although it undoubtedly does occur in some facultatively apomictic plants. Likewise, an extra premeiotic replication, which must be genetically determined, is an event that either occurs or does not—an intermediate condition is impossible.

Considerations such as these have suggested that some types of thelytoky must have arisen by single macromutations affecting the maturation divisions. Even so, we must expect that the genetic control of the ability of the egg to initiate development without having been fertilized will be separate from that of the maturation divisions. Thus, the switch from sexual to thelytokous reproduction is likely to be in all cases a process involving allelic changes at a number of different loci. In part, this is the answer to Williams's paradox. Even if a particular population could adapt more efficiently if it abandoned sexuality for thelytoky, there may be rather strong selection against imperfect, intermediate genetic systems.

In the case of thelytokous species with a hybrid origin, the problem is to explain how genetically determined changes affecting the behavior of the chromosomes in the oocyte will coincide or follow rapidly on the act of hybridization. Particularly, in those cases where a hybrid origin is accompanied by a mechanism involving an extra premeiotic replication, we must ask whether it is necessary to postulate the coincidence of two rare events—a hybridization and a cytological macromutation. Or may it be that in some cases the hybrid spontaneously acquires the extra replication cycle, which is not present in either parental form?

At the present stage, it would seem more profitable to examine the genetic architecture of thelytokous species and the circumstances under which they have arisen than to engage in further model-building on the adaptive advantages or disadvantages of sexual reproduction. Nevertheless, models can be very useful in suggesting what to look for. For example, it is possible that if, as suggested by Maynard Smith (1971), the advantages of sexual reproduction disappear below a certain population size (estimated by him at 10^6 individuals), then thelytoky should frequently have arisen in species populations that were small and isolated. A possible example of this would be the three thelytokous biotypes in cave crickets, referred to earlier, although no estimates of the size of their populations or those of their bisexual ancestors have been published. The origin of thelytokous weevils and *Solenobia* moths in the Alpine region during the Pleistocene glaciation may also have occurred in very small populations.

The long-term evolutionary limitations of thelytoky are most clear in animals, where (with a single exception) there is no single large genus or higher taxonomic group in which all biotypes reproduce asexually. The exception, which is poorly understood but undoubtedly genuine, is the bdelloid rotifers, an order that includes about 200 described species, classified in four families and about 19 genera. According to Hsu (1956a, b) the thelytoky of two species in different genera is apomictic; there are two maturation divisions that are mitotic in type, with no chromosomal synapsis. The karyotypes do not seem to consist of pairs of similar homologs; both species showed 13 chromosomes, including several that were unique in size. It is thus possible that these rotifers are cytologically haploid, as the males of non-bdelloid rotifers undoubtedly are. However, in one species of non-bdelloid rotifer that shows an alternation of parthenogenetic and sexual generations, i.e., one that periodically gives rise to haploid males as well as diploid females, selection experiments on exclusively parthenogenetic clones (supposedly with apomictic maturation divisions) led to metrical changes in phenotype (Badino and Robotti, 1975). It is thus possible that in rotifers in general (including Bdelloidea) some kind of nonmeiotic genetic recombination can occur, possibly according to the scheme proposed by Cognetti (1961; and see Cognetti and Pagliai, 1963) for aphids, which also typically have an alternation of sexual and parthenogenetic generations.

The genetic systems of normally self-fertilizing plant species, even though sexual, will bear some resemblance to those of asexually reproducing species that undergo diploidization, following meiosis, by doubling the chromosome number during the cleavage divisions of the embryo. Populations of such species will consist of essentially homozygous individuals but may include a large number of different genotypes. They may consequently show as much variability as those of outcrossing species (Kannenberg and Allard, 1967, for the grass *Festuca microstachys*). This type of population structure suggests that the individual genotypes are subject to frequency-dependent selection in nature.

Thelytokous species of hybrid origin (and these are probably the majority of all thelytokous species) are likely to be ones that are strongly dependent on heterosis and epistasis. Kimura and Ohta (1971b) have emphasized that recombination is disadvantageous under such circumstances. More relevant, perhaps, is the fact that it is only through hybrid thelytoky that very high levels of heterosis can be attained in the first place.

Clearly, there are some general differences between asexual genetic systems in the animal and plant kingdoms. The plant systems seem to have been much more liable to evolve facultative apomixis or coexistence of more than one reproductive mechanism in the same individuals. And the vast complexes of apomictic microspecies that occur in several families of higher plants have no real equivalents in the animal kingdom, although smaller complexes may exist in some earthworms (see p. 299), in a few flatworms such as *Dugesia benazzii,* and probably also in the brine shrimp *Artemia salina.* A number of diploid and polyploid thelytokous biotypes of *A. salina*—some apomictic, others probably automictic—coexist in waters of high salinity (from the Black Sea and the Mediterranean to the Great Salt Lake and the Gulf of California) with two sexually reproducing sibling species from which they must be derived (Artom, 1931; Goldschmidt, 1952; Barigozzi, 1957, 1974; Stefani, 1960; Halfer-Cerini *et al.,* 1968).

There is little agreement among recent writers as to the range of adaptation of asexually reproducing biotypes. Williams (1975) is impressed by their ecological adaptability. However, some of his evidence, based on plant genera such as *Taraxacum* and *Hieracium,* is suspect because it does not relate to individual biotypes but to whole complexes of microspecies, which originated independently and polyphyletically from bisexual ancestors. To state that dandelions grow luxuriantly in both Florida and Iceland is misleading if the microspecies are different. It is true that the tetraploid thelytokous biotype of *Solenobia triquetrella* has a great range of distribution in Northern Europe, compared with its bisexual ancestor. But if it colonized the land left bare by the retreating Würm ice sheet, this may have been a very uniform environment at the time. If we consider individual thelytokous biotypes rather than vast complexes, which may be wholly or partly polyphyletic, it seems likely that they are adaptively limited. The ranges occupied by the single biotypes of thelytokous *Cnemidophorus* lizards and of the thelytokous forms of the *Lacerta saxicola* group in the Caucasus are all fairly restricted. The grasshopper *Warramaba virgo* must once have occupied a very large territory in Australia, but it became extinct throughout the middle part of its range, an area about 1,600 kilometers wide. It shows deviant karyotypes at a number of localities, which must have arisen by chromosomal rearrangements from a "standard" ancestral karyotype, but there is no evidence that these are adaptive to their habitats where they occur, and they may owe their presence simply to drift or founder effects. The extinction of *W. virgo* over the central part of its range was probably due to its inability to adapt to Pleistocene or post-Pleistocene climatic changes. The species feeds exclusively on shrubs of the genus *Acacia,* but has been able to adapt to living on at least five main species of this genus and can also exist on a species of the related genus *Cassia*.

In theory, genetic systems including both thelytokous and sexual reproduction might be expected to combine the evolutionary advantages of each system. One combination is facultative apomixis, as it exists for example in *Hieracium aurantiacum* in Poland (see p. 314). Another is the alternation of thelytokous and sexual generations as shown in various ways by many aphids, gall wasps (Cynipidae), *Daphnia* (Crustacea), rotifers, and digenetic trematodes. This latter system has been called the aphid–rotifer model by Williams and Mitton (1973).

The only deep study of the genetics of a species reproducing by cyclical parthenogenesis is that of Hebert, Ward, and Gibson (1972; and see Hebert, 1974a, b, c) on the cladoceran crustracean *Daphnia magna* in England. Like many Cladoceran species, this one has a very wide distribution in Eurasia, North America, and southern Africa, with minimal geographic variation being evident. Natural populations consisting of apomictic females inhabit ponds and pools of fresh water. Under special circumstances of crowding and photoperiod, males are produced parthenogenetically. When sexual reproduction occurs the females lay two eggs in a resistant structure, the ephippium, which is capable of being transported passively (on the feet of water birds or in other ways) to new locations.

Hebert and co-workers showed that gene flow between populations of different ponds was on a very small scale. Large-scale differences in allelic frequencies were common between populations separated by a few meters and allelic arrays were often different in demes a few hundred meters apart. The data were interpreted as showing that founder effects (occasional transport of ephippia from pond to pond) had played a major role in determining the allelic composition of individual populations, although selective forces are responsible for the maintenance of variability. There were marked temporal fluctuations in the genotypic frequencies in the populations of "permanent" ponds and frequent departures from Hardy–Weinberg ratios, with heterozygotes generally in excess (an indication of the importance of heterosis). The overall level of polymorphism was rather low (populations being polymorphic, on the average, for 13 percent of their enzyme loci), as a result of inbreeding. Hebert suggests that the fragmentation of the species population into a large number of temporarily isolated demes has reduced the efficiency of genetic response to selection gradients, which explains the lack of geographic differentiation of *Daphnia magna* and other species having the same type of population structure.

Speciation in cladocerans has been discussed by Brooks (1957). He recognized 13 species of *Daphnia* in North America, but the taxonomy is much complicated by nongenetic variation. Brooks believed that much introgressive hybridization is occurring between the species; it would seem desirable that the material should be reinvestigated, using modern biochemical techniques.

Unlike pure thelytoky, which occurs sporadically in groups of sexually reproducing species, cyclical parthenogenesis characterizes whole groups of animals. It has been a long-term evolutionary success in aphids, cynipid wasps, cladocerans, and rotifers, and has permitted speciation to occur on a large scale. In most of these groups a few species have secondarily lost the sexual part of the cycle (e.g., the so-called "anholocyclic" aphid species) and become entirely thelytokous. The number of animal species with cyclical parthenogenesis far exceeds that of purely thelytokous ones and they occupy a far greater variety of habitats. This is true, even though two separate ecological niches (food plants or host species in the case of a parasite, or alternate seasons) are generally required for an alternation of generations. The cytogenetic tricks that have made this possible in each case (production of haploid males in rotifers and cynipids, loss of one X chromosome in eggs destined to give rise to male aphids) have been in existence for tens or even hundreds of millions of years. In a number of cases the switch from thelytokous to sexual reproduction only occurs under adverse conditions, such as desiccation. All these facts controvert those modern views that claim that genetic recombination in sexual reproduction is unimportant. There have been few studies of speciation in groups with alternation of sexual and thelytokous generations, except for the old work of Kinsey (1930, 1936) on the cynipids (see p. 239). Speciation in aphids seems to have been frequently accompanied by rather large karyotypic changes. Theoretically, an alternation of generations should provide favorable conditions for the establishment of chromosomal rearrangements. A rearrangement, such as a translocation, that would run into difficulties in the meiosis of a heterozygote can be multiplied many-fold in the course of the apomictic generations, thus greatly increasing the chance of homozygotes appearing in succeeding sexual generations. Mechanisms of chromosomal speciation may thus have been important in such groups.

Until recently, the study of asexual genetic systems in higher organisms was carried out almost exclusively with the techniques of chromosome cytology. With the advent of biochemical techniques for studying levels of polymorphism and heterozygosity it has taken on a new aspect. The greatest need at present would seem to be the identification and study of the bisexual ancestors of thelytokous biotypes. Only when their population dynamics and population genetics are properly understood will we be in a position to interpret the significance of evolutionary switches from sexuality to thelytoky.

10

Conclusions

Our survey of the evidence that has accumulated in recent years reveals that the processes involved in speciation are considerably more complex and also more varied than has been generally recognized in the past. The analytical study of speciation thus proves to be much more difficult but also far more interesting than has been realized until recently.

Not much attention has been paid in this book to certain discredited and out-of-date models of speciation, e.g., the "hopeful monster," "macromutation," and "chromosomal repatterning" concepts of Goldschmidt (1940). No purpose would be served by using these as straw men in order to discredit certain types of models for all time. The history of evolutionary ideas is undoubtedly an interesting field. But as far as theories of speciation are concerned, there simply was not enough information (of the type outlined on pp. 11–12) before about 1950 for them to be much more than inspired guesses as to what actually occurs. The study of speciation in a deep sense is thus essentially a modern branch of biology, one that has developed in the past quarter century. It may well be that our successors in the year 2,000 will consider the ideas in this book just as antiquated, unfounded, speculative, and superficial, as some of the concepts of the 1930s and 1940s appear to us today.

One great difficulty in speciation studies and theory is that they now require expertise in a wide range of fields—enzyme and protein biochemistry, molecular cytogenetics, population genetics, ecology, ethology, and biometry at least—and the most elaborate and sophisticated laboratory studies. They also require the kind of detailed knowledge of organisms in the field more commonly found among the naturalists of former generations than in the molecular biologists of today. The models of speciation of the future are likely to be framed largely in mathematical terms. This has not been attempted in the present work, because the author is not mathematically gifted. This book is hence only the forerunner of the definitive work on modes of speciation that should be written, about the year 2,000, by someone with a much more extensive knowledge of the basic facts and with the ability to incorporate them in mathematical models. Lewontin (1974) has stated, certainly overpessimistically, that "we know virtually nothing about the genetic changes that occur in species formation." That information, however, will only be accumulated and critically analyzed and evaluated if the attempt is generally felt to be worthwhile. One reason for writing this book was to stress the fact that speciation is a complex and wide-ranging field for inquiry, a legitimate and challenging area of biology, and one that requires a high level of technical as well as intellectual sophistication. It is emphatically not a field that should be encumbered by sterile semantic arguments as to the meaning and definition of terms, as has too often been the case in the past. And, at the present time, it would seem that sweeping generalizations about methods of speciation should be avoided as far as possible (it may be much more justifiable to generalize in 20 years time).

To some extent, Lewontin's very negative conclusion may be due to the fact that population geneticists have been asking the wrong questions. One of the main conclusions of this book is that over 90 percent (and perhaps over 98 percent) of all speciation events are accompanied by karyotypic changes, and that in the majority of these cases the structural chromosomal rearrangements have played a primary role in initiating divergence. In general, however, population geneticists have neglected karyotypic changes and have tended to think of speciation in terms of the building-up of distinct gene pools of alleles without considering structural rearrangements of the chromosomes or looking upon such changes as simply equivalent to point mutations at individual loci—Wallace (1959, 1966) is a conspicuous exception among population geneticists. Although the process of speciation is not exempt from the operations of the laws of population genetics, it almost certainly involves principles that do not operate in ordinary phyletic evolution.

The models of speciation considered in this book have been distinguished from one another, mainly on the basis of the geographic component, as allopatric, clinal, sympatric, stasipatric, etc. If we had a better understanding of the precise genetic changes involved in speciation in a wide variety of organisms, a classification of modes of speciation on purely genetic criteria, omitting the geographic aspects, might be possible. Since the present state of information about genetic changes in speciation is, as Lewontin insists, extremely incomplete, any purely genetic classification of modes of speciation would be premature.

One genetic model of speciation that has not been discussed in the earlier chapters of this book has recently been put forward by Carson (1975). According to this highly speculative hypothesis, diploid species possess two different systems of variability in their genetic architecture. One is an "open" system that contains a large number of polymorphic loci that can recombine freely without producing a serious loss of adaptation. The other is a "closed" system consisting of blocks of internally balanced and coadapted genes that normally cannot be disrupted by crossing-over without the production of greatly reduced viability (in some cases these blocks may be stabilized by chromosomal inversions, but this is not held to be necessary). When natural selection is relaxed, as during a population "flush," recombinants that would normally not survive may do so and selection may then act on the "perturbed" genetic system to produce new coadapted blocks, which then would play a primary role in speciation.

There seems little direct evidence for this model. Where a population is polymorphic for inversions there will undoubtedly be blocks of internally balanced and coadapted genes. But crossing-over between mutually inverted segments ordinarily leads to totally inviable acentric and dicentric chromosomes (in the case of paracentric inversions) or ones carrying duplications and deficiencies (where the inversions are pericentric). Such chromosomes will not survive even under the conditions of an extreme population expansion. The only types of recombinant chromosome that could be expected to arise and persist under a relaxation of natural selection would be ones produced by double crossing-over within the inversion, and these will only occur if the inversion is long enough.

One might postulate that Carson's coadapted blocks of genes would occur in species with extreme localization of chiasmata. On the whole, however, the evidence tends to show that if there is strict chiasma localization in one sex, the distribution of chaismata in the other sex is much more nearly random or even localized in the opposite sense, e.g., if there is proximal localization in one sex there may be distal localization in the other.

In general, it seems that there is rather little evidence for a sharp distinction between two kinds of adaptive genetic systems of the type postulated by Carson in the chromosomal architecture. This not to deny the reality of blocks of coadapted genes. But we have already given reasons for rejecting the view that population "flushes" have anything to do with speciation and, on the whole, it seems probable that if disruption of coadapted gene complexes and the origin of new ones plays a role in speciation, this is mainly where it is caused by structural rearrangements of the karyotype and not where it results from ordinary crossing-over.

It would be intellectually satisfying if all the various modes of speciation in higher organisms could be classified unequivocally into a limited number of discrete, nonoverlapping categories. At the present time, however, with so much vital information lacking, it is doubtful whether this is possible. Nevertheless, an attempt has been made by Bush (1975a), who recognizes three modes of speciation: allopatric, parapatric, and sympatric. This classification is, of course, based entirely on the geographic component. A classification based on the genetic component would have to include, at one end of the spectrum, homosequential speciation and, at the other end, those types of speciation in which a major chromosomal rearrangement plays the primary role. It is now fairly certain that the two classifications would only coincide very imperfectly, i.e., it is probable that speciation without major chromosomal rearrangements occurs in some sympatric cases as well as in some allopatric ones.

One school of evolutionists, composed especially of those who have worked on groups with conspicuous mating calls, such as anurans and crickets, or on ones with complex courtship displays, have tended to regard the acquisition of a premating ethological barrier to cross-mating as the essential and primary element in every case of speciation. The argument of Mecham (1961), that premating isolating mechanisms avoid wastage of gametes while postmating ones lead to wastage, is used to provide a theoretical basis for the assumption of the primacy of ethological isolation in speciation (Littlejohn, 1969). Moreover, the increasing knowledge of the importance of sexual pheromones in lepidopterans (Roelofs and Cardé, 1974; Cardé *et al.,* 1975), coleopterans (Lanier and Burkholder, 1974), and even in *Drosophila* (Averhoff and Richardson, 1974, 1976), where their existence was not formerly suspected, has reinforced this tendency. On the other hand, it is evident that there is nothing strictly equivalent to ethological isolation in the plant kingdom, although various types of prefertilization isolation at the level of pollen–stigma interaction do operate.

Although ostensibly formulated in Darwinian, i.e., selectionist, terms, the arguments of Mecham and Littlejohn seem to have a distinctly teleological flavor. There is no doubt that in such animals as crickets and frogs the overwhelming majority of species, including many cases of sibling species barely or not at all separable on external morphology, do differ in respect of their mating calls (Alexander, 1962a, b; Littlejohn, 1959). But cricket species usually also differ quite strikingly in karyotype. The question, then, is whether the karyotypic or the ethological difference evolved first, or whether they did so concurrently. It is certainly possible, but surely unlikely, that speciation mechanisms are fundamentally different in crickets and eumastacid grasshoppers, which are voiceless. Sibling species of crickets that are allochronic or entirely allopatric may have the same or virtually the same calling and courtship songs (Alexander 1962b). At least this is so for *Gryllus pennsylvanicus* and *G. veletis* (see p. 250), which have different seasonal cycles in eastern North America; for *G. firmus* (eastern North America); *G. bermudiensis* (Bermuda);

G. campestris (Europe); and for two seasonally isolated species of *Scapsipedus* in Japan (Masaki, 1961). This suggests that when other isolating factors (allopatry or allochrony) exist, ethological isolation does not play a primary role in cricket speciation. The great diversity of karyotypes would seem to indicate that chromosomal rearrangements are just as important in the initiation of speciation in crickets as they appear to be in morabine grasshoppers (see p. 54), which have no calling or courtship songs. Clearly, however, once two closely related species (or incipient species) of crickets are sympatric, one of two things can happen. Either they evolve different seasonal cycles or different calling songs (Walker, 1964). But these changes are secondary—they reinforce the primary isolation due to cytogenetic causes.

It is natural that as more and more ethological isolating mechanisms are discovered, based on sound production, special courtship rituals, or pheromones, the impression should grow that these mechanisms are the primary agents of speciation—aided perhaps by behavioral changes that assist the population to occupy and exploit a new ecological niche. From this viewpoint, speciation is seen as essentially an ethological change that is genetically caused. This point of view is well expressed by Brncic, Nair, and Wheeler (1971): "Speciation phenomena depend primarily on behavioral traits such as those that determine reproductive isolation, or a particular manner of niche exploitation and coexistence."

It is undeniable that closely related species of *Drosophila* almost always differ ethologically. The critical question is, however, whether these differences are usually or always the very first ones to develop in the speciation process. The idea that they are is strongly contradicted by the crucial cases of *Drosophila pseudoobscura* and its geographically isolated subspecies *bogotana* in the Colombian Andes (Figure 1, p. 34) and *D. willistoni* and its subspecies *quechua* (Prakash, 1972; Dobzhansky, 1975). It was originally believed that the population of *D. bogotana,* separated by approximately 2,000 kilometers from the nearest population of *D. pseudoobscura,* in Guatemala, was only about 25 years old. But recent studies (Dobzhansky, 1975) suggest that it is much older, i.e., its separation from the main body of the species took place thousands or even tens of thousands of years ago. It was on the basis of the former age estimate that Lewontin (1974) referred to this as the *only* case of the first stage of speciation to have been studied genetically. The subspecies *bogotana* has differentiated to such an extent that, although it is not ethologically isolated to any significant extent from the North American *D. pseudoobscura,* F_1 males from crosses of female *bogotana* to North American males are completely or almost completely sterile; the reciprocal cross, however, gives fully fertile hybrids, and female hybrids are fertile whichever way the cross is made.

The situation with *D. willistoni* and its subspecies *quechua* is essentially identical (Dobzhansky, 1975). Crosses of *quechua* females to *willistoni* males yield sterile sons and fertile daughters, while the reciprocal cross gives fertile offspring of both sexes. As in the former case, ethological isolation is minimal or nonexistent. In both cases the factors responsible for the male sterility are located on all the major chromosomes. Allozyme studies suggest that the "genetic distance" between these subspecies is about 0.20 (compared to 0.03 between local populations within a subspecies, 0.23 between semispecies, 0.58 between sibling species, and 1.06 between nonsibling species of the *willistoni* group of *Drosophila).*

The evidence from these cases thus suggests that in this genus of flies, quite contrary to the view of the "ethological speciationists," speciation is an essentially cytogenetic process; that the primary isolating mechanisms are often, or perhaps usually, likely to be of the postmating type; and that it is in stage two of the speciation process (Lewontin, 1974) that ethological isolation is built up, especially if the incipient species are sympatric at this stage.

Evidence from some other species of *Drosophila* presents a rather different picture, however. *Drosophila athabasca,* in the United States, consists of three races or semispecies, which differ cytologically in the shape of the Y chromosome and by a number of distinctive inversions in the X chromosome and several autosomes. These taxa show variable but fairly strong ethological isolation from one another, which is associated with, and presumably at least in part due to, differences in the male courtship sounds (Miller, Goldstein, and Patty, 1975). Hybrids between these races are generally fertile in both sexes, although a few sterile males have been obtained in some crosses. Differences in male courtship sounds have also been found by Ewing (1969) in the sibling species *D. pseudoobscura* and *D. persimilis.* In the *D. paulistorum* complex of semispecies, however, only one, the "Amazonian," showed a significantly different courtship song from the others (Ewing, 1970).

The view that if the products of hybridization are going to be eliminated by natural selection there should be selection for ethological isolation between the two forms (thereby diminishing and ultimately eliminating the wastage of gametes) has been most positively stated by Dobzhansky (1951): "Reproductive isolation diminishes the frequency of the appearance of hybrids, prevents the reproductive wastage, permits the populations of the incipient species gradually to invade each other's territories, and finally to become partly and wholly sympatric. It is during the latter stages of this process that the selection pressure bolstering the reproductive isolation becomes strongest, helping to complete the process of speciation." Here, it would seem, is a model of speciation that is both allopatric and sympatric—it starts in allopatry and is completed sympatrically. More recently, Dobzhansky (1975) has stated that "Postmating mechanisms . . . are by-products of genetic divergence: premating ones are contrived by natural selection to mitigate or eliminate the losses of fitness which result from hybridization of genetically divergent and differentially adapted forms." However, Coyne (1974) has considered some circumstances in which hybrid inviability could have arisen as a result of selection. Dobzhansky seems also to have neglected the possibility that premating isolating mechanisms may arise, in some instances, as incidential by-products of genetic divergence between allopatric populations.

However, it is more difficult than many evolutionists have thought to find clear evidence in zones of overlap between incipient species of character displacement in behavior patterns, such as mating calls in frogs and crickets. Walker (1974b) states that there is no convincing case known in insects having mating calls and Loftus-Hills (1975) considered that the instance of two members of the *Litoria ewingi* complex of Australian tree frogs was "the only relatively convincing, documented case among animals." Ball and Jameson (1966) could find no evidence in two California tree frogs, *Hyla regilla* and *H. californiae,* of reinforcement of call differences in the area where the two are sympatric.

It is true, nevertheless, that an allopatric population of a species can in some instances (but perhaps rare ones) acquire a different mating call from the parent population. In general, one would not think of an evolutionary change in the mating call of a population as being adaptive to a particular habitat. Certainly, mating calls are not likely to be adapted to temperature, humidity, salinity, source of food, etc. Nevertheless, Littlejohn (1969) has pointed out that they may well be influenced by the calls of other, related species that may be present in a particular habitat. Thus, if a species population becomes geographically fragmented into two isolated populations, one of which is sympatric with species A, B, and C, while the other is sympatric with species D, E, and F, then the calls of the two populations may be expected to diverge—that of one will become progressively more different from those of A, B, and C, that of the other will tend to evolve in such a way as to maximize its differences from those of D, E, and F. It is thus theoretically possible to imagine mating call differences in frogs or crickets being built up during a period of allopatric isolation to such an extent that if the two populations later become sympatric they would constitute an effective ethological isolating mechanism. Walker (1962) has shown that individual variation in calling songs in cricket populations is almost certainly genetic. He found little evidence of geographic variation in the calling songs of North American cricket species, but *Oecanthus fultoni* chirps at a faster rate on the west coast of the United States, where it is sympatric with a slow-chirping sibling species, *O. rileyi,* than in the east, where the latter species does not occur.

It is possible that the role in speciation of evolutionary changes in the organic molecules constituting sexual pheromones may be somewhat different to that of changes in mating calls. At least theoretically, it is easy to imagine a simple allelic substitution changing the chemical formula of a pheromone. Roelofs and Comeau (1969, 1970) suggested that sympatric speciation could result from mutations changing pheromone systems, but Shorey (1970) considered it more likely that changes in pheromone mechanisms would follow speciation rather than playing an active role in initiating divergence. In any case it is clear that any change in sex pheromone specificity must involve changes in two biochemical systems, since a sex pheromone is synthesized by one sex and recognized by the other.

Sex pheromone studies in beetles have been carried out on the genera *Trogoderma, Ips,* and *Dendroctonus* (Lanier and Burkholder, 1974). In *Trogoderma inclusum* the female pheromone is 14-methyl-*cis*-8-hexadecen-1-ol, while in *T. glabrum* it is the corresponding *trans* compound. Males of these and three other species respond to stimulation by either substance, but the sensitivity of *glabrum* males for the *trans* compound is greater than for the *cis* one, and vice versa for *inclusum* males. Males of *T. sternale* show a low level of response for either compound, or for extracts of females of other species of *Trogoderma*. In general, the sex pheromones of *Trogoderma* and other dermestid beetles seem to show a low level of species specificity but are quite genus-specific.

In the bark beetles of the genus *Ips* 10 species groups have been distinguished taxonomically. Members of the same species group are usually allopatric or parapatric, while members of different groups are frequently sympatric. Three sex pheromones have been identified in this group: *cis*-verbenol, 2-methyl-6-methylene-7-octen-4-ol ("ipsenol") and 2-methyl-6-methylene-2,7-octadien-4-ol ("ipsdienol"). In general, species of *Ips* belonging to the same species group are strongly cross-attractive, but species of different species groups are only weakly (or not at all) cross-attractive. Species that are not cross-reactive may share some pheromones, but the total bouquet is different and inhibitory substances may block the response in some cases.

The situation in another genus of bark beetles, *Dendroctonus,* is extremely complex. Three pheromones, brevicomin, frontalin, and *trans*-verbenol, have been identified in five species, and in *D. pseudotsugae* 3-methyl-2-cyclohexen-l-ol is present in addition; at least two of these substances are present in each sex of each species, but the combinations differ. Two additional compounds block the attraction in some species; also, brevicomin and *trans*-verbenol attract some species but inhibit the response in others. *D. rufipennis* and *D. pseudotsugae,* which both bore in white spruce (the latter also bores in douglas fir), were strongly cross-reactive. Lanier and Burkholder conclude that "neither production nor reception of attractant compounds from both the beetle and the tree is sufficient to enforce reproductive isolation among many of the species of sympatric *Dendroctonus*. . . . Specificity of pheromone systems is clearly not the principal mechanism enforcing reproductive isolation in *Trogoderma* and *Dendroctonus*." With regard to *Ips*, Hedden and co-workers (1976) conclude that "The mechanisms responsible for reproductive isolation in *I. avulsus* and *I. grandicollis* are unknown but are probably related to differences in habitat preference within the host tree, specificity of mating behaviour, and genital incompatibility. . . . Our findings do not support the hypothesis that specificity of pheromones among sympatric *Ips* species is an important mechanism in maintaining species integrity and preventing interspecific competition. . . ."

The situation in Lepidoptera seem to be rather different. Even though a number of species have been shown to produce secondary chemical components as well as a main attractant, a number of sibling species, semispecies, and races do seem to differ in the main pheromonal compound, the difference acting as a primary isolating factor. Thus, Roelofs and Comeau (1969) found that males of two closely related and sympatric moth species, *Bryotropha similis* and *B.* sp., were attracted to *cis*- and *trans*-9-tetradecenyl acetate, produced by their respective females. A mixture of two isomers attracts males of neither species, suggesting that each is inhibitory for the other. Similarly, two closely related apple tortricids with the same life cycle, *Graphalitra prunivora* and *G. packardi,* are isolated because the first is attracted to *cis*-8-dodecenyl acetate, the second to the *trans* isomer (Roelofs and Cardé, 1974). Some cases of cryptic sibling species have been revealed because they were attracted to different pheromones. However, not all sympatric sibling species have different sexual attractants. The western Canadian species *Choristoneura occidentalis* and *C. biennis* use the same aldehyde pheromone but are kept isolated by other factors; in eastern Canada *C. fumiferana* uses the same aldehyde, while the sympatric *G. pinus* uses an acetate (Freeman *et al.*, 1967; Sanders, 1971a, b).

Evolution of pheromonal systems is obviously occurring in several lepidopteran species at the present time. Thus, the European corn borer *Ostrinia nubilalis* is believed to have been introduced to the United States on a number of different occasions. In Iowa and Ontario the male is maximally attracted to a mixture of 96 percent *cis*-11-14-acetate and 4 percent of the *trans* isomer, while in New York the most potent attractant is a mixture of 98 percent of the *trans* isomer and 2 percent of the *cis* compound, and these mixtures are approximately what are produced by Ontario and New York females, respectively. In Pennsylvania there seem to be two distinct populations of European corn borers whose males are attracted to 97:3 and 2:98 mixtures of *cis:trans* isomers (Kochansky *et al.*, 1975; Cardé *et al.*, 1975). Only future studies will determine whether these two populations are effectively isolated genetically in nature (in which case they should be regarded as separate species). The matter is complicated by the fact that in Europe neither the *cis* nor the *trans* compound appears to be an effective attractant.

Related species that produce different relative amounts of the components of a two-component or multicomponent sex-attractive bouquet appear to be relatively common in Lepidoptera. Thus, the sympatric torticids *Adoxophyes orana* and *Clepsis spectrana* both produce *cis*-9-14 acetate and *cis*-11-14 acetate, but the former species is attracted when the first compound is present in excess, while the latter is attracted when high concentrations of the latter are present (Minks *et al.,* 1973).

In some cases two closely related species may utilize the same attractant but may release it at very different concentrations (Roelofs and Cardé, 1974), while in other cases they may liberate the same attractant at different times of the day. An example of the latter situation is provided by the gypsy moth *(Porthetria dispar)* and the closely related nun moth *(Lymantria monacha)* in Europe. Both use *cis*-7,8-epoxy-2-methyloctadecane as an attractant, but the former mates by day while the latter does so between 6 P.M. and 1 A.M. The gypsy moth was introduced into North America in 1868–1869, but the nun moth has never been introduced. Apparently the mating period in the United States of *P. dispar* was 9 A.M.–3 or 4 P.M. in 1896, 9 A.M.–6 P.M. in 1932 and 9 A.M.–8 P.M. in 1974 (Roelofs and Cardé, 1974); it has evolved a longer cycle over the past hundred years in the absence of *L. monacha.*

It now appears certain that pheromones have a significant role in the courtship of *Drosophila* (Averhoff and Richardson, 1974, 1976; Bennet-Clark *et al.,* 1976). However, the part played by chemical stimuli in the ethological isolation of closely related species of *Drosophila* is still very difficult to assess. Averhoff and Richardson believe that courtship in *D. melanogaster* includes genetically variable female pheromones that stimulate the male to court and variable male pheromones that induce the female to accept. The mechanism is envisaged as a force that acts to minimize inbreeding during population bottlenecks. The relative role of chemical and acoustic stimuli in the courtship of even the single species *D. melanogaster* is still controversial, and the earlier work of Spieth (1952) shows how complex the visible aspects of *Drosophila* courtship behavior are.

An important unresolved question in speciation theory is whether the primary role in initiating speciation is usually a premating isolating mechanism, such as a pheromonal or bio-acoustic difference between the two diverging populations, or a postmating one, whether due to chromosomal rearrangements or gene mutation. Only a series of investigations on suitable cases in various stages of speciation, in which both the ethological and the cytogenetic factors are intensively studied, will resolve this basic question. The answer will not necessarily be the same for all cases or for all groups of organisms. In all but a minute number of instances closely related species differ in karyotype. But at the same time, the great majority of closely related animal species differ in courtship behavior. Thus, most studies of sibling species are likely to be uninformative on this point, since both kinds of differences will exist. The cases that may help to resolve this problem will be those few instances of semispecies or sibling species that differ in a single respect only.

In addition to its undoubted role in allopolyploidy (Chapter 8) and in the origin of polyploid thelytokous biotypes (Chapter 9), hybridization has been said to have given rise to new species of diploid sexual organisms. Ross (1958) claimed that certain species of leafhoppers (Cicadellidae) had arisen in this manner. His case rests on the fact that certain species of the vast genus *Erythroneura* in eastern North America belong to one species group on the basis of most morphological characteristics but strongly resemble a member of a different species group in one character. No evidence was presented that this situation could not have arisen by mutation, which certainly seems a more probable explanation than a hybrid origin. Many other taxonomists have undoubtedly come on similar cases. Particularly where we are dealing with large numbers of closely related species (and the genus *Erythroneura* includes at least 500 described species) a certain amount of parallel and convergent evolution is only to be expected.

One species that is certainly of hybrid origin is the common European frog *Rana esculenta,* whose distribution extends from western France eastward to the Ural mountains. It is now known that this is a hybrid between *Rana ridibunda* and *R. lessonae* (Berger, 1973; Günther, 1973; Tunner, 1973, 1974; Hotz, 1974). However, the case is peculiar in a number of respects, not all of which have been fully clarified. Apparently, there are four biotypes of *R. esculenta* in central and eastern Europe, two diploid and two triploid. One diploid form, which coexists with *R. lessonae,* consists of both males and females; apparently, they discard the *lessonae* chromosome set at meiosis in both sexes, producing gametes with only *ridibunda* chromosomes. Effective crosses of this biotype are between female *esculenta* and male *lessonae,* the hybrid constitution being restored in the F_1. The other diploid biotype coexists with *R. ridibunda;* it consists mostly of males that lose the *ridibunda* chromosomes at meiosis. Again, the hybrid constitution is restored in the F_1 when a male mates with a *ridibunda* female. One triploid biotype carries two *ridibunda* sets of chromosomes and one *lessonae* set (RRL), while the other has two *lessonae* sets and one *ridibunda* one (LLR). Some triploid *esculenta* populations appear to exist in the absence of both parental species. It is obvious that this case is analogous to the "hybridogenesis" of some fishes in the genus *Poeciliopsis* in Mexico (p. 298). These exceptional cases should certainly not be used as an argument for hybridization as a normal or frequent mechanism of speciation in animals.

An entirely different type of hybridization is the rare instance of genetic "leakage" from the gene pool of one species into that of another. As far as animals are concerned, there can be no doubt that even in certain groups where premating isolating mechanisms of the ethological type have played a major part in speciation, occasional hybridization occurs and may lead to a significant level of genic "introgression" from one species into another. Mecham (1960) studied a case in which considerable hybridization was occurring between the tree frogs *Hyla cinerea* and *H. gratiosa* in an area of artificial ponds in Alabama, where the habitat had been greatly modified by human activity. The most interesting feature of this case is that there was apparently a strong tendency for hybrids to mate together, rather than backcross to either of the parental species; thus, the hybrids formed a breeding population partly isolated from both parent forms. Mecham (1975) also points out that in anurans of the United States (80 species) natural hybrids have been recorded for 27 combinations, and indirect evidence suggests that introgressive hybridization is a distinctive feature of this group and possibly more important than mutation or intraspecific gene flow as a source of new genetic variation. In spite of this, he concludes that there is no evidence that any existing species have arisen by recombination speciation, i.e. by hybridization.

The above general situation probably also exists in a number of groups of fishes (Hubbs, Hubbs, and Johnson, 1943; Hubbs, 1961), but is probably not widespread in animals. No evidence for introgressive hybridization on anything like the scale reported for anurans has been found in *Drosophila,* grasshoppers, lizards, or mammals. As far as *Drosophila* is concerned, Carson, Nair, and Sene (1975) have discussed two possible cases in the literature and have reported the existence of about two percent hybrid individuals in the mixed population of *D. setosimentum* and *D. ochrobasis* at one locality in Hawaii. Nevertheless, the species in this case maintain their distinctness, presumably because of the action of natural selection against adaptively inferior F_2 and backcross individuals. In the classical case of *Drosophila pseudoobscura* and *D. persimilis,* sibling species distinguishable in the male sex by a minor genitalic difference, Dobzhansky (1973) reported that in the course of 35 years work, involving the examination of the karyotypes of about 30,000 individuals in the polytene nuclei, he found only one hybrid female in the wild, which gave rise to an unmistakably backcross progeny. However, even this very low level of natural introgression would be expected to lead to the appearance of inversions characteristic of one species in populations of the other. The fact that this is not found indicates clearly that natural selection against hybrids, either in the F_1 or backcross generations, is sufficient to negate the effect of introgression. In the mosquitoes of the *Anopheles gambiae* complex, some evidence of effective interspecific hybridization has been published by G. B. White (1971).

A number of instances of natural hybridization on a substantial scale have been recorded in Lepidoptera. Sweadner (1937) long ago claimed that certain species of the moth genus *Platysamia* were hybridizing freely in a wide region of the United States. This case seems in need of reinvestigation by modern methods. Most instances of natural hybridization between animal species seem to represent failures of the genetic isolating mechanisms in particular geographic areas and do not lead to effective introgression because of selection against hybrids. This is almost certainly the case for the natural hybrids between the newts *Triturus c. cristatus* and *T. marmoratus* in France (Valleé, 1959) and for the hybrids between the butterfly species *Lysandra bellargus* (n = 45) and *L. coridon* (n = 87–88) in the Pyrenees and the Abruzzi (De Lesse, 1960). There is virtually no evidence to suggest that this type of hybridization has generated new species in any group of animals. It was claimed by some earlier authors that the Caucasian lizard species *Lacerta mixta* had arisen by hybridization between *L. derjugini* and a member of the *L. saxicola* complex. However, this interpretation has been rejected by Uzzell and Darevsky (1973) on electrophoretic evidence. Hybridization has played a major role in speciation in plants, of course, but essentially only where it leads to polyploidy. The only animal species of clearly hybrid origin are the thelytokous ones discussed in Chapter 9 and the very few "hybridogenetic" ones.

Obviously, whether we study karyotypes, ethology, external phenotype, or allozymes, we must expect to find numerous situations where three species A, B, and C form a linear series, with B more or less exactly intermediate between A and C. Such situations are expected on quite orthodox views of speciation and phylogeny, and do not necessarily imply that B is a species of hybrid origin.

The role of so-called *introgressive hybridization* has been much discussed. This is the persistent "leakage" of genetic material from one species into another, or between two species if it is mutual (unimpeded in both directions). Anderson (1949) elevated introgressive hybridization to the status of a major evolutionary force in the case of higher plants. There can be no doubt about the reality of introgression in many cases. At least in plants it seems to be especially prevalent in habitats that have been disturbed by human activity. But even granted it occurs on a fairly wide scale, it is not easy to evaluate the evolutionary role of introgression. If there is strong selection against segregants from the hybrids, the species may remain distinct even if introgression is quite frequent. This seems to be the most common situation in animals; even if hybridization occurs on a fairly large scale and the hybrids are fertile, the effective introgression may be zero if selection against the F_1 and backcross hybrids is severe.

It has been suggested by Levin (1969) that a plant population may undergo saltational speciation (he does not state whether chromosomal rearrangements would be involved) as a response to a threat to its genetic integrity posed by introgression from a related species. He cites only one case that he believes conforms to this model. *Phlox pilosa* and *P. glaberrima* grow in mesic and wet prairies respectively, in the mid-western United States. Some introgression has occurred between them, since there is a slight overlap in habitats and flowering seasons. *P. argillacea,* which occupies much more xeric habitats, is interpreted by Levin as a recent derivative from *P. pilosa,* which avoids introgression from *argillacea* by occupying a nonoverlapping environment. The main difficulty in accepting Levin's interpretation of this case would seem to be the fact that *P. pilosa* still persists in sympatry with *P. glaberrima.* It would seem simpler to suggest that the drier habitat merely provided a vacant niche for the evolution of a new species of *Phlox.*

The role of chromosomal rearrangements in speciation has been considered in Chapters 3 and 6. More and more, it appears as if such rearrangements, of many different types, have played the primary role in the majority of speciation events. It by no means follows, however, that their significance in speciation is always of the same type. In fact, each chromosomal rearrangement—whether fusion or dissociation, translocation, inversion, gain or loss of heterochromatin—must be regarded as a unique event whose consequences will be almost impossible to predict in the present stage of our knowledge. It is thus extremely difficult to incorporate chromosomal rearrangements into mathematical models of speciation and phyletic evolution, which may be one reason why they have been relatively neglected by many evolutionary geneticists. There are signs, however, that this situation is changing (Wilson, Sarich, and Maxson, 1974; Wilson, Bush, Case, and King, 1975).

In Chapter 6 we considered the role that many chromosomal rearrangements play in the stasipatric model of speciation in reducing the fecundity of structural heterozygotes, as a result of various kinds of meiotic "accidents." Another type of situation, which must now be discussed, is where rearrangements, especially those involving the sex chromosomes, lead to changes in the genetic mechanism of sex determination.

It seems fairly certain that the part played in speciation by changes in the balance system upon which sex determination depends has not been adequately appreciated in the past, and that even now the relationship of the sex determining mechanism to speciation is poorly understood. Perhaps the main difficulty is that there is no single relationship but several, and that in different cases one or other is predominant. Obviously, in the great majority of higher plants and in those groups of animals in which hermaphroditism is general or universal, balance mechanisms for the determination of sex are absent, so that these considerations do not arise.

In organisms with separate sexes there are no doubt many pairs of closely related species in which the genetic control of sex determination and differentiation is substantially the same in both. But the results of interspecific hybridization, in which one sex is frequently absent, rare, intersexual, or sterile in the F_1, indicate clearly that this is not always so. Long ago, Haldane (1922) pointed out that the sex affected in these various ways is frequently the heterogametic one. Although many exceptions to "Haldane's rule" have been reported since it was first put forward, it remains valid for the majority of cases in insects and vertebrates. The main point here, however, is that numerous species that are sufficiently closely related to be hybridizable, frequently and perhaps usually differ, quantitatively at least, in their sex-determining systems.

Many evolutionary changes in sex determination are not known to be due to chromosomal rearrangements, and may well be the result of allelic changes. The very extensive studies of Goldschmidt (1932) on the races of the gypsy moth, *Porthetria dispar,* are in this category. This is a very widespread species whose natural distribution extends from Japan to western Europe. All races have the male sex XX and the female sex XY, the X chromosome being male-determining and the Y chromosome female-determining (the role of the autosomes is somewhat unclear). However, the "potency" of the X as far as sex determination is concerned is highly variable in the different populations. Most European races have "weak" X chromosomes, as does the race *hokkaidoensis* from the northern island of Japan. On the other hand, those from the southern island of Japan are "medium" or "strong." Within each race the male-determining and female-determining chromosomes are in balance, so that only pure males and females are produced. When individuals of different races are crossed in the laboratory, however, intersexuality frequently results. The race *hokkaidoensis* has differentiated so far that its sex-determining mechanism constitutes an effective isolating mechanism; this race should probably be regarded as a semispecies or even a sibling species.

A number of marine invertebrate species polymorphic for hermaphroditism and unisexuality have been described. Thus, in the mollusc *Patella coerulea* at Naples most individuals are protandric hermaphrodites whose male phase may be long or short, depending on the genetic constitution; some individuals (about 8 percent of the total) are pure males or females that never undergo sex reversal (Bacci, 1955a, b; Montalenti, 1960). The starfish *Asterina gibbosa* has the same type of sexual polymorphism at Naples, but at Plymouth all individuals undergo sex reversal at the same stage of development. The worm *Ophryo-*

trocha puerilis normally undergoes a male to female sex reversal at a stage when it has 18 setigerous segments. By artificial selection in opposite directions, individuals were obtained that did not reverse their sex even though they had 26 setigerous segments (pure males); others were obtained that developed oocytes when they had only 11 segments, without passing through a male stage (Bacci, 1955b; Montalenti, 1960).

Quite different to the above are groups of organisms in which some females give rise only to sons and others only to daughters. Many species of sciarid midges are of this type, while others produce broods of mixed sex; yet others include some strains of one type and others of the opposite type. To what extent changes from one system to the other play a role in speciation is quite unknown. A somewhat different situation exists in certain isopod crustacea. In the freshwater species *Asellus aquaticus,* studied by Montalenti and Vitagliano-Tadini (1960), some pair-matings give rise to approximately equal numbers of sons and daughters (N matings), while others yield mainly sons (A matings) and others mainly daughters (T matings). A × A and T × T matings, even when not consanguineous, yield very few offspring; the system appears to strongly favor outbreeding.

It is conceivable that the changes in sex determination in the sciarid midges could lead to speciation, although this is certainly not proven. In the other cases cited above it seems unlikely that such changes could be directly associated with speciation. In *Patella* and *Orphryotrocha,* for example, one cannot select for purely hermaphroditic or purely bisexual strains, since one has to select in opposite directions to obtain pure males or females.

Structural changes in the sex chromosomes of a bisexual species present an entirely different case from those above. In organisms with polytene chromosomes there are now a very large number of instances where closely related species show far more structural differences in their sex chromosomes than in their autosomes. Thus, in the *Drosophila virilis* group (Stone, 1962) the species *ezoana, littoralis, montana,* and *laciocola* all have distinctive X chromosomes that have undergone such a radical process of structural change (presumably by multiple inversions) that the precise phylogeny is too complicated to be worked out, yet the inversions in the autosomes, although numerous, have been analyzed in detail. Just what has happened in these species is quite uncertain; they may have undergone "catastrophic" reconstructions of the X, based on multiple-break rearrangements. Alternatively, there may have been a large number of orthodox, two-break inversions that occurred seriatim. In any case, the relationship of the repatterning of the X chromosome to the speciation process by which these species arose is entirely obscure.

There are rather numerous cases of the same kind in other dipterans. It is apparent that much speciation in the black flies (Simuliidae) has been directly due to structural changes in the X and Y chromosomes (details in Chapter 3) and the same situation certainly exists in some chironomids. In the simuliids the formation of sibling species by structural rearrangements in the sex chromosomes seems to have taken place by a method entirely different to the stasipatric one, since it leads to sympatric rather than parapatric siblings.

Quite different to the above types of transformations of the sex chromosomes are fusions between sex chromosomes and autosomes. These have been studied most extensively in grasshoppers and other orthopteroid insects (details in White and Cheney, 1966, 1972; White, 1973, 1976). The primitive condition in these groups is for the male to be XO; sex determination must depend on a balance between female-determining factors in the X chromosome and male-determining ones in the autosomes (we use the term "factors" advisedly, since, with the exception of the H-Y gene for testicular differentiation in mammals, structural genes for sex determination have never been unequivocally demonstrated). When an X chromosome in an XO:XX species undergoes a fusion to an autosome, this leads to a neo-XY system, in which the unfused autosome, now confined to the male line, plays the role of a Y, pairing with the limb of the neo-X with which it is homologous at meiosis. Events of this kind have occurred repeatedly in Acrididae, Eumastacidae, Tettigoniidae, stick insects, and some other groups. They seem to be associated with processes of stasipatric speciation and, unlike the changes in simuliid sex chromosomes, lead to situations in which the XO and neo-XY taxa have parapatric distributions.

Originally, the neo-Y is entirely homologous to the chromosome arm that becomes incorporated, by fusion, into the neo-X. At this early stage the neo-Y is largely euchromatic and, being autosomal in origin, is presumably weakly male-determining. The neo-X must be preponderantly female-determining, but the limb that is homologous to the neo-Y is presumably weakly male-determining. Later changes in the neo-Y usually result in its becoming a largely heterochromatic chromosome that only retains a homology to the neo-X in a very small distal segment. The exact nature of the structural changes that lead to this "heterochromatinization" of the neo-Y are not understood, but they presumably involve repeated duplications of repetitive satellite-sequences and losses of unique sequences in the euchromatin. Whether the resulting neo-Y chromosomes are strongly (instead of weakly) male-determining is uncertain—some evidence from laboratory-produced triploid hybrids between the parthenogenetic grasshopper *Warramaba virgo* and related bisexual species (hybrids with and without neo-Y chromosomes) suggests, but without definitely proving, that heterochromatic neo-Y chromosomes are indeed strongly male-determining, and that they have in fact superseded the old X–autosome balance system of sex determination.

In a number of instances further fusions of autosomes to either the neo-Y or the neo-X have given rise to multiple sex-chromosome systems of the X_1X_2Y: $X_1X_1X_2X_2$ or XY_1Y_2:XX type. These changes seem to lead to parapatric taxa and may be presumed to accompany stasipatric speciation events.

It is very curious that, at least in the orthopteroid insects, fusions of sex chromosomes and autosomes seem to be strongly inhibitory of further speciation. Thus, most neo-XY species have only XO relatives, and although about 60 X–autosome fusions are now known in the acridid grasshoppers, there is not a single moderately sized genus (with 5 to 15 species) whose members all have neo-XY or X_1X_2Y mechanisms. The nearest approach in Acrididae would be *Paratylotropidia,* with one XY and two X_1X_2Y species; among the Australian morabine grasshoppers, *Culmacris* has two neo-XY and several X_1X_2Y species, while *Carnarvonella* has three neo-X and one X_1X_2Y species. The general situation in stick insects (Phasmatodea) seems to be similar (White, 1976). Only in the mantids do we find that an X_1X_2Y mechanism (which arose directly from an XO one by a mutual translocation and not by two successive fusions) has been an evolutionary success, since it occurs in several hundred species belonging to at least 17 genera, spread over all the warmer parts of the world.

The suggestion made above, that most fusions between sex chromosomes and autosomes inhibit further speciation is of course not the only conceivable explanation of the taxonomic situation described above—but it seems the most probable one. It is extremely unlikely that such changes have only recently started to occur in the phylogeny of grasshoppers and stick insects (i.e., in the last one or two million years). The inhibition of evolutionary potential by neo-XY and X_1X_2Y mechanisms signifies either that such species are doomed to rather rapid extinction or that their liability to undergo further speciation is severely reduced.

Much more complex systems of sex chromosomes have been built up in certain groups. Some of these undoubtedly are due to the inclusion of increasingly greater numbers of autosomal pairs into the sex chromosome mechanism, as a result of fusions or translocations; others are probably due to reduplication of parts of the original sex chromosomes, as independent elements in the karyotype. These systems are mentioned here because it seems likely that each change in the sex chromosome mechanism is associated with a speciation event. Systems involving five different X and four different Y chromosomes are known in a centipede species and 12 different X and six different Y chromosomes in the beetle *Blaps cribrosa.* In these groups the related species also possess multiple sex chromosome mechanisms but less complex ones; there seems to have been on inhibitory effect on the potential for speciation.

One group, Monotremata, shows an apparently extraordinary conservatism in having preserved a highly complex sex chromosome mechanism, probably from Mesozoic times. Although the full details are not yet published, it is known that earlier accounts of the sex chromosome mechanism are all incorrect and that three living species (a New Guinea *Zaglossus* and the Australian *Tachyglossus aculeatus* and *Ornithorhynchus anatinus*) all show long chains of sex chromosomes at the male meiosis. The first two (2n♂ = 63) are probably $X_1Y_1X_2Y_2X_3Y_3X_4Y_4X_5$, while the third (2n♂ = 52) has a chain of eight sex chromosomes at meiosis (Bick and Sharman, 1975). In addition, there seems to be a structurally heterozygous pair of chromosomes, probably in both sexes, which may function as a balanced lethal mechanism; its interrelationship (if any) with the chain of chromosomes is not understood. Whether this bizarre genetic system has inhibited speciation and evolutionary diversification and progress in the phylogeny of the monotremes can only be conjectured. Similar systems of multiple sex chromosomes have evolved in some species of mistletoes (Wiens and Barlow, 1965).

In summary, it appears that changes, both structural and functional, in the sex chromosome mechanism seem to have an important and intimate connection with speciation in many groups of animals. But this connection has been a variable one and has led to several quite different modes of speciation. Most modern writers on evolution seem to have largely or entirely neglected these changes in sex-determining systems as agents in speciation and phyletic change.

Some of Blackith's views (1973) on a possible relationship between methods of speciation and the number of gametes formed are almost certainly erroneous. His hypothesis is that in groups with stasipatric speciation, depending on chromosomal rearrangements, large numbers of sperm will be formed, making provision for wastage of large numbers of sperm carrying deleterious rearrangements. The whole concept rests on the supposition that "damaged" sperms will be attacked by enzymes in the female spermatheca—an almost certainly incorrect assumption in view of the great mass of evidence in *Drosophila* from the time of Muller and Settles (1927) onwards. Just what the different evolutionary consequences actually are of having large numbers of sperm (e.g., the subfamily Morabinae of the eumastacid grasshoppers) versus small numbers (e.g., the subfamilies Thericleinae and Mastacideinae, according to the data of White, 1970c) remains mysterious for the present; but the difference is almost certainly not directly related to stasipatric versus classical allopatric speciation.

When one compares the attitudes of biologists to problems of speciation, say, fifteen years ago and today, two important changes can be observed (in addition to the recognition that genetic polymorphism in natural populations is much higher than could have been predicted before the introduction of the technique of gel electrophoresis).

The first is an increasing awareness of the prevalence and importance of chromosomal rearrangements in speciation, an aspect that has been stressed especially in Chapters 3 and 6. Two factors tended to delay recognition of these facts. One was the discovery of a certain number of homosequential species groups in *Drosophila,* which, while it proved that speciation could occur without structural chromosomal rearrangements, was not seen as the exceptional case it undoubtedly is. The other was the view that in the genus *Drosophila* paracentric inversions appeared to have more of a role in population polymorphism than in speciation; in fact, their role in speciation seemed to be unclear and hence unimportant. Cytogenetic studies of other groups of animals have tended to correct these narrow views, and some of the most recent studies on the Hawaiian drosophilids are tending to change the picture, even for *Drosophila.*

The second aspect in which thinking has changed very greatly in recent years is in regard to closely related taxa with contiguous, i.e., parapatric, ranges. In 1967 Blair, in discussing a paper by Selander (published in 1969), could still take the view that such distribution patterns were rare and exceptional; it is doubtful if any informed biologist would uphold that idea today. This is due in part to the realization that many groups of organisms identified as species 10 or 20 years ago are in fact complexes of several sibling species or semispecies that conform to the biological concept of the species (or approximately, in the case of the semispecies). Groups formerly thought to be species include *Rana pipiens, Thomomys talpoides, Didymuria violescens, Drosophila paulistorum, Peromyscus maniculatus, Gryllotalpa gryllotalpa, Vandiemenella viatica, Simulium damnosum,* and *Anopheles gambiae* (examples from many animal groups, which have been discussed earlier in this book). When we look at the taxonomic units that make up these complexes, we find that parapatric distributions, instead of being rare, are in fact extremely common.

Closely related to this question of the frequency of occurrence of parapatric distributions is that of their origin and significance. So long as the allopatric model of speciation reigned supreme, as the only acceptable model, it was generally considered that when parapatric distributions did occur they were necessarily the result of secondary contacts between populations that had formerly been entirely disjunct. It now appears clear that many, but not all, of them are in fact primary, i.e., between populations that have never been out of contact with one another. Consequently, since parapatric zones of narrow overlap can arise in two different ways, and since in many instances it is difficult or impossible to be certain which interpretation is true, it seems undesirable on pragmatic and heuristic grounds to recognize the existence of a special category of parapatric speciation.

Other evolutionists have taken a different view. Thus Mayr (1970) defined parapatric speciation in the following terms:

> There is, however, the possibility that speciation can be completed in belts of secondary contact. If hybrids are of sufficiently lowered fitness, there will be a steady elimination of those genes and chromosome arrangements that permit the interbreeding of two incipient species . . . . This process of improving allopatrically acquired isolating mechanisms in a contact zone (a process corresponding to character displacement) can be called *parapatric speciation*.

This is close to the formulation of Dobzhansky (p. 328). Now, parapatric distributions are a fact of biogeography. Mayr clearly regards them as generally due to secondary contacts. However, in a majority of the cases cited in Chapter 6 there is no evidence for an earlier complete geographic separation of the two populations, and in such instances as *Vandiemenella, Didymuria, Sceloporus grammicus,* and the Apennine 2n = 22 races of *Mus musculus,* this seems in the highest degree improbable. However, some parapatric zones that we can observe at the present time are almost certainly secondary, e.g., the one that traverses continental Europe from Denmark to the Adriatic, which is almost the same one for the subspecies of *Mus musculus, Natrix natrix,* the hooded and carrion crows, and the eastern and western hedgehogs (see p. 165).

The question of the extent to which reinforcement of genetic isolating mechanisms in zones of overlap really does occur was discussed previously (p. 329). The best examples of this phenomenon, e.g., the isolation of *Drosophila miranda* from *D. pseudoobscura* and *D. persimilis* studied by Dobzhansky and Koller (1938) and the sexual isolation between populations of the *Drosophila willistoni* semispecies studied by Ehrman (1965), all seem to involve populations that are widely sympatric. The question is whether this kind of process is likely to occur in a very narrow zone of overlap with extensive populations of the pure races or semispecies on either side. On the whole, the answer seems to be no. The zone of secondary contact between *Mus m. musculus* and *M. m. domesticus* in Jutland (see pp. 206 and 208) has probably been in existence for at least 5,000 years, without ethological isolation developing to a degree that would prevent hybridization. It seems probable that if these two semispecies do eventually reach species level, i.e., if the genetic isolating mechanisms develop to a point where hybrids are no longer being produced in genetically significant numbers, this will more likely be due to the gradual buildup of genetic changes in sexual behavior in the huge populations on either side of the zone of overlap than to any genetic changes in the relatively minute populations of the two forms that are actually in contact (and which constitute perhaps less than one-millionth of the total population, in this case).

It seems probable, therefore, that the various cases of speciation that have been designated parapatric are actually a heterogeneous collection. Many of them are probably instances of stasipatric speciation. Others are due to area-effect speciation, clinal speciation, or allopatric processes. To group them all together, simply because in every case there is a very narrow zone of overlap, is more likely to lead to confusion than clarification.

Remington (1968) claimed that parapatric hybrid zones (which, along with Mayr, he regards as zones of secondary contact) are rather recent (up to a few thousand years old) and that the frequency of hybridization declines "rather rapidly" as a result of selection for premating isolating mechanisms. These views have been criticized by Hall and Selander (1973), who point to the strong evidence that most avian hybridization zones in the northern hemisphere are quite ancient, having been in existence since Wisconsin times (60,000–10,000 years ago) or even earlier (Short, 1970).

Selection for reproductive isolation between two sympatric taxa of plants, in contrast to animals, has been considered by Levin (1970), who points out that such selection may be due to the loss of reproductive potential associated with heterospecific matings or to competition for a resource, e.g., a pollinator, necessary for mating. He concludes that this kind of selection is more likely to be important in animals than in plants because the reproductive handicap per mating will be greater in animals. Selection for reproductive isolation in plants will usually be due to competition for pollinators.

Many very narrow zones of overlap between parapatric taxa are ones where free hybridization is occurring. Since the zones may often be only a few hundred meters wide and since the distribution of the two taxa is unlikely to be uniform throughout the zone, it is usually quite impossible to ascertain whether mating is actually random, but it certainly approximates to it in a number of insect populations. Zaslavsky (1967) studied such situations in the species pairs *Eremochorus elongatus–E. oppositus* and *Phytonomus variabilis–P. transylvanicus* in the U.S.S.R., and Mrongovius (1975) investigated similarly narrow or even narrower hybrid zones in the case of morabine grasshoppers of the genus *Vandiemenella* on Kangaroo Island, Australia (following the earlier work of White, Carson, and Cheney, 1964; White *et al.*, 1967; and White *et al.*, 1969). Selection against hybrid genotypes (F_1 and backcrosses) explains the narrowness of the hybrid zones; this is called "reproductive self-destruction" by Zaslavsky. This author makes the important point that the selection does not have to be particularly strong. As he puts it: ". . . a low intensity of reproductive self-destruction is able to determine the stability of the contact zones and to make it impossible for allopatric species to be distributed more widely. The decisive factor here is the fact that the descendants of immigrants from the overlapping zone into a pure population of another species are doomed to extinction."

In animals with complex social organization factors other than, or additional to, the simple genetic ones may stabilize narrow hybrid zones. Gabow (1975) has described such a zone in Ethiopa where the baboon species *Papio anubis* and *P. hamadryas* come into contact. *Hamadryas* males herd females (of either species) and incorporate them into their harems, but *anubis* males do not, and hybrid males herd only inefficiently. Because of the behavioral inferiority of the hybrids, increases in the number of hybrid males in the overlap zone act as a brake on hybridization. Moreover, as more *anubis* females are incorporated into *hamadryas* harems, the proportion of pure *hamadryas* males in the population falls, so that competition for females is lessened and raiding of *anubis* groups slows down or ceases.

An attempt to clarify the distinctions between different types of hybrid zones has been made by Woodruff (1973), who recognizes the categories of allopatric hybridization, parapatric hybridization, and sympatric hybridization (the latter may be "peripheral," "localized," or "widespread"). The most significant hybrid zones, from the standpoint of speciation, would appear to be those that fall in Woodruff's category of parapatric hybrid zones (the "tension zones" of Key, 1968, 1974).

In general, it is clear that in such parapatric hybrid zones the hybrids are at an adaptive disadvantage. A contrary opinion has been expressed by Littlejohn and Watson (1973) in the case of a hybrid zone between the Australian frog species *Crinia laevis* and *C. victoriana*. But since these authors had previously shown that hybrids have a lowered survival rate to metamorphosis (Littlejohn, Watson, and Loftus-Hills, 1971), their later conclusion seems doubtfully based on theoretical considerations. The idea that such hybrid zones occur in ecotonal belts of terrain within which hybrids are adaptively superior does not seem soundly based, in this case or in others of the same type. The hybrid zone studied by Littlejohn and Watson (1973, 1974) was in a region where a narrow salient inhabited by *C. laevis* extends into territory occupied by *C. victoriana*. The hybrid zone is at most 11 kilometers wide and probably narrower in most places.

A hybrid zone between two other Australian frog species, *Litoria ewingi* and *L. paraewingi*, has been studied by Gartside (1972a, b). It is about the same width as the *Crinia* one. In this case there is marked zygotic lethality; most embryos from the cross *L. ewingi* ♀ × *L. paraewingi* ♂ have developmental anomalies. In spite of this, presumed F_1 and backcross hybrids exist in nature, and evidence from the transferrin phenotypes suggests that introgression is occurring.

Most parapatric hybrid zones do not seem to correspond to any obvious ecological discontinuities or ecotonal transitions. Where such a correlation has been claimed, the evidence frequently seems unconvincing, in view of our almost total ignorance, in every case, of what particular factors in the environment are really of importance in determining the presence of the species under consideration.

Again and again we encounter evidence that should warn us against generalizations on mechanisms of speciation based on single groups of organisms, however thoroughly studied. For example, evidence was presented in Chapter 2 that sibling species of *Drosophila* are relatively similar as far as their allozymes are concerned (Table 2-7, p. 41). But in a totally different group of animals, the marine polychete worms, Grassle and Grassle (1976) have shown that the "species" *Capitella capitata* is actually a complex of six sibling species, which are all sympatric in the Cape Cod area and which have virtually no allozyme alleles in common. Possibly this information tells us nothing about the speciation mechanisms involved—it may convey information only about genetic differentiation following the actual speciation process. But, clearly, sibling species in *Capitella* are a somewhat different genetical phenomenon from the *Drosophila* sibling forms that have been so extensively studied.

We have repeatedly emphasized the role of high or low vagility in determining the modes of speciation in particular groups of organisms. Bush (1975a) has gone somewhat further than this by considering the effects of so-called K- and r-selection on modes of speciation. These concepts, introduced into population biology by MacArthur and Wilson (1967), are based on the K and r parameters of the logistic equation of population growth (Volterra, 1926; Pearl, 1930). It is supposed that certain species are K-strategists and others r-strategists. The characteristics of such species have been tabulated by Pianka (1970) and Force (1975). K-strategists have low fecundity, long generation time, tend to be large in size, are highly competitive for resources, and hence are efficient competitors. By contrast, r-strategists have high fecundity, short generation time, are usually small, short-lived, and dependent on abundant food supplies; they are therefore poor competitors and exist opportunistically, being more liable to density-independent control of population size. Bush plausibly suggests that, at least in vertebrates, K-strategists, such as Carnivora, are likely to speciate allopatrically, while r-strategists may do so stasipatrically.

Ayala (1975a) has considered the relationship between rates of speciation and rates of phyletic change in individual groups of organisms. He formulates the question: "Do highly cladogenetic or species-rich lineages evolve genetically at a faster rate per unit time than species-poor lineages?" He considers two mathematical models. In the first the answer to the question is no; in the second it is yes. On the whole, the evidence favors the first model, but Ayala concludes that the available evidence, on several groups of fishes and on the Hawaiian drosophilids, is insufficient to warrant a firm conclusion.

Paleontologists are, of course, concerned to explain evolutionary changes, including speciation, in terms of changes that have occurred in the environment. It was suggested by Bretsky and Lorenz (1970) that species of organisms inhabiting very stable environments would show reduced variability: "a homozygotic genome would be selected for (homoselection) by species subjected to long-term environmental stability, and it is this selection that would ultimately prove disastrous for survival in a later, only slightly less stable regime."

There seem to be sound theoretical reasons for rejecting this hypothesis of a correlation between levels of genetic variability and the *temporal* heterogeneity of the environment (as opposed to *spatial* heterogeneity, with which there is certainly a correlation). And the evidence from deepwater marine organisms living in an environment that, whatever its spatial heterogeneity, must be extremely stable in time, suggests that they have quite high levels of genetic polymorphism (Gooch and Schopf, 1972; Ayala and Valentine, 1974). And, as we have seen in Chapter 2, the black rat, a species that is exposed to highly unstable environments, is characterized by very low levels of genetic polymorphism.

There can be no doubt that a physicist would approach the subject of speciation mechanisms in a manner different from most biologists. He would demand to know first of all the basic parameters—the number of existing species, the number of years (or generations) they have been in existence, and the area each occupies, on the average, compared to the total area of the earth's surface available for occupation by living organisms. Only then would he feel able to approach the more complex problems of the genetic mechanisms involved in speciation. There can be no doubt that most biologists, being specialists on relatively species-poor groups of organisms, tend to forget the vast number of species of invertebrate animals, and especially insects, when formulating general ideas on speciation. Much of the over-emphasis on geographic isolation as a factor in speciation, which has characterized the literature in recent decades, is due to a failure to appreciate the sheer numbers of species on limited areas of the earth's surface. It has been stated by several authorities that the number of described species of animals is about a million and a quarter, of which about 800,000 to 900,000 are insects. But this is clearly far short of the actual number that exist. In the first place, this total takes little or no account of sibling species that are now being discovered in group after group, as they come under the scrutiny of cytogeneticists, biochemists, and other specialists employing modern techniques. Second, it is clear that in the tropics, where the greatest species diversity exists, less than half (perhaps less than a quarter) of the total number of species of insects and other terrestrial arthropods have been described. C. B. Williams (1960) presented some evidence that the total number of insects in the world may be approximately three million (a low estimate would be two million, a high one five million). But even these estimates are of clearly distinguishable morphospecies, recognizable by a museum taxonomist using conventional criteria. If the total estimate included all sibling species, which are the real biological entities, the figure might be almost twice as great.

Comparable estimates for the plant kingdom are not readily available, but there can be no doubt that large numbers of species of fungi and algae remain to be discovered. A figure of ten million for all forms of life does not seem unreasonable.

The insistence that all speciation is strictly allopatric (in the sense of the dumbbell model) will probably be seen in the future as belonging to the same period in the history of biology as classical population genetics. As pointed out by Murray (1972) and Lewontin (1973), both the discovery that the genetic polymorphism of most natural populations is far greater than earlier imagined and the realization of the importance of frequency-dependent selection in natural populations have rendered much of the classical mathematical treatment obsolete when applied to actual populations in the real world.

For a full understanding of speciation mechanisms it would be necessary to have independent evidence of the antiquity of subspecies, semispecies, and sibling species in those groups that have been intensively studied genetically and cytogenetically. Unfortunately, this kind of information is almost totally lacking. Thus, we really have no idea how long ago such well-studied species pairs as *Drosophila melangogaster* and *D. simulans* and the siblings *D. pseudoobscura* and *D. persimilis* diverged from one another in evolution. Epling (1944) claimed, on bioclimatic grounds, that some of the third-chromosome inversions in *pseudoobscura* must be as old as the Miocene, but his arguments are not convincing (Mayr, 1945). Estimates of ages of the Hawaiian species of *Drosophila* are much more secure, since the age of the individual islands in the Hawaiian archipelago is known with considerable accuracy. Carson (1976) has estimated the time of origin of seven species of the *D. planitibia* group on the basis of their genetic distance from *D. heteroneura,* considered to be the most recent species. The ages range from a few hundred thousand years for *D. silvestris* (endemic to Hawaii) to three million years for *D. picticornis* (on the oldest island, Kauai). Carson suggests, on the basis of the geological, geographical, and chromosomal data, that the genetic distance *(D,* as measured by gel-electrophoresis studies on allozymes) increases at a rate of one percent per 20,000 years. He points out that this rate is about 15 times faster than the rate estimated by Yang, Soulé, and Gorman (1974) for the *Anolis* lizards of the eastern Caribbean. These flies are slow-breeding by comparison with the better-known species of *Drosophila,* and may even have only two generations a year.

Evolutionary studies on living organisms may be carried out in two different ways. One way is to take a broad look at the patterns of distribution, adaptation, and variability on a large scale as they are manifested by entire species or groups of species. In such studies one is concerned with populations of thousands or millions of individuals sampled at a limited number of localities distributed over a territory whose total area may be thousands or millions of square kilometers. Most of the data of classical taxonomy and biogeography have been obtained in this way. The picture one obtains is "evolution through a telescope."

The other way to study evolution is through the investigation of natural populations in small critical areas, such as the edge of a species' distribution or a zone of parapatric overlap between two taxa. Studies of this kind (Bigelow, 1965; Vaughan, 1967; Watson, 1972; Jackson, 1973; Woodruff, 1973, 1974) have become increasingly frequent in recent years, although their importance was recognized as long ago as 1938 by Jordan. In such investigations no detail is irrelevant, and we are concerned with the fine-grained structure of the environment (differences in soil type, vegetational cover, microclimate, etc.) and with the operation of mechanisms of genetic adaptation (including genetic isolating mechanisms between different taxa) in relation to it. This is "evolution under the microscope."

In order to better understand the mechanisms of speciation, research of the latter kind is especially needed—on a great variety of biological materials, using the latest sophisticated techniques of biochemistry, ethology, and cytogenetics.

In evolutionary theory the "modern synthesis" of the 1940s is now 30 years old, and some attempts to revise it have not taken into account sufficiently the vast increases in knowledge that have occurred in recent years. The result is that, as far as speciation is concerned, most students of biology are presented with a sterotyped dogma that leaves them with the impression that all the basic problems have been solved. The main aim of this book has been to correct this situation, continuously pointing out how much still needs to be discovered before we can confidently construct the "new synthesis" of evolutionary theory some 25 or 30 years from now—a synthesis that may indeed be the *final* one in this field of knowledge. Nevertheless, however much evolutionists of the future may synthesize in the field of speciation, we can be confident that the diversity of living organisms is such that their evolutionary mechanisms cannot be forced into the straitjacket of any narrow, universal dogma.

Bibliography

Abramoff, P., R. M. Darnell, and J. S. Balsano.
1968 Electrophoretic demonstration of the hybrid origin of the gynogenetic teleost *Poecilia formosa*. *Amer. Nat.*, **102:**555–558.

Absher, J. K.
1975 Genetic variability in a population of *Musca domestica* L. *Genetics*, **80:**S8.

Acton, A. B.
1958 The gene contents of over-lapping inversions. *Amer. Nat.*, **92:**57–58.

Adams, R. P.
1975 Gene flow versus selection pressure and ancestral differentiation in the composition of species: Analysis of populational variation of *Juniperus ashei* Buch. using terpenoid data. *J. Mol. Evol.*, **5:**177–185.

Adensamer, W.
1937 Ein Beitrag zu Art- und Rassenstudien an mitteleuropäischen Muscheln. *Zool. Jahrb. Abt. Syst.*, **70:**227–242.

Aeppli, E.
1952 Natürliche Polyploidie bei den Planarien *Dendrocoelum lacteum* (Müller) und *Dendrocoelum infernale* (Steinmann). *Z. ind. Abst.-u. Vererbl.*, **84:**182–212.

Afzelius, K.
1932 Zur Kenntnis der Fortpflanzungsverhältnisse und Chromosomenzahlen bei *Nigritella nigra*. *Svensk. Bot. Tidskr.*, **26:**365–369.

Ahmed, M.
1974 Chromosome variation in three species of the marine gastropod *Nucella*. *Cytologia*, **39:**597–607.

Albignac, R., Y. Rumpler, and J. J. Petter.
1971 Hybridation des lémuriens de Madagascar. *Mammalia*, **35:**358–368.

Alexander, M. L.
1952 The effect of two pericentric inversions upon crossing over in *Drosophila melanogaster*. *Univ. Texas Publ.* No. 5204, pp. 219–226.

Alexander, R. D.
1962a The role of behavioral study in cricket classification. *Syst. Zool.*, **11:**53–72.
1962b Evolutionary change in cricket acoustical communication. *Evolution*, **16:**443–467.
1963 Animal species, evolution, and geographic isolation. *Syst. Zool.*, **12:**202–204.
1968 Life cycle, origins, speciation and related phenomena in crickets (Orthoptera, Gryllidae). *Quart. Rev. Biol.*, **43:**1–41.

Alexander, R. D., and R. S. Bigelow.
1960 Allochronic speciation in field crickets and a new species, *Acheta veletis*. *Evolution*, **14:**334–346.

Alexander, R. D., and T. E. Moore.
1962 The evolutionary relationships of 17-year and 13-year cicadas, and three new species (Homoptera, Cicadidae, *Magicicada*). *Misc. Publ. Mus. Zool. Univ. Mich.* No. 121, pp. 1–59.

Allee, W. C., A. E. Emerson, O. Park, T. Park, and K. P. Schmidt.
1949 *Principles of Animal Ecology.* Philadelphia: Saunders.

Altukhov, Yu. P., E. A. Salmenkova, V. T. Omelchenko, G. D. Satchko, and V. I. Slynko.
1972 The number of monomorphic and polymorphic loci in populations of the salmon, *Oncorhynchus keta*—one of the tetraploid species. (In Russian) *Genetika*, **8:**67–75.

Amadon, D.
1950 The Hawaiian honeycreepers (Aves, Drepaniidae). *Bull. Amer. Mus. Nat. Hist.*, **95:**151–262.

Anderson, E.
1949 *Introgressive Hybridization.* New York: Wiley.

Antonovics, J.
1971. The effects of a heterogeneous environment on the genetics of natural populations. *Amer. Sci.*, **59:**593–599.

Antonovics, J., and A. D. Bradshaw.
1970 Evolution in closely adjacent plant populations, VIII: Clinal patterns at a mine boundary. *Heredity*, **25:**349–362.

Antonovics, J., A. D. Bradshaw, and R. G. Turner.
1971 Heavy metal tolerance in plants. *Advanc. Ecol. Res.*, **7:**2–58.

Anway, J. C.
1969 The evolution and taxonomy of *Calectasia cyanea* R. Br. (Xanthorrhoeaceae) in terms of its present-day variation and cytogenetics. *Austral. J. Bot.*, **17:**147–159.

Árnason, Ú.
1972 The role of chromosomal rearrangement in mammalian speciation with special reference to Cetacea and Pinnipedia. *Hereditas*, **70:**113–118.
1974a Comparative chromosome studies in Pinnipedia. *Hereditas*, **76:**179–226.

1974b Comparative chromosome studies in Cetacea. *Hereditas,* **77:**1–36.
1974c Phylogeny and speciation in Pinnipedia and Cetacea: A cytogenetic study. Printed by Carl Bloms Boktryckeri (8 pp.)

Artom, C.
1931 L'origine e l'evoluzione della partenogenesi attraverso i differenti biotipi di una specie collettiva *(Artemia salina* L.*)* con speciale riferimento al biotipo diploide partenogenetico di Sète. *Mem. R. Accad. Ital. Fis. Mat. Nat.,* **2:**1–57.

Askew, R. R.
1968 Considerations on speciation in Chalcidoidea (Hymenoptera). *Evolution,* **22:**642–645.
1971 *Parasitic Insects.* London: Heinemann.

Astaurov, B. L.
1972 Experimental model of the origin of bisexual polyploid species in animals (the hypothesis of indirect origin of polyploid animals via parthenogenesis and hybridization). *Biol. Zentralbl.,* **90:**137–150.

Atchley, W. R.
1974 Morphometric differentiation in chromosomally characterized parapatric races of morabine grasshoppers (Orthoptera: Eumastacidae). *Austral. J. Zool.,* **22:**25–37.

Atchley, W. R., and J. Cheney.
1974 Morphometric differentiation in the *viatica* group of morabine grasshoppers (Orthoptera, Eumastacidae). *Syst. Zool.,* **23:**400–415.

Atchley, W. R., and D. A. Hensleigh.
1974 The congruence of morphometric shape in relation to genetic divergence in four races of morabine grasshoppers (Orthoptera: Eumastacidae). *Evolution,* **28:**416–427.

Averhoff, W. W., and R. H. Richardson.
1974 Pheromonal control of mating patterns in *Drosophila melanogaster. Behavior Genetics,* **4:**207–225.
1976 Multiple-pheromone system controlling mating in *Drosophila melanogaster. Proc. Nat. Acad. Sci. U.S.A.,* **73:**591–593.

Avise, J. C., and R. K. Selander.
1972 Evolutionary genetics of cave-dwelling fishes of the genus *Astyanax. Evolution,* **26:**1–19.

Avise, J. C., and M. H. Smith.
1974a Biochemical genetics of sunfish, I: Geographic variation and subspecific intergradation in the bluegill, *Lepomis macrochirus. Evolution,* **28:**42–56.
1974b Biochemical genetics of sunfish, II: Genic similarity between hybridizing species. *Amer. Nat.,* **108:**458–472.

Avise, J. C., M. H. Smith, R. K. Selander, T. E. Lawlor, and P. R. Ramsey.
1974 Biochemical polymorphism and systematics in the genus *Peromyscus,* V: Insular and mainland species of the subgenus *Hapolomyomys. Syst. Zool.,* **23:**226–238.

Ayala, F. J.
1972a Darwinian versus non-Darwinian evolution in natural populations of *Drosophila. Proc. Sixth Berkeley Symp. Math. Stat. and Prob. Vol. V: Darwinian, Neo-Darwinian and Non-Darwinian Evolution,* pp. 211–236.
1972b Enzyme variability in the *Drosophila willistoni* group, IV: Genic variation in natural populations of *Drosophila willistoni. Genetics,* **70:**113–139.
1975a Genetic differentiation during the speciation process. *Evol. Biol.,* **8:**1–78.
1975b Scientific hypotheses, natural selection, and the neutrality theory of protein evolution. In *The Role of Natural Selection in Human Evolution,* F. M. Salzano, ed., Amsterdam: North-Holland Publishing Co., pp. 19–42.

Ayala, F. J., D. Hedgecock, G. S. Zumwalt, and J. W. Valentine.
1973 Genetic variation in *Tridacna maxima,* an ecological analog of some unsuccessful evolutionary lineages. *Evolution,* **27:**177–191.

Ayala, F. J., C. A. Mourão, S. Pérez-Salas, R. Richmond, and Th. Dobzhansky.
1970 Enzyme variability in the *D. willistoni group,* I: Genetic variation among sibling species. *Proc. Nat. Acad. Sci. U.S.A.,* **67:**225–232.

Ayala, F. J., and J. R. Powell.
1972 Allozymes as diagnostic characters of sibling species of *Drosophila. Proc. Nat. Acad. U.S.A.,* **69:**1094–1096.

Ayala, F. J., J. R. Powell, and Th. Dobzhansky.
1971 Enzyme variability in the *Drosophila willistoni group,* II: Polymorphisms in continental and island populations of *Drosophila willistoni. Proc. Nat. Acad. Sci. U.S.A.,* **68:**2480–2483.

Ayala, F. J., J. R. Powell, and M. L. Tracey.
1972 Enzyme variability in the *Drosophila willistoni group,* V: Genic variation in natural populations of *Drosophila equinoxialis. Genet. Res.,* **20:**19–42.

Ayala, F. J., J. R. Powell, M. L. Tracey, C. A. Mourão, and S. Pérez-Salas.
1972 Enzyme variability in the *Drosophila willistoni* group, IV: Genic variation in natural populations of *Drosophila willistoni. Genetics,* **70:**113–139.

Ayala, F. J., and M. L. Tracey.
1973 Enzyme variability in the *Drosophila willistoni* group, VIII: Genetic differentiation and reproductive isolation between two subspecies. *J. Hered.,* **64:**120–124.
1974 Genetic differentiation within and between species of the *Drosophila willistoni group. Proc. Nat. Acad. Sci. U.S.A.,* **71:**999–1003.

Ayala, F. J., M. L. Tracey, D. Hedgecock, and R. C. Richmond.
1974 Genetic differentiation during the speciation process in *Drosophila. Evolution,* **28:**576–592.

Ayala, F. J., and J. W. Valentine.
1974 Genetic variability in the cosmopolitan deep-water ophiuran *Ophiomusium lymani. Mar. Biol.* (Berlin), **27:**51–57.

Ayala, F. J., J. W. Valentine, L. G. Barr, and G. S. Zumwalt.
1974 Genetic variability in a temperate intertidal phoronid, *Phoronopsis viridis. Biochem. Genet.,* **11:**413–427.

Ayala, F. J., J. W. Valentine, T. E. De Laca, and G. S. Zumwalt.
1975 Genetic variability of the antarctic brachiopod *Liothyrella notorcadensis* and its bearing on mass extinction hypotheses. *J. Palaeont.*, **49:**1–9.

Babcock, E. B.
1942 Systematics, cytogenetics, and evolution in *Crepis*. *Bot. Rev.*, **8:**139–190.
1947 The genus *Crepis*, I and II. *Univ. Calif. Publ. Bot.* No. 21–22 (1,030 pp).

Babcock, E. B., and J. A. Jenkins.
1943 Chromosomes and phylogeny in *Crepis*, III: The relationships of 113 species. *Univ. Calif. Publ. Bot.* 18, pp. 241–292.

Babcock, E. B., and G. L. Stebbins.
1938 The American species of *Crepis:* Their interrelationships and distribution as affected by polyploidy and apomixis. *Carnegie Inst. Wash. Publ.* No. 504 (200 pp).

Baccetti, B.
1960 Cariologia di popolazioni partenogenetiche e bisessuate di *Troglophilus cavicola* Koll. (Inst. Orthopt.). *Verh. XI Int. Kongr. Entomol. Wien,* **1:**418–422.

Baccetti, B., and F. Capra.
1969 Notulae orthopterologicae, XXVI: Osservazioni faunistiche e cariologiche sui *Troglophilus* italiani Rhaphidophoridae. *Rassegna Speleol. Ital.*, **21:**1–15.

Bacci, G.
1955a Variabilità sessuale di popolazioni, razze e specie ermafrodite. *Ric. Sci., Suppl.*, **25:**1–6.
1955b La variabilità dei genotipi sessuali negli animali ermafroditi. *Pubbl. Staz. Zool. Napoli,* **26:**110–137.

Bachmann, K.
1970 Specific nuclear amounts in toads of the genus *Bufo*. *Chromosoma,* **29:**365–374.

Badino, G., and C. Robotti.
1975 Selection in parthenogenetic lines of *Asplanchna sieboldi* (Leydig) 1854 (Rotatoria). *Experientia,* **31:**298–299.

Bailey, D. W.
1956 Re-examination of the diversity in *Partula taeniata*. *Evolution,* **10:**360–366.

Baird, D. E., J. H. Dickson, M. W. Holdgate, and N. M. Wace.
1965 The biological report of the Royal Society expedition to Tristan da Cunha, 1962. *Phil. Trans. Roy. Soc. Lond. (B),* **249:**257–434.

Baker, H. G.
1961 *Ficus* and *Blastophaga*. *Evolution,* **15:**378–379.

Baker, R. J., W. R. Atchley, and V. R. McDaniel.
1972 Karyology and morphometrics of Peters' tent-making bat, *Uroderma bilobatum* Peters (Chiroptera, Phyllostomatidae). *Syst. Zool.*, **21:**414–429.

Baldwin, J. T. Jr.
1941 *Galax:* The genus and its chromosomes. *J. Hered.*, **32:**249–254.
Ball, R. W., and D. L. Jameson.
1966 Premating isolating mechanisms in sympatric and allopatric *Hyla californiae. Evolution,* **20:**533–551.
Bantock, C. R., and W. C. Cockayne.
1975 Chromosomal polymorphism in *Nucella lapillus. Heredity,* **34:**231–245.
Barber, H. N., and W. D. Jackson.
1957 Natural selection in action in *Eucalyptus. Nature,* **179:**1267–1269.
Barigozzi, C.
1933a Die Chromosomen-garnitur der Maulwurfsgrille und ihre systematische Bedeutung. *Z. Zellforsch.*, **18:**626–641.
1933b L'unicità della specie *Gryllotalpa gryllotalpa* L. e il suo ciclo biologico. *Boll. Lab. Zool. Portici,* **27:**145–155.
1942 Sulla struttura dei cromosomi in nuclei iperploidi di *Gryllotalpa gryllotalpa* L. *Chromosoma,* **2:**345–366.
1957 Différentiation des génotypes et distribution géographique d'*Artemia salina* Leach: données et problèmes. *Ann. Biol.*, **33:**241–250.
1974 *Artemia:* a survey of its significance in genetic problems. *Evol. Biol.*, **7:**221–252.
Barker, J. S. F., and L. J. Cummins.
1969 Disruptive selection for sternopleural bristle number in *Drosophila melanogaster. Genetics,* **61:**697–712.
Barnes, H. F.
1946–1951 *Gall Midges of Economic Importance,* 5 vols. London: Crosby Lockwood.
Barrio, A., and P. Rinaldi de Chieri.
1970a Estudios citogenéticos sobre el género *Pleurodema* y sus consecuencias evolutivas (Amphibia, Anura, Leptodactylidae). *Physis* (Buenos Aires), **30:**309–319.
1970b Relaciones cariosistemáticas de los Ceratophryidae de la Argentina. *Physis,* **30:**321–329.
Barsacchi, G., L. Bussotti, and G. Mancino.
1970 The maps of the lampbrush chromosomes of *Triturus* (Amphibia Urodela), IV: *Triturus vulgaris meridionalis. Chromosoma,* **31:**255–279.
Bartholomew, B., L. C. Eaton, and P. H. Raven.
1973 *Clarkia rubicunda:* a model of plant evolution in semiarid regions. *Evolution,* **27:**505–517.
Basilewsky, P., *et al.*
1972 *La Faune Terrestre del l'Isle de Sainte Helène.* Deuxieme partie, II: Insectes. 9. Coleoptera. *Ann. Mus. Roy. Afr. Cent. Ser. Octavo Zool.*, No. 192 (530 pp).
Basrur, V. R., and K. H. Rothfels.
1959 Triploidy in natural populations of the blackfly *Cnephia mutata* (Malloch). *Canad. J. Zool.*, **37:**571–589.

Bates, M.
1940 The nomenclature and taxonomic status of the mosquitoes of the *Anopheles maculipennis* complex. *Ann. Ent. Soc. Amer.,* **33:**343–356.

Bateson, W.
1922 Evolutionary faith and modern doubts. *Science,* **55:**55–61.

Batistic, R. F., M. Soma, M. L. Beçak, and W. Beçak.
1975 Further studies on polyploid amphibians: A diploid population of *Phyllomedusa burmeisteri. J. Hered.,* **66:**160–162.

Baylis, H. A.
1938 Helminths and evolution. In *Evolution: Essays on Aspects of Evolutionary Biology Presented to Professor E. S. Goodrich,* Oxford: Clarendon Press, pp. 249–270.

Bazykin, A. D.
1965 On the possibility of sympatric species formation. *Bull. Moscow Soc. Nat. Biol. Div.,* **70:**(1):161–165.
1969 Hypothetical mechanism of speciation. *Evolution,* **23:**685–687.

Beçak, M. L., and W. Beçak.
1974 Diploidization in *Eleutherodactylus* (Leptodactylidae–Amphibia). *Experientia,* **30:**624–625.

Beçak, M. L., W. Beçak, and M. N. Rabello.
1966 Cytological evidence of constant tetraploidy in the bisexual South American frog *Odontophrynus americanus. Chromosoma* **19:**188–193.

Beçak, M. L., W. Beçak, and L. D. Vizotto.
1970 A diploid population of the polyploid amphibian *Odontophrynus americanus* and an artificial interspecific triploid hybrid. *Experientia,* **26:**545–546.

Beçak, M. L., L. Denaro, and W. Beçak.
1970 Polyploidy and mechanisms of karyotypic diversification in Amphibia. *Cytogenetics,* **9:**225–238.

Beçak, W.
1969 Genic action and polymorphism in polyploid species of Amphibians. *Genetics,* **61** (Suppl.):183–190.

Beçak, W., and M. L. Beçak.
1973 Evolution and differentiation of polyploid amphibians. *Genetics,* **74** (Suppl.): S17–S18.

Beçak, W., M. L. Beçak, D. Lavalle, and G. Schreiber.
1967 Further studies on polyploid amphibians (Ceratophrydidae), II: DNA content and nuclear volume. *Chromosoma,* **23:**14–23.

Beçak, W., and G. Goissis.
1971 DNA and RNA content in diploid and tetraploid amphibians. *Experientia,* **27:**345–346.

Beçak, W., and M. T. Pueyo.
1970 Gene regulation in the polyploid amphibian *Odontophrynus americanus. Exp. Cell Res.,* **63:**448–451.

Beermann, W.
1955 Geschlechtsbestimmung und Evolution der genetischen Y-Chromosomen bei *Chironomus*. *Biol. Zbl.*, **74:**525–544.
1956 Inversionsheterozygotie und Fertilität der Mannchen von *Chironomus*. *Chromosoma,* **8:**1–11.
Bell, A. W.
1959 *Enchytraeus fragmentosus,* a new species of naturally fragmenting oligochaete worm. *Science,* **129:**1278.
Benazzi, M.
1957 Considerazioni sulla evoluzione cromosomica negli animali. *Boll. di Zool.*, **24:**373–409.
1960 Evoluzione cromosomica e differenziamento razziale specifico nei tricladi. *Accad. Naz. Lincei,* Quaderno 47, "Evoluzione e Genetica," pp. 273–297.
1963 Il problema sistematico delle *Polycelis* del gruppo *nigratenuis* alla luce di ricerche citologiche e genetiche. *Monit. Zool. Ital.*, **70–71:**288–300.
1967 Nuovi dati sul differenziamento citologico e genetico di planarie delle isole tirreniche. *Atti. Accad. Naz. Lincei Rend. Cl. Sci. Fis. Mat. Nat.*, Ser. VIII, **42:**469–472.
1968 Popolazioni di *Dugesia benazii* della Sardegna e della Corsica di probabile origine ibrida. *Atti Assoc. Genet. Ital.*, **13:**117–124.
1969 Annotazioni citosistematiche sui Tricladi di alcune isole tirreniche. *Atti Accad. Naz. Lincei Rend. Cl. Sci. Fis. Mat. Nat.*, Ser. VIII, **46:**605–609.
1974 Fissioning in planarians from a genetic standpoint. In *Biology of the Turbellaria: Libbie H. Hyman Memorial Volume,* pp. 476–492. New York: McGraw-Hill.
Benazzi, M., J. Baguñá, and R. Ballester.
1970 First report on an asexual form of the planarian *Dugesia lugubris* s.l. *Atti Accad. Naz. Lincei Rend. Cl. Sci. Fis. Mat. Nat.*, Ser. VIII, **48:**42–44.
Benazzi, M., and E. Giannini.
1970 Terzo contributo alla conoscenza di popolazioni della planaria *Dugesia benazzii* con due tipi di ovociti: Analasi statistica delle frequenze e considerazioni d'ordine genetico. *Atti Assoc. Genet. Ital.*, **15:**72–74.
Benazzi, M., I. Pulcinelli, and R. Del Papa.
1970 The planarians of the *Dugesia lugubris–polychroa* group: Taxonomic inferences based on cytogenetic and morphologic data. *Atti Accad. Naz. Lincei Rend. Cl. Sci. Fis. Mat. Nat.*, Ser. VIII, **48:**369–376.
Benazzi, M., and V. Scali.
1964 Modalità riproduttive della popolazione di *Bacillus rossius* (Rossi) dei dintorni di Pisa. *Atti Accad. Naz. Lincei Rend. Sci. Fis. Mat. Nat.*, Ser. VIII, **36:**311–314.
Benazzi-Lentati, G.
1966 Amphimixis and pseudogamy in freshwater triclads: Experimental reconstruction of polyploid pseudogamic biotypes. *Chromosoma,* **20:**1–14.

1970 Gametogenesis and egg fertilization in Planarians. *Internat. Rev. Cytol.*, **27:**101–179.

Bennet-Clark, H. C.

1970 A new French mole cricket, differing in song and morphology from *Gryllotalpa gryllotalpa* L. (Orthoptera, Gryllotalpidae). *Proc. Roy. Entomol. Soc. Lond. Ser. Taxon.*, **39:**125–132.

Bennet-Clark, H. C., M. Dow, A. W. Ewing, A. Manning, and F. Von Schilcher.

1976 Courtship stimuli in *Drosophila melanogaster. Behav. Genet.*, **6:**93–95.

Berger, L.

1973 Systematics and hybridization in European green frogs of *Rana esculenta* complex *J. Herpetol.*, **7:**1–10.

Bergman, B.

1941 Studies on the embryo sac mother cell and its development in *Hieracium* subg. *Archieracium. Svensk. Bot. Tidskr.*, **35:**1–42.

Bernstein, S. C., L. H. Throckmorton, and J. L. Hubby.

1973 Still more genetic variability in natural populations. *Proc. Nat. Acad. Sci. U.S.A.*, **70:**3928–3931.

Berry, D. L., and R. J. Baker.

1971 Apparent convergence of karyotypes in two species of pocket gophers of the genus *Thomomys* (Mammalia, Rodentia). *Cytogenetics*, **10:**1–9.

Berry, R. J.

1969 History in the evolution of *Apodemus sylvaticus* (Mammalia) at one edge of its range. *J. Zool. Lond.*, **159:**311–328.

1973 Chance and change in British long-tailed field mice *(Apodemus sylvaticus). J. Zool. Lond.*, **170:**351–366.

1975 On the nature of genetical distance and island races of *Apodemus sylvaticus. J. Zool. Lond.*, **176:**292–295.

Betts, E.

1961 Outbreaks of the African migratory locust (*Locusta migratoria migratorioides* R. and F.) since 1871. *Anti-Locust Memoir* No. 6 (25 pp).

Bick, B. A. E., and G. B. Sharman.

1975 The chromosomes of the platypus (*Ornithorhynchus:* Monotremata). *Cytobios*, **14:**17–28.

Biemont, M.-C., and C. Laurent.

1974 Essai de phylogénie d'espèces proches: Cheval, âne et mulet, par une technique de marquage, *C.R. Acad. Sci. Paris*, **279** D:323–326.

Bigelow, R. S.

1965 Hybrid zones and reproductive isolation. *Evolution*, **19:**449–458.

Birch, L. C.

1961 Natural selection between two species of Tephritid fruit fly of the genus *Dacus*. *Evolution*, **15:**360–374.

Birch, L. C.
1965 Evolutionary opportunity for insects and marsupials in Australia. In *The Genetics of Colonizing Species,* H. G. Baker and G. L. Stebbins, eds., pp. 197–211. New York: Academic Press.

Birch, L. C., and W. G. Vogt.
1970 Plasticity of taxonomic characters of the Queensland fruit flies *Dacus tryoni* and *Dacus neohumeralis* (Tephritidae). *Evolution,* **24:**320–343.

Blackith, R. E.
1973 Clues to the Mesozoic evolution of the Eumastacidae *Acrida,* **2:**V-XVIII.

Blackwood, M.
1953 Chromosomes of *Phylloglossum drummondii* Kunze. *Nature,* **172:**591.

Blair, W. F.
1947 Variation in shade of pelage of local populations of the cactus-mouse *(Peromyscus eremicus)* in the Tularosa basin and adjacent areas of southern New Mexico. *Contrib. Lab. Vert. Biol. Univ. Mich.,* **37:**1–7.
1950 Ecological factors in speciation of *Peromyscus. Evolution,* **4:**253–275.
1953 Factors affecting gene exchange between populations in the *Peromyscus maniculatus* group. *Texas J. Sci.,* **5:**17–33.
1955 Mating call and stage of speciation in the *Microhyla olivacea–M. carolinensis* complex. *Evolution,* **9:**469–480.
1958 Mating call in the speciation of anuran amphibians. *Amer. Nat.,* **92:**27–51.
1964 Isolating mechanisms and interspecies interactions in anuran amphibians. *Quart. Rev. Biol.,* **39:**334–344.
1974 Character displacement in frogs. *Amer. Zool.,* **14:**1119–1125.

Block, K.
1975 Chromosomal variation in Agromyzidae (Diptera), IV: Further observations on natural hybridization between two semispecies within *Phytomyza abdominalis. Hereditas,* **79:**199–208.

Bocquet, C.
1953 Recherches sur le polymorphisme naturel des *Jaera marina* (Fabr.) (Isopodes Asellotes). Essai de systématique évolutive. *Arch. Zool. Exp. Gén.,* **90:**187–450.
1969 Le problème des forms apparentées à distribution contigüe. *Bull. Soc. Zool. Fr.,* **94:**517–526.

Bogart, J. P.
1967 Chromosomes of the South American amphibian family Ceratophridae, with a reconsideration of the taxonomic status of *Odontophrynus americanus. Canad. J. Genet. Cytol.,* **9:**531–542.
1970a Systematic problems in the amphibian family Leptodactylidae (Anura) as indicated by karyotypic analysis. *Cytogenetics,* **9:**369–383.
1970b Los cromosomas de amfibios anuros del género *Eleutherodactylus. IV Congr. Latinoamericano Zool. Caracas,* 1968.
1972 Karyotypes. In *Evolution in the Genus Bufo,* W. F. Blair, ed. Austin: University of Texas Press, pp. 171–175.

1974 A karyosystematic study of frogs in the genus *Leptodactylus* (Anura· Leptodactylidae). *Copeia,* 1974:724–737.

Bogart, J. P., and A. O. Wasserman.
1972 Diploid-tetraploid cryptic species pairs: A possible clue to evolution by polyploidization in anuran amphibians. *Cytogenetics,* **11:**7–24.

Boller, E. F., and G. L. Bush.
1973 The population biology of the European cherry fruit fly, *Rhagoletis cerasi* L. (Diptera: Tephritidae), I: Evidence for genetic variation based on physiological parameters and hybridization experiments. *Entomol. Exp. Appl.,* **17:**279–293.

Bonnell, M. L., and R. K. Selander.
1974 Elephant seals: Genetic variation and near extinction. *Science,* **184:**908–909.

Bonnet, D.A.
1950 The hybridization of *Aedes aegypti* L. and *Aedes albopictus* Skuse in the Territory of Hawaii. *Proc. Hawaiian Ent. Soc.,* **14:**35–39.

Bowers, J. H., R. J. Baker, and M. H. Smith.
1973 Chromosomal, electrophoretic, and breeding studies of selected populations of deer mice *(Peromyscus maniculatus)* and black-eared mice *(P. melanotis). Evolution,* **27:**378–386.

Bowman, R. I.
1961 Morphological differentiation and adaptation in the Galápagos Finches. *Univ. Calif. Publ. Zool.* 58, 302 pp.

Boyes, J. M.
1967 The cytology of muscoid flies. In *Genetics of Insect Vectors of Disease,* J. W. Wright and R. Pal, eds., Amsterdam: Elsevier, pp. 371–384.

Bradshaw, A. D.
1952 Populations of *Agrostis tenuis* resistant to lead and zinc poisoning. *Nature,* **169:**1098.

Bradshaw, A. D., T. S. McNeilly, R. P. G. Gregory.
1965 Industrialization, evolution, and the development of heavy metal tolerance in plants. *Brit. Ecol. Soc. Symp.,* **6:**327–343.

Bradshaw, W. N., and T. C. Hsu.
1972 Chromosomes of *Peromyscus* (Rodentia, Cricetidae), III: Polymorphism in *Peromyscus maniculatus. Cytogenetics,* **11:**436–451.

Bretsky, P. W., and D. M. Lorenz.
1970 An essay on genetic adaptive strategies and mass extinctions. *Bull. Geol. Soc. Amer.,* **81:**2449–2456.

Britten, R. J., and D. E. Kohne.
1967 Nucleotide sequence repetition in DNA. *Carnegie Inst. Wash. Yearb.,* **65:**78–106.

Brncic, D., P. S. Nair, and M. R. Wheeler.
1971 Cytogenetic relationships within the *mesophragmatica* species group of *Drosophila. Univ. Texas Publ.* No. 7103, pp. 1–16.

Brock, R. D., and J. A. M. Brown.
1961 Cytotaxonomy of Australian *Danthonia*. *Austral. J. Bot.*, **9:**62–91.
Brooks, J. L.
1950 Speciation in ancient lakes. *Quart. Rev. Biol.*, **25:**30–60 and 131–176.
1957 The species problem in freshwater animals. In *The Species Problem*, E. Mayr, ed. *Amer. Assoc. Adv. Sci. Publ.* No. 50, pp. 81–123.
Brower, L. P., and J. van Z. Brower.
1962 The relative abundance of model and mimic butterflies in natural populations of the *Baltus philenor* mimicry complex. *Ecology*, **43:**154–158.
Brown, D. D., and K. Sugimoto.
1973 The structure and evolution of ribosomal and 5s DNA in *Xenopus laevis* and *Xenopus mulleri*. *Cold Spr. Harb. Symp. Quant. Biol.*, **38:**501–505.
Brown, D. S., and C. A. Wright.
1972 On a polyploid complex of freshwater snails (Planorbidae: *Bulinus*) in Ethiopia. *J. Zool.*, **167:**97–132.
Brown, K. S., Jr., P. M. Sheppard, and J. R. G. Turner.
1974 Quaternary refugia in tropical America: Evidence from race formation in *Heliconius* butterflies. *Proc. Roy. Soc. Lond. (B)*, **187:**369–378.
Brown, L. E.
1973 Speciation in the *Rana pipiens* complex. *Amer. Zool.*, **13:**73–79.
Brown, L. E., and J. R. Brown.
1972 Call types of the *Rana pipiens* complex in Illinois. *Science*, **176:**928–929.
Brown, S. W.
1965 Chromosomal survey of the armored and Palm scale insects (Coccoidea: Diaspididae and Phoenicoccidae). *Hilgardia*, **36:**189–294.
Brown, W. L. Jr.
1957 Centrifugal speciation. *Quart. Rev. Biol.*, **32:**247–277.
Brown, W. L., and E. O. Wilson.
1956 Character displacement. *Systemat. Zool.*, **5:**49–64.
Brown, W. V.
1948 A cytological study in the Gramineae. *Amer. J. Bot.*, **35:**382–395.
Brown, W. V., and W. H. P. Emery.
1957 Apomixis in the Gramineae, tribe Andropogoneae: *Themeda triandra* and *Bothriochloa ischaemum*. *Bot. Gaz.*, **118:**246–253.
Brues, C. T.
1924 The specificity of food plants in the evolution of phytophagous insects. *Amer. Nat.*, **58:**127–144.
Brutlag, D. L., and W. J. Peacock.
1975 Sequences of highly repeated DNA in *Drosophila melanogaster*. In *The Eukaryote Chromosome*, W. J. Peacock and R. D. Brock, eds. Canberra: Australian National University Press, pp. 35–45.
Bullini, L.
1964 Richerche sul rapporto sessi in *Bacillus rossii* (Fab.), *Atti Accad. Naz. Lincei Rend. Sci. Fis. Mat. Nat.*, Ser. VIII, **36:**897–902.

1965 Research into the biological characteristics of amphigony and parthenogenesis in a bisexual population of *Bacillus rossius* (Rossi) (Cheleutoptera, Phasmoidea). *Rivista di Biol.*, **58:**189–244.

Bullini, L., and M. Coluzzi.
1972 Natural selection and genetic drift in protein polymorphism. *Nature*, **239:**160–161.

Burch, J. B.
1964 Cytological studies of Planorbidae (Gastropoda: Basommatophora), I: The African subgenus *Bulinus* s.s. *Malacologia*, **1:**387–400.

Burch, J. B., P. F. Basch, and L. L. Bush.
1960 Chromosome numbers in Ancylid snails. *Rev. Portuguesa Zool. Biol. Geral*, **2:**199–204.

Bush, G. L.
1966 Taxonomy, cytology, and evolution of the genus *Rhagoletis* in North America (Diptera: Tephritidae). *Bull. Mus. Comp. Zool. Harvard Univ.*, **134:**431–562.
1969a Sympatric host race formation and speciation in frugivorous flies of the genus *Rhagoletis*. *Evolution*, **23:**237–251.
1969b Mating behavior, host specificity, and the ecological significance of sibling species in frugivorous flies of the genus *Rhagoletis* (Diptera, Tephritidae). *Amer. Nat.*, **103:**669–672.
1974 The mechanism of sympatric host race formation in the true fruit flies (Tephritidae). In *Genetic Mechanisms of Speciation in Insects*, M. J. D. White, ed., Sydney: Australia and New Zealand Book Co., pp. 3–23.
1975a Modes of animal speciation. *Ann. Rev. Ecol. Systemat.*, **6:**339–364.
1975b Sympatric speciation in phytophagous parasitic insects. In *Evolutionary Strategies of Parasitic Insects*, P. W. Price, ed., London: Plenum Press, pp. 187–206.

Bush, G. L., S. M. Case, A. C. Wilson, and J. L. Patton.
1977 Rapid speciation and chromosomal evolution in mammals. *Proc. Nat. Acad. Sci. U.S.A.*, in press.

Buxton, P. A.
1939 *The Louse*. London: Edward Arnold and Co.

Cain, A. J.
1971 Color and banding morphs in subfossil samples of the snail *Cepaea*. In *Ecological Genetics and Evolution*, R. Creed, ed., Oxford: Blackwell, pp. 65–92.

Cain, A. J., and J. D. Currey.
1963a Area effects in *Cepaea*. *Phil. Trans. Roy. Soc. Lond. (B)*, **246:**1–181.
1963b The causes of area effects. *Heredity*, **18:**467–471.

Cain, A. J., and P. M. Sheppard.
1950 Selection in the polymorphic land snail *Cepaea nemoralis*. *Heredity*, **4:**275–294.

Cain, A. J., and P. M. Sheppard
1954 Natural selection in *Cepaea*. *Genetics,* **39:**89–116.
1957 Some breeding experiments with *Cepaea nemoralis* (L.). *J. Genet.* **55:** 195–199.

Cain, A. J., P. M. Sheppard, J. M. B. King, M. A. Carter, J. D. Currey, B. Clarke, C. Diver, J. Murray, and R. W. Arnold.
1968 Studies on *Cepaea,* I–VII. *Phil. Trans. Roy. Soc. Lond. (B),* **253:**383–595.

Caire, W., and E. G. Zimmerman.
1975 Chromosomal and morphological variation and circular overlap in the deer mouse, *Peromyscus maniculatus,* in Texas and Oklahoma. *Syst. Zool.,* **24:**89–95.

Callan, H. G.
1972 Replication of DNA in the chromosomes of eukaryotes. *Proc. Roy. Soc. Lond. (B),* **181:**19–41.

Callan, H. G., and L. Lloyd.
1960 Lampbrush chromosomes of crested newts, *Triturus cristatus* (Laurenti). *Phil. Trans. Roy. Soc. Lond. (B),* **243:**135–219.

Callan, H. G., and H. Spurway.
1951 A study of meiosis in interracial hybrids of the newt *Triturus cristatus. J. Genet.,* **50:**235–249.

Calton, M. S., and T. E. Denton.
1974 Chromosomes of the chocolate gourami: A cytogenetic anomaly. *Science,* **185:**618–619.

Canberra.
1975 *The Eukaryote Chromosome.* Papers presented at the Canberra, 1974 Eukaryote Chromosome Conference, W. J. Peacock and R. D. Brock, eds., Australian National University Press.

Capanna, E.
1973 Concluding remarks. In *Cytotaxonomy and Vertebrate Evolution,* A. B. Chiarelli and E. Capanna, eds., London: Academic Press, pp. 681–695.

Capanna, E., M. V. Civitelli, and M. Cristaldi.
1973 Chromosomal polymorphism in an alpine population of *Mus musculus* L. *Boll. di Zool.,* **40:**379–383.
1974 Una popolazione appenninica di *Mus musculus* L. caratterizzata da un cariotipo a 22 cromosomi. *Atti Accad. Naz. Lincei Rend. Cl. Sci. Fis. Mat. Nat.,* Ser. VIII, **54:**981–984.

Capanna, E., M. Cristaldi, P. Perticone, and M. Rizzoni.
1975 Identification of chromosomes involved in the nine Robertsonian fusions of the Apennine mouse with a 22-chromosome karyotype. *Experientia,* **31:**294–296.

Capanna, E., A. Gropp, H. Winking, G. Noack, and M.-V. Civitelli.
1976 Robertsonian metacentrics in the mouse. *Chromosoma,* **58:**341–353.

Capanna, E., and M. G. Manfredi Romanini.
1971 Nuclear DNA content and morphology of the karyotype in certain palearctic microchiroptera. *Caryologia,* **24:**471–482.

Carano, E.
1926 Ulteriori osservazioni su *Euphorbia dulcis* L. in rapporto col suo comportamento apomittico. *Ann. Bot. (Roma),* **17:**50–79.

Cardé, R. T., J. Kochansky, J. F. Stimmel, A. G. Wheeler, Jr., and W. L. Roelofs.
1975 Sex pheromones of the European corn borer *Ostrinia nubilalis: cis* and *trans* responding males in Pennsylvania. *Environ. Entomol.,* **4:**413–414.

Carson, H. L.
1959 Genetic conditions which promote or retard the formation of species. *Cold Spr. Harb. Symp. Quart. Biol.,* **24:**87–105.
1962 Fixed heterozygosity in a parthenogenetic species of *Drosophila.* In *Studies in Genetics II,* Univ. Texas Publ. No. 6205, pp. 55–62.
1965 Chromosomal morphism in geographically widespread species of Drosophila. In *Genetics of Colonizing Species,* H. G. Baker and G. L. Stebbins, eds., New York: Academic Press, pp. 503–531.
1967 Selection for parthenogenesis in *Drosophila mercatorum. Genetics,* **55:**157–171.
1968 The population flush and its genetic consequences. In *Population Biology and Evolution,* R. C. Lewontin, ed., Syracuse University Press, pp. 123–137.
1969 Parallel polymorphisms in different species of Hawaiian *Drosophila. Amer. Nat.,* **103:**323–329.
1970a Chromosome tracers of the origin of species. *Science,* **168:**1414–1418.
1970b Chromosomal tracers of founders events. *Biotropica,* **2:**3–6.
1971 Speciation and the founder principle. *Stadler Genetics Symposia,* **3:**51–70.
1973a Ancient chromosomal polymorphism in Hawaiian *Drosophila. Nature,* **241:**200–202.
1973b Reorganization of the gene pool during speciation. In "Genetic Structure of Populations," N. E. Morton, ed., *Population Genetics Monographs Vol. III.* University of Hawaii Press, pp. 274–280.
1973c The genetic system in parthenogenetic strains of *Drosophila mercatorum. Proc. Nat. Acad. Sci. U.S.A.,* **70:**1772–1774.
1974 Patterns of speciation in Hawaiian *Drosophila* inferred from ancient chromosomal polymorphism. In *Genetic Mechanisms of Speciation in Insects,* M. J. D. White, ed., Sydney: Australia and New Zealand Book Co., pp. 81–93.
1975 The genetics of speciation at the diploid level. *Amer. Nat.,* **109:**83–92.
1976 Inference of the time of origin of some *Drosophila* species. *Nature,* **259:**395–396.

Carson, H. L., and W. C. Blight.
1952 Sex chromosome polymorphism in a population of *Drosophila americana. Genetics,* **37:**572.

Carson, H. L., F. E. Clayton, and H. D. Stalker.
1967 Karyotypic stability and speciation in Hawaiian *Drosophila. Proc. Nat. Acad. Sci. U.S.A.,* **57:**1280–1285.

Carson, H. L., D. E. Hardy, H. T. Spieth, and W. S. Stone.
1970 The evolutionary biology of the Hawaiian Drosophilidae. In *Essays in*

Evolution and Genetics in Honor of Theodosius Dobzhansky, M. K. Hecht and W. C. Steere, eds., New York: Appleton Century Crofts, pp. 437–543.

Carson, H. L., P. S. Nair, and F. M. Sene.
1975 *Drosophila* hybrids in nature: Proof of gene exchange between sympatric species. *Science,* **189:**806–807.

Carson, H. L., and J. E. Sato.
1969 Microevolution within three species of Hawaiian *Drosophila. Evolution,* **23:**493–501.

Carson, H. L., and H. D. Stalker.
1969 Polytene chromosome relationships in Hawaiian species of *Drosophila,* IV: The *D. primaeva* subgroup. *Studies in Genetics V,* Univ. Texas Publ. No. 6918, pp. 85–93.

Carson, H. L., I. Y. Wei, and J. A. Neiderkon, Jr.
1969 Isogenicity in parthenogenetic strains of *Drosophila mercatorum. Genetics,* **63:**619–628.

Carson, H. L., M. R. Wheeler, and W. B. Heed.
1957 A parthenogenetic strain of *Drosophila mangabeirai* Malogolowkin. *Univ. Texas Publ.* No. 5721, pp. 115–122.

Carter, C. R., and S. Smith-White.
1972 The cytology of *Brachycome lineariloba,* III: Accessory chromosomes. *Chromosoma,* **39:**361–379.

Carter, C. R., S. Smith-White, and D. W. Kyhos.
1974 The cytology of *Brachycome lineariloba,* IV: The 10-chromosome quasi-diploid. *Chromosoma,* **44:**439–456.

Caspari, E., and G. S. Watson.
1959 On the evolutionary importance of cytoplasmic sterility in mosquitoes. *Evolution,* **13:**568–570.

Cassagnau, P.
1974 Les chromosomes polytènes de *Neanura monticola* Cassagnau (Collemboles), I: Polymorphisme écologique du chromosome X. *Chromosoma,* **46:**343–363.

Cavalier-Smith, T.
1974 Palindromic base sequences and replication of eukaryote chromosome ends. *Nature,* **250:**467–470.

Chabora, A. J.
1968 Disruptive selection for sternopleural chaeta number in various strains of *Drosophila melanogaster. Amer. Nat.,* **102:**525–532.

Chandley, A. C., R. Jones, H. M. Dott, W. R. Allen, and R. V. Short.
1974 Meiosis in interspecific equine hybrids, I: The male mule *(E. asinus × E. caballus)* and hinny *(E. caballus × E. asinus). Cytogenetics Cell Genet.,* **13:**330–341.

Chiarugi, A., and E. Francini.
1930 Apomissia in *"Ochna serrulata"* Walp. *Nuovo G. Bot. Ital.* (N.S.), **37:**1–250.

Chooi, W. Y.

1971a Variation in nuclear DNA content in the genus *Vicia*. *Genetics,* **68:**195–211.

1971b Comparison of the DNA of six *Vicia* species by the method of DNA-DNA hybridization. *Genetics,* **68:**213–230.

Christensen, B.

1961 Studies on cytotaxonomy and reproduction in the Enchytraeidae, with notes on parthenogenesis and polyploidy in the animal kingdom. *Hereditas,* **47:**387–450.

Christiansen, M. P.

1942 The *Taraxacum* flora of Iceland. In *The Botany of Iceland,* Vol. 3, Part III, L. K. Rosenvinge and E. Warming, eds., Copenhagen: J. Frimodt, pp. 235–343.

Chu, E. H. Y., and M. A. Bender.

1961 Chromosome cytology and evolution in primates. *Science,* **133:**1399–1405.

1962 Cytogenetics and evolution of primates. *Ann. New York Acad. Sci.,* **102:**253–266.

Chu, E. H. Y., and B. A. Swomley.

1961 Chromosomes of lemurine lemurs. *Science,* **133:**1925–1926.

Chubareva, L. A.

1968a Triploidy in natural populations of blackflies, *Prosimulium macropyga* Lundstr. (Simuliidae, Diptera). *Tsitologia,* **10:**750–754.

1968b On polyploidy in Diptera, family Simuliidae. *Tsitologia i Genetika,* **2:**456–465.

Clarke, B.

1966 The evolution of morph-ratio clines. *Amer. Nat.,* **100:**389–402.

1968 Balanced polymorphism and regional differentiation in land snails. In *Evolution and Environment,* E. T. Drake, ed., Yale University Press, pp. 351–368.

1969 Discussion. In *Systematic Biology,* National Academy of Sciences, U.S.A., pp. 242–244.

1974 Book review of R. C. Lewontin's *The Genetic Basis of Evolutionary Change*. *Science,* **186:**524–525.

Clarke, B., and J. Murray.

1969 Ecological genetics and speciation in land snails of the genus *Partula*. *Biol. J. Linn. Soc.,* **1:**31–42.

1971 Polymorphism in a polynesian land snail *Partula suturalis vexillum*. In *Ecological Genetics and Evolution,* R. Creed, ed., pp. 51–64. Oxford and Edinburgh: Blackwell Scientific Publications.

Clarke, C. A., and P. M. Sheppard.

1960 The genetics of *Papilio dardanus* Brown, III: Race *antinorii* from Abyssinia and race *meriones* from Madagascar. *Genetics,* **45:**683–698.

1975 The genetics of the mimetic butterfly *Hypolimnas bolina* (L.). *Phil. Trans. Roy. Soc. Lond. (B),* **272:**229–265.

Clausen, J., D. D. Keck, and W. M. Hiesey.
1945 Experimental studies on the nature of species, II: Plant evolution through amphiploidy and autoploidy, with examples from the Madiinae. *Carnegie Inst. Wash. Publ.* No. 564 (174 pp).

Clay, T.
1949 Some problems in the evolution of a group of ectoparasites. *Evolution,* **3:**279–299.

Clayton, F. E., H. L. Carson, and J. E. Sato.
1972 Polytene chromosome relationships in Hawaiian species of *Drosophila,* VI: Supplementary data on metaphases and gene sequences. In *Studies in Genetics, VII,* Univ. Texas Publ. No. 7213, pp. 163–177.

Cleland, R. E.
1950 Studies in *Oenothera* cytogenetics and phylogeny. *Indiana Univ. Publ. Sci. Ser.* No. 16.
1962 The cytogenetics of *Oenothera. Adv. in Genet.,* **11:**147–237.
1964 The evolutionary history of the North American evening primroses of the *"biennis* group." *Proc. Amer. Phil. Soc.,* **108:**88–98.

Cognetti, G.
1961 Citogenetica della partenogenesi negli afidi. *Arch. Zool. Ital.,* **46:**89–122.

Cognetti, G., and A. M. Pagliai.
1963 Razze sessuali in *Brevicoryne brassicae* L. (Homoptera Aphididae). *Arch. Zool. Ital.,* **48:**329–337.

Cold Spring Harbor 1973.
1974 *"Chromosome Structure and Function." Cold Spr. Harb. Symp. Quart. Biol.,* **38** (1,010 pp).

Cole, C. J.
1971 Karyotypes of the five monotypic species groups of lizards in the genus *Sceloporus. Amer. Mus. Novit.,* **2450:**1–17.
1972 Chromosome variation in North American fence lizards (genus *Sceloporus; undulatus* species group). *System. Zool.,* **21:**357–363.
1974 Chromosome evolution in selected tree frogs, including casque-headed species *(Pternotyla, Triprion, Hyla,* and *Smilisca). Amer. Mus. Novit.* No. 2541 (10 pp).

Coluzzi, M.
1964 Morphological divergences in the *Anopheles gambiae* complex. *Riv. Malariol.,* **43:**197–232.
1966 Osservazioni comparative sul cromosoma X nelle specie A e B del complesso *Anopheles gambiae. Atti Accad. Naz. Lincei Rend. Cl. Sci. Fis. Mat. Nat.,* Ser. VIII, **40:**671–677.
1970 Sibling species in *Anopheles* and their importance in malariology. *Misc. Publ. Ent. Soc. Amer.,* **7:**63–77.

Coluzzi, M., and L. Bullini.
1971 Enzyme variants as markers in the study of pre-copulatory isolating mechanisms. *Nature,* **231:**455–456.

Coluzzi, M., A. M. Gironi, and D. A. Muir.
1970 Ulteriori esperimenti d'incrocio tra le forme del complesso *mariae* del genere *Aedes*. *Parassitol.*, **12:**119–123.
Coluzzi, M., and A. Sabatini.
1967 Cytogenetic observations on species A and B of the *Anopheles gambiae* complex. *Parassitologia,* **9:**73–88.
1968a Divergenze morphologiche e barriere di sterilità nel complesso *Aedes mariae* (Diptera, Culicidae). *Riv. Parassitol.*, **29:**49–70.
1968b Cytogenetic observations on species C of the *Anopheles gambiae* complex. *Parassitol.*, **10:**155–165.
1969 Cytogenetic observations on the salt water species, *Anopheles merus* and *Anopheles melas,* of the *gambiae* complex. *Parassitol.*, **11:**177–187.
Corbet, G. B.
1970 Patterns of subspecific variation. *Symp. Zool. Soc. Lond.*, **26:**105–116.
Cordeiro, M., L. Wheeler, C. S. Lee, C. D. Kastritsis, and R. H. Richardson.
1975 Heterochromatic chromosomes and satellite DNA's of *Drosophila nasutoides*. *Chromosoma,* **51:**65–73.
Coyne, J. A.
1974 The evolutionary origin of hybrid inviability. *Evolution,* **28:**505–506.
Craddock, E. M.
1970 Chromosome number variation in a stick insect, *Didymuria violescens* (Leach). *Science,* **167:**1380–1382.
1974a Chromosomal evolution and speciation in *Didymuria*. In *Genetic Mechanisms of Speciation in Insects,* M. J. D. White, ed., Sydney: Australia and New Zealand Book Co., pp. 24–42.
1974b Degrees of reproductive isolation between closely related species of Hawaiian *Drosophila*. In *ibid.*, pp. 111–139.
1974c Reproductive relationships between homosequential species of Hawaiian *Drosophila*. *Evolution,* **28:**593–606.
1975 Intraspecific karyotypic differentiation in the Australian phasmatid *Didymuria violescens* (Leach), I: The chromosome races and their structural and evolutionary relationships. *Chromosoma,* **53:**1–24.
Crampton, H. E.
1916 Studies on the variation, distribution, and evolution of the genus *Partula:* The species inhabiting Tahiti. *Carnegie Inst. Wash. Publ.*, No. 228 (331 pp.).
1932 Studies on the variation, distribution, and evolution of the genus *Partula:* The species inhabiting Moorea. *Carnegie Inst. Wash. Publ.*, No. 410 (335 pp.).
Creed, E. R., W. H. Dowdeswell, E. B. Ford, and K. G. McWhirter.
1970 Evolutionary studies on *Maniola jurtina* (Lepidoptera, Satyridae): The boundary phenomenon in southern England, 1961 to 1968. In *Essays in Evolution and Genetics in Honor of Theodosius Dobzhansky,* M. K. Hecht and W. C. Steere, eds., New York: Appleton-Century-Crofts, pp. 263–287.
Croizat, L., G. Nelson, and D. E. Rosen.
1974 Centers of origin and related concepts. *Syst. Zool.*, **23:**265–287.

Crosby, J. L.
1970 The evolution of genetic discontinuity: Computer models of the selection of barriers to interbreeding between subspecies. *Heredity,* **25:**253–297.
Crow, J. F., and M. Kimura.
1965 Evolution in sexual and asexual populations. *Amer. Nat.* **99:**439–450.
Crozier, R. H.
1973 Apparent differential selection at an isozyme locus between queens and workers of the ant *Aphaenogaster rudis. Genetics,* **73:**313–318.
1975 *Hymenoptera.* Vol. 3, Insecta 7, of *Animal Cytogenetics,* B. John, ed., Berlin: Borntraeger (95 pp).
Cuellar, O.
1971 Reproduction and the mechanism of meiotic restitution in the parthenogenetic lizard *Cnemidophorus uniparens. J. Morph.,* **133:**139–165.
1974 On the origin of parthenogenesis in vertebrates: The cytogenetic factors. *Amer. Nat.,* **108:**625–648.
Cuellar, O., and A. G. Kluge.
1972 Natural parthenogenesis in the gekkonid lizard *Lepidodactylus lugubris. J. Genet.,* **61:**14–26.

Da Cunha, A. B., H. Burla, and Th. Dobzhansky.
1950 Adaptive chromosomal polymorphism in *Drosophila willistoni. Evolution,* **4:**212–235.
Da Cunha, A. B., Th. Dobzhansky, O. Pavlovsky, and B. Spassky.
1959 Genetics of natural populations, XXVIII: Supplementary data on the chromosomal polymorphism in *Drosophila willistoni* in its relation to the environment. *Evolution,* **13:**389–404.
Darevski, I. S.
1966 Natural parthogenesis in a polymorphic group of Caucasian rock lizards related to *Lacerta saxicola* Eversmann. *J. Ohio Herpetol. Soc.,* **5:**115–152.
Darevski, I. S., and V. N. Kulikova.
1961 Natürliche Parthenogenese in der polymorphen Gruppe der Kaukasichen Felseidechse *(Lacerta saxicola Eversmann). Zool. Jahrb. Syst.,* **89:**119–176.
1964 Natural triploidy in a polymorphic group of Caucasian rock lizards *(Lacerta saxicola* Eversmann*)* resulting from hybridization between bisexual and parthenogenetic forms of this species. (Russian) *Dokl. Akad. Nauk S.S.S.R.,* **158:**202–205.
Darlington, C. D., and G. W. Shaw.
1959 Parallel polymorphism in the heterochromatin of *Trillium* species. *Heredity,* **13:**89–121.
Davidson, E. H., G. A. Galau, R. C. Angerer, and R. J. Britten.
1975 Comparative aspects of DNA organization in Metazoa. *Chromosoma,* **51:**253–259.

Davidson, G., and R. H. Hunt.
1973 The crossing and chromosome characteristics of a new, sixth species in the *Anopheles gambiae* complex. *Parassitologia,* **15:**121–128.

Davidson, G., H. E. Paterson, M. Coluzzi, G. F. Mason, and D. W. Micks.
1967 The *Anopheles gambiae* complex. In *Genetics of Insect Vectors of Disease,* J. W. Wright and R. Pal, eds., Amsterdam: Elsevier, pp. 211–250.

Davies, E.
1956 Cytology, evolution, and origin of the aneuploid series in the genus *Carex. Hereditas,* **42:**349–365.

Davis, B. L., and R. J. Baker.
1974 Morphometrics, evolution, and cytotaxonomy of mainland bats of the genus *Macrotus* (Chiroptera, Phyllostomatidae). *Syst. Zool.,* **23:**26–39.

Day, A.
1965 The evolution of a pair of sibling allotetraploid species of Cobwebby *Gilias* (Polemoniaceae). *Aliso,* **6:**25–75.

De Lesse, H.
1952 Quelques formules chromosomiques chez les *Lycaenidae* (Lépidoptères, Rhopalocères). *C.R. Acad. Sci. Paris,* **235:**1692–1694.
1953 Formules chromosomiques nouvelles chez les Lycaenidae (Lepid., Rhopal.). *C.R. Acad. Sci. Paris,* **237:**1781–1783.
1954 Formules chromosomiques nouvelles chez les Lycaenidae (Lepid., Rhopal.). *C.R. Acad. Sci. Paris,* **238:**514–516.
1955 Une nouvelle formule chromosomique dans le groupe d'*Erebia tyndarus* Esp. (Lepidoptères, Satyrinae). *C.R. Acad. Sci. Paris,* **241:**1505–1507.
1957 Description de deux nouvelles espèces d'*Agrodiaetus* (Lep., Lycaenidae) separées à la suite de la decouverte de leurs formules chromosomiques. *Lambillionea,* **57:**65–71.
1959a Separation specifique d'un *Lysandra* d'Afrique du Nord a la suite de la decouverte de sa formule chromosomique (Lycaenidae). *Alexanor,* **1:**61–64.
1959b Note sur deux espèces d'*Agrodiaetus* (Lep., Lycaenidae) recemment separées d'après leurs formules chromosomiques. *Lambillionea,* **59:**5–10.
1960 Spéciation et variation chromosomique chez les Lépidoptères Rhopalocères. *Ann. Sci. Nat. Zool.,* **12** Ser. 2:1–223.
1964 Les nombres de chromosomes chez quelques *Erebia* femelles (Lep., Satyrinae). *Rev. Fr. Entomol.,* **31:**112–115.
1969 Les nombres de chromosomes dans le groupe de *Lysandra coridon* (Lep., Lycaenidae). *Ann. Soc. Entomol. Fr.* (N.S.), **5:**469–522.
1970 Les nombres de chromosomes dans le groupe de *Lysandra argester* et leur incidence sur sa taxonomie. *Bull. Soc. Entomol. Fr.,* **75:**64–68.

Descamps, M.
1973 Révision des Eumastacoidea (Orthoptera) aux échelons des familles et des sous-familles (genitalia, répartition, phylogenie). *Acrida,* **2:**161–298.

Dev, V. G., D. A. Miller, R. Tantrarahi, R. R. Schreck, T. H. Roderick, B. F. Erlanger, and O. J. Miller.

1975 Chromosome markers in *Mus musculus:* Differences in C-banding between the subspecies *M. m. musculus* and *M. m. molossinus*. *Chromosoma,* **53:**335–344.

De Vries, G. F., H. F. de France, and J. A. M. Scheners.

1975 Identical banding patterns of two *Macaca* species: *Macaca mulatta* and *M. fascicularis*. *Cytogenet. Cell Genet.,* **14:**26–33.

De Wet, J. M. J.

1954 The genus *Danthonia* in grass phylogeny. *Amer. J. Bot.,* **41:**204–211.

1968 Diploid-tetraploid-haploid cycles and the origin of variability in *Dichanthium* agamospecies. *Evolution,* **22:**394–397.

Dhaliwal, H. S., and B. L. Johnson.

1976 Anther morphology and the origin of the tetraploid wheats. *Amer. J. Bot.,* **63:**363–368.

Dickinson, H., and J. Antonovics.

1973 Theoretical considerations of sympatric divergence. *Amer. Nat.,* **107:**256–274.

Dirsh, V. M.

1970 La faune terrestre de l'ile de Sainte-Helène (première partie), 7. Orthoptera, c. Acridoidea. *Ann. Mus. Roy. Afr. Cent. Ser. Octavo Zool.,* **181:** 201–210.

1974 The genus *Schistocerca* (Acridomorpha, Insecta). The Hague: W. Junk.

Dobzhansky, Th.

1937a Genetic nature of species differences. *Amer. Nat.,* **71:**404–420.

1937b *Genetics and the Origin of Species,* 1st ed., Columbia University Press.

1951 *Genetics and the Origin of Species,* 3rd ed., Columbia University Press.

1955 A review of some fundamental concepts and problems of population genetics. *Cold Spr. Harb. Symp. Quart. Biol.,* **20:**1–15.

1957 Genetics of natural populations, XXVI: Chromosomal variability in island and continental populations of *Drosophila willistoni* from Central America and the West Indies. *Evolution,* **11:**280–293.

1970 *Genetics of the Evolutionary Process*. Columbia University Press.

1972 Species of *Drosophila*. *Science,* **177:**664–669.

1973 Is there gene exchange between *Drosophila pseudobscura* and *Drosophila persimilis* in their natural habitats? *Amer. Nat.,* **107:**312–314.

1975 Analysis of incipient reproductive isolation within a species of *Drosophila*. *Proc. Nat. Acad. Sci. U.S.A.,* **72:**3638–3641.

Dobzhansky, Th., L. Ehrman, O. Pavlovsky, and B. Spassky.

1964 The superspecies *Drosophila paulistorum*. *Proc. Nat. Acad. Sci. U.S.A.,* **51:**3–9.

Dobzhansky, Th., and P. C. Koller.

1938 An experimental study of sexual isolation in *Drosophila*. *Biol. Zentralbl.* **58:**589–607.

Dobzhansky, Th., and O. Pavlovsky.
1962 A comparative study of the chromosomes in the incipient species of the *Drosophila paulistorum* complex. *Chromosoma,* **13:**196–218.
1966 Spontaneous origin of an incipient species in the *Drosophila paulistorum* complex. *Proc. Nat. Acad. Sci. U.S.A.,* **55:**727–733.
1967 Experiments on the incipient species of the *Drosophila paulistorum* complex. *Genetics,* **55:**141–156.
1971 Experimentally created incipient species of *Drosophila. Nature,* **230:**289–292.
1975 Unstable intermediates between Orinocan and Interior semispecies of *Drosophila paulistorum. Evolution,* **29:**242–248.
Dobzhansky, Th., O. Pavlovsky, and L. Ehrman.
1969 Transitional populations of *Drosophila paulistorum. Evolution,* **23:**482–492.
Dobzhansky, Th., and B. Spassky.
1959 *Drosophila paulistorum,* a cluster of species in statu nascendi. *Proc. Nat. Acad. Sci. U.S.A.,* **45:**419–428.
Duffey, P. A.
1972 Chromosome variation in *Peromyscus:* A new mechanism. *Science,* **176:** 1333–1334.
Dunbar, R. W.
1959 The salivary gland chromosomes of seven forms of black fly included in *Eusimulium aureum* Fries. *Canad. J. Zool.,* **37:**495–525.
1972 Speciation in the *Simulium (Edwardsellum) damnosum* complex (Diptera: Simuliidae). *Abstracts XIV Internat. Congr. Entomol., Canberra,* p. 286.
Dunsmuir, P.
1976 Satellite DNA in the kangaroo *Macropus rufogriseus. Chromosoma,* **56:**111–125.
Du Rietz, G. E.
1930 The fundamental units of biological taxonomy. *Svensk Bot. Tidskr.,* **24:**333–428.
Dutrillaux, B., M.-O. Rethoré, M. Prieur, and J. Lejeune.
1973 Analyse de la structure fine des chromosomes du Gorille *(Gorilla gorilla).* Comparaison avec *Homo sapiens* et *Pan troglodytes. Humangenetik,* **20:**343–354.
Dwyer, P. D.
1966 The population pattern of *Miniopterus schreibersii* (Chiroptera) in north-eastern New South Wales. *Austral. J. Zool.,* **14:**1073–1137.

Egozcue, J.
1975 *Mammalia II: Placentalia 5, Primates.* Vol. 4, Chordata 4, of *Animal Cytogenetics,* B. John, ed., Berlin: Borntraeger (74 pp).
Ehrendorfer, F.
1949 Zur Phylogenie der Gattung *Galium* L., I: Polyploidie und geographisch-ökologische Einheiten in der Gruppe des *Galium pumilum* Murray (Sekt.

Leptogalium Lange sensu Rouy) im österreichischen Alpenraum. *Öst. Bot. Z.*, **96:**109–138.
1964 Cytologie, Taxonomie, und Evolution bei Samenpflanzen. *Vistas in Botany,* **4:**99–186.
1965 Dispersal mechanisms, genetic systems, and colonizing abilities in some flowering plant families. In *The Genetics of Colonizing Species,* H. G. Baker and G. L. Stebbins, eds., New York: Academic Press, pp. 331–352.

Ehrlich, P.
1961 Has the biological species concept outlived its usefulness? *Syst. Zool.*, **10:**167–176.

Ehrlich, P., and R. W. Holm.
1962 Patterns and populations. *Science,* **137:**652–657.
1963 *The Process of Evolution.* New York: McGraw-Hill.

Ehrlich, P., and P. H. Raven.
1969 Differentiation of populations. *Science,* **165:**1228–1232.

Ehrman, L.
1965 Direct observation of sexual isolation between allopatric and between sympatric strains of the different *Drosophila paulistorum* races. *Evolution,* **19:**459–464.

Ehrman, L., and D. L. Williamson.
1965 Transmission by injection of hybrid sterility to nonhybrid males in *Drosophila paulistorum:* Preliminary report. *Proc. Nat. Acad. Sci. U.S.A.*, **54:** 481–483.
1969 On the etiology of the sterility of hybrids between certain strains of *Drosophila paulistorum. Genetics,* **62:**193–199.

Endler, J. A.
1973 Gene flow and population differentiation. *Science,* **179:**243–250.

Epling, C.
1944 The historical background. In "Contributions to the Genetics, Taxonomy, and Ecology of *Drosophila pseudoobscura* and its Relatives," *Carnegie Inst. Wash. Publ.*, No. 554, pp. 145–183.

Evans, J. W.
1962 Evolution in the Homoptera. In *The Evolution of Living Organisms,* G. W. Leeper, ed., Melbourne University Press, pp. 250–259.

Ewing, A. W.
1969 The genetic basis of sound-production in *Drosophila pseudobscura* and *D. persimilis. Anim. Behav.*, **17:**555–560.
1970 The evolution of courtship songs in *Drosophila. Rev. Comport. Anim.*, **4**(4):3–8.

Falleroni, D.
1926 Fauna anofelica italiana e suo "habitat." *Riv. Malariol.*, **5:**553–593.

Favarger, C.
1957 Sur le pourcentage des polyploides dans la flore de l'étage nival des alpes suisses. *Proc. VII Internat. Bot. Congr.*, pp. 51–58.

Felt, E. P.

1920 34th report of the state entomologist on injurious and other insects of the state of New York, 1918. *N.Y. State Mus. Bull.* Nos. 231–232 (288 pp).

1925 Key to gall midges (A resumé of studies I–VII, Itonididae). *N.Y. State Mus. Bull.* No. 257 (239 pp).

Felsenstein, J.

1974 The evolutionary advantage of recombination. *Genetics,* **78:**737–756.

1975 The genetic basis of evolutionary change. *Evolution,* **29:**587–590.

Fisher, R. A.

1930 *The Genetical Theory of Natural Selection.* Oxford: Clarendon Press.

1958 *The Genetical Theory of Natural Selection,* second revised edition. New York: Dover Publications.

Forbes, W. T. M.

1928 Variation in *Junonia lavinia* (Lep., Nymphalidae). *J. N.Y. Ent. Soc.,* **36:**305–320.

Force, D. C.

1975 Succession of *r* and *K* strategists in parasitoids. In *Evolutionary Strategies of Parasitic Insects,* P. W. Price, ed., London: Plenum Press, pp. 112–129.

Ford, C. E., and E. P. Evans.

1973 Robertsonian translocations in mice: Segregational irregularities in male heterozygotes and zygotic unbalance. *Chromosomes Today,* **4:**387–397.

Ford, C. E., and J. L. Hamerton.

1970 Chromosome polymorphism in the common shrew, *Sorex araneus. Symp. Zool. Soc. Lond.,* **26:**223–226.

Ford, C. E., J. L. Hamerton, and G. B. Sharman.

1957 Chromosome polymorphism in the common shrew. *Nature,* **180:**392–393.

Forde, M. B., and D. G. Faris.

1962 Effect of introgression on the serpentine endemism of *Quercus durata. Evolution,* **16:**338–347.

Fredga, K., A. Gropp, H. Winking, and F. Frank.

1976 Fertile XX- and XY-type females in the wood lemming *Myopus schisticolor. Nature,* **261:**225–227.

Freeman, T. N., G. Stehr, G. T. Harvey, and I. M. Campbell

1967 On coniferophagous species of *Choristoneura* (Lepidoptera: Torticidae) in North America. *Canad. Ent.,* **99:**449–506.

Friauf, J. J.

1957 Clarification of the species in the genus *Dendrotettix* (Orthoptera: Acrididae, Cyrtacanthacrinae). *Florida Ent.,* **40:**127–139.

Fritts, T. H.

1969 The systematics of the parthenogenetic lizards of the *Cnemidophorus cozumela* complex. *Copeia,* 1969:519–535.

Frizzi, G.

1947 Cromosomi salivari in *Anopheles maculipennis. Sci. Genet.,* **3:**67–68.

1949 Genetica di popolazioni in *Anopheles maculipennis. Ric. Sci.,* **19:**544–552.

Frizzi, G.

1950 Studio sulla sterilità degli ibridi nel genere *Anopheles,* I: Sterilità nel incrocio fra *Anopheles mac. atroparvus* ed *Anopheles mac. typicus* e nel reincrocio dei cromosomi salivari. *Sci. Genet.,* **3:**260–270.

1951 Dimorfismo cromosomico in *Anopheles maculipennis messeae. Sci. Genet.,* **4:**79–93.

1953 Étude cytogénétique d'*Anopheles maculipennis* en Italie. Extension des recherches à d'autres espèces d'anopheles. *Bull. World Health Org.,* **9:**335–344.

Frizzi, G., and L. De Carli.

1954 Studio preliminare comparativo genetico e citogenetico fra alcun specie nordamericane di *Anopheles maculipennis* e l'*Anopheles maculipennis atroparvus* Italiano. *Symp. Genet. Pavia,* **2:**184–206.

Frost, J. S., and J. T. Bagnara.

1976 A new species of Leopard Frog (*Rana pipiens* complex) from northwestern Mexico. *Copeia,* **1976:**332–338.

Frydenberg, O., D. Møller, G. Naevdal, and K. Sick.

1965 Haemoglobin polymorphism in Norwegian cod populations. *Hereditas,* **53:**257–271.

Fryer, G.

1959a Some aspects of evolution in Lake Nyasa. *Evolution,* **13:**440–451.

1959b The trophic interrelationships and ecology of some littoral communities of Lake Nyasa with especial reference to the fishes, and a discussion of the evolution of a group of rock-frequenting Cichlidae. *Proc. Zool. Soc. London,* **132:**153–281.

1960a Some controversial aspects of speciation in African cichlid fishes. *Proc. Zool. Soc. London,* **135:**569–578.

1960b Evolution of fishes in Lake Nyasa. *Evolution,* **14:**396–400.

Fryer, G., and T. D. Isles.

1972 *The Cichlid Fishes of the Great Lakes of Africa: Their Biology and Evolution.* Edinburgh: Oliver and Boyd.

Fryxell, P. A.

1957 Mode of reproduction in higher plants. *Bot. Rev.,* **23:**135–233.

Fujino, K.

1970 Immunological and biochemical genetics of tunas. *Trans. Amer. Fish Soc.,* **99:**152–178.

Fukuda, I.

1967 The biosystematics of *Achlys. Taxon,* **16:**308–316.

Gabbutt, P. D.

1959 The bionomics of the wood cricket, *Nemobius sylvestris* (Orthoptera: Gryllidae). *J. Anim. Ecol.,* **28:**15–42.

Gabow, S. A.

1975 Behavioral stabilization of a baboon hybrid zone. *Amer. Nat.,* **109:**701–712.

Gajewski, W.

1957 A cytogenetic study on the genus *Geum* L. *Monogr. Botanicae* (Warszawa), **4:**1–415.

1959 Evolution in the genus *Geum*. *Evolution,* **13:**378–388.

Gall, J. G., and D. D. Atherton.

1974 Satellite DNA sequences in *Drosophila virilis*. *J. Mol. Biol.,* **85:**633–664.

Gartside, D. F.

1972a The *Litoria ewingi* complex (Anura: Hylidae) in south-eastern Australia, II: Genetic incompatibility and delimitation of a narrow hybrid zone between *L. ewingi* and *L. paraewingi*. *Austral. J. Zool.,* **20:**423–433.

1972b The *Litoria ewingi* complex (Anura: Hylidae) in southeastern Australia, III: Blood protein variation across a narrow hybrid zone between *L. ewingi* and *L. paraewingi*. *Austral. J. Zool.,* **20:**435–443.

Gileva, É. A.

1973 Chromosomes, unusual inheritance of sex chromosomes, and sex ratio in the arctic lemming (*Dicrostonyx torquatus torquatus* Pall. 1779). *Dokl. Akad. Nauk S.S.S.R.,* **213:**952–955.

1975 Karyotypes of *Dicrostonyx torquatus chionopaes* Allen and an unusual chromosomal mechanism of sex determination in palearctic hooved lemmings. *Dokl. Akad. Nauk S.S.S.R.,* **224:**697–700.

Gill, B. S., and G. Kimber.

1974 Giemsa C-banding and the evolution of wheat. *Proc. Nat. Acad. Sci. U.S.A.,* **71:**4086–4090.

Gillespie, J. H., and K. Kojima.

1968 The degree of polymorphism in enzymes involved in energy production compared to that in non-specific enzymes in two *Drosophila ananassae* populations. *Proc. Nat. Acad. Sci. U.S.A.,* **61:**582–585.

Goldring, E. S., D. L. Brutlag, and W. J. Peacock.

1975 Arrangement of the highly repeated DNA of *Drosophila melanogaster*. In *The Eukaryote Chromosome,* W. J. Peacock and R. D. Brock, eds., pp. 47–59. Canberra: Australian National University Press.

Goldschmidt, E.

1952 Fluctuation in chromosome number in *Artemia salina*. *J. Morph.,* **91:**111–134.

1956 Structural polymorphism in the Israel race of *Drosophila subobscura*. *Bull. Res. Council Israel,* **5**B:150.

Goldschmidt, R.

1932 Untersuchungen zur Genetik der geographischen Variation, III: Abschliessendes über die Geschlechtsrassen von *Lymantria dispar*. *Roux' Arch. Entwicklungsm. Organ.,* **126:**277–324.

1940 *The Material Basis of Evolution*. Yale University Press.

1955 *Theoretical Genetics*. University of California Press.

Göltenboth, F.

1973 DNA-Replikation und Chromosomenstruktur von *Mesostoma* (Turbellaria). *Chromosoma,* **44:**147–181.

Gooch, J. L., and T. J. M. Schopf.
1972 Genetic variability in the deep sea: Relation to environmental variability. *Evolution,* **26:**545–552.
Goodhart, C. B.
1963 "Area effects" and non-adaptive variation between populations of *Cepaea* (Mollusca). *Heredity,* **18:**459–465.
Gorman, G. C.
1973 The chromosomes of the Reptilia: A cytotaxonomic interpretation. In *Vertebrate Cytotaxonomy,* E. Capanna and B. Chiarelli, eds., London: Academic Press, pp. 349–424.
Gorman, G. C., and L. Atkins.
1968 New karyotypic data for 16 species of *Anolis* (Sauria, Iguanidae) from Cuba, Jamaica, and the Cayman Islands. *Herpetologica,* **24:**13–21.
1969 The zoogeography of Lesser Antillean *Anolis* lizards: An analysis based upon chromosomes and lactic dehydrogenases. *Bull. Mus. Comp. Zool. Harvard Univ.,* **138:**53–80.
Gottlieb, L. D.
1973 Enzyme differentiation and phylogeny in *Clarkia franciscana, C. rubicunda,* and *C. amoena. Evolution,* **27:**205–214.
1974 Genetic confirmation of the origin of *Clarkia lingulata. Evolution,* **28:** 244–250.
Gould, S. J., D. S. Woodruff, and J. P. Martin.
1974 Genetics and morphometrics of *Cerion* at Pongo Carpet: A new systematic approach to this enigmatic land snail. *Syst. Zool.,* **23:**518–535.
Grandi, G.
1963a Catalogo ragionato degli Agaonidi del mondo descritti fino a oggi (57° contributo alla conoscenza degli insetti dei fichi). *Boll. Ist. Ent. Univ. Bologna,* **26:**319–373.
1963b The hymenopterous insects of the superfamily Chalcidoidea developing within the receptacles of figs: Their life history, symbioses, and morphological adaptations. *Boll. Ist. Ent. Univ. Bologna,* **26:**i-xiii.
Grant, V.
1949 Pollination systems as isolating mechanisms. *Evolution,* **3:**82–97.
1963 *The Origin of Adaptations.* Columbia University Press.
1964a Genetic and taxonomic studies in *Gilia,* XII: Fertility relationships of the polyploid cobwebby *Gilias. Aliso,* **5:**479–507.
1964b *The Architecture of the Germ Plasm.* New York: Wiley.
1971 *Plant Speciation.* Columbia University Press.
Grant, V., and A. Grant.
1960 Genetic and taxonomic studies in *Gilia,* XI: Fertility relationships of the diploid cobwebby *Gilias. Aliso,* **4:**435–481.
Grassle, J. P., and J. F. Grassle.
1976 Sibling species in the marine pollution indicator *Capitella* (Polychaeta). *Science,* **192:**567–569.

Gropp, A., P. Citoler, and M. Geisler.
1969 Karyotypvariation und Heterochromatinmuster bei Igeln *(Erinaceus* und *Hemiechinus). Chromosoma,* **27:**288–307.

Gropp, A., V. Tettenborn, and E. von Lehman.
1970 Chromosomenvariationen vom Robertson'schen Typus bei der Tabakmaus *M. poschiavinus,* und ihren Hybriden mit der Laboratoriumsmaus. *Cytogenetics,* **9:**9–23.

Gropp, A., H. Winking, L. Zech, and H. Müller.
1972 Robertsonian chromosomal variation and identification of metacentric chromosomes in feral mice. *Chromosoma,* **39:**265–288.

Grouchy, J. de, C. Turleau, M. Roubin, and M. Klein.
1972 Evolutions cariotypiques de l'homme et du chimpanzé: Etude comparative des topographies de bandes après dénaturation menagée. *Ann. Genet.,* **15:**79–84.

Günther, R.
1973 Über die verwandschaftlichen Beziehungen zwischen den europäischen Grünfröschen und den Bastardcharakter von *Rana esculenta* L. (Anura, Amphibia). *Biol. Zbl.,* **89:**327–342.

Gustafsson, Å.
1946–1947 Apomixis in the higher plants, I, II, and III. *Lunds Univ. Arsskr.* (N.F. Avd. 2), **42**(3):1–67; **43**(2):71–178; and **44**(2):183–370.

Haas, F.
1940 A tentative classification of the palearctic unionids. *Zool. Ser. Field Mus. Nat. Hist.,* **24:**115–141.

Haffer, J.
1969 Speciation in Amazonian forest birds. *Science,* **165:**131–137.

Haga, T.
1969 Structure and dynamics of natural populations of a diploid *Trillium. Chromosomes Today,* **2:**207–217.
1974 Cytogenetic structures of 15 natural populations of *Trillium kamtschaticum* Pallas: A compilation with 15 tables and five figures. Cytogenetics Laboratory, Kyushu University, Fukuoka, Japan (19 pp).

Hair, J. B., and E. J. Buizenberg.
1958 Chromosomal evolution in the Podocarpaceae. *Nature,* **181:**1584–1586.

Haldane, J. B. S.
1922 Sex-ratio and unisexual sterility in hybrid animals. *J. Genet.,* **12:**101–109.
1948 The theory of a cline. *J. Genet.,* **48:**277–284.
1964 A defense of beanbag genetics. *Perspectives in Biology and Medicine,* **7:**343–359.

Halfer-Cervini, A. M., M. Piccinelli, P. Prosdocimi, and L. Baratelli-Zambruni.
1968 Sibling species in *Artemia* (Crustacea: Branchiopoda). *Evolution,* **22:**373–381.

Halkka, L., O. Halkka, U. Skarén, and V. Soderlund.
1974 Chromosome banding pattern in a polymorphic population of *Sorex araneus* from northeastern Finland. *Hereditas,* **76:**305–314.

Hall, W. P.
1970 Three probable cases of parthenogenesis in lizards (Agamidae, Chamaeleontidae, Gekkonidae). *Experientia,* **26:**1271–1273.
1973 *Comparative Population Cytogenetics, Speciation, and Evolution of the Iguanid Lizard Genus* Sceloporus. Ph.D. thesis, Harvard University.

Hall, W. P., and R. K. Selander.
1973 Hybridization in karyotypically differentiated populations of the *Sceloporus grammicus* complex (Iguanidae). *Evolution,* **27:**226–242.

Hamilton, T. H., and I. Rubinoff.
1963 Isolation, endemism, and multiplication of species in the Darwin finches. *Evolution,* **17:**388–403.
1964 On models predicting the abundance of species and endemics of the Darwin finches in the Galápagos archipelago. *Evolution,* **18:**339–342.
1967 On predicting insular variation in endemism and sympatry for the Darwin finches in the Galápagos archipelago. *Amer. Nat.,* **101:**161–172.

Hanelt, P.
1966 Polyploidie-Frequenz und geographische Verbreitung bei höheren Pflanzen. *Biol. Rundsch.,* **4:**183–196.

Hardy, D. E.
1974 Introduction and background information. In *Genetic Mechanisms of Speciation in Insects,* M. J. D. White, ed., Sydney: Australia and New Zealand Book Co., pp. 71–80.

Harris, H., and D. A. Hopkinson.
1972 Average heterozygosity in man. *J. Human. Genet.,* **36:**9–20.

Haskins, C. P., E. F. Haskins, and R. E. Hewitt.
1960 Pseudogamy as an evolutionary factor in the Poeciliid fish *Mollienisia formosa. Evolution,* **14:**473–483.

Hatch, F. T., A. J. Bodner, J. A. Mazrimas, and D. H. Moore.
1976 Satellite DNA and cytogenetic evolution: DNA quantity, satellite DNA, and karyotypic variations in kangaroo rats (genus *Dipodomys*). *Chromosoma,* **58:**155–165.

Hayman, D. L.
1960 The distribution and cytology of the chromosome races of *Themeda australis* in southern Australia. *Austral. J. Bot.,* **8:**58–68.

Hayman, D. L., and P. G. Martin
1974 *Mammalia I: Monotremata and Marsupialia.* Vol. 4, chordata 4, of *Animal Cytogenetics,* B. John, ed., Berlin: Borntraeger (110 pp).

Hebert, P. D. N.
1974a Enzyme variability in natural populations of *Daphnia magna,* I: Population structure in East Anglia. *Evolution,* **28:**546–556.

1974b Enzyme variability in natural populations of *Daphnia magna,* II: Genotypic frequencies in permanent populations. *Genetics,* **77:**323–334.

1974c Enzyme variability in natural populations of *Daphnia magna, III:* Genotypic frequencies in intermittent populations. *Genetics,* **77:**335–341.

Hebert, P. D. N., R. D. Ward, and J. B. Gibson.

1972 Natural selection for enzyme variants among parthenogenetic *Daphnia magna. Genet. Res.,* **19:**173–176.

Hedden, R., J. P. Vite, and K. Mori.

1976 Synergistic effect of a pheromone and a kairomone on host selection and colonization by *Ips avulsus. Nature,* **261:**696–697.

Hedgecock, D., and F. J. Ayala.

1974 Evolutionary divergence in the genus *Taricha* (Salamandridae). *Copeia,* 1974:738–747.

Helwig, E. R.

1955 Spermatogenesis in hybrids between *Circotettix verruculatus* and *Trimerotropis suffusa* (Orthoptera: Oedipodidae). *Univ. Colorado Study Series in Biol.* No. 3, pp. 47–64.

Hendrickson, J. R.

1966 The Galápagos tortoises, *Geochelone* Fitzinger (*Testudo* Linnaeus 1758 in part). *Proc. Symp. Galápagos Int. Sci. Project,* University of California Press, pp. 252–257.

Henning, V., and C. Yanovsky.

1963 An electrophoretic study of mutationally altered A proteins of the tryptophane synthetase of *Escherichia coli. J. Mol. Biol.,* **6:**16–21.

Herre, A.W.C.T.

1933 The fishes of Lake Lanao: A problem in evolution. *Amer. Nat.,* **67:**154–162.

Hewitt, G. M.

1973 (1974) The integration of supernumerary chromosomes into the Orthopteran genome. *Cold Spr. Harb. Symp. Quant. Biol.,* **38:**183–194.

1975 A new hypothesis for the origin of the parthenogenetic grasshopper *Moraba virgo. Heredity,* **34:**117–123.

Hill, R.

1937 Morphometry of the cisco, *Leucichthys artedi* (Le Suer), in the lakes of the northeastern highlands, Wisconsin. *Intern. Rev. Ges. Hydrol. Hydrog.,* **36:**57–130.

Hinegardner, R.

1968 Evolution of cellular DNA content in teleost fishes. *Amer. Nat.,* **102:**517–523.

Hoar, C. S.

1931 Meiosis in *Hypericum punctatum* Lam. *Bot. Gaz.,* **92:**396–406.

Hogben, L.

1940 Problems of the origin of species. In *The New Systematics* J. Huxley, ed., Oxford University Press, pp. 269–286.

Hopkins, G. H. E.
1949 The host-associations of the lice of mammals. *Proc. Zool. Soc. London,* **191:**387–604.
Hopkins, G. H. E., and T. Clay.
1952 *A Check List of the Genera and Species of Mallophaga.* London: British Museum (Natural History).
Hotz, H.
1974 Ein Problem aus vielen Fragen—europäischen Grünfrösche *(Rana esculenta* Komplex) und ihre Verbreitung. *Nat. Mus.,* **104:**262–272.
Hsu, S. W.
1956a Oogenesis in the bdelloid rotifer *Philodina roseola. La Cellule,* **57:**283–296.
1956b Oogenesis in *Habrotrocha tridens* (Milne). *Biol. Bull Woods Hole,* **111:**364–374.
Hsu, T. C.
1973 Longitudinal differentiation of chromosomes. *Ann. Rev. Genet.,* **7:**153–176.
Hsu, T. C., and F. E. Arrighi.
1966 Chromosomal evolution in the genus *Peromyscus* (Cricetidae, Rodentia). *Cytogenetics,* **5:**355–359.
1968 Chromosomes of *Peromyscus* (Rodentia, Cricetidae): Evolutionary trends in 20 species. *Cytogenetics,* **7:**417–446.
Hubbell, T. H.
1956 Some aspects of geographic variation in insects. *Ann. Rev. Entomol.,* **1:**71–88.
Hubbell, T. H., and R. W. Walker.
1928 A new shrub-inhabiting species of *Schistocerca* from central Florida. *Occ. Pap. Mus. Zool. Univ. Mich.* No. 197, pp. 1–12.
Hubbs, C. L.
1961 Isolating mechanisms in the speciation of fishes. In *Vertebrate Speciation,* W. F. Blair, ed., University of Texas Press, pp. 5–23.
Hubbs, C. L., and L. C. Hubbs.
1932 Apparent parthenogenesis in nature, in a form of fish or hybrid origin. *Science,* **76:**628–630.
Hubbs, C. L., L. C. Hubbs, and R. E. Johnson.
1943 Hybridization in nature between species of catostomid fishes. *Contrib. Lab. Vert. Biol. Univ. Mich.* No. **22:**1–76.
Hubbs, C. L., and R. R. Miller.
1948 Correlation between fish distribution and hydrographic history in the desert basins of the western United States. In "The Great Basin, With Emphasis on Glacial and Postglacial Times," *Bull. Univ. Utah,* **38:**18–166.
Hubby, J. L., and R. C. Lewontin.
1966 A molecular approach to the study of genic heterozygosity in natural populations, I: The number of alleles at different loci in *Drosophila pseudoobscura. Genetics,* **54:**577–594.

Hubby, J. L., and L. H. Throckmorton.
1965 Protein differences in *Drosophila,* II: Comparative species genetics and evolutionary problems. *Genetics,* **52:**203–215.
1968 Protein differences in *Drosophila,* IV: A study of sibling species. *Amer. Nat.,* **102:**193–205.

Hubendick, B.
1960 The Ancylidae of Lake Ochrid and their bearing on intralacustrine speciation. *Proc. Zool. Soc. London,* **133:**497–529.

Huettel, M.D., and G. L. Bush.
1972 The genetics of host selection and its bearing on sympatric speciation in *Procecidochares* (Diptera: Tephritidae). *Ent. Exp. Appl.,* **15:**465–480.

Huey, R. B., and E. R. Pianka.
1974 Ecological character displacement in a lizard. *Amer. Zoologist,* **14:**1127–1136.

Hughes-Schrader, S., and F. Schrader.
1956 Polyteny as a factor in the chromosomal evolution of the Pentatomini (Hemiptera). *Chromosoma,* **8:**135–151.

Hughes-Schrader, S., and E. Tremblay.
1966 *Gueriniella* and the cytotaxonomy of iceryine coccids (Coccoidea: Margarodidae). *Chromosoma,* **19:**1–13.

Hunt, W. G., and R. K. Selander.
1973 Biochemical genetics of hybridization in European house mice. *Heredity,* **31:**11–33.

Husted, L., and T. K. Ruebush.
1940 A comparative study of *Mesostoma ehrenbergii ehrenbergii* and *Mesostoma ehrenbergii wardii. J. Morph.,* **67:**387–410.

Hutchinson, G. E.
1959 Homage to Santa Rosalia, or why are there so many kinds of animals. *Amer. Nat.,* **93:**145–159.

Huxley, J. S.
1939 Clines: An auxiliary method in taxonomy. *Bijdr. Dierk.,* **27:**491–520.
1942 *Evolution: The Modern Synthesis.* London: George Allen and Unwin.

Jackson, J. F.
1973 The phenetics and ecology of a narrow hybrid zone. *Evolution,* **27:**58–68.

Jackson, R. C.
1957 New low chromosome number for plants. *Science,* **128:**1115–1116.
1962 Interspecific hybridization in *Haplopappus* and its bearing on chromosome evolution in the *Blepharodon* section. *Amer. J. Bot.,* **49:**119–132.
1965 A cytogenetic study of a three-paired race of *Haplopappus gracilis. Amer. J. Bot.,* **52:**946–953.
1973 Chromosomal evolution in *Haplopappus gracilis:* A centric transposition race. *Evolution,* **27:**243–284.

Jain, S. K., and A. D. Bradshaw.
1966 Evolutionary divergence among adjacent plant populations, I: The evidence and its theoretical analysis. *Heredity,* **21:**407–441.

James, S. H.
1965 Complex hybridity of *Isotoma petraea,* I: The occurrence of interchange heterozygosity, autogamy, and a balanced lethal system. *Heredity,* **20:**341–353.
1970a Complex hybridity in *Isotoma petraea,* II: Components and operation of a possible evolutionary mechanism. *Heredity,* **25:**53–78.
1970b A demonstration of a possible mechanism of sympatric divergence using simulation techniques. *Heredity,* **25:**241–252.

Janzen, D. H.
1968 Host plants as islands in evolutionary and contemporary time. *Amer. Nat.,* **102:**592–594.

Jaretsky, R.
1928 Histologische and karyologische Studien an Polygonaceen. *Jahrb. wiss. Bot.,* **69:**357–490.

Jensen, H. W.
1936 Meiosis in *Rumex,* I: Polyploidy and the origin of new species. *Cytologia,* **7:**1–22.

Jha, A. G.
1975 Cytogenetics, evolution, and systematics of digenea (Trematoda: Platyhelminthes). *Egypt. J. Genet. Cytol.,* **4:**201–233.

John, B., and G. M. Hewitt.
1966 Karyotype stability and DNA variability in the Acrididae. *Chromosoma,* **20:**155–172.
1968 Patterns and pathways of chromosome evolution within the Orthoptera. *Chromosoma,* **25:**40–74.

John, B., and K. R. Lewis.
1957 Studies on *Periplaneta americana,* I: Experimental analysis of male meiosis. *Heredity,* **11:**1–9.
1958 Studies on *Periplaneta americana,* III: Selection for heterozygosity. *Heredity,* **12:**185–97.
1959 Selection for interchange heterozygosity in an inbred culture of *Blaberus discoidalis* (Serville). *Genetics,* **44:**251–267.
1965 Genetic speciation in the grasshopper *Eyprepocnemis plorans. Chromosoma,* **16:**308–344.

Johnson, B. L.
1975 Identification of the apparent B-genome donor of wheat. *Canad. J. Genetics Cytol.,* **17:**21–39.

Johnson, F. M., H. E. Schaffer, J. E. Gillaspy, and E. S. Rockwood.
1969 Isozyme genotype–environment relationships in natural populations of the harvester ant, *Pogonomyrmex barbatus,* from Texas. *Biochem. Genet.,* **3:**429–450.

Johnson, L. A. S., and B. G. Briggs.
1963 Evolution in the Proteaceae. *Aust. J. Bot.*, **11:**21–61.
Johnson, M. S.
1976 Allozymes and area effects in *Cepaea nemoralis* on the western Berkshire Downs. *Heredity,* **36:**105–121.
Johnson, W. E., and R. K. Selander.
1971 Protein variation and systematics in kangaroo rats (genus *Dipodomys*). *Syst. Zool.*, **20:**377–405.
Johnson, W. E., R. K. Selander, M. H. Smith, and Y. J. Kim.
1972 Biochemical genetics of sibling species of the cotton rat *(Sigmodon). Studies in Genetics VII,* Univ. Texas Publ. No. 7213, pp. 297–305.
Jones, A. W.
1945 Studies in cestode cytology. *J. Parasitol.*, **31:**213–235.
Jordan, K.
1938 Where subspecies meet. *Novit. Zool.*, **41:**103–111.
Jotterand, M.
1972 Le polymorphisme chromosomique des *Mus* (Leggadas) africains: Cytogénétique, zoogéographie, évolution. *Rev. Suisse Zool.*, **79:**287–359.
1975 The African *Mus* (pigmy mice): The role of chromosomal polymorphism in speciation. *Caryologia,* **28:**335–344.

Kallman, K. D.
1962 Population genetics of the gynogenetic teleost, *Mollienisia formosa* (Girard). *Evolution,* **16:**497–504.
Kaneshiro, K. Y., H. L. Carson, F. E. Clayton, and W. B. Heed.
1973 Niche separation in a pair of homosequential *Drosophila* species from the island of Hawaii. *Amer. Nat.*, **107:**766–774.
Kannenberg, L. G., and R. W. Allard.
1967 Population studies in predominantly self-pollinated species, VIII: Genetic variability in the *Festuca microstachys* complex. *Evolution,* **21:**227–240.
Karpechenko, G. D.
1927 Polyploid hybrids of *Raphanus sativus* L. × *Brassica oleracea* L. *Bull. Appl. Bot. Genet. Plant Breeding, Leningrad,* **17:**305–310.
Kastritsis, C. D.
1969 The chromosomes of some species of the *guarani* group of *Drosophila. J. Hered.*, **60:**50–57.
Kastritsis, C. D., and Th. Dobzhansky.
1967 *Drosophila pavlovskiana,* a race or a species? *Amer. Midl. Nat.*, **78:**244–247.
Kayano, H., and H. Watanabe.
1970 Panmixis in natural populations of *Trillium kamtschaticum. Jap. J. Genet.*, **45:**59–69.
Kemp, D. J.
1975 Unique and repetitive sequences in multiple genes for feather keratin. *Nature,* **254:**573–577.

Kernaghan, R. P., and L. Ehrman.
1970 An electron microscopic study of the etiology of hybrid sterility in *Drosophila paulistorum,* I: Mycoplasma-like inclusions in the testes of sterile males. *Chromosoma,* **29:**291–304.

Kerr, W. E., and Z. V. da Silveira.
1972 Karyotypic evolution of bees and corresponding taxonomic implications. *Evolution,* **26:**197–202.

Kessler, S.
1962 Courtship rituals and reproductive isolation between the races or incipient species of *Drosophila paulistorum. Amer. Nat.,* **96:**117–121.

Key, K. H. L.
1954 *The Taxonomy, Phases, and Distribution of* Chortoicetes *and* Austroicetes, Canberra: Commonwealth Scientific and Industrial Research Organization (237 pp).
1968 The concept of stasipatric speciation. *Systemat. Zool.,* **17:**14–22.
1970 Principles of classification and nomenclature. In *Insects of Australia,* Melbourne University Press, pp. 141–151.
1972 A revision of the Psednurini (Orthoptera: Pyrgomorphidae). *Austral. J. Zool., Suppl. Ser.,* No. 14 (72 pp).
1974 Speciation in the Australian morabine grasshoppers: Taxonomy and ecology. In *Genetic Mechanisms of Speciation in Insects,* M. J. D. White, ed., Sydney: Australia and New Zealand Book Co., pp. 43–56.
1976 A generic and suprageneric classification of the Morabinae (Orthoptera: Eumastacidae) with description of the type species and a bibliography of the subfamily. *Austral. J. Zool., Suppl. Ser.,* No. 37 (185 pp).

Key, K. H. L., and J. Balderson.
1972 Distributional relations of two species of *Psednura* (Orthoptera: Pyrgomorphidae) in the Evans Head area of New South Wales. *Austral. J. Zool.,* **20:**411–422.

Keyl, H.-G.
1962 Chromosomenevolution bei Chironomus, II: Chromosomenumbauten und phylogenetische Beziehung der Arten. *Chromosoma,* **13:**464–514.
1965 Duplikationen von Untereinheiten der chromosomale DNS während der Evolution von *Chironomus thummi. Chromosoma,* **17:**139–180.
1966 Lokale DNS-Replikationen in Riesenchromosomen. In *Probleme der biologischen Reduplikation,* P. Sitte, ed., Berlin: Springer Verlag.

Khoshoo, T. N.
1959 Polyploidy in gymnosperms. *Evolution,* **13:**24–39.
1960 Chromosome numbers in gymnosperms. *Silvae Genet.,* **10:**1–9.
1969 Chromosome evolution in cycads. *Chromosomes Today,* **2:**236–240.

Khoshoo, T. N., and M. R. Ahuja.
1963 The chromosomes and relationships of *Welwitschia mirabilis. Chromosoma,* **14:**522–532.

Kihara, H., and T. Ono.

1923 Cytological studies on *Rumex,* II: On the relation of chromosome number and sexes in *Rumex acetosa* L. *Bot. Mag. Tokyo,* **37:**147–149.

1925 The sex chromosomes of *Rumex Acetosa. Z. indukt. Abst. Vererbl.,* **39:**1–7.

1926 Chromosomenzahlen und systematische Gruppierung der Rumex-Arten. *Z. Zellf.,* **4:**475–481.

Kiknadze, I. I., and L. V. Vysotskaya.

1970 Measurement of DNA mass per nucleus in the grasshopper species with different numbers of chromosomes. *Tsitologia,* **12:**1100–1107.

Kimber, G.

1974 A reassessment of the origin of the polyploid wheats. *Genetics,* **78:**487–492.

Kimura, M.

1968 Evolutionary rate at the molecular level. *Nature,* **217:**624–626.

1969 The rate of molecular evolution considered from the standpoint of population genetics. *Proc. Nat. Acad. Sci. U.S.A.,* **63:**1181–1188.

Kimura, M., and T. Ohta.

1971a Protein polymorphism as a phase of molecular evolution. *Nature,* **299:**467–469.

1971b *Theoretical Aspects of Population Genetics.* Princeton University Press.

1974 On some principles governing molecular evolution. *Proc. Nat. Acad. Sci. U.S.A.,* **71:**2848–2852.

King, J. L.

1967 Continuously distributed factors affecting fitness. *Genetics,* **55:**483–492.

King, M., and D. King.

1975 Chromosomal evolution in the lizard genus *Varanus* (Reptilia). *Austr. J. Biol. Sci.,* **28:**89–108.

King, M.-C., and A. C. Wilson.

1975 Evolution at two levels in humans and chimpanzees. *Science,* **188:**107–116.

Kinsey, A. C.

1930 The gall wasp genus *Cynips:* A study in the origin of species. *Indiana Univ. Studies* No. 16 (577 pp).

1936 The origin of higher categories in *Cynips. Indiana Univ. Publ., Science Series* No. 4 (334 pp).

Kitzmiller, J. B., G. Frizzi, and R. H. Baker.

1967 Evolution and speciation within the *maculipennis* complex of the genus *Anopheles.* In *Genetics of Insect Vectors of Disease,* J. W. Wright and R. Pal, eds., Amsterdam: Elsevier, pp. 151–210.

Klingstedt, H.

1939 Taxonomic and cytological studies on grasshopper hybrids, I: Morphology and spermatogenesis of *Chorthippus bicolor* Charp. × *Ch. biguttulus* L. *J. Genet.,* **37:**389–420.

Kluge, A. G., and M. J. Eckardt.
1969 *Hemidactylus garnotii* Duméril and Bibron: A triploid all-female species of gekkonid lizard. *Copeia,* 1969:651–664.
Kochansky, J., R. T. Cardé, J. Liebherr, and W. L. Roelofs.
1975 Sex pheromones of the European corn borer *(Ostrinia nubilalis)* in New York. *J. Chem. Ecol.,* **1:**225–231.
Kojima, K., J. Gillespie, and Y. N. Tobari.
1970 A profile of *Drosophila* species enzymes assayed by electrophoresis, I: Number of alleles, heterozygosities, and linkage disequilibrium in glucose-metabolizing systems and some other enzymes. *Biochem. Genet.,* **4:**627–637.
Koryakov, E. A.
1955 On the fertility and type of the spawning population of *Comephorus. Dokl. Acad. Sci. U.S.S.R.,* **101:**965–967.
1956 Some ecological adaptive features in the propagation of *Comephorus. Dokl. Acad. Sci. U.S.S.R.,* **111:**1111–1114.
Kozhov, M.
1963 *Lake Baikal and Its Life.* The Hague: W. Junk.
Kozlovsky, A. I.
1970 Chromosome polymorphism in eastern-Siberian populations of the common shrew *Sorex araneus* L. *Tsitologia,* **12:**1459–1464.
1974 Karyological differentiation of the northeastern subspecies of the hooved lemmings. *Dokl. Akad. Nauk. U.S.S.R.,* **219:**981–984.
Krepp, S. R., and M. H. Smith.
1974 Genic heterozygosity in the 13-year cicada, *Magicicada. Evolution,* **28:** 396–401.
Krimbas, K. B.
1960 Ta didyma eidi *Gryllotalpa* en Helladi: Kylfarologiki kai morfologiki ereyna, Athens: privately printed (27 pp).
Kupka, E.
1948 Chromosomale Verschiedenheiten bei schweizerischen Coregonen (Felchen). *Rev. Suisse Zool.,* **55:**285–293.
Kurabayashi, M.
1958 Evolution and variation in Japanese species of *Trillium. Evolution,* **12:** 286–310.
1963 Karyotype differentiation in *Trillium sessile* and *T. ovatum* in the western United States. *Evolution,* **17:**296–306.
Kushnir, T.
1948 Chromosomal evolution in the European mole-cricket. *Nature,* **161:**531.
1952 Heterochromatic polysomy in *Gryllotalpa gryllotalpa* L. *J. Genet.,* **50:**361–383.

Lack, D.
1947 *Darwin's Finches.* Cambridge University Press.
1969 Subspecies and sympatry in Darwin's finches. *Evolution,* **23:**252–263.

LaCroix, J. C.

1968 Étude descriptive des chromosomes en écouvillon dans le genre *Pleurodeles* (Amphibien, Urodele). *Ann. Embryol. Morph.* **1:**179–202.

1970 Mise en évidence sur les chromosomes en écouvillon de *Pleurodeles poireti* Gervais, Amphibien urodèle, d'une structure liée au sexe, identifiant le bivalent sexuel et marquant le chromosome W. *C. R. Acad. Sci. Paris,* **271:**102–104.

Lagowski, J. M., M. Y. W. Yu, H. S. Forrest, and C. D. Laird.

1973 Dispersity of repeat DNA sequences in *Oncopeltus fasciatus,* an organism with diffuse centromeres. *Chromosoma,* **43:**349–373.

Lakovaara, S., and A. Saura.

1971a Genetic variation in natural populations of *D. obscura. Genetics,* **69:** 377–384.

1971b Genic variation in marginal populations of *D. subobscura. Hereditas,* **69:**77–82.

Lakovaara, S., A. Saura, and C. T. Falk.

1972 Genetic distance and evolutionary relationships in the *Drosophila obscura* group. *Evolution,* **26:**177–184.

Lamb, R. Y.

UNPUB. Cytogenetic studies of parthenogenetic and related bisexual populations in two species of cave cricket (*Hadenoecus,* Rhaphidophoridae, Orthoptera).

Lamb, R. Y., and R. B. Willey.

1975 The first parthenogenetic populations of Orthoptera Saltatoria to be reported from North America. *Ann. Entomol. Soc. Amer.,* **68:**771–772.

Lamotte, M.

1951 Recherches sur la structure génétique des populations naturelles de *Cepaea nemoralis* (L). *Bull. Biol. Fr. Belg. Suppl.,* **35:**1–238.

1954 Distribution en France des divers systèmes de bandes chez *Cepaea nemoralis* L. *J. Conchyl.,* **94:**125–147.

1959 Polymorphism of natural populations of *Cepaea nemoralis. Cold Spr. Harb. Symp. Quant. Biol.,* **24:**65–86.

1960 La théorie actuelle des mecanismes de l'évolution. *Arch. Philos.,* **23:**8–57.

Landau, R.

1962 Four forms of *Simulium tuberosum* (Lundstr.) in southern Ontario: A salivary gland chromosome study. *Canad. J. Zool.,* **40:**921–939.

Lanier, G. N. and W. E. Burkholder.

1974 Pheromones in speciation of Coleoptera. In *Pheromones,* M. C. Birch, ed., Amsterdam: North-Holland Publishing Co.

Laven, H.

1959 Speciation by cytoplasmic isolation in the *Culex pipiens* complex. *Cold Spr. Harb. Symp. Quant. Biol.,* **24:**166–175.

1967 Speciation and evolution in *Culex pipiens.* In *Genetics of Insect Vectors of Disease,* J. W. Wright and R. Pal, eds., Amsterdam: Elsevier, pp. 251–275.

Lawlor, T. E.
1974 Chromosomal evolution in *Peromyscus*. *Evolution,* **28:**689–692.
Lea, A.
1969 The distribution and abundance of brown locusts, *Locustana pardalina* (Walker), between 1954 and 1965. *J. Ent. Soc. S. Afr.,* **32:**367–398.
Lécher, P.
1962 Etude de la formule chromosomique de l'Isopode asellote *Jaera marina posthirsuta* Forsman. *C.R. Acad. Sci. Paris,* **254:**561–563.
1964 Recherches complémentaires sur le polytypisme de la super-espèce *Jaera albifrons* Leach = *Jaera marina* (Fabricius), III: Etude chromosomique de differentes populations de *Jaera (albifrons) syei* Bocquet. *Bull. Biol. Fr. Belg.,* **98:**415–431.
1967 Cytogénétique de l'hybridization expérimentale et naturelle chez l'isopode *Jaera (albifrons) syei* Bocquet. *Arch. Zool. Exp. Gén.,* **108:**633–698.
1968 Polymorphisme chromosomique dans les populations baltes et scandinaves de l'Isopode *Jaera (albifrons) Syei* Bocquet. *Arch. Zool. Exp. Gén.,* **109:**211–227.
Lécher, P., and G. Prunus.
1972 Caryologie et taxinomie de *Jaera (albifrons) albifrons* (Crustacé, Isopode), populations des côtes bretonnes. *Arch. Zool. Exp. Gén.,* **112:**715–730.
Lécher, P., and M. Solignac.
1972 Etude caryologique des *Jaera (albifrons) ischiosetosa* (Crustacés, Isopodes), I: Polymorphisme chromosomique robertsonien dans trois populations d'Islande. *Arch. Zool. Exp. Gén.,* **113:**439–450.
Lee, A. K.
1967 Studies in Australian amphibia, II: Taxonomy, ecology, and evolution of the genus *Heleioporus* Gray (Anura: Leptodactylidae). *Austral. J. Zool.,* **15:** 367–439.
Levin, D.A.
1969 The challenge from a related species: A stimulus for saltational speciation. *Amer. Nat.,* **103:**316–322.
1970 Reinforcement of reproductive isolation: Plants versus animals. *Amer. Nat.,* **104:**571–581.
Levitan, M., H. L. Carson, and H. D. Stalker
1954 Triads of overlapping inversions in *Drosophila robusta*. *Amer. Nat.,* **88:** 113–114.
Lewis, K. R., and B. John.
1957 Studies on *Periplaneta americana,* II: Interchange heterozygosity in isolated populations. *Heredity,* **11:**11–22.
Lewis, H.
1953a The mechanism of evolution in the genus *Clarkia*. *Evolution,* **7:**1–20.
1953b Chromosome phylogeny and habitat preference in *Clarkia*. *Evolution,* **7:**102–109
1962 Catastrophic selection as a factor in speciation. *Evolution,* **16:**257–271

1966 Speciation in flowering plants. *Science,* **152:**167–172.

1973 The origin of diploid neospecies in *Clarkia. Amer. Nat.,* **107:**161–170.

Lewis, H., and P. H. Raven.

1958 Rapid evolution in *Clarkia. Evolution,* **12:**319–336.

Lewis, H., and M. R. Roberts.

1956 The origin of *Clarkia lingulata. Evolution,* **10:**126–138.

Lewis, W. H., R. L. Oliver, and Y. Suda.

1967 Cytogeography of *Claytonia virginica* and its allies. *Ann. Missouri Bot. Garden,* **54:**153–171.

Lewontin, R. C.

1957 The adaptations of populations to varying environments. *Cold Spr. Harb. Symp. Quant. Biol.,* **22:**395–408.

1967 An estimate of average heterozygosity in man. *Amer. J. Human Genet.,* **19:**681–685.

1973 Population genetics. *Ann. Rev. Genet.,* **7:**1–17.

1974 *The Genetic Basis of Evolutionary Change.* Columbia University Press.

Lewontin, R. C., and J. L. Hubby.

1966 A molecular approach to the study of genic heterozygosity in natural populations, II: Amount of variation and degree of heterozygosity in natural populations of *Drosophila pseudoobscura. Genetics,* **54:**595–609.

Li, H.-L.

1953 Present distribution and habitats of the conifers and Taxads. *Evolution,* **7:**245–261.

Lim, H.-C., V. R. Vickery, and D. K. McE. Kevan.

1973 Cytogenetic studies in relation to taxonomy within the family Gryllidae (Orthoptera), I: Subfamily Gryllinae. *Canad. J. Zool.,* **51:**179–186.

Linder, D., and J. Power.

1970 Further evidence for postmeiotic origin of teratomas in the human female. *Ann. Hum. Genet.,* **34:**21–30.

Littlejohn, M.J.

1959 Call differentiation in a complex of seven species of *Crinia* (Anura, Leptodactylidae). *Evolution,* **13:**452–468.

1965 Premating isolation in the *Hyla ewingi* complex (Anura: Hylidae). *Evolution,* **19:**234–243.

1969 The systematic significance of isolating mechanisms. In *Systematic Biology, Proceedings of an International Conference,* National Academy of Sciences U.S.A., pp. 459–482.

Littlejohn, M. J., and J. J. Loftus-Hills.

1968 An experimental evaluation of premating isolation in the *Hyla ewingi* complex (Anura: Hylidae). *Evolution,* **22:**659–663.

Littlejohn, M. J., and R. S. Oldham.

1968 *Rana pipiens* complex: Mating call structure and taxonomy. *Science,* **162:**1003–1005.

Littlejohn, M. J., and J. D. Roberts.
1975 Acoustic analysis of an intergrade zone between two call races of the *Limnodynastes tasmaniensis* complex (Anura: Leptodactylidae) in southeastern Australia. *Aust. J. Zool.,* **23:**113–122.
Littlejohn, M. J., and G. F. Watson.
1973 Mating-call variation across a narrow hybrid zone between *Crinia laevis* and *C. victoriana* (Anura: Leptodactylidae). *Aust. J. Zool.,* **21:**277–284.
1974 Mating-call discrimination and phonotaxis by females of the *Crinia laevis* complex (Anura: Leptodactylidae). *Copeia,* 1974:171–175.
Littlejohn, M. J., G. F. Watson, and J. J. Loftus-Hills.
1971 Contact hybridization in the *Crinia laevis* complex (Anura: Leptodactylidae). *Austral. J. Zool.,* **19:**85–100.
Lloyd, M., and H. S. Dybas.
1966a The periodical cicada problem, I: Population ecology. *Evolution,* **20:** 133–149.
1966b The periodical cicada problem, II: Evolution. *Evolution,* **20:**466–505.
Lo Bianco, S.
1909 Notizie biologiche riguardanti specialmente il periodo di maturità sessuale negli animali viventi nel Golfo di Napoli. *Mitt. Zool. Sta. Neapel,* **19:**513–761.
Loftus-Hills, J. J.
1975 The evidence for reproductive character displacement between the toads *Bufo americanus* and *B. woodhousii fowleri. Evolution,* **29:**368–369.
Lokki, J., E. Suomalainen, A. Saura, and P. Lankinen.
1975 Genetic polymorphism and evolution in parthenogenetic animals. II: Diploid and polyploid *Solenobia triquetrella* (Lepidoptera: Psychidae). *Genetics,* **79:**513–525.
Lorković, Z.
1958a Die Merkmale der unvollständigen Speciationsstufe und die Frage der Einführung der Semispezies in der Systematik. *Uppsala Univ. Arsskrift,* **6:**159–168.
1958b Some peculiarities of spatially and sexually restricted gene exchange in the *Erebia tyndarus* group. *Cold Spring Harb. Symp. Quant. Biol.,* **23:**319–325.
Löve, Á.
1943 Cytogenetic studies in *Rumex* subgenus *Acetosella. Hereditas,* **30:**1–136.
1957 Sex determination in *Rumex. Proc. Genet. Soc. Canada,* **2:**31–36.
1969 Conservative sex chromosomes in *Acetosa. Chromosomes Today,* **2:**166–171.
Löve, Á., and V. Evenson.
1967 The taxonomical status of *Rumex pauciflorus. Taxon,* **16:**423–425.
Löve, Á., and B. M. Kapoor.
1966 An allopolyploid *Ophioglossum. The Nucleus,* **9:**132–138.
1967 The highest plant chromosome number in Europe. *Svensk Bot. Tidskr.,* **61:**29–32.

Löve, Á., and D. Löve.
1957 Arctic polyploidy. *Proc. Genet. Soc. Canada,* **2:**23–27.
Löve, Á., and N. Sarkar.
1956 Cytotaxonomy and sex determination of *Rumex paucifolius. Canad. J. Bot.,* **34:**261–268.
Lowe, C. H., C. J. Cole, and J. L. Patton.
1967 Karyotype evolution and speciation in lizards (genus *Sceloporus)* during evolution of the North American desert. *Syst. Zool.,* **16:**296–300.
Lowe, C. H., and J. W. Wright.
1966a Evolution of parthenogenetic species of *Cnemidophorus* (Whiptail lizards) in western North America. *J. Arizona Acad. Sci.,* **4:**81–87.
1966b Chromosomes and karyotypes of cnemidophorine teiid lizards, *Mamm. Chrom. Newsl.,* **22:**199–200.

MacArthur, R. H.
1972 *Geographical Ecology: Patterns in the Distributions of Species.* New York: Harper and Row.
MacArthur, R. H., and E. O. Wilson.
1967 *The Theory of Island Biogeography.* Princeton University Press.
Macdonald, G. A., and A. T. Abbott.
1970 *Volcanoes in the Sea—The Geology of Hawaii.* Honolulu: University Hawaii Press.
Macgregor, H. C., H. Horner, C. A. Owen, and I. Parker.
1973 Observations on centromeric heterochromatin and satellite DNA in salamanders of the genus *Plethodon. Chromosoma,* **43:**328–348.
Macgregor, H. C., and T. M. Uzzell, Jr.
1964 Gynogenesis in salamanders related to *Ambystoma jeffersonianum. Science,* **143:**1043–1045.
Mackerras, I. M.
1970 Composition and distribution of the fauna. In *Insects of Australia,* Melbourne University Press, pp. 187–203.
Madahar, D. P.
1969 The salivary gland chromosomes of seven taxa in the subgenus *Stegopterna* (Diptera, Simuliidae, *Cnephia). Canad. J. Zool.,* **47:**115–119.
Maeki, K., and S. A. Ae.
1966 A chromosome study of 28 species of Himalayan butterflies (Papilionidae, Pieridae). *Spec. Bull. Lep. Soc. Japan,* pp. 107–119.
Maeki, K., and C. L. Remington.
1960a Studies of the chromosomes of North American Rhopalocera, I: Papilionidae. *J. Lepidopt. Soc.,* **13:**193–203.
1960b Studies of the chromosomes of North American Rhopalocera, II: Hesperiidae, Megathymidae, and Pieridae. *J. Lepidopt. Soc.,* **14:**37–57.

Main, A. R., A. K. Lee, and M. J. Littlejohn.
1958 Evolution in three genera of Australian frogs. *Evolution,* **12:**224–233.

Main, B. Y.
1957 Biology of aganippine trapdoor spiders (Mygalomorphae: Ctenizidae). *Austral. J. Zool.* **5:**402–473.
1962 Adaptive responses and speciation in the spider genus *Aganippe* Cambridge. In *The Evolution of Living Organisms,* G. W. Leeper, ed., University of Melbourne Press, pp. 359–369.

Malheiros-Gardé, N., and A. Gardé.
1951 Agmatoploidia no genero *Luzula* DC. *Genética Iberica,* **3:**155–176.

Mancino, G.
1963 La minuta struttura di un tratto del lampbrush chromosome XII di *Triturus cristatus cristatus. Rend. Accad. Naz. Lincei, Cl. Sci. Fis. Mat. Nat.* Ser. VIII, **34:**65–67.
1968 Sulla validità specifica di *Triturus italicus* (Peracca 1898) (Anfibi Urodeli). *Rend. Accad. Naz. Lincei, Cl. Sci. Fis. Mat. Nat.* Ser. VIII, **54:** 697–700.

Mancino, G., and G. Barsacchi.
1965 Le mappe dei cromosomi *lampbrush* di *Triturus,* I: *Triturus alpestris apuanus. Caryologia,* **18:**637–665.
1966 The maps of the lampbrush chromosomes of *Triturus* (Amphibia, Urodela), II: *Triturus helveticus helveticus. Riv. Biologia,* **59:**311–351.
1969 The maps of the lampbrush chromosomes of *Triturus,* III: *Triturus italicus. Ann. Embr. Morph.,* **2:**355–377.

Mancino, G., G. Barsacchi, and I. Nardi.
1969 The lampbrush chromosomes of *Salamandra salamandra* (L.) (Amphibia, Urodela). *Chromosoma,* **26:**365–387.

Mancino, G., and I. Nardi.
1971 Chromosomal heteromorphism and female heterogamety in the marbled newt *Triturus marmoratus* (Latreille, 1800). *Experientia,* **27:**821–822.

Manfredi-Romanini, M. G.
1973 The DNA nuclear content and the evolution of vertebrates. In *Cytotaxonomy and Vertebrate Evolution,* A. B. Chiarelli and E. Capanna, eds., London: Academic Press, pp. 39–81.

Manton, I.
1950 *Problems of Cytology and Evolution in the Pteridophyta.* Cambridge University Press.
1951 Cytology of *Polypodium* in America. *Nature,* **167:**37.
1953 The cytological evolution of the fern flora of Ceylon. *Symp. Soc. Exp. Biol.* No. 7, *Evolution,* pp. 174–185.

Manton, I., and W. A. Sledge.
1954 Observations on the cytology and taxonomy of the pteridophyte flora of Ceylon. *Phil. Trans. Roy. Soc. Lond. (B)* **238:**125–185.

Marks, G. E. and D. Schweizer.
1974 Giemsa banding: Karyotype differences in some species of *Anemone* and *Hepatica nobilis*. *Chromosoma,* **44:**405–416.

Marlier, G.
1959 Observations sur la biologie littorale du lac Tanganika. *Rev. Zool. Bot. Afr.* **59:**164–183.

Martin, J.
1967 Meiosis in inversion heterozygotes in Chironomidae. *Canad. J. Genet. Cytol.,* **9:**255–268.

Martin, R. D.
1972 Adaptive radiation and behaviour of the Malagasy lemurs. *Phil. Trans. Roy. Soc. Lond. (B)* **264:**295–352.

Masaki, S.
1961 Geographic variation of diapause in insects. *Bull. Fac. Agric. Hirosaki Univ.* **7:**66–98.

Mascarello, J. T., A. D. Stock, and S. Pathak.
1975 Conservatism in the arrangement of genetic material in rodents. *J. Mammal.,* **55:**695–704.

Mascarello, J. T., J. W. Warner, and R. J. Baker.
1975 A chromosome banding analysis of the mechanisms involved in the karyological divergence of *Neotoma phenax* (Merriam) and *Neotoma micropus* Baird. *J. Mammal.,* **55:**831–834.

Maslin, T. P.
1967 Skin-grafting in the bisexual teiid lizard *Cnemidophorus sexlineatus* and in the unisexual *C. tesselatus. J. Exp. Zool.,* **166:**137–150.

Matthes, H.
1962 Poissons nouveaux ou interessants du lac Tanganika et du Ruanda. *Ann. Mus. Roy. Afr. Cent. Ser. Octavo Zool.,* **111:**27–88.

Matthey, R.
1941 Etude biologique et cytologique de *Saga pedo* Pallas (Orthoptères: Tettigoniidae). *Rev. Suisse Zool.,* **48:**91–102.
1949 *Les Chromosomes des Vertébrés*. Lausanne: Rouge.
1957 Cytologie comparée et taxonomie des Chamaeleontidae (Reptilia, Lacertilia). *Rev. Suisse Zool.,* **64:**709–732.
1961 La formule chromosomique et la position systématique de *Chamaeleo gallus* Günther (Lacertilia). *Zool. Anz.,* **166:**153–159.
1964 La signification des mutations chromosomiques dans les processus de speciation: Etude cytologique du sous-genre *Leggada* Gray (Mammalia–Muridae). *Arch. Biol.,* **75:**169–206.
1965 Le probleme de la détermination du sexe chez *Acomys selousi* de Winton: Cytogénétique du genre *Acomys* (Rodentia, Muridae). *Rev. Suisse Zool.,* **72:**119–144.
1966 Le polymorphisme chromosomique des *Mus* africains du sous-genre *Leggada. Rev. Suisse Zool.,* **73:**585–607.

Matthey, R.
1968a Un nouveau systeme chromosomique polymorphe chez des *Leggada* africaines du groupe *tennellus* (Rodentia, Muridae). *Genetica,* **38:**211–226.
1968b Cytogénétique et taxonomie du genre *Acomys:* A. *percivali* Dollman et *A. wilsoni* Thomas, espèces d'Abyssinie. *Mammalia,* **32:**621–627.
1970 L' "éventail robertsonien" chez les *Mus (Leggada)* africains du groupe *minutoides – musculoides. Rev. Suisse Zool.,* **77:**625–629.
Matthey, R., and M. Jotterand.
1970 Nouveau système polymorphe non-robertsonien chez les Leggadas *(Mus sp.)* de république Centrafricaine. *Rev. Suisse Zool.,* **77:**630–665.
Matthey, R., and J. van Brink.
1956 Note préliminaire sur la cytologie chromosomique comparée des Cameleons. *R. Suisse Zool.,* **63:**241–246.
1960 Nouvelle contribution à la cytologie comparée des Chamaeleontidae (Reptilia, Lacertilia), *Bull. Soc. Vaud. Sci. Nat.,* **67:**333–348.
Maxson, L. R., A. C. Wilson, and V. M. Sarich.
1974 A comparison of chromosomal, protein and organismal evolution. *Genetics,* **77:**S41–S42.
Maynard-Smith, J.
1962 Disruptive selection, polymorphism, and sympatric speciation. *Nature,* **195:**60–62.
1966 Sympatric speciation. *Amer. Nat.,* **100:**637–650.
1970 The causes of polymorphism. *Symp. Zool. Soc. Lond.,* **26:**371–383.
1971 What use is sex? *J. Theor. Biol.,* **30:**319–335.
1974 Recombination and the rate of evolution. *Genetics,* **78:**299–304.
Mayr, E.
1940 Speciation phenomena in birds. *Amer. Nat.,* **74:**249–278.
1942 *Systematics and the Origin of Species.* Columbia University Press.
1945 Introduction to Symposium on Age and Distribution Patterns of Gene Arrangements in *Drosophila pseudo-obscura. Lloydia* (Cincinnati), **8:**69–83.
1947 Ecological factors in speciation. *Evolution,* **1:**263–288.
1954a Geographic speciation in tropical echinoids. *Evolution,* **8:**1–18.
1954b Change of genetic environment and evolution. In *Evolution as a Process,* J. Huxley, A. C. Hardy, and E. B. Ford, eds., pp. 157–180. London: Allen and Unwin.
1957 Die denkmöglichen Formen der Artentstehung. *Rev. Suisse Zool.,* **64:**219–235.
1963a *Animal Species and Evolution.* Harvard University Press.
1963b Reply to criticism by R. D. Alexander. *Syst. Zool.,* **12:**204–206.
1969a Species, speciation, and chromosomes. In *Comparative Mammalian Cytogenetics,* K. Benirschke, ed., New York: Springer Verlag, pp. 1–7.
1969b *Principles of Systematic Zoology.* New York: McGraw-Hill.
1970 *Populations, Species, and Evolution.* Harvard University Press.

Mayr, E., and C. B. Rosen.
1956 Geographic variation and hybridization in populations of Bahama snails *(Cerion)*. *Amer. Mus. Novitates* No. 1806, pp. 1–48.

Mazrimas, J. A., and F. T. Hatch.
1972 A possible relationship between satellite DNA and the evolution of kangaroo rat species (genus *Dipodomys*). *Nature New Biol.*, **240:**102–105.

Mecham, J. S.
1960 Introgressive hybridization between two southeastern treefrogs. *Evolution,* **14:**445–457.
1961 Isolating mechanisms in anuran amphibians. In *Vertebrate Speciation,* W. F. Blair, ed., University of Texas Press, pp. 24–61.
1975 Incidence and significance of introgressive hybridization in anuran amphibians. *Amer. Zoologist,* **15:**831.

Mecham, J. S., M. J. Littlejohn, R. S. Oldham, L. E. Brown, and J. R. Brown.
1973 A new species of leopard frog (*Rana pipiens* complex) from the plains of the central United States. *Occas. Pap. Mus. Texas Tech. Univ.* No. 18 (11 pp).

Mehra, P. N.
1961 Cytological evolution of ferns with particular reference to Himalayan forms. *Proc. 48th Indian Sci. Congr.,* **2:**1–24.

Meise, W.
1928 Die Verbreitung der Aaskrähe (Formenkreis *Corvus corone* L.). *J. Ornithol.,* **76:**1–203.

Mello, M. L. S., and Z. V. da Silveira.
1970 Somatic polyploidy in larval malpighian tubules of *Melipona quinquefasciata* Lep. *Nucleus,* **13:**59–61.

Mello-Sampayo, T.
1971 Genetic regulation of meiotic chromosome pairings by chromosome 3D of *Triticum aestivum*. *Nature New Biol.,* **230:**22–23.

Mestriner, M. A.
1969 Biochemical polymorphism in bees *(Apis mellifera ligustica)*. *Nature,* **223:**188–189.

Mettler, L. E., and J. J. Nagle.
1966 Corroboratory evidence for the concept of the sympatric origin of isolating mechanisms. *Dros. Inf. Serv.,* **40:**82–83.

Meyer, H.
1938 Investigations concerning the reproductive behavior of *Mollienisia "formosa."* *J. Genet.,* **36:**327–366.

Meylan, A.
1964 Le polymorphisme chromosomique de *Sorex araneus* L. (Mamm.–Insectivora). *Rev. Suisse Zool.,* **71:**903–983.
1965 Répartition géographique des races chromosomiques de *Sorex araneus* L. en Europe (Mamm.–Insectivora). *Rev. Suisse Zool.,* **72:**636–646.

Meylan, A., and J. Hausser.
1973 Les chromosomes des *Sorex* du groupe *araneus–arcticus* (Mammalia, Insectivora). *Z. Säugetierk.,* **38:**143–158.

Mikulska, I.
1960 New data to the cytology of the parthenogenetic weevils of the genus *Otiorrhynchus* Germ. (Curculionidae, Coleoptera) from Poland. *Cytologia,* **25:**322–333.

Milani, R.
1967 The genetics of *Musca domestica* and of other muscoid flies. In *Genetics of Insect Vectors of Disease,* J. W. Wright and R. Pal, eds., Amsterdam: Elsevier, pp. 315–369.

Milkman, R. D.
1967 Heterosis as a major cause of heterozygosity in nature. *Genetics,* **55:**493–495.

Miller, A. H.
1956 Ecologic factors that accelerate formation of races and species of terrestrial vertebrates. *Evolution,* **10:**262–277.

Miller, D. D., R. B. Goldstein, and R. A. Patty.
1975 Semispecies of *Drosophila athabasca* distinguishable by male courtship sounds. *Evolution,* **29:**531–544.

Miller, R. R.
1948 The cyprinodont fishes of the Death Valley system of eastern California and southwestern Nevada. *Misc. Publ. Mus. Zool. Univ. Mich.* No. 68, pp. 1–155.
1950 Speciation in fishes of the genera *Cyprinodon* and *Empetrichthys* inhabiting the Death Valley region. *Evolution,* **4:**155–163.

Minks, A. K., W. L. Roelofs, F. J. Ritter, and C. J. Persoons.
1973 Reproductive isolation of two tortricid moth species by different ratios of a two-component sex attractant. *Science,* **180:**1073.

Missiroli, A.
1939 The varieties of *Anopheles maculipennis* and the malaria problem in Italy. *Verh. VII Int. Kongr. Entomol. 1938,* **3:**1619–1640.

Mizuno, S., and H. C. Macgregor.
1974 Chromosomes, DNA sequences, and evolution in salamanders of the genus *Plethodon. Chromosoma,* **48:**239–296.

Montalenti, G.
1960 Alcune considerazioni sull' evoluzione della determinazione del sesso. *Accad. Naz. Lincei Quaderno* No. 47, pp. 153–181.

Montalenti, G., and L. Fratini.
1959 Observations on the spermatogenesis of *Bacillus rossius* (Phasmoidea). *Proc. XV Int. Congr. Zool. London,* pp. 749–750.

Montalenti, G., and G. Vitagliano-Tardini.
1960 Osservazioni sull'idoneità (fitness) di alcuni incroci di *Asellus. Atti Assoc. Genet. Ital.,* **5:**207–216.

Moorehead, A.

1969 *Darwin and the Beagle.* London: Hamish Hamilton.

Mooring, J.

1958 A cytogenetic study of *Clarkia unguiculata,* I: Translocations. *Amer. J. Bot.,* **45:**233–242.

Morales, A. E.

1940 Los Gryllotalpinae de España. *Bol. Pat. Veg. Ent. Agric. (Madrid),* **9:**212–233.

Morescalchi, A.

1973 Amphibia. In *Cytotaxonomy and Vertebrate Evolution,* A. B. Chiarelli and E. Capanna, eds., London: Academic Press, pp. 233–348.

Moriwaki, K., H. Kato, H. Imai, K. Tsuchiya, and T. Yosida.

1975 Geographical distribution of 12 transferrin alleles in black rats of Asia and Oceania. *Genetics,* **79:**295–304.

Mosquin, T.

1962 *Clarkia stellata,* a new species from California. *Leafl. Western Bot.,* **9:**215–216.

1967 Evidence for autopolyploidy in *Epilobium angustifolium* (Onagraceae). *Evolution,* **21:**713–719.

Mrongovius, M.

1975 *Studies on Hybrids Between Members of the* viatica *Group of Morabine Grasshoppers.* Ph.D. thesis, University of Melbourne.

Muller, C. H.

1952 Ecological control of hybridization in *Quercus:* A factor in the mechanism of evolution. *Evolution,* **6:**147–161.

Muller, H. J.

1925 Why polyploidy is rarer in animals than in plants. *Amer. Nat.,* **59:**346–353.

1932 Some genetic aspects of sex. *Amer. Nat.,* **66:**118–138.

1950 Our load of mutations. *Amer. J. Hum. Genet.,* **2:**111–176.

Muller, H. J., and F. Settles.

1927 The non-functioning of genes in spermatozoa. *Z. indukt. Abstamm. Vererbungsl.,* **43:**285–312.

Müller, P.

1973 The dispersal centres of terrestrial vertebrates in the neotropical realm. *Biogeographica,* **2:**1–244.

Müntzing, A.

1932 Cytogenetic investigations on synthetic *Galeopsis tetrahit. Hereditas,* **16:**105–154.

1936 The evolutionary significance of autopolyploidy. *Hereditas,* **21:**263–378.

1938 Sterility and chromosome pairing in intraspecific *Galeopsis* hybrids. *Hereditas,* **24:**117–188.

1943 Characteristics of two haploid twins in *Dactylis glomerata. Hereditas,* **29:**134–140.

Murdy, W. H., and H. L. Carson.
1959 Parthenogenesis in *Drosophila mangabeirai* Malog. *Amer. Nat.,* **93:**355–363.

Murray, J.
1972 *Genetic Diversity and Natural Selection.* Edinburgh: Oliver and Boyd.

Murray, J., and B. Clarke.
1966 The inheritance of polymorphic shell characters in *Partula* (Gastropoda). *Genetics,* **54:**1261–1277.
1968 Partial reproductive isolation in the genus *Partula* (Gastropoda) on Moorea. *Evolution,* **22:**684–698.

Myers, G. S.
1960a The endemic fish fauna of Lake Lanao, and the evolution of higher taxonomic categories. *Evolution,* **14:**323–333.
1960b Fish evolution in Lake Nyasa. *Evolution,* **14:**394–396.

Nadler, C. F., and D. M. Lane.
1967 Chromosomes of some species of *Meriones* (Mammalia: Rodentia). *Zeits. Säugetierk.,* **32:**285–291.

Nair, P. S., D. Brncic, and K. Kojima.
1971 Isozyme variations and evolutionary relationships in the *mesophragmatica* species group of *Drosophila. Studies in Genetics VI,* Univ. Texas Publ. No. 7103, pp. 17–28.

Nankivell, R. N.
1967 A terminal association of two pericentric inversions in first metaphase cells of the Australian grasshopper *Austroicetes interioris* (Acrididae). *Chromosoma,* **22:**42–68.

Narbel-Hoftstetter, M.
1964 *Les Altérations de la Méiose chez les Animaux Parthénogénétiques. Protoplasmatologia,* Vol. VI, F2.

Nardi, I., M. Ragghianti, and G. Mancino.
1972 Characterization of the lampbrush chromosomes of the marbled newt *Triturus marmoratus* (Latreille 1800). *Chromosoma,* **37:**1–22.

Navashin, M.
1932 The dislocation hypothesis of evolution of chromosome numbers. *Z. indukt. Abstamm.- u. Vererbl.,* **63:**224–231.

Neaves, W. B.
1969 Adenosine deaminase phenotypes among sexual and parthenogenetic lizards in the genus *Cnemidophorus* (Teiidae). *J. Exp. Zool.,* **171:**175–184.

Nei, M.
1972 Genetic distance between populations. *Amer. Nat.,* **106:**283–292.
1975 *Molecular Population Genetics and Evolution.* Amsterdam: Elsevier.

Nevo, E.
1969 Mole rat *Spalax ehrenbergi:* Mating behavior and its evolutionary significance. *Science,* **163:**484–486.

1973 Test of selection and neutrality in natural populations. *Nature,* **244:**573–575.

Nevo, E., and S. A. Blondheim.
1972 Acoustic isolation in the speciation of mole crickets. *Ann. Ent. Soc. Amer.,* **65:**980–981.

Nevo, E., Y. J. Kim, C. R. Shaw, and C. S. Thaeler, Jr.
1974 Genetic variation, selection and speciation in *Thomomys talpoides* pocket gophers. *Evolution* **28:**1–23.

Nevo, E., G. Naftali, and R. Guttman.
1975 Aggression patterns and speciation. *Proc. Nat. Acad. Sci. U.S.A.,* **72:** 3250–3254.

Nevo, E., and V. Sarich.
1973 Immunology and evolution in mole rats, *Spalax. Israel J. Zool.,* **23:**210.

Nevo, E., and C. R. Shaw.
1972 Genetic variation in a subterranean mammal, *Spalax ehrenbergi. Biochem. Genet.,* **7:**235–241.

Nevo, E., and A. Shkolnik.
1974 Adaptive metabolic variation of chromosome forms in mole rats, *Spalax. Experientia,* **30:**724–726.

Newton, W. C. F., and C. Pellew.
1929 *Primula kewensis* and its derivatives. *J. Genet.,* **20:**405–467.

Ninan, C. A.
1958 Studies on the cytology and phylogeny of the pteridophytes, VI: Observations on the Ophioglossaceae. *Cytologia,* **23:**291–316.

Nobs, M. A.
1963 Experimental studies on species relationships in *Ceanothus. Carnegie Inst. Wash. Publ.* No. 623, pp. 1–94.

Nolte, D. J.
1968 Strain crosses in the Desert Locust. *Proc. 3rd Congr. S. Afr. Genet. Soc.* (1966), pp. 17–22.

Nordenskiöld, H.
1951 Cytotaxonomical studies in the genus *Luzula,* I: *Hereditas,* **37:**324–355.

Nowakowski, J. T.
1962 Introduction to a systematic revision of the family Agromyzidae (Diptera) with some remarks on host plant selection by these flies. *Anales Zoologici,* **20:**68–183.

Nur, U.
1963 Meiotic parthenogenesis and heterochromatization in a soft scale, *Pulvinaria hydangeae* (Coccoidea: Homoptera). *Chromosoma,* **14:**123–139.
1971 Parthenogenesis in coccids (Homoptera). *Amer. Zool.,* **11:**301–308.

Nursall, J. R.
1974 Character displacement and fish behavior, especially in coral reef communities. *Amer. Zoologist,* **14:**1099–1118.

Nygren, A.
1954 Apomixis in the Angiosperms, II. *Bot. Rev.,* **20:**577–649.

Ohno, S.
1970 *Evolution by Gene Duplication*. London: Allen and Unwin.
1974 *Protochordata, Cyclostomata, and Pisces*. Vol. 4, Chordata 1, of *Animal Cytogenetics* B. John, ed., Berlin: Borntraeger, (92 pp).
Ohno, S., and N. B. Atkin.
1966 Comparative DNA values and chromosome complements of eight species of fishes. *Chromosoma,* **18:**455–466.
Ohno, S., J. Muramoto, L. Christian, and N. B. Atkin.
1967 Diploid-tetraploid relationship among Old World members of the fish family Cyprinidae. *Chromosoma,* **23:**1–9.
Ohno, S., J. Muramoto, J. Klein, and N. B. Atkin.
1969 Diploid-tetraploid relationship in clupeoid and salmonoid fish. *Chromosomes Today,* **2:**139–147.
Ohno, S., U. Wolf, and N. B. Atkin.
1968 Evolution from fish to mammals by gene duplication. *Hereditas,* **59:**169–187.
Ojima, Y., and S. Hitotsumachi.
1969 Cytogenetical studies in loaches (Pisces, Cobitidae). *Zool. Mag. Tokyo,* **78:**4.
Oliver, J. H. Jr.
1971 Parthenogenesis in mites and ticks (Arachnida: Acari). *Amer. Zool.,* **11:** 283–299.
Olmo, E.
1976 Genome size in some reptiles. *J. Exp. Zool.,* **195:**305–310.
Olmo, E., and A. Morescalchi.
1975 Evolution of the genome and cell sizes in salamanders. *Experientia,* **31:**804–806.
Omodeo, P.
1951 Gametogenesi e sistematica come problemi connessi con la poliploidia nei Lumbricidae. *Atti Soc. Toscana Sci. Nat. (B),* **58:**1–12.
1952 Cariologia dei Lumbricidae, I. *Caryologia,* **4:**173–274.
1953 Specie e razze poliploidi nei lombrichi. *Convegno di Genetica, La Rivista Scientifica,* Suppl., **23:**43–49.
1954 Aspetti biogeografici della speciazione. *Boll. di Zool.,* **21:**1–56.
1955 Cariologia dei Lumbricidae, II. *Caryologia,* **8:**135–178.
Ortiz, E.
1951 Los cariotipos de *Gryllotalpa gryllotalpa* (L) de la peninsula ibérica. *Bol. Soc. Esp. Hist. Nat.,* **49:**155–158.
1958 El valor taxonomico de las llamadas razas cromosómicas de *Gryllotalpa gryllotalpa* L. *Publ. Inst. Biol. Apl. Barcelona,* **27:**181–194.
Ott, J.
1968 Nachweis natürlicher reproduktiver Isolation zwischen *Sorex gemellus* sp. n. und *Sorex araneus* Linnaeus 1758 in der Schweiz. *Rev. Suisse Zool.,* **75:**53–75.

Ottonen, P. O.
1966 The salivary gland chromosomes of six species in the IIIS-1 group of *Prosimulium* Roub. (Diptera: Simuliidae), *Canad. J. Zool.,* **44:**677–701.
Ottonen, P. O., and R. Nambiar.
1969 The salivary gland chromosomes of *Prosimulium multidentatum* Twinn and three forms included in *Prosimulium magnum* (Dyar and Shannon) (Diptera: Simuliidae). *Canad. J. Zool.,* **47:**943–949.

Pace, A. E.
1974 Systematic and biological studies of the leopard frogs *(Rana pipiens* complex) of the United States. *Misc. Publ. Mus. Zool. Univ. Michigan,* **148:**1–140.
Pardue, M. L.
1973 Localization of repeated DNA sequences in *Xenopus* chromosomes. *Cold Spr. Harb. Symp. Quant. Biol.,* **38:**475–482.
Pardue, M. L., D. D. Brown, and M. L. Birnstiel.
1973 Location of the genes for 5s ribosomal RNA in *Xenopus laevis. Chromosoma* **42:**191–203.
Parker, E. D. Jr., and R. K. Selander.
1976 The organization of genetic diversity in the parthenogenetic lizard *Cnemidophorus tesselatus. Genetics,* **84:**791–805.
Parsons, P. A.
1975 The comparative evolutionary biology of the sibling species *Drosophila melanogaster* and *D. simulans. Quant. Rev. Biol.,* **50:**151–169.
Pasternak, J.
1964 Chromosome polymorphism in the blackfly *Simulium vittatum* (Zett.). *Canad. J. Zool.,* **42:**135–158.
Paterson, H. E., and S. H. James.
1973 Animal and plant speciation studies in Western Australia. *J. Roy. Soc. West. Aust.,* **56:**31–43.
Pathak, S., T. C. Hsu, and F. E. Arrighi.
1973 Chromosomes of *Peromyscus* (Rodentia, Cricetidae), IV: The role of heterochomatin in karyotypic evolution. *Cytogenet. Cell Genet.,* **12:**315–326.
Patterson, J. T., and W. S. Stone.
1952 *Evolution in the Genus Drosophila.* New York: Macmillan.
Patton, J. L.
1969 Chromosome evolution in the pocket mouse, *Perognathus goldmani* Osgood. *Evolution,* **23:**645–662.
1972 Patterns of geographic variation in karyotype in the pocket gopher, *Thomomys bottae* (Eydoux and Gervois). *Evolution,* **26:**574–586.
1973 An analysis of natural hybridization between the pocket gophers, *Thomomys bottae* and *Thomomys umbrinus,* in Arizona. *J. Mammal.,* **54:**561–584.

Patton, J. L., and R. E. Dingman.
1968 Chromosome studies of pocket gophers, genus *Thomomys,* I: The specific status of *Thomomys umbrinus* (Richardson) in Arizona. *J. Mammal.,* **49:**1–13.
1970 Chromosome studies of pocket gophers, genus *Thomomys,* III. Variation in *T. bottae* in the American southwest. *Cytogenetics,* **9:**139–151.

Patton, J. L., and P. Myers.
1974 Chromosomal identity of Black Rats *(Rattus rattus)* from the Galápagos Islands, Ecuador. *Experientia,* **30:**1140–1142.

Patton, J. L. R. K. Selander, and M. H. Smith.
1972 Genic variation in hybridizing populations of gophers (genus *Thomomys*). *Syst. Zool.,* **21:**263–270.

Patton, J. L., S. Y. Yang, and P. Myers.
1975 Genetic and and morphologic divergence among introduced rat populations *(Rattus rattus)* of the Galápagos Archipelago, Ecuador. *Syst. Zool.,* **24:** 296–310.

Pazy, B., and D. Zohary.
1965 The process of introgression between *Aegilops* polyploids: Natural hybridization between *A. variabilis, A. ovata,* and *A. biuncialis. Evolution,* **19:** 385–394.

Peacock, A. D., and U. Weidman.
1961 Recent work on the cytology of animal parthenogenesis. *Przegląd Zoologiczny,* **5:**5–27, 101–122.

Peacock, W. J., D. Brutlag, E. Goldring, R. Appels, C. W. Hinton, and D. L. Lindsley.
1974 The organization of highly repeated DNA sequences in *Drosophila melanogaster* chromosomes. *Cold Spr. Harb. Symp. Quant. Biol.,* **38:**405–416.

Pearl, R.
1930 *The Biology of Population Growth.* New York: Knopf.

Peccinini-Seale, D., and O. Frota-Pessoa.
1974 Structural heterozygosity in parthenogenetic populations of *Cnemidophorus lemniscatus* (Sauria, Teiidae) from the Amazonas valley. *Chromosoma,* **47:**439–451.

Pederson, R. A.
1971 DNA content, ribosomal gene multiplicity, and cell size in fish. *J. Exp. Zool.,* **177:**65–78.

Pell, P. E., and D. I. Southern.
1975 Symbionts in the female tsetse fly *Glossina morsitans morsitans. Experientia,* **31:**650–651.

Petryszak, B.
1975 Chromosome number of *Trachyphloeus scabriculus* (L.) and *T. aristatus* (Gyll.) (Coleoptera, Curculionidae). *Acta Biol. Cracov.,* **18:**91–95.
Pianka, E. R.
1970 On *r*- and *K*-selection. *Amer. Nat.,* **104:**592–597.
Pickford, R.
1953 A two-year life cycle in grasshoppers (Orthoptera: Acrididae) overwintering as eggs and nymphs. *Canad. Entomol.,* **85:**9–14.
Pijnacker, L. P.
1969 Automictic parthenogenesis in the stick insect *Bacillus rossius* Rossi (Cheleutoptera, Phasmidae). *Genetica,* **40:**393–399.
Pimentel, D., G. J. C. Smith and J. Soans.
1967 A population model of sympatric speciation. *Amer. Nat.,* **101:**493–504.
Piza, S. de T.
1943 Meiosis in the male of the Brazilian scorpion, *Tityus bahiensis. Rev. Agric. S. Paulo,* **18:**247–276.
1947 Interessante comportamento dos cromossômios na espermatogenese do escorpiao *Isometrus maculatus* de Geer. *An. Esc. Sup. Agric. L. de Queiroz,* **4:**177–182.
Platt, A. P., and L. P. Brower.
1968 Mimetic versus disruptive coloration in intergrading populations of *Limenitis arthemis* and *astyanax* butterflies. *Evolution,* **22:**699–718.
Platz, J. E.
1972 Sympatric interaction between two forms of Leopard Frog *(Rana pipiens* complex) in Texas. *Copeia,* 1972:232–240.
Poliński, W.
1932 Die reliktäre Gastropodenfauna des Ochrida-Sees. *Zool. Jahrb. Abt. Syst.,* **62:**611–666.
Poll, M.
1956 Poissons Cichlidae. *Résult. Scient. Explor. Hydrobiol. Lac Tanganika* (1946-1947) **3,** Fasc. 4b, pp. 1–619.
Porter, D. L.
1971 Oogenesis and chromosomal heterozygosity in the thelytokous midge, *Lundstroemia parthenogenetica* (Diptera, Chironomidae). *Chromosoma,* **32:**333–342.
Powell, J. R.
1971 Genetic polymorphisms in a varied environment. *Science,* **174:**1035–1036.
Prakash, S.
1969 Genic variation in a natural population of *Drosophila persimilis. Proc. Nat. Acad. Sci. U.S.A.,* **62:**778–784.
1972 Origin of reproductive isolation in the absence of apparent genic differentiation in a geographic isolate of *Drosophila pseudoobscura. Genetics,* **72:** 143–155.

Prakash, S.
1973 Patterns of gene variation in central and marginal populations of *Drosophila robusta. Genetics,* **75:**347–369.
Prakash, S., R. C. Lewontin, and J. L. Hubby.
1969 A molecular approach to the study of genic heterozygosity in natural populations, IV: Patterns of genic variation in central, marginal, and isolated populations of *Drosophila pseudoobscura. Genetics,* **61:**841–858.
Pryor, L. D.
1959 Species distribution and association in *Eucalyptus.* In *Biogeography and Ecology in Australia* (Monographiae Biologicae 8), The Hague: Junk, pp. 461–471.

Ragge, D. R.
1970 La faune terrestre de l'ile de Sainte Helene (première partie), 7. Orthoptera, a. Tettigonioidea. *Ann. Mus. Roy. Afr. Cent. Ser. Octavo Zool.,* **181:**183–190.
Raicu, P., D. Duma, and S. Torcea.
1973 Chromosomal polymorphism in the lesser mole rat, *Spalax leucodon. Chromosomes Today,* **4:**383–386.
Raicu, P., and E. Taisescu.
1972 *Misgurnus fossilis,* a tetraploid fish species. *J. Hered.,* **63:**92–94.
Raikow, R.
1973 Puffing in salivary gland chromosomes of picture-winged Hawaiian *Drosophila. Chromosoma,* **41:**221–230.
Rajasekarasetty, M. R., and C. V. Ramanamurthy.
1964 Analysis of male meiosis in *Periplaneta* sp. *La Cellule,* **65:**83–91.
Ramírez, B. W.
1970 Host specificity of fig wasps (Agaonidae). *Evolution,* **24:**680–691.
Rasch, E. M., R. M. Darnell, K. D. Kallman, and P. Abramoff.
1965 Cytophotometric evidence for triploidy in hybrids of the gynogenetic fish, *Poecilia formosa. J. Exp. Zool.,* **160:**155–170.
Rasch, E. M., L. M. Prehn, and R. W. Rasch.
1970 Cytogenetic studies of *Poecilia* (Pisces), II: Triploidy and DNA levels in naturally occurring populations associated with the gynogenetic teleost *Poecilia formosa* (Girard). *Chromosoma,* **31:**18–40.
Raven, P. H., and H. J. Thompson.
1964 Haploidy and angiosperm evolution. *Amer. Nat.,* **98:**251–252.
Rees, H.
1964 The question of polyploidy in the Salmonidae. *Chromosoma,* **15:**275–279.
Rees, H., and M. H. Hazarika.
1969 Chromosome evolution in *Lathyrus. Chromosomes Today,* **2:**158–165.
Rees, H., and R. N. Jones.
1972 The origin of the wide species variation in nuclear DNA content. *Internat. Rev. Cytol.,* **32:**53–92.

Reig, O. A., O. Angel Spotorno, and D. R. Fernandez.
1972 A preliminary survey of chromosomes in populations of the Chilean burrowing octodont rodent *Spalacopus cyanus* Molina (Caviomorpha, Octodontidae). *Biol. J. Linn. Soc. Lond.*, **4:**29–38.

Reig, O. A., and P. Kiblisky.
1969 Chromosome multiformity in the genus *Ctenomys* (Rodentia, Octodontidae). *Chromosoma,* **28:**211–244.

Reig, O. A., and M. Useche.
UNPUB. Diversidad cariotípica y sistemática en poblaciones venezolanas de *Proechimys,* con datos adicionales sobre poblaciones de Perú y Colombia.

Remington, C. L.
1968 Suture zones of hybrid interaction between recently joined biotas. *Evolutionary Biology,* **2:**321–428.

Renner, O.
1925 Untersuchungen über die faktorielle Konstitution einiger komplexheterozygotischer *Oenotheren. Bibliotheca Genetica,* **9:**1–169.

Richardson, R. H.
1974 Effects of dispersal, habitat selection, and competition on a speciation pattern of *Drosophila* endemic to Hawaii. In *Genetic Mechanisms of Speciation in Insects,* M. J. D. White, ed., Sydney: Australia and New Zealand Book Co., pp. 140–164.

Richmond, R. C.
1972 Enzyme variability in the *Drosophila willistoni* group, III: Amounts of variability in the superspecies *D. paulistorum. Genetics,* **70:**87–112.

Riley, R.
1966 The genetic regulation of meiotic behaviour in wheat and its relatives. *Proc. 2nd Internat. Wheat Genet. Symp.,* **2:**395–408.
1974 Cytogenetics of chromosome pairing in wheat. *Genetics,* **78:**193–203.

Riley, R., and V. Chapman.
1958 Genetic control of the cytologically diploid behaviour of hexaploid wheat. *Nature,* **182:**713–715.

Rivas, L. R.
1964 A reinterpretation of the concepts "sympatric" and "allopatric" with proposal of the additional terms "syntopic" and "allotopic." *Syst. Zool.,* **13:**42–43.

Robertson, A.
1970 A note on disruptive selection experiments in *Drosophila. Amer. Nat.,* **104:**561–569.
1975 The genetic basis of evolutionary change (book review of R. C. Lewontin's *Genetic Basis of Evolutionary Change). Nature,* **254:**367.

Rockwood, E. S., C. G. Kanapi, M. R. Wheeler, and W. S. Stone.
1971 Allozyme changes during the evolution of Hawaiian *Drosophila. Studies in Genetics VI,* Univ. Texas Publ. No. 7103, pp. 193–212.

Roelofs, W. L., and R. T. Cardé.

1974 Sex pheromones in the reproductive isolation of lepidopterous species. In *Pheromones,* M. C. Birch, ed., Amsterdam: North-Holland Publishing Co., pp. 96–114.

Roelofs, W. L., and A. Comeau.

1969 Sex pheromone specificity: Taxonomic and evolutionary aspects in Lepidoptera. *Science,* **165:**398–400.

1970 Lepidopterous sex attractants discovered by field screening tests. *J. Econ. Entomol.,* **63:**969–974.

Rogers, J. S.

1972 Measures of genetic similarity and genetic distance. *Studies in Genetics VII,* Univ. Texas Publ. No. 7213, pp. 145–153.

Rogers, W. P.

1962 *The Nature of Parasitism: The Relationship of some Metazoan Parasites to their Hosts.* New York: Academic Press.

Rollins, R.

1944 Evidence for natural hybridity between guayule *(Parthenium argentatum)* and mariola *(P. incanum). Amer. J. Bot.,* **31:**93–99.

1945 Interspecific hybridization in *Parthenium,* I: Crosses between guayule *(P. argentatum)* and mariola *(P. incanum). Amer. J. Bot.,* **32:**395–404.

1946 Interspecific hybridization in *Parthenium,* II: Crosses involving *P. argentatum, P. incanum, P. stramonium, P. tomentosum,* and *P. hysterophorus. Amer. J. Bot.,* **33:**21–30.

Ross, H. H.

1958 Evidence suggesting a hybrid origin for certain leafhopper species. *Evolution,* **12:**337–446.

Roth, L. M., and S. H. Cohen.

1968 Chromosomes of the *Pycnoscelus indicus* and *P. surinamensis* complex. *Psyche,* **75:**54–76.

Rothfels, K.

UNPUB. *Simuliidae.* Vol. 3, Insecta 3, Diptera IV, of *Animal Cytogenetics,* B. John, ed., Stuttgart: Borntraeger.

Rothfels, K., E. Sexsmith, M. Heimburger, and M. O. Krause.

1966 Chromosome size and DNA content of species of *Anemone* L. and related genera (Ranunculaceae). *Chromosoma,* **20:**54–74.

Rousi, A.

1967 Cytological observations on some species and hybrids of *Vaccinium.* Züchter, **36:**352–359.

Rubini, P. G.

1964 Polimorfismo cromosomico in *Musca domestica* L. *Boll. Zool.,* **31:**679–694.

Rubini, P. G., and M. G. Franco.

1965 Osservazioni sugli eterocromosomi e considerazioni sulla determinazione del sesso in *Musca domestica* L. (Nota preliminare). *Boll. Zool.,* **32:**823–827.

Ruebush, T. K.

1937 The genus *Dalyellia* in America. *Zool. Anz.,* **119:**237–256.

1938 A comparative study of Turbellarian chromosomes. *Zool. Anz.,* **122:**321–329.

Rumpler, Y., and R. Albignac.

1969a Etude cytogénétique de deux Lémuriens, *Lemur macaco macaco* Linné 1766, et *Lemur fulvus rufus* (Audebert 1800) et d'un hybride *macaco macaco/fulvus rufus. C. R. Séanc. Soc. Biol.,* **163:**1247–1250.

Existence d'une variabilité chromosomique intraspécifique chez certains Lémuriens. *C. R. Séanc. Soc. Biol.,* **163:**1989–1992.

1970 Evolution chromosomique des lémuriens malgaches. *Ann. Univ. Madagascar, Médicine et Biologie,* **12–13:**123–134.

Saccá, G.

1953 Contributo alla conoscenza tassonomica del "gruppo" *domestica* (Diptera, Muscidae). *Rend. Ist. Sup. Sanità,* **16:**442–464.

1957 Ricerche sulla speciazione nelle mosche domestiche, IV: Esperienze sull' isolamento sessuale fra le sub-specie di *Musca domestica* L. *Rend. Ist. Sup. Sanità,* **20:**702–712.

1958 Ricerche sulla speciazione nelle mosche domestiche, VI: Ibridismo naturale e ibridismo sperimentale fra le subspecie di *Musca domestica* L. *Rend. Ist. Sup. Sanità,* **21:**1170–1184.

1967 Speciation in *Musca.* In *Genetics of Insect Vectors of Disease,* J. W. Wright and R. Pal, eds., Amsterdam: Elsevier, pp. 385–415.

Saccá, G., and L. Rivosecchi.

1958 Ricerche sulla speciazione nelle mosche domestiche. V. L'areale di distribuzione delle subspecie di *Musca domestica* L. *(Diptera, Muscidae). Rend. Ist. Sup. Sanità,* **21:**1149–1169.

Saez, F. A.

1956 Caso extraordinario de un ortóptero acridido con ocho cromosomas diploides y mecanismo sexual XY. *Biologica, Santiago-Chile,* **22:**27–30.

1957 An extreme karyotype in an orthopteran insect. *Amer. Nat.,* **91:**259–264.

Saez, F. A., and N. Brum.

1959 Citogenética de anfibios anuros de America del sur. *Anales Fac. Med. Montevideo,* **44:**414–423.

1960 Chromosomes of South American amphibians. *Nature,* **175:**945.

Saez, F. A., and N. Brum-Zorrilla.

1966 Karyotype variation in some species of the genus *Odontophrynus* (Amphibia-Anura). *Caryologia,* **19:**55–63.

Saez, F. A., and B. de N. Zorrilla.

1963 Cytogenetics of South American amphibians. *Proc. XI Int. Congr. Genet.,* **1:**141.

Sanders, C. J.
1971a Daily activity patterns and sex pheromone specificity as sexual isolating mechanisms in two species of *Choristoneura* (Lepidoptera: Tortricidae). *Canad. Entomol.*, **103:**498–502.
1971b Sex pheromone specificity and taxonomy of budworm moths (*Choristoneura*). *Science*, **171:**911–913.
Sax, K.
1931 Chromosome ring formation in *Rhoeo discolor. Cytologia*, **3:**36–53.
Scali, V.
1968 Biologia riproduttiva del *Bacillus rossius* (Rossi) nei dintorni di Pisa con particolare riferimento all' influenza del fotoperiodo. *Atti Soc. Tosc. Sci. Nat. Mem. (B)*, **75:**108–139.
Scharloo, W.
1964 The effect of disruptive and stabilizing selection on a cubitus interruptus mutant in *Drosophila*. *Genetics*, **50:**553–562.
1971 Reproductive isolation by disruptive selection: Did it occur? *Amer. Nat.*, **105:**83–86.
Scharloo, W., M. den Boer, and M. S. Hoogmoed.
1967 Disruptive selection on sternopleural chaetae number. *Genet. Res.*, **9:**115–118.
Scharloo, W., M. S. Hoogmoed, and A. ter Kuile.
1967 Stabilizing and disruptive selection on a mutant character in *Drosophila*, I: The phenotypic variance and its components. *Genetics*, **56:**709–726.
Schnetter, M.
1951 Veränderungen der genetischen Konstitution in natürlichen Populationen der polymorphen Bänderschnecken. *Zool. Anz., Suppl.*, **15:**192–206.
Schnitter, H.
1922 Die Najaden der Schweiz. *Rev. Hydrobiol.*, Suppl., **2:**1–200.
Scholl, H.
1956 Die Chromosomen parthenogenetischer Mücken. *Naturwiss.*, **43:**91–92.
1960 Die Oogenese einiger parthenogenetischer Orthocladiinen (Diptera). *Chromosoma*, **11:**380–401.
Schultz, R. J.
1961 Reproductive mechanism of unisexual and bisexual strains of the viviparous fish *Poeciliopsis*. *Evolution*, **15:**302–325.
1966 Hybridization experiments with an all-female fish of the genus *Poeciliopsis*. *Biol. Bull. Woods Hole*, **130:**415–429.
1967 Gynogenesis and triploidy in the viviparous fish *Poeciliopsis*. *Science*, **157:**1564–1567.
1969 Hybridization, unisexuality, and polyploidy in the teleost *Poeciliopsis* (Poeciliidae) and other vertebrates. *Amer. Nat.*, **103:**605–619.
1971 Special adaptive problems associated with unisexuality in fish. *Amer. Zool.*, **11:**351–360.

Sears, E. R., and M. Okamoto.
1958 Interogenomic relationships in hexaploid wheat. *Proc. X. Internat. Congr. Genet.*, **2**:258–259.

Sedlmair, H.
1956 Verhaltens-, Resistenz-, und Gehäuseunterschiede bei den polymorphen Bänderschnecken *Cepaea hortensis* (Müll.) and *Cepaea nemoralis* (L.). *Biol. Zentr.*, **75**:281–313.

Seiler, J.
1961 Untersuchungen über die Entstehung der Parthenogenese bei *Solenobia triquetrella* F.R. (Lepidoptera, Psychidae), III: Die geographische Verbreitung der drei Rassen von Solenobia *triquetrella* (bisexuell, diploid, und tetraploid parthenogenetisch) in der Schweiz und in den angrenzenden Ländern und die Beziehung zur Eiszeit. Bemerkungen über die Entstehung der Parthenogenese. *Z. Vererb.*, **92**:261–316.

Selander, R. K.
1969 The ecological aspects of the systematics of animals. In *Systematic Biology, Proceedings of an International Conference*, pp. 213–247. National Academy of Sciences, U.S.A.

Selander, R. K., W. G. Hunt, and S. Y. Yang.
1969 Protein polymorphism and genic heterozygosity in two European subspecies of the house mouse. *Evolution*, **23**:379–390.

Selander, R. K., and W. E. Johnson.
1973 Genetic variation among vertebrate species. *Ann. Rev. Ecol. Systemat.*, **4**:75–91.

Selander, R. K., D. W. Kaufman, R. J. Baker, and S. L. Williams.
1974 Genic and chromosomal differentiation in pocket gophers of the *Geomys bursarius* group. *Evolution*, **28**:557–564.

Selander, R. K., M. H. Smith, S. Y. Yang, W. E. Johnson, and J. B. Gentry.
1971 Biochemical polymorphism and systematics in the genus *Peromyscus*, I: Variation in the old-field mouse *(Peromyscus polionotus)*. *Studies in Genetics VI*, Univ. Texas Publ. No. 7103, pp. 49–90.

Selander, R. K., and S. Y. Yang.
1969 Protein polymorphism and genic heterozygosity in a wild population of the house mouse *(Mus musculus)*. *Genetics*, **63**:653–667.

Selander, R. K., S. Y. Yang, R. C. Lewontin, and W. E. Johnson.
1970 Genetic variation in the horseshoe crab *(Limulus polyphemus)*, a phylogenetic "relic." *Evolution*, **24**:402–414.

Senna, A.
1911 La spermatogenesi di *Gryllotalpa vulgaris* Latr. *Monit. Zool. Ital.*, **22**: 65–77.

Serov, O. L.
1972 Monomorphism of some structural genes in populations of wild and laboratory rats. *Isozyme Bull.*, **5**:38.

Sharma, A. K.
1969 Evolution and taxonomy of monocotyledons. *Chromosomes Today,* **2:**241–249.
Sharma, G. P., R. Parshad, and M. G. Joneja.
1959 Chromosome mechanism in the males of three species of scorpions (Scorpiones, Buthidae). *Res. Bull.* (N.S.) *Panjab Univ.,* **10:**197–207.
Sharman, G. B.
1956 Chromosomes of the common shrew. *Nature,* **177:**941–942.
Sherman, M.
1946 Karyotype evolution: A cytogenetic study of seven species and six interspecific hybrids of *Crepis. Univ. Calif. Publ. Bot.,* **18:**369–408.
Shorey, H. H.
1970 Sex pheromones of Lepidoptera. In *Control of Insect Behaviour by Natural Products,* D. L. Wood, R. M. Silverstein, and M. Nakajima, eds., New York: Academic Press, pp. 249–281.
Short, L. L. Jr.
1970 A reply to Uzzell and Ashmole on "Suture zones: An alternate view." *Syst. Zool.,* **19:**199–202.
Short, R. V., A. C. Chandley, R. C. Jones, and W. R. Allen.
1974 Meiosis in interspecific equine hybrids, II: The Przewalski horse × domestic horse hybrid *(Equus przewalskii × E. caballus). Cytogenet. Cell Genet.,* **13:**465–478.
Siew-Ngo, L.
1973 Meiotic instability in the plant genus *Globba. Chromosomes Today,* **4:**321–334.
Skalinska, M.
1967 Cytological analysis of some *Hieracium* species, subg. *Pilosella* from mountains of southern Poland. *Acta Biol. Cracoviensa, Bot.,* **10:**127–141.
Slatkin, M.
1974 Cascading speciation. *Nature,* **252:**701–702.
Small, E.
1971 The evolution of reproductive isolation in *Clarkia,* section Myxocarpa. *Evolution,* **25:**330–346.
Smith, B. W.
1969 Evolution of sex-determining mechanisms in *Rumex. Chromosomes Today,* **2:**172–182.
Smith, H. S.
1941 Racial segregation in insect populations and its significance in applied entomology. *J. Entomol.,* **34:**1–12.
Smith, M. H., R. K. Selander, and W. E. Johnson.
1973 Biochemical polymorphism and systematics in the genus *Peromyscus,* III: Variation in the Florida deer mouse *(Peromyscus floridanus),* a Pleistocene relict. *J. Mammal.,* **54:**1–13.

Smith, S. G.
1966 Natural hybridization in the coccinellid genus *Chilocorus*. *Chromosoma,* **18:**380–406.

Smith-White, S.
1959 Cytological evolution in the Australian flora. *Cold Spr. Harb. Symp. Quant. Biol.,* **24:**273–289.
1968 *Brachycome lineariloba:* A species for experimental cytogenetics. *Chromosoma,* **23:**359–364.

Smith-White, S., and C. R. Carter.
1970 The cytology of *Brachycome lineariloba,* II: The chromosome species and their relationships. *Chromosoma,* **30:**129–153.

Smith-White, S., C. R. Carter, and H. M. Stace.
1970 The cytology of *Brachycome,* I: The subgenus *Eubrachycome,* a general survey. *Aust. J. Bot.,* **18:**99–125.

Smith-White, S., and A. R. Woodhill.
1954 The nature and significance of nonreciprocal fertility in *Aedes scutellaris* and other mosquitoes. *Proc. Linn. Soc. N.S.W.,* **79:**163–176.
1955 The nature and significance of nonreciprocal fertility in *Aedes scutellaris* and other mosquitoes. *Proc. Linn. Soc. N.S.W.,* **79:**163–176.

Smreczyński, S.
1966 Ryjkowe–Curculionidae. Podrodziny Otiorrhynchinae, Brachyderinae. *Klucze do Oznaczania Owadów Polski,* **19:**98b, 1–130.

Snaydon, R. W.
1970 Rapid population differentiation in a mosaic environment, I: Response of *Anthoxanthum odoratum* populations to soils. *Evolution,* **24:**257–269.

Snyder, T. P.
1974 Lack of allozymic variability in three bee species. *Evolution,* **28:**687–689.

Sokal, R. R., and T. J. Crovello.
1970 The biological species concept: A critical evaluation. *Amer. Nat.,* **104:** 127–153.

Soldatović, B., S. Zivković, I. Savić, and M. Milosević.
1967 Vergleichende Analyse der Morphologie und der Anzahl der Chromosomen zwischen verschiedenen Populationen von *Spalax leucodon* Nordmann, 1840. *Z. Säugertierk.,* **32:**238–245.

Solignac, M.
1969a Distributions contiguës, sympatrie et hybridation naturelle chez la superespèce *Jaera albifrons* Leach (Isopodes Asellotes). *C. R. Acad. Sci. Paris,* **268:**1610–1612.
1969b Hybridation introgressive dans la population complexe de *Jaera albifrons* de Luc-sur-Mer. *Arch. Zool. Exp. Gén.,* **110:**629–652.
1972 Comparaison des comportments sexuels specifiques dans la super-espece *Jaera albifrons* (Isopodes Asellotes). *C. R. Acad. Sci. Paris,* **274:**2236–2239.

Spassky, B., R. C. Richmond, S. Pérez-Salas, O. Pavlovsky, C. A. Mourão, A. S. Hunter, H. Hoenigsberg, Th. Dobzhansky, and F. J. Ayala.
1971 Geography of the sibling species related to *Drosophila willistoni,* and of the semispecies of the *Drosophila paulistorum* complex. *Evolution,* **25:**129–143.

Spieth, H. T.
1952 Mating behavior within the genus *Drosophila* (Diptera). *Bull. Amer. Mus. Nat. Hist.* **99:**397–474.
1968 Evolutionary implications of sexual behavior in *Drosophila. Evol. Biol.* **2:**157–193.
1974 Mating behavior and evolution of the Hawaiian *Drosophila.* In *Genetic Mechanisms of Speciation in Insects,* M. J. D. White, ed., Sydney: Australia and New Zealand Book Co., pp. 94–101.

Staiger, H.
1954 Der Chromosomendimorphismus beim Prosobranchier *Purpura lapillus* in Beziehung zur Ökologie der Art. *Chromosoma,* **6:**419–478.
1955 Reziproke Translokationen in natürlichen Populationen von *Purpura lapillus* (Prosobranchia). *Chromosoma,* **7:**181–197.

Staiger, H., and C. Bocquet.
1954 Cytological demonstration of female heterogamety in Isopods. *Experientia,* **10:**64–66.
1956 Les chromosomes de la super-espèce *Jaera marina* (F.) et de quelques autres Janiridae (Isopodes Asellotes). *Bull. Biol. Fr. Belg.,* **90:**1–32.

Stalker, H. D.
1954 Parthenogenesis in *Drosophila. Genetics,* **39:**4–34.
1956 On the evolution of parthenogenesis in *Lonchoptera* (Diptera). *Evolution,* **10:**345–359.
1972 Intergroup phylogenies in *Drosophila* as determined by comparison of salivary banding patterns. *Genetics,* **70:**457–474.

Stanković, S.
1932 Die Fauna des Ochridsees und ihre Herkunft. *Arch. Hydrobiol.,* **23:**557–616.
1960 *The Balkan Lake Ochrid and Its Living World.* The Hague: Junk.

Stearns, H. T.
1966 *Geology of the State of Hawaii.* Palo Alto, Calif.: Pacific Books.

Stebbins, G. L.
1938 Cytological characteristics associated with the different growth habits in the dicotyledons. *Amer. J. Bot.,* **25:**189–198.
1950 *Variation and Evolution in Plants.* Columbia University Press.
1963 Perspectives, I. *Amer. Scientist,* **51:**362–370.
1966 *Processes of Organic Evolution.* Englewood Cliffs, New Jersey: Prentice Hall.
1970 Variation and evolution in plants: Progress during the past twenty years. In *Essays in Evolution and Genetics in Honor of Theodosius Dobzhansky,*

M. K. Hecht and W. C. Steere, eds., pp. 173–208. Appleton Century Crofts.
1971 *Chromosomal Evolution in Higher Plants.* London: Arnold.

Stebbins, G. L., and E. B. Babcock.
1939 The effect of polyploidy and apomixis on the evolution of species in *Crepis. J. Hered.,* **30:**519–530.

Stebbins, G. L., and D. Zohary.
1959 Cytogenetic and evolutionary studies in the genus *Dactylis,* I: Morphology, distribution, and interrelationships of the diploid subspecies. *Univ. Calif. Publ. Bot.,* **31:**1–39.

Stefani, R.
1956 Il problema della partenogenesi in *"Haploembia solieri"* Ramb. (Embioptera, Oligotomidae). *Atti Accad. Naz. Lincei Mem. Cl. Sci. Fis. Mat. Nat.* Sez. IIIa, **5:**127–201.
1959 Secondo contributo alla conocenza della cariologia negli insetti Embiotteri: il corredo cromosomico nelle specie dell' Europa meridionale. *Atti Accad. Naz. Lincei Rend. Sci. Fis. Mat. Nat.* Ser. VIII, **26:**396–399.
1960 *L'Artemia salina* partenogenetica a Cagliari. *Riv. di Biol.,* **53:**463–491.

Steopoe, I.
1939 Nouvelles recherches sur la spermatogénèse chez *Gryllotalpa vulgaris* de Roumanie. *Arch. Zool. Exp. Gén.,* **80:**445–464.

Stern, R.
1972 Satellite DNA's of *Xenopus mulleri. Carnegie Inst. Wash. Yearbook,* **71:**22.

Stock, A. D.
1971 Chromosome evolution in the genus *Dipodomys* and its phylogenetic implications. *Mammalian Chrom. Newsl.,* **12:**122–128.

Stock, A. D., and T. C. Hsu.
1973 Evolutionary conservation in arrangement of genetic material: A comparative analysis of chromosome banding between the rhesus macaque (2n = 42, 84 arms) and the African green monkey (2n = 60, 120 arms). *Chromosoma,* **43:**211–224.

Stone, W. S.
1955 Genetic and chromosomal variability in *Drosophila. Cold Spr. Harb. Symp. Quant. Biol.,* **20:**256–270.
1962 The dominance of natural selection and the reality of superspecies (species groups) in the evolution of *Drosophila. Studies in Genetics II,* Univ. Texas Publ. No. 6205, pp. 507–537.

Stone, W. S., W. C. Guest, and F. D. Wilson.
1960 The evolutionary implications of the cytological polymorphism and phylogeny of the *virilis* group of *Drosophila. Proc. Nat. Acad. Sci. U.S.A.,* **46:**350–361.

Straw, R. M.
1955 Hybridization, homogamy, and sympatric speciation. *Evolution,* **9:**441–444.

Stromberg, P. C., and J. L. Crites.
1974 Specialization, body volume, and geographical distribution of Camallanidae (Nematoda). *Syst. Zool.,* **23:**189–201.

Suomalainen, E.
1958 On polyploidy in animals. *Proc. Finn. Acad. Sci. Letters.* **1:**105–119.
1961 On morphological differences and evolution of different polyploid parthenogenetic weevil populations. *Hereditas,* **47:**309–341.
1963 On the chromosomes of the Geometrid moths *Cidaria. Proc. XIth Internat. Genet. Congr.,* **1:**137–138.
1965 On the chromosomes of the Geometrid moth genus *Cidaria. Chromosoma,* **16:**166–184.
1969 Evolution in parthenogenetic Curculionidae. *Evol. Biol.,* **3:**261–296.

Suomalainen, E., and A. Saura.
1973 Genetic polymorphism and evolution in parthenogenetic animals, I: Polyploid Curculionidae. *Genetics,* **74:**489–508.

Sutton, W. D., and M. McCallum.
1972 Related satellite DNA's in the genus *Mus. J. Mol. Biol.,* **71:**633–656.

Svärdson, G.
1945 Chromosome studies on Salmonidae. *Rep. Swed. State Inst. Freshw. Fish Res. Drottingholm,* **23:**1–151.

Sved, J. A., T. E. Reed, and W. F. Bodmer.
1967 The number of balanced polymorphisms that can be maintained in a natural population. *Genetics,* **55:**469–481.

Sweadner, W. R.
1937 Hybridization and the phylogeny of the genus *Platysamia. Carnegie Mus. Ann.,* **25:**163–242.

Świetlińska, Z.
1963 Cytogenetic relationships among *Rumex acetosa, Rumex arifolius,* and *Rumex thyrsiflorus. Acta Soc. Bot. Pol.,* **32:**215–279.

Takenouchi, Y.
1969 A further study on the chromosomes of the parthenogenetic weevil, *Listroderes costirostris* Schönherr, from Japan. *Cytologia,* **34:**360–368.
1972 A chromosome study of a new polyploid parthenogenetic weevil, *Myllocerus nipponicus* Zumpt (Coleoptera: Curculionidae). *Kontyû,* **40:**121–123.

Taylor, K. M., D. A. Hungerford, R. L. Snyder, and F. A. Ulmer.
1968 Uniformity of karyotypes in the Camelidae. *Cytogenetics,* **7:**8–15.

Templeton, A. R., H. L. Carson, and C. F. Sing.
1976 The population genetics of parthenogenetic strains of *Drosophila mercatorum,* II: The capacity for parthenogenesis in a natural, bisexual population. *Genetics,* **82:**527–542.

Tettenborn, U., and A. Gropp.
1970 Meiotic nondisjunction in mice and mouse hybrids. *Cytogenetics,* **9:**272–283.

Thaeler, C. S. Jr.
1968a Karyotypes of 16 populations of the *Thomomys talpoides* complex of pocket gophers (Rodentia, Geomyidae). *Chromosoma,* **25:**172–183.
1968b An analysis of the distribution of pocket gopher species in northeastern California. *Univ. Calif. Publ. Zool.,* **86:**1–46.
1968c An analysis of three hybrid populations of pocket gophers (genus *Thomomys*). *Evolution,* **22:**543–555.
1974 Four contacts between ranges of different chromosome forms of the *Thomomys talpoides* complex (Rodentia: Geomyidae). *Systemat. Zool.,* **23:**343–354.

Theim, H.
1933 Beitrag zur Parthenogenese und Phänologie der Geschlechter von *Lecanium corni* Bouche (Coccidae). *Z. Morph. Ökol. Tiere,* **27:**294–324.

Thoday, J. M.
1965 Effects of selection for genetic diversity. *Proc. XI. Int. Congr. Genet.,* **3:**533–540.
1972 Disruptive selection. *Proc. Roy. Soc. Lond. (B),* **182:**109–143.

Thoday, J. M., and T. B. Boam.
1959 Effects of disruptive selection, III: Polymorphism and divergence without isolation. *Heredity,* **13:**205–218.

Thoday, J. M., and J. B. Gibson.
1962 Isolation by disruptive selection. *Nature,* **193:**1164–1166.
1970 The probability of isolation by disruptive selection. *Amer. Nat.,* **104:**219–230.
1971 Reply to Scharloo. *Amer. Nat.,* **105:**86–88.

Thomas, B.
1973 Evolutionary implications of karyotypic variation in some insular *Peromyscus* from British Columbia, Canada. *Cytologia,* **38:**485–495.

Thomas, P. T.
1940 Reproductive versatility in *Rubus,* II: The chromosomes and development. *J. Genet.,* **40:**119–128.

Thompson, P. E.
1971 Male and female heterogamety in populations of *Chironomus tentans* (Diptera: Chironomidae). *Canad. Entomol.,* **103:**369–372.

Thompson, P. E., and J. S. Bowen.
1972 Interactions of differentiated primary sex factors in *Chironomus tentans. Genetics,* **70:**491–493.

Thompson, V.
1976 Does sex accelerate evolution? *Evol. Theory,* **1:**131–156.

Thomsen, M.
1927 Studien über die Parthenogenese bei einigen Cocciden und Aleurodiden. *Z. Zellforsch.*, **5:**1–116.

Thomson, J. A., K. R. Radok, D. C. Shaw, M. J. Whitten, G. G. Foster, and L. M. Birt.
1976 Genetics of lucilin, a storage protein from the sheep blowfly *Lucilia cuprina* (Calliphoridae). *Biochem. Genet.*, **14:**145–160.

Thornton, I. W. B.
1971 *Darwin's Islands: A Natural History of the Galápagos*. New York: Natural History Press.

Thorpe, R. S.
1975 Biometric analysis of incipient speciation in the Ringed Snake, *Natrix natrix* (L.). *Experientia*, **31:**180–182.

Thorpe, W. H.
1930 Biological races in insects and allied groups. *Biol. Rev.*, **5:**177–212.
1940 Ecology and the future of systematics. In *The New Systematics*, J. Huxley, ed., Oxford University Press, pp. 341–364.
1945 The evolutionary significance of habitat selection. *J. Anim. Ecol.*, **14:**67–70.

Throckmorton, L. H.
1966 The relationships of the endemic Hawaiian Drosophilidae. Univ. Texas Publ. No. 6615, pp. 335–396.

Tobgy, H. A.
1943 A cytological study of *Crepis fuliginosa*, *C. neglecta*, and their F_1 hybrid, and its bearing on the mechanism of phylogenetic reduction in chromosome number. *J. Genet.*, **45:**67–111.

Todd, N. B.
1970 Karyotypic fissioning and canid phylogeny. *J. Theor. Biol.*, **26:**445–480.

Tosi, M.
1959 The chromosome sets of *Gryllotalpa gryllotalpa* L. *Caryologia*, **12:**189–198.

Toumanoff, C.
1950 L'intercroisement de l'*Aëdes (Stegomyia) aegypti* et *Aëdes (Stegomyia) albopictus* Skuse. *Bull. Soc. Path. Exot.*, **43:**234–240.

Townsend, J. I.
1952 Genetics of marginal populations of *Drosophila willistoni*. *Evolution*, **6:**428–442.
1958 Chromosomal polymorphism in Caribbean Island populations of *Drosophila willistoni*. *Proc. Nat. Acad. Sci. U.S.A.*, **44:**38–42.

Tucker, J. M., and C. H. Muller.
1956 The geographic history of *Quercus ajoensis*. *Evolution*, **10:**157–175.

Tunner, H. G.
1973 Demonstration of the hybrid origin of the common green frog *Rana esculenta* L. *Naturwiss.*, **60:**481–482.

1974 Die klonale Struktur einer Wasserfroschpopulation. *Z. Zool. Syst. Evolutionsf.,* **12:**309–314.

Turleau, C., J. de Grouchy, and M. Klein.

1972 Phylogénie chromosomique de l'homme et des primates hominiens *(Pan troglodytes, Gorilla gorilla, Pongo pygmaeus.* Essai de reconstitution du caryotype de l'ancêtre commun. *Ann. Génét.,* **15:**225–240.

Turner, B. J.

1974 Genetic divergence of Death Valley pupfish species: Biochemical versus morphological evidence. *Evolution,* **28:**281–294.

Turner, J. R. G.

1964 Evolution of complex polymorphism and mimicry in distasteful South American butterflies. *Proc. XII Int. Congr. Entomol. London,* p. 267.

1971a Studies of Müllerian mimicry and its evolution in burnet moths and heliconid butterflies. In *Ecological Genetics and Evolution,* E. R. Creed, ed., Oxford: Blackwell, pp. 224–260.

1971b Two thousand generations of hybridization in a *Heliconius* butterfly. *Evolution,* **25:**471–482.

1972 The genetics of some polymorphic forms of the butterflies *Heliconius melpomene* (Linnaeus) and *H. erato* (Linnaeus), II: The hybridization of subspecies of *H. melpomene* from Surinam and Trinidad. *Zoologica* (N.Y.), **56:**125–157.

Turner, J. R. G., and J. Crane.

1962 The genetics of some polymorphic forms of the butterflies *Heliconius melpomene* Linnaeus and *H. erato* Linnaeus, I: Major genes. *Zoologica* (N.Y.), **47:**141–152.

Tymowska, J., and M. Fischberg.

1973 Chromosome complements of the genus *Xenopus. Chromosoma,* **44:**335–342.

Ueshima, N.

1963 Chromosome behavior in the *Cimex pilosellus* complex (Cimicidae: Hemiptera). *Chromosoma,* **14:**511–521.

Upcott, M.

1939 The nature of tetraploidy in *Primula kewensis. J. Genet.,* **39:**79–100.

Ursin, E.

1952 Occurrence of voles, mice and rates (Muridae) in Denmark, with a special note on a zone of intergradation between two subspecies of the house mouse *(Mus musculus* L.*). Vid. Medd. Dansk Naturhist. Foren.,* **114:**217–244.

Utter, F. M., F. W. Allendorf, H. O. Hodgins, and A. G. Johnson.

1973 Letter to the editors. *Genetics,* **73:**159.

Uzzell, T. M.

1963 Natural triploids in salamanders related to *Ambystoma jeffersonianum. Science,* **139:**113–114.

Uzzell, T. M.
1964 Relations of the diploid and triploid species of the *Ambystoma jeffersonianum* complex (Amphibia, Caudata). *Copeia,* 1964:257–300.
1970 Meiotic mechanisms of naturally occurring unisexual vertebrates. *Amer. Nat.,* **104:**433–445.
Uzzell, T. M., and I. S. Darevski.
1973 Electrophoretic examination of *Lacerta mixta,* a possible hybrid species (Sauria, Lacertidae). *J. Herpetol.,* **7:**11–15.
Uzzell, T. M., and S. M. Goldblatt.
1967 Serum proteins of salamanders of the *Ambystoma jeffersonianum* complex, and the origin of the triploid species of this group. *Evolution,* **21:**345–354.

Vaarama, A.
1950 Studies on chromosome numbers and certain meiotic features of several Finnish moss species. *Bot. Notiser,* 1950:239–256.
1953 Some chromosome numbers of Californian and Finnish moss species. *Bryologist,* **56:**169–177.
1956 A contribution to the cytology of some mosses of the British Isles. *Ir. Nat. J.,* **12:**30–40.
Valkanov, A.
1938 Cytologische Untersuchungen über die Rhabdocoelen. *Jb. Univ. Sofia (Phys.-Math. Fak.),* **34:**321–402.
Vallée, L.
1959 Recherches sur *Triturus blasii* de l'Isle, hybride naturel de *Triturus cristatus* Latr. × *Triturus marmoratus* Latr. *Mém. Soc. Zool. Fr.,* **31:**1–96 (offprint).
Vandel, A.
1928 La parthénogénèse géographique: Contribution à l'étude biologique et cytologique de la parthénogénèse naturelle, I. *Bull. Biol. Fr. Belg.,* **68:**419–463.
Van der Pijl, L.
1934 Über die Polyembryonie bei *Eugenia. Rec. Trav. Bot. Nederl.,* **31:**113–187.
Vanzolini, P. E., and E. E. Williams.
1970 South American anoles: Geographic differentiation and evolution of *Anolis chrysolepsis* species group (Sauria, Iguanidae). *Arq. Zool., São Paulo,* **19:**1–298.
Vasek, F. C.
1958 The relationship of *Clarkia exilis* to *Clarkia unguiculata. Amer. J. Bot.,* **45:**150–162.
1960 A cytogenetic study of *Clarkia exilis. Evolution,* **14:**88–97.
1964 The evolution of *Clarkia unguiculata* derivatives adapted to relatively xeric environments. *Evolution,* **18:**26–42.
1968 The relationships of two ecologically marginal, sympatric *Clarkia* populations. *Amer. Nat.,* **102:**25–40.

Vaughan, T. A.

1967 Two parapatric species of pocket gophers. *Evolution,* **21:**148–158.

Vlad, M., and H. C. Macgregor.

1975 Chromomere number and its genetic significance in lampbrush chromosomes. *Chromosoma,* **50:**327–347.

Volterra, V.

1926 Variazioni e fluttuazioni del numero d'individui in specie animali conviventi. *Atti Acad. Naz. Lincei Mem. Cl. Sci. Fis. Mat. Nat.,* Ser. 6, **2:**31–113.

Vrijenhoek, R. C., and R. J. Schultz.

1974 Evolution of a trihybrid unisexual fish *(Poeciliopsis,* Poeciliidae). *Evolution,* **28:**306–319.

Wagner, M.

1889 *Die Entstehung der Arten durch räumliche Sonderung.* Basel: Benno Schwalbe.

Wahrman, J.

1972 Hybridization in nature between two chromosome forms of spiny mice. *Chromosomes Today,* **3:**294.

Wahrman, J., and R. Goitein.

1972 Hybridization in nature between two chromosome forms of spiny mice. *Chromosomes Today,* **3:**228–237.

Wahrman, J., R. Goitein, and E. Nevo.

1969a Geographic variation of chromosome forms in *Spalax,* a subterranean mammal of restricted mobility. In *Comparative Mammalian Cytogenetics,* K. Benirschke, ed., New York: Springer Verlag, pp. 30–48.

1969b Mole rat *Spalax:* Evolutionary significance of chromosome variation. *Science,* **164:**82–84.

Wahrman, J., and P. Gourevitz.

1973 Extreme chromosome variability in a colonizing rodent. *Chromosomes Today,* **4:**399–424.

Wahrman, J., and A. Zahavi.

1955 Cytological conclusions to the phylogeny and classification of the rodent genus *Gerbillus. Nature,* **175:**600–602.

1958 Cytogenetic analysis of mammalian sibling species by means of hybridization. *Proc. Xth Int. Genet. Congr.,* **2:**304–305.

Waines, J. G.

1976 A model for the origin of diploidizing mechanisms in polyploid species. *Amer. Nat.,* **110:**415–430.

Walker, T. G.

1962 Cytology and evolution in the fern genus *Pteris* L. *Evolution* **16:**27–43.

Walker, T. J.

1962 Factors responsible for intraspecific variation in the calling songs of crickets. *Evolution,* **16:**407–428.

Walker, T. J.
1964 Cryptic species among sound-producing ensiferan Orthoptera (Gryllidae and Tettigoniidae). *Quant. Rev. Biol.*, **39:**345–355.
1974a *Gryllus ovisopis* n. sp., a taciturn cricket with a life cycle suggesting allochronic speciation. *Fla. Entomol.*, **57:**13–22.
1974b Character displacement and acoustic insects. *Amer. Zool.*, **14:**1137–1150.
Walknowska, J.
1963 Les chromosomes chez *Spalax leucodon* Nordm. *Folia Biol. (Kraków)*, **11:**293–309.
Wallace, B.
1953 On coadaptation in *Drosophila*. *Amer. Nat.*, **87:**343–358.
1959 The influence of genetic systems on geographical distribution. *Cold Spr. Harb. Symp. Quant. Biol.*, **20:**16–24.
1966 *Chromosomes, Giant Molecules, and Evolution*. New York: Norton.
Waloff, Z.
1966 The upsurges and recessions of the desert locust plague: An historical survey. *Anti-Locust Memoir* 8, London: Anti-Locust Research Centre (111 pp).
Walters, J. L.
1942 Distribution of structural hybrids in *Paeonia californica*. *Amer. J. Bot.*, **29:**270–275.
Ward, B. L., and W. B. Heed.
1970 Chromosome phylogeny of *Drosophila pachea* and related species. *J. Hered.*, **61:**248–258.
Wasserman, A. O.
1970 Polyploidy in the common tree toad, *Hyla versicolor* Le Conte. *Science*, **167:**385–386.
Wasserman, M.
1962 Cytological studies of the *repleta* group of the genus *Drosophila*, V: The *mulleri* subgroup. *Studies in Genetics II*, Univ. Texas Publ. No. 6205, pp. 119–134.
1963 Cytology and phylogeny of *Drosophila*. *Amer. Nat.*, **97:**333–352.
Wassif, K., R. G. Lutfy, and S. Wassif.
1969 Morphological, cytological, and taxonomical studies of the rodent genera *Gerbillus* and *Dipodillus* from Egypt. *Proc. Egypt. Acad. Sci.*, **22:**77–93.
Watson, G. F.
1972 The *Litoria ewingi* complex (Anura: Hylidae) in southeastern Australia, II: Genetic incompatibility and delimitation of a narrow hybrid zone between *L.ewingi* and *L.paraewingi*. *Austral. J. Zool.*, **20:**423–433.
Watson, G. F., J. J. Loftus-Hills, and M. J. Littlejohn.
1971 The *Litoria ewingi* complex (Anura: Hylidae) in south-eastern Australia, I: A new species from Victoria. *Austral. J. Zool.*, **19:**401–416.
Webb, G. C., and M. J. D. White.
1975 Heterochromatin and timing of DNA replication in morabine grasshoppers. In *The Eukaryote Chromosome*, W. J. Peacock and R. D. Brock, eds., Canberra: Australian National University Press, pp. 395–408.

Webster, T. P., W. P. Hall, and E. E. Williams.
1972 Fission in the evolution of a lizard karyotype. *Science,* **177:**611–613.

Webster, T. P., R. K. Selander, and S. Y. Yang.
1972 Genetic variability and similarity in the *Anolis* lizards of Bimini. *Evolution* **26:**523–535.

Westergaard, M.
1958 The mechanism of sex determination in dioecious flowering plants. *Adv. Genet.,* **9:**217–281.

White, G. B.
1971 Chromosomal evidence for natural interspecific hybridization by mosquitoes of the *Anopheles gambiae* complex. *Nature,* **231:**184–185.

White, M. J. D.
1948 The chromosomes of the parthenogenetic mantid *Brunneria borealis. Evolution,* **2:**90–93.

1957 Cytogenetics of the grasshopper *Moraba scurra,* I: Meiosis of interracial and interpopulation hybrids. *Austral. J. Zool.,* **5:**285–304.

1963 Cytogenetics of the grasshopper *Moraba scurra,* VIII: A complex spontaneous translocation. *Chromosoma,* **14:**140–145.

1966 Further studies on the cytology and distribution of the Australian parthenogenetic grasshopper, *Moraba virgo. Rev. Suisse Zool.,* **73:**383–398.

1968 Models of speciation. *Science,* **159:**1065–1070.

1969 Chromosomal rearrangements and speciation in animals. *Ann. Rev. Genet.,* **3:**75–98.

1970a Cytogenetics of speciation. *J. Austral. Entomol. Soc.,* **9:**1–6.

1970b Heterozygosity and genetic polymorphism in parthenogenetic animals. In *Essays in Evolution and Genetics in Honor of Theodosius Dobzhansky,* M. K. Hecht and W. C. Steere, eds., New York: Appleton-Century-Crofts, pp. 237–262.

1970c Karyotypes and meiotic mechanisms of some Eumastacid grasshoppers from East Africa, Madagascar, India, and South America. *Chromosoma,* **30:**62–97.

1973 *Animal Cytology and Evolution,* 3d ed., Cambridge University Press.

1974 Speciation in the Australian morabine grasshoppers: The cytogenetic evidence. In *Genetic Mechanisms of Speciation in Insects,* M. J. D. White, ed., Sydney: Australia and New Zealand Book Co, pp. 57–68.

1975 Chromosomal repatterning: Regularities and restrictions. *Genetics,* **79:** 63–72.

1976 *Blattodea, Mantodea, Isoptera, Grylloblattodea, Phasmatodea, Dermaptera, and Embioptera.* Vol. 3, Insecta 2, of *Animal Cytogenetics,* B. John, ed., Berlin: Borntraeger (75 pp).

White, M. J. D., R. E. Blackith, R. M. Blackith, and J. Cheney.
1967 Cytogenetics of the *viatica* group of morabine grasshoppers, I: The "coastal" species. *Austral. J. Zool.,* **15:**263–302.

White, M. J. D., H. L. Carson, and J. Cheney.
1964 Chromosomal races in the grasshopper *Moraba viatica* in a zone of geographic overlap. *Evolution,* **18:**417–429.

White, M. J. D., and J. Cheney.
1966 Cytogenetics of the *cultrata* group of morabine grasshoppers, I: A group of species with XY and X_1X_2Y sex chromosome mechanisms. *Aust. J. Zool.* **14:**821–834.
1972 Cytogenetics of a group of morabine grasshoppers with XY and X_1X_2Y males. *Chromosomes Today,* **3:**177–196.

White, M. J. D., J. Cheney, and K. H. L. Key.
1963 A parthenogenetic species of grasshopper with complex structural heterozygosity (Orthoptera: Acridoidea). *Austral. J. Zool.,* **11:**1–19.

White, M. J. D., N. Contreras, J. Cheney, and G. C. Webb.
1977 Cytogenetics of the parthenogenetic grasshopper *Warramaba* (formerly *Moraba*) *virgo* and its bisexual relatives, II: Hybridization studies. *Chromosoma,* **61:**127–148.

White, M. J. D., K. H. L. Key, M. Andre, and J. Cheney.
1969 Cytogenetics of the *viatica* group of morabine grasshoppers, II: Kangaroo Island populations. *Austral. J. Zool.,* **17:**313–328.

White, M. J. D., and G. C. Webb.
1968 Origin and evolution of parthenogenetic reproduction in the grasshopper *Moraba virgo* (Eumastacidae: Morabinae). *Aust. J. Zool.,* **16:**647–671.

White, M. J. D., G. C. Webb, and J. Cheney.
1973 Cytogenetics of the parthenogenetic grasshopper *Moraba virgo* and its bisexual relatives; I: A new species of the *virgo* group with a unique sex chromosome mechanism. *Chromosoma,* **40:**199–212.

Wiebes, J. T.
1965 Host specificity of fig wasps (Hymenoptera, Chalcidoidea, Agaonidae). *Proc. XIIth Int. Congr. Entomol., London,* pp. 95–96.
1966 Provisional host catalog of fig wasps (Hymenoptera, Chalcidoidea). *Zool. Verh.,* **83:**1–44.

Wiens, D., and B. A. Barlow.
1975 Permanent translocation heterozygosity and sex determination in East African mistletoes. *Science,* **187:**1208–1209.

Williams, C. B.
1960 The range and pattern of insect abundance. *Amer. Nat.,* **94:**137–151.

Williams, G. C.
1975 *Sex and Evolution.* Princeton University Press.

Williams, G. C., R. K. Koehn, and J. B. Mitton.
1973 Genetic differentiation without isolation in the American eel, *Anguilla rostrata. Evolution,* **27:**192–204.

Williams, G. C., and J. B. Mitton.
1973 Why reproduce sexually? *J. Theor. Biol.,* **39:**545–554.

Williamson, D. L., and L. Ehrman.
1968 Infectious hybrid sterility and the "New Llanos" strain of *Drosophila paulistorum*. *Nature,* **219:**1266–1267.

Wilson, A. C., G. L. Bush, S. M. Case, and M.-C. King.
1975 Social structuring of mammalian populations and rate of chromosomal evolution. *Proc. Nat. Acad. Sci. U.S.A.,* **72:**5061–5065.

Wilson, A. C., V. M. Sarich, and L. R. Maxon.
1974 The importance of gene rearrangement in evolution: Evidence from studies on rates of chromosomal, protein and anatomical evolution. *Proc. Nat. Acad. Sci. U.S.A.,* **71:**3028–3030.

Wolda, H.
1963 Natural populations of the polymorphic land snail *Cepaea nemoralis* (L.). *Arch. Néerlandaises Zool.,* **15:**381–471.
1967a The effect of temperature of reproduction in some morphs of the land snail *Cepaea nemoralis* (L.). *Evolution,* **21:**117–129.
1967b Reproductive isolation between two closely related species of the Queensland fruit fly, *Dacus tryoni* (Frogg.) and *D. neohumeralis* Hardy (Diptera: Tephritidae), I and II. *Austral. J. Zool.,* **15:**501–513 and 515–539.

Woltereck, R.
1931 Wie ensteht eine endemische Rasse oder Art? *Biol. Zentralbl.,* **51:**231–253.

Woodruff, D. S.
1973 Natural hybridization and hybrid zones. *Syst. Zool.,* **22:**213–218.
1975 Allozyme variation and genic heterozygosity in the Bahaman pulmonate snail *Cerion bendalli. Malacol. Rev.,* **8:**47–55.

Worthington, E. B.
1940 Geographical differentiation in fresh waters with special reference to fish. In *The New Systematics,* J. Huxley, ed., Oxford: Clarendon Press, pp. 287–302.

Wright, J. W., and R. Pal, eds.
1967 *Genetics of Insect Vectors of Disease.* Amsterdam: Elsevier.

Wülker, W., J. E. Sublette, and J. Martin.
1968 Zur Cytotaxionomie nordamerikanischer *Chironomus*-Arten. *Ann. Zool. Fenn.,* **5:**155–158.

Wurster, D. H., and N. B. Atkin.
1972 Muntjac chromosomes: A new karyotype for *Muntiacus muntjac. Experientia,* **28:**972–973.

Wurster, D. H., and K. Benirschke.
1970 Indian Muntjac, *Muntiacus muntjac:* A deer with a low diploid number. *Science,* **168:**1364–1366.

Wurster-Hill, D. H.
1973 Chromosomes of eight species from five families of Carnivora. *J. Mammal.,* **54:**753–760.

Wurster-Hill, D. H., and C. W. Gray.
1973 Giemsa banding patterns in the chromosomes of 12 species of cats (Felidae). *Cytogenet. Cell Genet.*, **12:**377–397.

Yang, S. Y., M. Soulé, and G. C. Gorman.
1974 *Anolis* lizards of the eastern Caribbean: A case study in evolution, I: Genetic relationships, phylogeny, and colonization sequence of the *roquet* group. *Syst. Zool.*, **23:**387–399.

Yang, S. Y., L. L. Wheeler, and I. R. Bock.
1972 Isozyme variations and phylogenetic relationships in the *Drosophila bipectinata* species complex. In *Studies in Genetics VII,* Univ. Texas Publ. No. 7213, pp. 213–227.

Yang, T. W.
1967 Chromosome numbers in populations of creosote bush *(Larrea divaricata)* in the Chihuahuan and Sonoran subdivisions of the North American desert. *J. Arizona Acad. Sci.*, **4:**183–184.
1968 A new race of *Larrea divaricata* in Arizona. Western Reserve Academy Natural History Museum Special Publication No. 2, pp. 1–4.

Yen, J. H., and A. R. Barr.
1974 Incompatibility in *Culex pipiens*. In *The Use of Genetics in Insect Control,* R. Pal and M. J. Whitten, eds., Amsterdam: Elsevier/North-Holland, pp. 97–118.

Yosida, T. H.
1973 Evolution of karyotypes and differentiation in *Rattus* species. *Chromosoma,* **40:**285–297.

Zahavi, A., and J. Wahrman.
1957 The cytotaxonomy, ecology, and evolution of the gerbils and jirds of Israel (Rodentia: Gerbillinae). *Mammalia,* **21:**341–380.

Zaslavsky, V. A.
1963 Hybrid sterility as a limiting factor in the distribution of allopatric species. (Russian) *Dokl. Akad. Nauk S.S.S.R.*, **149:**470–471.
1967a Reproductive self-destruction as an ecological factor (ecological consequences of genetical interaction between populations). (Russian) *J. Gen. Biol.*, **28:**3–11.
1967b Reproductive isolation in some freely intermating species of insects. (Russian) *Dokl. Akad. Nauk S.S.S.R.*, **174:**1433–1434.

Zimmering, S., L. Sandler, and B. Nicoletti.
1970 Mechanisms of meiotic drive. *Ann. Rev. Genet.*, **4:**409–436.

Zimmerman, E. C.
1938 Cryptorhynchinae of Rapa. *Bernice P. Bishop Mus. Bull.*, **151:**1–75.
1960 Possible evidence of rapid evolution in Hawaiian moths. *Evolution,* **14:**137–138.

Zimmerman, E. G.
1970 Karyology, systematics and chromosomal evolution in the rodent genus *Sigmodon*. *Publ. Mus. Michigan State Univ. Biol. Ser.*, **4**(9):385–454.
Zimmerman, E. G., and M. R. Lee.
1968 Variation in chromosomes of the cotton rat, *Sigmodon hispidus*. *Chromosoma*, **24:**243–250.
Zohary, D., and M. Feldman.
1962 Hybridization between amphidiploids and the evolution of polyploids in the wheat *(Aegilops–Triticum)* group. *Evolution*, **16:**44–61.
Zouros, E., C. B. Krimbas, S. Tsakas, and M. Loukas.
1974 Genic versus chromosomal variation in natural populations of *Drosophila subobscura*. *Genetics*, **78:**1223–1244.
Zuckerman, S.
1953 The breeding seasons of mammals in captivity. *Proc. Zool. Soc. Lond.*, **122:**827–950.
Zúk, J.
1963 An investigation on polyploidy and sex determination within the genus *Rumex*. *Acta Soc. Bot. Pol.*, **32:**5–67.
Zweifel, R. G.
1965 Variation in and distribution of the unisexual lizard, *Cnemidophorus tesselatus*. *Amer. Mus. Novit.*, **2235:**1–49.

Name Index

Subject Index

Passerella melodia, 17
Patella coerulea, 337
Pattern effect theory, 224
Pediculus humanus, 244
Pedogenesis, 117
Perdita, 238
Pericentric inversions. *See* Inversions.